Macworld® DVD Studio Pro™ Bible

Macworld® DVD Studio Pro™ Bible

Todd Kelsey and Chad Fahs

Best-Selling Books • Digital Downloads • e-Books • Answer Networks • e-Newsletters • Branded Web Sites • e-Learning

New York, NY ✦ Cleveland, OH ✦ Indianapolis, IN

Macworld® DVD Studio Pro™ Bible
Published by
Hungry Minds, Inc.
909 Third Avenue
New York, NY 10022
www.hungryminds.com

Copyright © 2002 Hungry Minds, Inc. All rights reserved. No part of this book, including interior design, cover design, and icons, may be reproduced or transmitted in any form, by any means (electronic, photocopying, recording, or otherwise) without the prior written permission of the publisher.

Library of Congress Control Number: 2001094148

ISBN: 0-7645-3633-8

Printed in the United States of America

10 9 8 7 6 5 4 3 2 1

1B/TQ/RR/QR/IN

Distributed in the United States by Hungry Minds, Inc.

Distributed by CDG Books Canada Inc. for Canada; by Transworld Publishers Limited in the United Kingdom; by IDG Norge Books for Norway; by IDG Sweden Books for Sweden; by IDG Books Australia Publishing Corporation Pty. Ltd. for Australia and New Zealand; by TransQuest Publishers Pte Ltd. for Singapore, Malaysia, Thailand, Indonesia, and Hong Kong; by Gotop Information Inc. for Taiwan; by ICG Muse, Inc. for Japan; by Intersoft for South Africa; by Eyrolles for France; by International Thomson Publishing for Germany, Austria, and Switzerland; by Distribuidora Cuspide for Argentina; by LR International for Brazil; by Galileo Libros for Chile; by Ediciones ZETA S.C.R. Ltda. for Peru; by WS Computer Publishing Corporation, Inc., for the Philippines; by Contemporanea de Ediciones for Venezuela; by Express Computer Distributors for the Caribbean and West Indies; by Micronesia Media Distributor, Inc. for Micronesia; by Chips Computadoras S.A. de C.V. for Mexico; by Editorial Norma de Panama S.A. for Panama; by American Bookshops for Finland.

For general information on Hungry Minds' products and services please contact our Customer Care department within the U.S. at 800-762-2974, outside the U.S. at 317-572-3993 or fax 317-572-4002.

For sales inquiries and reseller information, including discounts, premium and bulk quantity sales, and foreign-language translations, please contact our Customer Care department at 800-434-3422, fax 317-572-4002 or write to Hungry Minds, Inc., Attn: Customer Care Department, 10475 Crosspoint Boulevard, Indianapolis, IN 46256.

For information on licensing foreign or domestic rights, please contact our Sub-Rights Customer Care department at 212-884-5000.

For information on using Hungry Minds' products and services in the classroom or for ordering examination copies, please contact our Educational Sales department at 800-434-2086 or fax 317-572-4005.

For press review copies, author interviews, or other publicity information, please contact our Public Relations department at 317-572-3168 or fax 317-572-4168.

For authorization to photocopy items for corporate, personal, or educational use, please contact Copyright Clearance Center, 222 Rosewood Drive, Danvers, MA 01923, or fax 978-750-4470.

LIMIT OF LIABILITY/DISCLAIMER OF WARRANTY: THE PUBLISHER AND AUTHOR HAVE USED THEIR BEST EFFORTS IN PREPARING THIS BOOK. THE PUBLISHER AND AUTHOR MAKE NO REPRESENTATIONS OR WARRANTIES WITH RESPECT TO THE ACCURACY OR COMPLETENESS OF THE CONTENTS OF THIS BOOK AND SPECIFICALLY DISCLAIM ANY IMPLIED WARRANTIES OF MERCHANTABILITY OR FITNESS FOR A PARTICULAR PURPOSE. THERE ARE NO WARRANTIES WHICH EXTEND BEYOND THE DESCRIPTIONS CONTAINED IN THIS PARAGRAPH. NO WARRANTY MAY BE CREATED OR EXTENDED BY SALES REPRESENTATIVES OR WRITTEN SALES MATERIALS. THE ACCURACY AND COMPLETENESS OF THE INFORMATION PROVIDED HEREIN AND THE OPINIONS STATED HEREIN ARE NOT GUARANTEED OR WARRANTED TO PRODUCE ANY PARTICULAR RESULTS, AND THE ADVICE AND STRATEGIES CONTAINED HEREIN MAY NOT BE SUITABLE FOR EVERY INDIVIDUAL. NEITHER THE PUBLISHER NOR AUTHOR SHALL BE LIABLE FOR ANY LOSS OF PROFIT OR ANY OTHER COMMERCIAL DAMAGES, INCLUDING BUT NOT LIMITED TO SPECIAL, INCIDENTAL, CONSEQUENTIAL, OR OTHER DAMAGES.

Trademarks: Macworld is a registered trademark or trademark of International Data Group, Inc. DVD Studio Pro is a registered trademark or trademark of Apple Computer, Inc. All other trademarks are the property of their respective owners. Hungry Minds, Inc., is not associated with any product or vendor mentioned in this book.

Incubus stills courtesy of Contempo III and Anthony M. Taylor.

 is a trademark of Hungry Minds, Inc.

Credits

Acquisitions Editor
Michael Roney

Project Editor
Mica Johnson

Development Editors
Katie Dvorak
Kenyon Brown

Technical Editor
Dennis Short

Copy Editors
Beth Parlon
Suzanne Thomas

Editorial Manager
Rev Mengle

Project Coordinator
Regina Snyder

Graphics and Production Specialists
Sean Decker, Brian Drumm,
Joyce Haughey, Jackie Nicholas,
Laurie Petrone, Betty Schutte,
Rashell Smith, Jeremey Unger

Quality Control Technicians
Laura Albert, Vickie Broyles,
David Faust, John Greenough,
Carl Pierce

Permissions Editor
Carmen Krikorian

Media Development Specialist
Travis Silvers

Media Development Coordinator
Marisa Pearman

Illustrator
Anthony Bunyan

Proofreading and Indexing
TECHBOOKS Production Services

Cover Image
Murder By Design

About the Authors

Todd Kelsey is the owner of Selendrian Group, Inc., a digital media company that has created music and video projects that received national attention. One such project he produced with is "GLF," collaborating with Jim Cooper of the Detholz, and Skipworth of Skipwave Productions. GLF a cross between Alvin and the Chipmunks and Toontzes the Cat. GLF started as a joke, but then spawned a CD (and video) that received airplay on college and commercial radio stations throughout the U.S. and Canada. The video was a hit on IMNTV, an independent music video channel that gave it play in Virginia, Florida, and California. Todd has recently created a promotional DVD to shop GLF to television.

Todd has a BA in Literature and an MA in Communications, with a background in desktop publishing and editorial work. He is also a former performing member and online manager of the major label recording act "Sister Soleil," whose first release on Universal Records was recorded at Realworld Studios in England, featuring a cameo by Peter Gabriel.

His current pursuits include independent DVD authoring, Flash animation, and a focus on exploring the possibilities of DVD music video compilations. He is forming a band in the Chicago area and will be touring in the Midwest region and releasing an independent album.

His dream is to get a ride in a Huey helicopter. Another dream is to play Hendrix's version of the National Anthem at the 100th Anniversary of Woodstock in 2069. Along the way, he wants to play a lot of shows to keep his fingers limber, get a Ph.D. in Communications, and teach helicopter flying lessons on the side.

Chad Fahs knew he wanted to tell stories since he was very young. In his mind, moving images were the best medium to accomplish this dream. Whether making home movies or spending countless hours in a movie theater, it was all done with a single goal in sight. This passion soon led to the formation of his own production company in Chicago in 1998, using an Avid system to edit video.

During this time, he traveled the globe in search of new inspiration, working on documentary projects in countries as far apart as Japan and France. At the same time, he worked on a variety of other projects, many of which were for local and national bands, in order to feed his musical interests and need for visual experimentation.

After working as an Avid editor, Chad decided to expand his digital palette. Eventually, Final Cut Pro became a part of the mix and DVD soon followed. At the same time, he developed an interest in streaming video and interactive elements as a means of telling his stories on the Web.

Currently, Chad is working at one of the top internet sites, while continuing to seek out new projects which merge video, streaming media applications, and interactive content. He is also working on a screenplay for an independent film project.

To my parents, Fred and Linda, and my brother Mark, for putting up with me over the years, and to my good friend Howie Beno, for all your help in the music industry.

— Todd

To those who have inspired and supported me all these years. To my parents, Berit and Gerald, my sister Pia, and my closest friend, Meghann.

— Chad

Foreword

Back in December of 1995, few people had a clue what DVD was and how it would change the world. There were, however, a few hints in the air. Multimedia CD-ROMs were becoming an increasingly popular way to communicate and entertain, while the explosive growth of the Internet revealed a pent up demand for interactive content. It was in this light that DVD drew its first tentative breath.

Since its first commercial introduction in 1997, DVD has enjoyed a faster adoption rate than any other consumer electronic device in history. Consumers adopted DVD players at a greater rate than television sets, radios, VHS decks, or even toasters for that matter. The number of available DVD movies grew at an exponential rate. DVD movies aren't just a newer twist on VHS tape. Many popular Hollywood titles take full advantage of DVD's impressive feature set, including motion menus, multiple camera angle scenes, and multi-channel audio tracks.

At first, DVD movies were incredibly expensive to produce. DVD authoring tools started at around $60,000 and rose sharply from there. Creating DVD movies was a black art reserved for the most skillful compressionists and programmers. Gradually, however, DVD movies became cheaper and easier to produce. In the spring of 2000, Apple bought a little known authoring tool called DVDirector, and officially joined the ranks of the DVD community. By January 2001, Apple rechristened the tool as DVD Studio Pro and dropped its price to $999, a new low for a full-featured, professional DVD authoring tool.

In many ways, DVD Studio Pro represented a democratization of the DVD authoring process. Previously, professional DVD authoring tools were geared specifically towards highly technical users, and were very difficult to master. Eventually, low-priced authoring tools came to the market, but they were universally feature-bare and didn't include many of DVD's most compelling features. In order for DVD authoring to catch on with creative professionals, it had to be affordable, powerful, and easy to use. Along with it's low price point, DVD Studio Pro appeals its users because it uses the type of workflow that creative professionals are already used to. However, DVD Studio Pro's ease of use did not come at the price of features. Nearly every interactive feature of the DVD video specification is supported, and it is implemented with the ease of use and intuitiveness that Apple is known for.

This unique combination of affordability and ease of use has proven to be a winning formula for a professional DVD authoring application, especially when you consider that DVD authoring is no longer just for the Hollywood types. DVD Studio Pro represents a significant lowering of the financial bar for those interested in getting into DVD authoring. This means that for a moderate sum, you can stand on the same footing as those who previously spent hundreds of thousands of dollars for the ability to create DVD movies. Armed with these capabilities, a world of infinite possibilities awaits. You might use DVDs to preserve your memories, tell your story, or as a means to turn your home-grown feature film into the next blockbuster. Whatever your choice, this book will help you understand the capabilities of this powerful new medium.

The *Macworld DVD Studio Pro Bible* is another remarkable step down the road to bringing DVD authoring to the masses. Inside this book, you will find nearly everything you need to clearly and easily understand both DVD authoring and the use of DVD Studio Pro. Within these pages is a comprehensive understanding of the core principals surrounding DVD Studio Pro's workflow and production models. Make the most out of this book, and you will master DVD Studio Pro in no time.

Have fun!

Tony Knight
Product Marketing Manager
DVD Studio Pro, Apple Computer, Inc.

Calling All DVD Studio Pro Users!

We wanted to take this moment to extend an invitation to submit a DVD project, and to visit/participate in the official Web site. In the next version of this book, we'd like to include the readers.

We are keeping our eyes open for a few good DVD projects, and we'll select from the submitted material and choose a few to feature in the next revision of the *DVD Studio Pro Bible,* and on the official Web site.

If you have an interesting project that you've done in DVD Studio Pro, whether it's a Hollywood release or an independent production, send us a copy.

Things that might be interesting:

- A special technique you developed to achieve a certain effect
- Overcoming a certain technical challenge you faced, such as space limits
- Use of 3D animation in motion menus
- Educational and/or training projects
- Something unique or appealing about the underlying visual or audio content
- Something interesting about the way the content was put together
- Use of multiple languages, angles or audio streams
- No use of written/spoken language whatsoever, completely visual/symbolic with sound or music
- Projects involving underwater, microscopic or interstellar videography/photography
- Projects from countries other than the U.S.
- Projects involving science, or science fiction
- DVD-based games
- A student/class project
- Projects used in the context of raising awareness for non-profit organizations

- A project from a setting where DVD Studio Pro is being taught/used
- Projects made by or involving famous people
- Projects where video was made with the Fisher Price Pixelvision camera
- Cultural/history projects, such as Native American, Celtic/Irish or Japanese
- Creative use of features in ways they weren't necessarily intended
- DVD or DVD-R based dating service?
- Interesting discoveries of business opportunities for DVD
- Interesting packaging
- Anything having to do with helicopters
- A project involving fashion/modeling/acting
- Portfolio projects, such as demo reels or acting/modeling portfolios
- Memoir DVDs with ancestral information or just memories
- DVDs that were made as a gift for a specific person
- Making 3D interfaces that you can see with those funky two-color glasses
- If you didn't know how to use a feature like how to record Surround, then figured it out

Special Categories:

- Use of DVD Studio Pro in the music industry
- Use of DVD Studio Pro in Hollywood
- Use of DVD Studio Pro in film schools
- Use of DVD Studio Pro by former Sonic/Spruce users
- Use of Surround Sound
- Motion menus and loop point techniques
- Using Final Cut Pro and DVD Studio Pro together
- Scripting
- DVD-ROM content that connects to the Internet
- DVD-ROM content using Flash, Director or other programs
- Projects made by children, or from very mature DVD Studio Pro users

Note: The project doesn't necessarily have to be released commercially. A finished project would be better, but it could even be an element of a project, such as an experimental use of scripting. The project doesn't necessarily have to be successful either; as long as you learned something from it. Basically, don't rule yourself out. If you like the project, or learned something from it, send it in.

Limits: You can send ANYTHING that is not hateful or child porn, but keep in mind that for things that are on the edge, we might not be able to use it in the book. Also, there is the risk that if we get a ton of submissions, we might not be able to check every disc out, so feel free to be creative with presentation.

For more information on where and how to send submissions, visit www.dvdspa.com and look for the DVD Studio Pro Gallery link.

Questions about DVD Studio Pro

If you have a question about DVD Studio Pro or DVD in general, try the following:

1. Make sure you've read the *DVD Studio Pro Bible,* and the latest DVD Studio Pro manual. If you have DVD Studio Pro 1.0, you can download the free upgrade to 1.1 from www.apple.com/dvdstudiopro/update, and it comes with an expanded manual.
2. Try going to www.dvdspa.com and see if someone has already asked the same question and see the answer.
3. Visit www.dvddemystified.com to see if the answer is in the DVD FAQ. The DVD FAQ is also on the DVD-ROM that goes with this book, in HTML format.
4. Try going to http://discussions.info.apple.com/ or www.apple.com/support/ and look through the latest topics and postings.

   If you can't find the answer, feel free to submit a question at www.dvdspa.com

The DVD Studio Pro site (www.dvdspa.com)

There will be some interesting stuff on the site, some of it coming from readers.

DVD Studio Pro enhancements

No program, no matter how cool, is perfect. If you have discovered a bug in the program, feel free to go to www.dvdspa.com and check in the Bugs area to see if it's been reported, and if a workaround has been developed. If you don't see it there, feel free to email in, and we'll give you credit for it, and possibly use it in the next book.

Errors and complaints

No author or their writing, no matter how cool, is perfect. If you have discovered a bug in one of the authors, feel free to go to www.dvdspa.com and check in the Errata area, to see if the inaccuracy has been reported.

DVD-RW (Reading and Watching)

There will be reviews of DVD-related books and videos on the site, to encourage further exploration and learning. (Note: If you read or write an interesting DVD-related book, or found/produced an interesting training video, see if it's already listed/reviewed, and if it isn't, go on the site in the DVD-RW area and submit a request.)

Manufacturing/Replication/Media/Equipment

There will be a list of some DVD related companies on the site that provide equipment, software, or services that would be of interest to DVD authors. (Note: If your company or product is not there, feel free to visit the Resources area of the site and submit a request, even if you're in another country.)

Find a DVD author

If we have enough time and resources, we will see if we can put together a way for independent and corporate DVD authoring services to be listed by region. Check the DVD Authoring Services area for the latest.

Acknowledgments

We'd like to thank Hungry Minds for making this book a reality, not just a possibility. Most of all, we'd like to thank Michael Roney and Mica Johnson for being there every step of the way. Their guidance is what got us through the book in one piece. Special thanks to David Fugate, and all the other folks at the Waterside Literary Agency.

Our gratitude extends to everyone who helped with the book. For everyone who generously contributed material and gave us assistance along the way.

Chad — Personal thanks to Gerald, Berit, and Pia Fahs as well as Meghann Matwichuk for their encouragement, inspiration, and continued support. Also, thanks to all the films, filmmakers, artists, and people I've met along the way who have inspired me to pursue a dream.

To Brian Dressel of OVT Visuals, Jason Zada of Evolution Bureau, Chris Greene of Alien Sound and Arts, Alan Martin, Paul Feith, and Dan Fenster for their expertise and generous contributions to the book. Also, thanks to Anthony Taylor for the use of stills from the movie *Incubus*.

Todd — Thanks to Dad and Mom for encouraging me with computers when I was starting out with the Radio Shack TRS-80 color computer with 4k of ram, back when it wasn't hip to be a computer nerd. To my brother Mark, for inspiring me with his sense of professionalism.

This book wouldn't be half of what it is without the assistance, materials and information provided by friends and colleagues. To fellow authors Lee Purcell and Jim Taylor, for patiently answering questions and taking the time to discuss complex technical issues. To Tim Dop and the folks at Metatec, for providing information and assistance with DVD manufacturing. To David Chou at EMVUSA, for answering questions and providing guidance on DVD manufacturing. To Craig Muller and Rich Ludwig of the Warm Blankets Foundation, for helping kids in Cambodia, and letting us invade their offices and use material for the case study. To Martin Baumgaertner of Angle Park Productions, for inspiration and an example of creative professionalism. To Paul Holtz of Timeline Video, for advice and expert assistance with MPEG-2 compression.

To Brian Skipworth of Skipwave Productions, for generous assistance with preparation of source material.

To Beba, Mirsada, and Alma, of the Rat Patrol.

To Andrew, Ben, and Dennis, who share the house and put up with all the late nights and phone calls. To Katie, for being a patient friend. To Liz, my first DVD student.

Most importantly, thanks to Chad, my fellow author, for telling me about DVD Studio Pro.

Contents at a Glance

Foreword ix
Calling All DVD Studio Pro Users! xi
Acknowledgments xv
Introduction xxxi

Part I: Working with DVD Studio Pro 1
Quick Start: Up and Running with DVD Studio Pro 3
Chapter 1: Understanding DVD Studio Pro 17
Chapter 2: Inside the DVD Studio Pro Workspace 29
Chapter 3: Working with Assets 53

Part II: Creating Interactivity 73
Chapter 4: Planning and Prototyping 75
Chapter 5: Working with Buttons 99
Chapter 6: Working with Menus 125
Chapter 7: Using Subtitles 149
Chapter 8: Exploring Interface Design 173

Part III: Using Assets in a Project 203
Chapter 9: Understanding DVD Audio 205
Chapter 10: Working with DVD Audio 223
Chapter 11: Understanding DVD Video 249
Chapter 12: Working with DVD Video 269
Chapter 13: Working with Slideshows 295

Part IV: Bringing It All Together 315
Chapter 14: Building a Project 317
Chapter 15: Preparing for Output 339
Chapter 16: Outputting a Project 351

Part V: Advanced Interactivity 371
Chapter 17: Understanding Scripting 373
Chapter 18: Adding Multiple Angles to a Project 395
Chapter 19: Working with Multiple Languages 415

Part VI: Exploring DVD Content Delivery **443**
Chapter 20: Evaluating DVD Player Options 445
Chapter 21: Developing DVD-ROM Content 457
Chapter 22: Cross-Developing for DVD, CD-ROM, and the Internet 475

Part VII: Case Studies **503**
Chapter 23: Case Study: Warm Blankets 505
Chapter 24: Case Study: Atomic Paintbrush Studios 513
Chapter 25: Case Study: Pioneer Electronics' SuperDrive 523
Chapter 26: Case Study: Simple Motion Menus 533
Chapter 27: Case Study: Metatec — Manufacturing a DVD 543
Chapter 28: Case Study: Real World Promotional DVD 559

Part VIII: Appendixes **571**
Appendix A: DVD Studio Pro Installation and System Requirements 573
Appendix B: Resources 577
Appendix C: Glossary and Terms 581
Appendix D: About the DVD-ROM 585

Index 589
End-User License Agreement 615

Contents

Foreword ix

Calling All DVD Studio Pro Users! xi

Acknowledgments xv

Introduction xxxi

Part I: Working with DVD Studio Pro — 1

Quick Start: Up and Running with DVD Studio Pro 3
Starting a DVD Project 3
Importing Assets 6
Setting up Menus and Tracks 8
    Setting up a menu 9
    Setting Up a Track 10
    Previewing Tiles 11
Creating and Linking Buttons 12
Burning a DVD 14

Chapter 1: Understanding DVD Studio Pro 17
DVD Studio Pro and DVD Technology 17
    DVD storage capacity 18
    DVD menus 18
    DVD video 18
    DVD audio 19
    Multiple languages 19
    Subtitles 20
DVD Studio Pro Projects 21
    DVD Studio Pro building blocks 22
    Tracks in DVD Studio Pro 22
    Menus in DVD Studio Pro 23
Working with Project Files 25
    Starting a project 25
    Saving a project 26
    File management 27
Summary 28

Chapter 2: Inside the DVD Studio Pro Workspace 29

Understanding the Graphical View 30
    Getting to know Tiles 30
    Exploring the drop-down menus 32
    Project element buttons 34
Understanding the Project View Tabs 38
The Assets Container 39
Using the Property Inspector 40
    Disc properties 40
    General properties 40
    Disc menu settings 42
    Variable Names 43
    Remote-Control 43
Track properties 44
Menu properties 45
Script properties 45
Slideshow properties 45
Additional Ways to View Projects 46
    Using the editors 46
    The Matrix views 48
    The Troubleshooting Windows 51
Summary 52

Chapter 3: Working with Assets 53

Understanding Compatible File Formats 53
    Understanding video compatibility: MPEG and MPEG-2 54
    Considering audio compatibility 54
    Considering graphics compatibility: Photoshop 54
    Understanding file extensions 55
Importing assets 56
    Importing through the File menu 56
    Importing with the drag-and-drop method 58
    Importing by creating a new asset 59
Organizing and Controlling Your Project Assets 59
    Understanding the Assets Container 59
    Understanding the Asset Matrix 63
    Understanding the Asset Files window 66
Summary 70

Part II: Creating Interactivity   73

Chapter 4: Planning and Prototyping 75

Considering the Audience 75
    Considering enjoyability 76
    Asking the questions 77
Storyboarding a DVD Project 80

Prototyping DVD Interactivity 82
    Considering a DVD-R Prototype 83
    Prototyping with Microsoft Word 83
    Prototyping with HTML in Dreamweaver 86
Summary 97

Chapter 5: Working with Buttons 99

Creating Buttons in Photoshop 100
    Preparing layers for use in a still menu 100
    Matching colors in Photoshop 105
    Using Photoshop shapes and styles 106
    Using text in Photoshop 110
Creating Buttons in ImageReady 111
Understanding Square versus Non-Square Pixels 113
Considering Workflow Possibilities 114
Setting Button Properties 118
    Creating overlay images in Photoshop 119
    Working with Button Hilites 120
    Using buttons as interactive markers 122
    Understanding the Matrix Views 123
Summary 123

Chapter 6: Working with Menus 125

Working with Layered Image Files 126
    Working with backgrounds 126
    Working with foregrounds 130
Exploring Motion Menus 131
    Preparing video with Final Cut Pro 132
    Looping video 134
    Considering animations 135
Enabling Menus 135
    Using the Property Inspector with menus 136
    Linking assets to menu buttons 140
    Testing completed menus 144
Summary 147

Chapter 7: Using Subtitles 149

Working with the Subtitle Editor 150
    Selecting project settings 150
    Exploring the Subtitle Editor's windows 156
Creating Subtitles with Key Frames 161
    Creating subtitles with a Project Movie 161
    Importing subtitles 167
    Positioning subtitle text in a frame 168
Working with Fonts 169
    Selecting fonts and styles for subtitles 169
    Considering fonts and the user experience 169

Importing External Subtitles 170
    Acquiring subtitles created by another Subtitle Editor 170
    Obtaining subtitles from a service bureau 170
Summary 171

Chapter 8: Exploring Interface Design 173

Understanding the Principles of Interface Design 173
    Enjoyability 174
    Usability 175
Suggestions for Designing DVD Interfaces 175
Evaluating DVD Interface Designs 181
Technical Considerations When Designing for DVD 183
Looking at Third-Party Applications for Compositing,
    Animation, and Special Effects 186
    After Effects 187
    Flash 195
    Freehand 199
    Director 200
Summary 202

Part III: Using Assets in a Project   203

Chapter 9: Understanding DVD Audio 205

Investigating Audio Formats 206
    Linear PCM 206
    AC-3/Dolby Digital 210
    MPEG audio 214
    DTS 214
    SDDS 215
    DVD-Audio 215
Mixing for AC-3/Dolby Digital 216
Monitoring 5.1 Channels of Audio 218
Summary 220

Chapter 10: Working with DVD Audio 223

Evaluating Compatible Audio Formats 224
Preparing Audio in Third-Party Applications 224
    Considering multitrack recording 225
Experimenting with Dolby Digital 227
Investigating the A.Pack Interface 228
    The Instant Encoder window 228
    The Batch Encoder window 235
Encoding Audio for AC-3/Dolby Digital Using A.Pack 237
    Mixing audio 237
    Preparing audio for import 237
    Using the Instant Encoder to produce AC-3/Dolby Digital streams . . . 238

Contents

        Encoding AC-3/Dolby Digital tracks in batches 240
        Monitoring AC-3/Dolby Digital audio 242
    Using Audio in DVD Studio Pro 242
        Importing audio 243
        Managing files 244
        Adding multiple audio streams to a project 245
    Summary 248

Chapter 11: Understanding DVD Video 249

    Reviewing Digital Video Basics 249
        Looking at frame rates 250
        Comparing film with video 250
        Understanding progressive versus interlaced scanning 253
        Evaluating screen size 254
        Considering chroma and luminance for digital video 260
    Exploring the DVD Video Standard 260
        Evaluating image quality 261
        Evaluating sound quality 261
    Investigating the File Structure of a DVD 264
        Video Title Set 265
        Presentation files (VOB) 265
        Navigation files (IFO) 265
        Backup files (BUP) 265
    Examining the MPEG-2 Format 265
        Compressing video with MPEG 266
        Intra frames 266
        Predicted frames 267
        Bidirectional frames 267
    Variable Bit-rate Encoding 267
    Summary 268

Chapter 12: Working with DVD Video 269

    Evaluating Compatible Video Formats 270
    Preparing Video in Third-Party Applications 270
        Final Cut Pro 271
        Cleaner 271
        Toast Titanium 276
    Encoding Video for DVD 277
        Encoding MPEG-2 with QuickTime Pro 277
        Encoding MPEG-2 with Cleaner 279
        Hardware-based MPEG-2 encoding 282
    Using Video in DVD Studio Pro 283
        Importing video 283
        Managing video files 286
        Placing markers in a video track 288
        Creating stories with markers 290
        Previewing video 293
    Summary 293

Chapter 13: Working with Slideshows 295

Using Still Frame Images 296
    Scanning still images for use in a DVD project 296
    Determining scanner resolution and bit depth 297
    Detecting and eliminating moiré patterns 297
Resizing Images in Photoshop 300
Exploring the Slideshow Editor 303
Understanding the Slide Area of the Slideshow Editor 304
Creating a Slideshow 307
Using Languages 311
Summary 313

Part IV: Bringing It All Together   315

Chapter 14: Building a Project 317

Preparing the DVD Project File 317
    Creating the tiles 318
    Associating the assets 319
Creating the Buttons 322
    Creating the buttons for the Main Menu 322
    Creating buttons for the Detholz Menu 324
    Creating the band member menus 327
    Understanding the beauty of the Duplicate function 327
    Cloning the Rick tile 328
Linking and Thinking 328
    Linking the Main Menu 329
    Linking the Detholz Menu 331
    Linking the Rick Menu 332
    Cloning the Rick Menu and altering its genes 333
    Completing the Detholz Menu 334
    Setting the Track properties 335
    Setting the Disc properties 336
Summary 338

Chapter 15: Preparing for Output 339

Testing Interactivity 339
    Reviewing remote-control settings 340
    Reviewing interactivity using the Preview button 340
    Checking Interactive settings in the Property Inspector 341
    Developing a Checklist 344
    Multiplexing 345
Multiplexing with Build Disc 346
Disc Previewing 348
Summary 349

Chapter 16: Outputting a Project 351

Understanding Multiplexing 352
    Using the Build Disc command 352
    Using the Build Disc and Format command 352
Burning a DVD-R in DVD Studio Pro 354
Burning a DVD-R in Toast Titanium 356
Outputting to DLT 361
    Manufacturing a DVD 361
    Outputting to DLT in DVD Studio Pro 362
DVD-R Media 363
DVD-R Media in DVD Players 366
    Trusting no one 366
    Sharing DVD-R-based projects 367
Outputting a Project to CD-ROM 367
Summary 370

Part V: Advanced Interactivity   371

Chapter 17: Understanding Scripting 373

Understanding the Concept of Scripting 373
    Comprehending variables 374
    Playing with menus 375
    Assigning scripts 378
    Understanding IF-THEN statements 378
Understanding the Script Editor 379
    Understanding the general drop-down menus 379
    Understanding the Helpers drop-down menu 381
    Befriending the Log window 391
Scripting Capability 392
Summary 393

Chapter 18: Adding Multiple Angles to a Project 395

The Importance of Multiple Angles 395
Preparing Tracks 396
    Managing multiple video files in DVD Studio Pro 396
    Preparing angles in Final Cut Pro 399
Creating Multiple Angles in DVD Studio Pro 406
Adjusting Bit-rate 411
Previewing a Multi-Angle Project 411
Summary 412

Chapter 19: Working with Multiple Languages 415

Understanding DVD Multiple Language Capability 416
    Choosing a default language 417
    Considering various multilingual approaches 418
Managing Multiple Languages in DVD Studio Pro 421

Making a Multiple Language DVD Project 422
    Adding a new language 422
    Assigning assets for the new language 426
Making an Alternate Reality Multiple Language DVD Project 431
    Setting up the Spanish assets 433
    Redirecting buttons in duplicated tiles 433
    Summary 440

Part VI: Exploring DVD Content Delivery   443

Chapter 20: Evaluating DVD Player Options 445

Investigating Copy Protection 445
    Debating the use of copy protection 446
    Contents Scrambling System 447
    Macrovision 448
Understanding Region Coding 448
Justification for Region Coding 450
Evaluating the Apple DVD Player 451
    Modifying Player Preferences 451
    Using the controller 454
Summary 455

Chapter 21: Developing DVD-ROM Content 457

What is DVD-ROM? 457
    How CD-ROM influenced DVD-ROM 457
    CD-R discs 458
DVD-ROM Capabilities 458
    Standalone content 458
    Player-based content 460
    Considering platform compatibility 460
Cross-Platforming Standalone Content 462
    Comparing Flash and Director 463
    Fireworks 466
    LiveMotion 467
    Adding player-based Flash on a DVD-ROM 467
Preparing DVD-ROM Content 467
    Manufacturing a DVD with DVD-ROM content 468
    Enhanced DVD-ROM? 469
    UDF Bridge 469
Exploring DVD-ROM Options 469
    Previewing cross-platform content with Virtual PC 469
    Evaluating PCFriendly/InterActual Player 471
Experimenting with Multiple DVD 472
Summary 473

Contents   xxvii

Chapter 22: Cross-Developing for DVD, CD-ROM, and the Internet 475

Investigating Delivery Options 475
    Paying passenger and stowaway content 476
    Perception and delivery 477
Considering Disc Delivery 479
    DVD 479
    DVD-ROM 479
    CD-ROM 480
    Video CD 481
Considering Internet Delivery 482
    DVD and Internet parallels 483
    Streaming media 485
    Streaming animation 486
    HTML delivery 487
Managing Workflow Between Applications 488
Real World Examples — Simulating DVD 489
    Simulating DVD in Flash 490
    Using QuickTime with Flash to simulate DVD 495
    Simulating DVD with Adobe Acrobat (PDF) 497
Summary 502

Part VII: Case Studies   503

Chapter 23: Case Study: Warm Blankets 505

Background 505
Project Planning 506
Development Issues 508
Motion Menu Loop Points with QuickTime Pro 508
Summary 511

Chapter 24: Case Study: Atomic Paintbrush Studios 513

Conversation with Atomic Paintbrush Studios 513

Chapter 25: Case Study: Pioneer Electronics' SuperDrive 523

Before the SuperDrive: Pioneer DVR-S201 523
The SuperDrive A03 Mechanism 523
DVD-R media 525
External FireWire DVD-R burners 525
Conversation with Pioneer Electronics 526

Chapter 26: Case Study: Simple Motion Menus 533
Creating a Motion Menu Using Adobe's After Effects 533
    Prepping the clips 533
    Creating the After Effects composition 534
    Adjusting layers in the Timeline 535
    Positioning the clips 536
    Exporting as QuickTime 536
    Encoding into MPEG-2 537
    Creating the motion menu in DVD Studio Pro 538
    Jubilation 539
Summary 540

Chapter 27: Case Study: Metatec — Manufacturing a DVD 543
Background 543
DVD Manufacturing 544
    Premastering 544
    Mastering 546
    Replication 547
    Printing 548
    Packaging 549
Pictorial Case Study: The GLF DVD 550
    Step 1: Ordering online — Getting a quote 550
    Step 2: Choosing a quantity 551
    Step 3: Choosing disc capacity 551
    Step 4: Choosing art colors 552
    Step 5: Choosing packaging 552
    Step 6: Finishing up 554
    Step 7: Preparing the art 554
    Step 8: Submitting the project and waiting 556
Discount for Readers 558
Summary 558

Chapter 28: Case Study: Real World Promotional DVD 559
Visual Tour of the Real World DVD Project 560
    Main menu 564
    Graphical View window 564
    General settings 565
    Photoshop layer arrangement 566
    Design guides in Photoshop 566
    Buttons in the Menu Editor 567
    Adjusting button display settings 568
    Preview 570
Summary 570

Part VIII: Appendixes — 571

Appendix A: DVD Studio Pro Installation and System Requirements 573

System Requirements for DVD Studio Pro 573
    Hardware requirements 573
    Software requirements 574
Installing and Configuring DVD Studio Pro 574

Appendix B: Resources 577

Appendix C: Glossary and Terms 581

Appedix D: About the DVD-ROM 585

What's on the DVD? 585
    Tutorial files 586
    Applications 586
    Case studies 586
    Goodies 586
    Full-Color electronic version of Macworld DVD Studio Pro Bible . . . 587
Using the DVD with the Mac OS 587
Troubleshooting 587

Index 589

End-User License Agreement 615

Introduction

Welcome to the world of DVD authoring!

Apple Computer is starting a DVD authoring revolution with DVD Studio Pro, which began shipping in February of 2001, following in the tradition of their successful Final Cut Pro video editing software. With the introduction of the SuperDrive, Apple brought affordable DVD burning to the mass market for first time. An excellent combination of professional DVD creation software and hardware is now available for Mac users, ushering in an era or virtually limitless creative possibilities for producers of both digital video and multimedia content.

Combined with a SuperDrive or external DVD-R burner, you will be able to produce a disc that actually plays back on your set-top DVD player. Or, using DVD Studio Pro, you can send a disc or DLT tape off for manufacturing a larger quantity. Whether you are a beginner or an expert, you will find that DVD Studio Pro is easy to use, yet powerful enough to meet the demands of advanced projects.

Our approach in writing this book was to combine step-by-step examples with in depth explanations to provide guidance for readers of any experience level. Through a series of chapter-based tutorials, sidebars and diagrams, case studies of several projects, as well as an accompanying hybrid DVD/DVD-ROM (which can be put in a regular DVD player or computer), we hope to provide the reader with valuable information that will assist them in the creation of their own DVD project.

With DVD Studio Pro, producing a DVD is not only easy but fun as well. With this book, we hope to give you a resource that will help you to complete the DVD project of your dreams, which (until DVD Studio Pro arrived) seemed nearly impossible.

As an added bonus to the material we have provided within the chapters, we sprinkled several chapters with commentaries or sidebars of interviews conducted with a fascinating array of movers and shakers in the audiovisual world. The people we spoke to are the following:

Chapter 8
- Justin Kuzmanich (Ad2) — Interface Design
- Alan Martin — Interface Design Considerations
- Brian Dressel (OVT Visuals) — Motion Menu Backgrounds
- Dan Fenster — Designing for DVD
- Jason Zada (Evolution | Bureau) — Design considerations, designing for TV

Chapter 9
- Nika Aldrich (Sweetwater Sound) — Mixing Surround Sound

Chapter 11
- Martin Baumgaertner (Angle Park Productions) — General commentary on DVD and video

Chapter 12
- Paul Feith (Consultant) — Encoding for Visual Quality

Chapter 16
- Jim Taylor (author of DVD Demystified) — Q/A about DVD
- Lee Purcell (author of CD-R/DVD Disc Recording Demystified) — Article on DVD-R
- Andy Marken (Markencom, a PR firm) — Viewpoints on DVD and DVD-R

Chapter 21
- Brad Mooberry (Ad2) — Dynamic CD

Chapter 28
- York Tillyer (Director of Real World Interactive) — Commentary on Real World initiatives

We hope you find the interviews and commentary to be insightful and inspiring as you immerse in the world of DVD authoring and DVD Studio Pro.

Chicago, Illinois

Todd Kelsey
dvdchicago@aol.com

Chad Fahs
dvdstudiopro@aol.com

Working with DVD Studio Pro

PART

I

In This Part

Quick Start
Up and Running with DVD Studio Pro

Chapter 1
Understanding DVD Studio Pro

Chapter 2
Inside the DVD Studio Pro Workspace

Chapter 3
Working with Assets

Up and Running with DVD Studio Pro

In This Chapter

Starting the DVD project

Importing assets

Setting up menus and tracks

Creating and linking buttons

Burning a DVD

Welcome to the world of DVD Studio Pro! In this quick chapter, you get a whirlwind tour of the program and are taken through the process of creating a DVD from start to finish. You can use your own files, or follow along and use files that are already provided for you from the DVD-ROM.

If you plan to use the files from the disc, we recommend that you make a copy of the PS Records DVD Assets folder, located in the Tutorial section of the DVD-ROM, by dragging it to your hard drive.

Before you do anything, make sure you have upgraded to the latest version of DVD Studio Pro, especially if you are using Version 1.0. Version 1.1 adds several enhancements and is a free upgrade. If you are not sure which version you have, open DVD Studio Pro, select the Apple menu in the upper left-hand corner of the screen, and choose About DVD Studio Pro. (This is also a convenient time to write down the serial number, which you will need to download the free upgrade from www.apple.com/dvdstudiopro/update/.)

So without further ado, let's begin!

Starting a DVD Project

The overall process of making the DVD will consist of creating a new project file, importing the assets (the graphic, video, and audio files), creating the project elements such as menus and video tracks, and linking everything together.

To start a project, follow these steps:

1. **Start DVD Studio Pro by clicking the application icon located in the DVD Studio Pro folder. (See Figure QS-1.)**

4   Part I ✦ Working with DVD Studio Pro

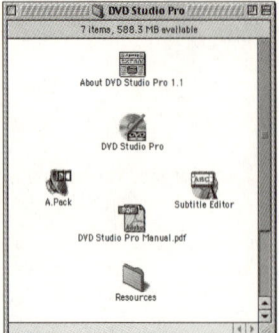

Figure QS-1: The DVD Studio Pro folder, including DVD Studio Pro and the supporting programs, A.Pack and Subtitle Editor

When the program opens up, take a moment to review the workspace. The Graphical View window and the Property Inspector (See Figure QS-2) will be the areas where you will be doing most of the adjustments to the project.

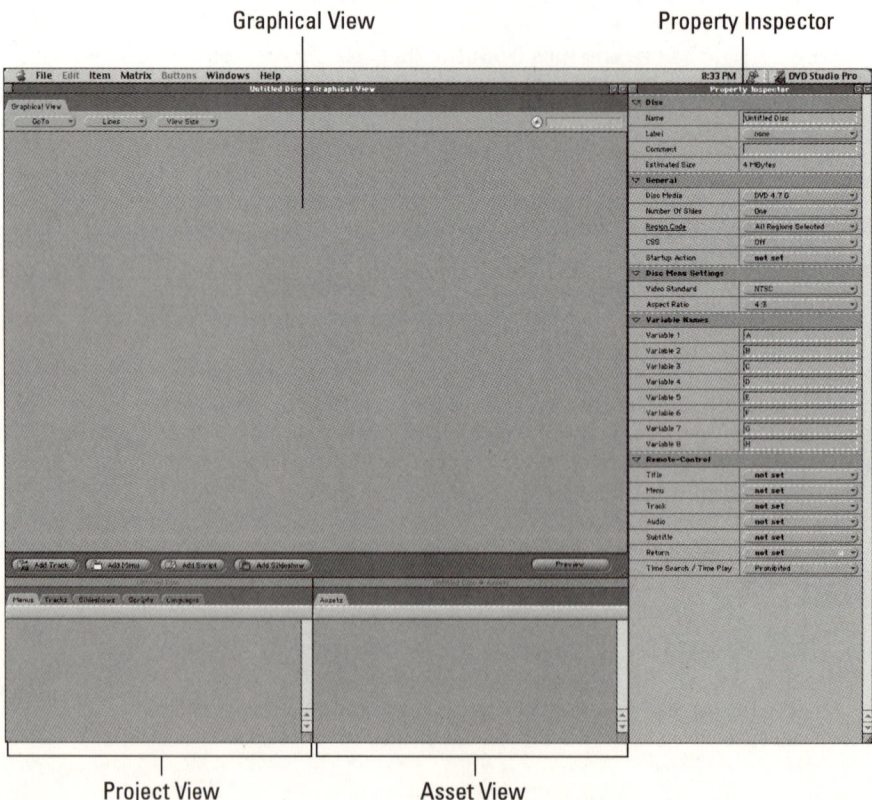

Figure QS-2: Bird's-eye view of the DVD Studio Pro workspace

2. **When the program starts up, it automatically creates a new file for you.** If you already have the program open and want to create a new file, choose File–>New.

3. **Create a new menu by clicking the Add Menu button at the bottom of the Graphical View window.** A blue tile appears. (See Figure QS-3.) A *tile* is the basic building block of DVD Studio Pro, and each of the main project elements has its own respective color.

    ✦ **Menu tiles:** blue
    ✦ **Track tile:** green
    ✦ **Slideshow tiles:** dark gray
    ✦ **Script tiles:** orange

   It is necessary to add something to a DVD Studio Pro project before you can save it.

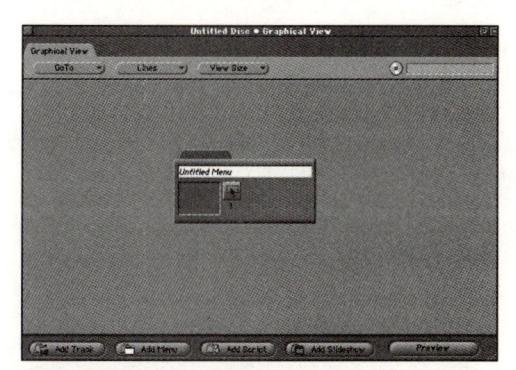

Figure QS-3: The new, untitled menu tile. The name of the tile appears in italics until assets are associated with it.

4. **Save the DVD Studio Pro project by choosing File–>Save.** The Save dialog box appears (Figure QS-4). If you haven't already, create a folder for your DVD by clicking on the New (folder) button. Choose a name for your project and click the Save button.

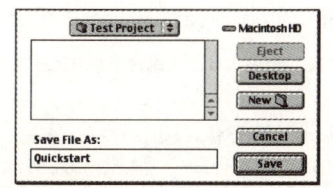

Figure QS-4: The Save dialog box

5. **Click the menu tile to select it and choose a name for the menu (See Figure QS-5).** You can change the name by clicking in the white area at the top of the menu tile, or by changing the name in the Property Inspector.

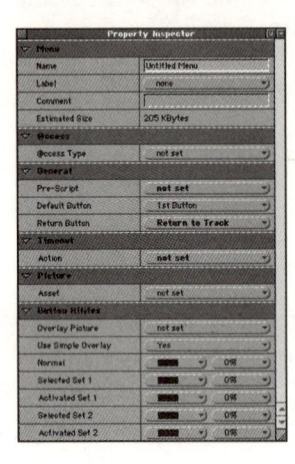

Figure QS-5: Menu properties. When a project element tile is selected, its properties show in the Property Inspector.

You have completed the first step in setting up your DVD. The next step is to import the graphic, video, and audio assets for your project.

Importing Assets

To import files, they must be in a format that is compatible with DVD Studio Pro. For example, video must be in DVD-compatible MPEG format. There are more choices with audio, but in many cases, you will probably end up with an audio file when you encode your video into MPEG format, using a program like QuickTime Pro. There is more information on compatible file formats in Chapter 3.

For menus, DVD Studio Pro is designed to work with Photoshop .PSD files that have multiple layers. In most cases, you will want to have separate layers for each aspect of a menu, such as buttons, foreground images, and some kind of background.

To import assets for your project, follow these steps:

1. **Choose File–>Import and open the folder where your assets are located.**
2. **In the Import dialog box, click each asset that you want to import and then click the Add button or click Add All.**
3. **When you have added the files you want, click the Import button (see Figure QS-6).** If you are using the files from the DVD-ROM, locate the PS Records DVD Assets folder that you copied to your hard drive, and add everything except bandmembers.psd and detholzmenu.psd.

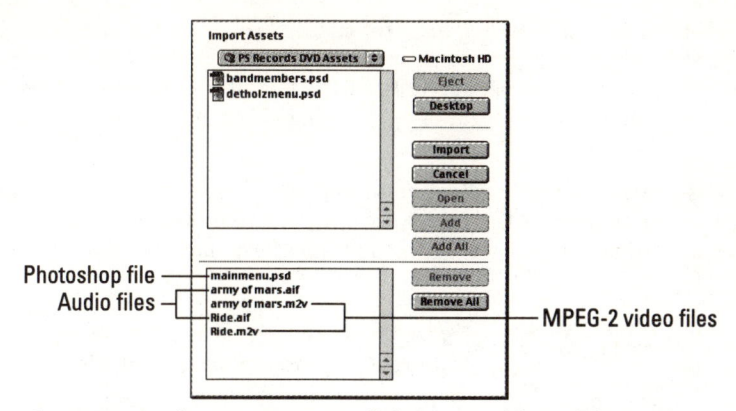

Figure QS-6: The Import Assets dialog box. Add your files, and click Import.

When you have imported the files into the program, the files appear in the Asset View window (see Figure QS-7).

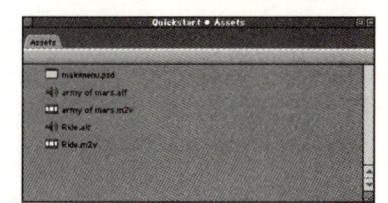

Figure QS-7: The Asset View window, showing imported assets

Don't be alarmed if you notice that when you click on the background in the Graphical View window, something different appears in the Property Inspector than when you select the menu tile you just created. What appears is the overall disc properties.

Before you set up the rest of your DVD, it is a good idea to set the Startup Action. The Startup Action is the first thing your DVD does when it is inserted in a player. For example, if you do an entire project and burn a disc without setting the Startup Action, the DVD may not do anything at all.

The rest of the disc properties can often be left as they are for simple projects, but you will always want to set a Startup Action. Using the Startup Action is a good habit to get into, and an easy thing to forget. (Author's Confession: This happened to me the first time I used DVD Studio Pro.)

To set the Startup Action, follow these steps:

1. **Click anywhere in the background of the Graphical View window to bring up the disc properties in the Property Inspector.** If you like, you can choose a name for your disc.
2. **In the General area of the Property Inspector, click the Startup Action drop-down menu and select the menu tile that you previously created.** This menu is the first thing you will see when the DVD starts playing. (See Figure QS-8.)

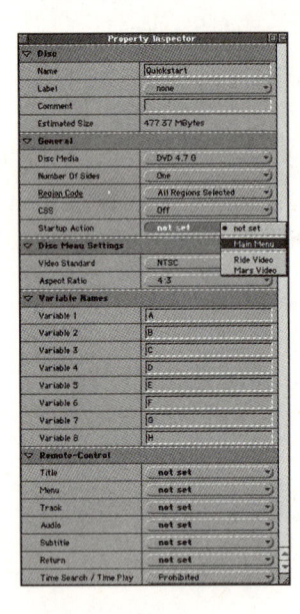

Figure QS-8: The disc properties, showing the Startup Action set to go to the Main Menu tile

Now that the assets are imported and the Startup Action is set, you are ready to begin setting up project elements such as menus and tracks.

Setting up Menus and Tracks

When you insert a DVD into a DVD player or in your computer, a DVD menu comes up, unless the disc is set up to go to something else. DVD menus allow you to access the various parts of a DVD.

In DVD Studio Pro, menu tiles hold what you want to appear in your menu, and the menus end up being connected to track tiles, which hold the video and/or audio. A menu can either be a *still menu*, where a Photoshop file is used, and the menu itself has no video or sound, or it can be a *motion menu*, where you have some kind of video or animation running underneath. In this example, we are setting up a still menu.

Setting up a menu

To set up a menu, you need to tell the menu tile which asset you want it to display. Follow these steps:

1. **Click on the menu tile you created earlier to bring up the menu properties in the Property Inspector.**
2. **In the Picture area of the Property Inspector, click the Asset drop-down menu and select the Photoshop file that you want to use for the menu.** (If you are using the Tutorial files, select mainmenu.psd.) When you select a Picture Asset, a new option appears underneath the Asset section, named Layers (Always Visible). This is where you select which layers of the Photoshop file you want to be visible.
3. **Click on the Layers (Always Visible) drop-down menu and select the layers you want to appear in the menu.** (For the mainmenu.psd, select all five layers.) To select a layer, drag the cursor down to an individual layer, click the drop-down menu again, select another layer, and so on. The black indicates a layer is selected.
4. **In the Button Hilites area of the Property Inspector, click the Selected Set 1 drop-down menus to select a color and level of transparency.** This setting affects the way a DVD menu button will look when selected. (For the Tutorial files, select gray and a value of 33%.) (See Figure QS-9.)

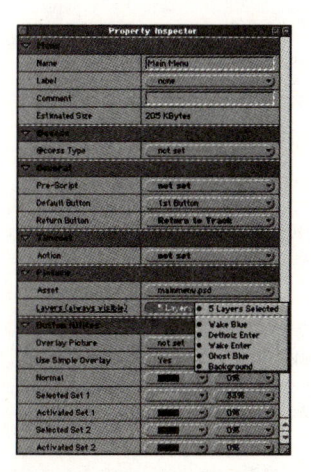

Figure QS-9: The menu properties, featuring the Layers (Always Visible) menu

Now that you have set up your menu, you are ready to move on to setting up a track.

Setting Up a Track

Creating and setting up a track is very similar to creating a menu. For this exercise, you will want to create track tiles to correspond to each of the video clips you imported. Follow these steps:

1. **To create a track tile, click the Add Track button at the bottom of the Graphical View window and name it in a way that corresponds to the video you will be using.** If you are using the Tutorial files, create two track tiles, naming them Mars Video and Ride Video.

2. **Select each track tile to bring up the track's properties in the Property Inspector and in the Video area, select the video asset you want to use for the track.** For the Tutorial files, select army of mars.m2v for the Mars Video track, and Ride.m2v for the Ride Video track.

3. **In the General area of each track, set the Jump When Finished value to the menu tile you created (Main Menu).** This setting determines where a viewer will be taken when the video segment in the track is done playing. (See Figure QS-10.)

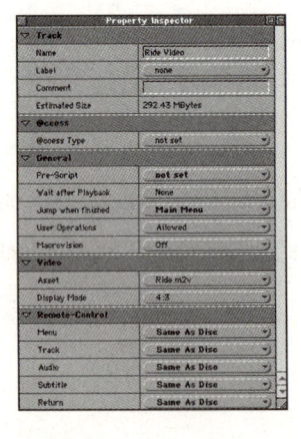

Figure QS-10: The track properties in the Property Inspector

4. **To add audio to each track, click on the appropriate audio asset in the Asset View window and drag it to the corresponding track tile.** For the Tutorial files, drag army of mars.aif to the Mars Video tile and drag Ride.aif to the Ride Video tile.

If you want DVD Studio Pro to show you the connections between your project elements, you can click on the Lines drop-down menu at the top of the Graphical View window and set it to Always. This will make lines appear between tiles to indicate any connections. (See Figure QS-11.)

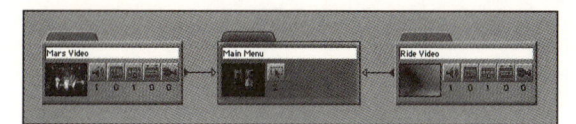

Figure QS-11: A close-up view of a track tile connected to a menu tile. In this example, the arrow indicates that the Jump When Finished value for the tracks are set to lead back to the menu.

Previewing Tiles

At this point, you may wish to preview some of the tiles. To preview a menu or track tile, simply select it in the Graphical View window and then click the Preview button at the bottom-right corner of the Graphical View window (see Figure QS-12).

Figure QS-12: The Preview window, showing a video segment from the Ride Video track tile

The buttons at the bottom of the Preview window (see Figure QS-13) simulate the actions that are available with a DVD remote control. When you are done previewing, click the Stop button (the one with the solid square on it) at the bottom of the Preview window.

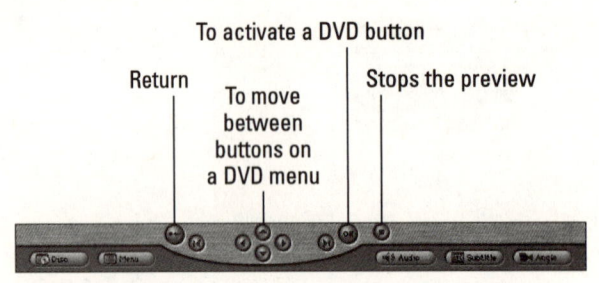

Figure QS-13: The controls at the bottom of the Preview window

Now that you have set up your menu and track tiles, you are ready to create some buttons!

Creating and Linking Buttons

Buttons in a DVD menu allow you to link a menu to another project element, such as a track or even another menu.

To create a button in a menu, follow these steps:

1. **Double-click on the thumbnail image area in a menu tile to open up the Menu Editor.** A new button automatically appears in the upper-left corner of the screen.

2. **Move the cursor over the button so that the hand appears (see Figure QS-14).** Click on the menu button, hold the mouse down, and drag the button over the part of the image you want to make the button for. (For the Tutorial files, move the button over the Enter text on the left.)

3. **Place the cursor on the corner of the menu button and adjust it to the shape you want (see Figure QS-15).** The shape of the menu button determines the area that will appear highlighted when the button is selected.

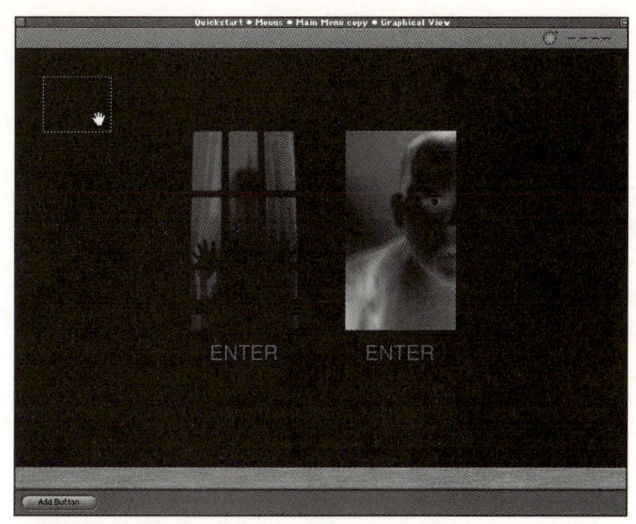

Quick Start ✦ **Up and Running with DVD Studio Pro** 13

Figure QS-14: The Menu Editor, showing the new button in the upper left-hand corner.

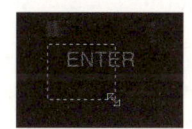

Figure QS-15: Adjusting the size of the button

4. **Add as many buttons as necessary by clicking Add Button at the bottom of the Menu Editor screen, moving the buttons to the appropriate area and adjusting the size.** (For the Tutorial file, create another button and put it over the remaining Enter text.)

5. **Select each button and in the Property Inspector set the Jump When Activated value (see Figure QS-16) to reflect which track tile you want the button to lead to.** (For the Tutorial file, select Mars Video for the Enter button on the left and Ride Video for the other button.)

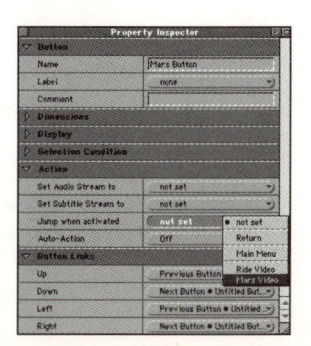

Figure QS-16: Selecting the Jump When Activated setting in the Property Inspector, with a button selected in the menu editor

6. Close the Menu Editor by clicking in the square in the upper left-hand corner of the Menu Editor window.

Congratulations, your DVD is ready to go! (See Figure QS-17.)

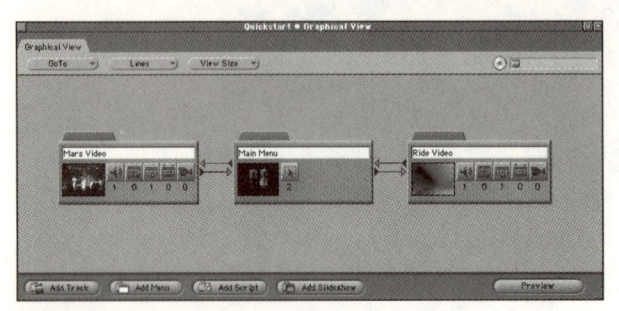

Figure QS-17: Menu tile connected to the track tiles. The finished DVD.

To preview your DVD, click on the background in the Graphical View window, and click on the Preview button at the bottom of the window. If things don't seem to make sense, go back and make sure you addressed each one of the settings in these step-by-step examples. For example, if you don't see any hilites on the screen for the buttons, you may have forgotten to set the Selected Set 1 values for the menu, which can be seen in Figure QS-9.

If you are using the SuperDrive or if you have an external DVD-R burner connected, you may wish to try burning a disc. Before you do, go back and test everything, and make sure you set the Startup Action!

Burning a DVD

When you're ready to burn the DVD, unwrap a beautiful, tasty DVD-R disc, and if you've never done it before, take a moment to gaze at its wondrous, revolutionary circular nature. Ahhhhh.

After you have appreciated it for a while, make sure that it's a DVD-R General disc. If you got the disc from Apple, it is definitely the right kind. Most people will need to use the DVD-R General disc, unless you happen to have a DVD-R burner which uses DVD-R Authoring media.

Tip: If you are using an external FireWire DVD-R burner, upgrade to DVD Studio Pro Version 1.1. The free upgrade adds support for some external FireWire DVD-R burners.

Quick Start ✦ **Up and Running with DVD Studio Pro** 15

To burn the DVD, follow these steps:

1. **Insert the disc in your DVD-R burner, and choose File–>Build and Format Disc.** Choose a location where the files for your DVD will be written, and click the Select button at the bottom of the window. (See Figure QS-18.) The VIDEO_TS and AUDIO_TS folders are written to your hard drive, during a process known as multiplexing, where all the information in the DVD project is encoded into a format that a DVD player will understand.

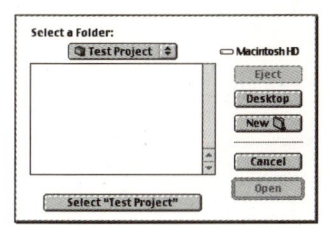

Figure QS-18: The Select a Folder dialog box

2. **In the Format window, select the DVD-R burner and click the OK button (Figure QS-19).**

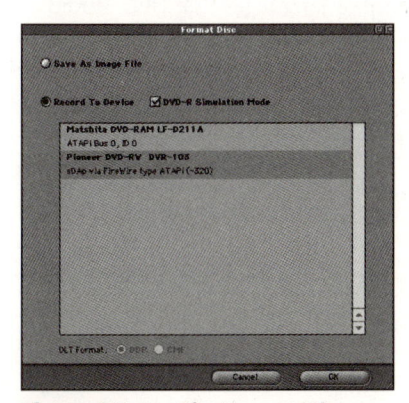

Figure QS-19: The Format Disc window. This is where you choose the output device.

> **Note** If you are using an external drive and it doesn't show up in the Format window, make sure you are using at least version 1.1 of DVD Studio Pro and check with your drive manufacturer to see what software you have to have installed to be able to use it. You may only be able to burn the DVD with a program like Roxio's Toast. In this case, simply choose File-->Build Disc, and when you have the VIDEO_TS and AUDIO_TS folders, burn them to the disc with a program other than DVD Studio Pro.

3. Kick back with a bottle of Jolt Cola and watch the mysterious multiplexing window (Figure QS-20).

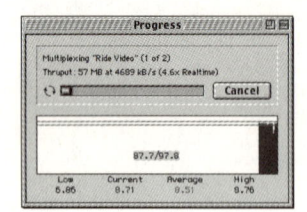

Figure QS-20: The Multiplexing Progress window

4. **Take the disc and put it in a DVD player. Woohoo!**

> **Note** If you didn't exclaim "Woohoo!" when you inserted the DVD disc in the player, and instead you cursed in any one of the many languages that DVD Studio Pro understands because the disc didn't play, check the compatibility lists for DVD-R media. Not all DVD players are compatible with DVD-R media. For some links, check www.dvdspa.com.

Well, that wraps it up for the Quickstart.

If you're completely new to DVD, don't worry if you have a lot of questions. In Chapter 1, you will get a more extended introduction to the program as you get to know more about its capabilities and some of the related concepts.

Welcome to the wacky world of DVD authoring, and keep on truckin'! By the time you reach the end of this book, you'll be able to say, what a long, strange trip it's been!

✦ ✦ ✦

Understanding DVD Studio Pro

CHAPTER 1

✦ ✦ ✦ ✦

In This Chapter

An introduction to DVD Studio Pro Version 1.0

DVD Studio Pro projects

Working with project files

✦ ✦ ✦ ✦

This chapter introduces the program, the new terminology, and the basic building blocks that you use to construct DVD projects. In addition, we describe how to work with project files and provide some suggestions on file management to ensure a smooth ride.

The power of DVD Studio Pro is the way that it allows you to combine files from other programs (such as image files from Photoshop and video files from Final Cut Pro or Premiere) and incorporate them into an overall project file, which ultimately becomes a DVD. When you use DVD Studio Pro to author a DVD, DVD Studio Pro generates a series of files that are burned onto the disc, which the computer inside the DVD player translates to create the interactive experience.

DVD Studio Pro and DVD Technology

Users of DVD Studio Pro are taking part in a revolutionary breakthrough in technology, with an easy-to-use software environment that has a low learning curve. This program is excellent right out of the box and includes a helpful suite of tools that takes the headache out of DVD authoring so that you can jump right in. DVD Studio Pro harnesses the power of DVD interactivity and helps you organize the related elements of your project.

Before delving into the depths of DVD Studio Pro, you'll want to review some DVD standards, because DVD Studio Pro's capabilities are tied closely to the capabilities of DVDs in general.

DVD storage capacity

The storage capacity of a standard, single-sided DVD, also known as the DVD-5 standard, is 4.7 gigabytes. At first glance, 4.7 gigabytes seems like a lot of room to work with. However, when you start incorporating all of the features of DVD technology, the space fills up. Roughly speaking, video in a DVD project uses less than one megabyte per second, and depending on the compression and what kind of audio is used, a single-sided DVD can hold about two hours of video. Adding such features as multiple audio and video tracks takes up even more space.

> **Tip** As you become more familiar with DVD features and the relative file sizes, practice estimating storage requirements so that when you start planning for larger projects, you will have a better idea of what the size consequence of using certain features may be. For more information on project planning, see Chapter 4.

DVD menus

All DVDs are based around *menus* — a selection of features (each assigned to an individual button) that the user selects to view or listen to content (see Figure 1-1).

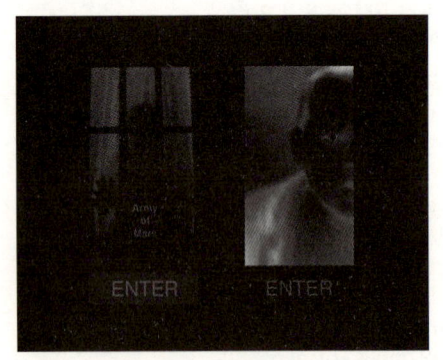

Figure 1-1: An example of a simple DVD menu, with the Enter button selected

All DVD players share common features, such as the way you navigate through menus with the remote control. All of these features are accessible when you develop a project in DVD Studio Pro. With DVD Studio Pro, you can customize the user experience to a certain extent by disabling specific keys on the remote or programming special functions when certain keys are pressed. However, the function of each key remains the same from project to project.

DVD video

All DVDs use a standard DVD video format — the MPEG-2. *MPEG* stands for Moving Picture Experts Group, which is a series of standards developed to help people encode in a compressed audio and video format. In general, MPEG compression allows for better quality at smaller file sizes than other compression techniques. Before video can be incorporated into a project, it must be encoded into this format. DVD Studio Pro includes a capable software-based MPEG-2 encoder that we discuss in detail in Chapter 12. DVD Studio Pro contains no video editing capability, so video needs to be prepared in another application such as Apple's Final Cut Pro or Adobe Premiere.

> **Note** Of all the various tasks related to making a DVD project, encoding video content can often be the most time-intensive. Even with today's speedy processors, it can take some time but does provide you with an opportunity for a break!

DVD audio

DVD Studio Pro supports a wide variety of audio formats, giving you the flexibility of working with anything from basic mono to full-blown Dolby 5.1 Surround Sound. As with video, the recording of audio is up to you outside of the DVD Studio Pro environment. DVD Studio Pro has no sound input or editing capability. In many cases, you may decide to use the original audio that was recorded along with the video through the video camera, but for more advanced projects an application such as Bias' Peak may be appropriate.

> **Cross-Reference** For more detail on DVD audio, see Chapter 9 and Chapter 10.

Multiple languages

It's the 21st Century, and it's time to get multilingual! An advanced feature of DVD technology is the use of multiple languages, and DVD Studio Pro shines in this area. For example, one entertaining challenge you can undertake when learning DVD Studio Pro is to do an entire project in Esperanto. (Esperanto was an attempt at creating a universal language, popular in the '70s, by using a mixture of various languages, including English, Spanish, and German. The language never really caught on, but Esperanto buffs may be amused to note that William Shatner starred in a movie entitled *Incubus* that was filmed entirely in Esperanto.)

The use of multiple languages in DVD Studio Pro is accomplished in a variety of ways, but the most typical plan for a multilingual project is to use multiple audio streams to represent each given language with a video track.

DVD Studio Pro enables you to associate project elements with just about any language in the world (see Figure 1-2), so you can have custom visual and audio content automatically appear after a user selects a language. You can even set the project up to automatically select which language to use and play the appropriate underlying audio stream.

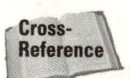

> **Cross-Reference** See Chapter 19 for more detail on the use of multiple languages.

Figure 1-2: Available languages in DVD Studio Pro

Subtitles

Budding kung fu directors and other multilingualists can add and edit subtitles with the DVD Studio Pro Subtitle Editor. The Subtitle Editor is a program that comes with DVD Studio Pro, but it is a separate application, launched separately from DVD Studio Pro. The Subtitle Editor enables you to type in text or import from an external source and preview the video content to properly position the subtitle (see Figure 1-3).

Figure 1-3: The Subtitle Editor

When you are working with subtitles, the original video file is not affected. DVD Studio Pro, also known as DVDSP, subtitles the story through generating a *subtitle stream,* or data that a DVD player interprets and uses to display text over the video on a television. The text overlay is generated on a television in a process that is similar to closed-captioning.

Cross-Reference: See Chapter 7 for further discussion of subtitles.

DVD Studio Pro Projects

The process of creating a project in DVD Studio begins with the preparation of source material, also referred to as *assets*. Examples of assets might include a Photoshop file or a video file in the MPEG-2 format. This preparation is where you shoot, edit, and encode your video content and also prepare any graphics that are used in the project in an image editing program, such as Adobe Photoshop.

Cross-Reference: See Chapter 3 for more information on how to prepare source material.

This book concentrates on the second phase of a project — where you bring your various assets together in DVD Studio Pro to set up the interactive experience. DVD Studio Pro is considered to be an authoring environment. An *authoring environment* gives you the ability to set up relationships between all files that you import into the program. With an authoring environment, a series of tools is also provided to help define the characteristics of the finished project. (Doing so enables you to go to parties and say that you are an author.)

If you are just beginning to learn DVD authoring, keep in mind that you will probably end up doing most of the work for a DVD project outside of DVD Studio Pro (see Figure 1-4). You move between various applications when you are starting a new project in DVD Studio Pro, importing graphics and content, and then going back into programs, such as Photoshop or Final Cut Pro to readjust source material or create new content. As you become more familiar with how DVD authoring works, you develop the ability to forecast what you need, so that DVD Studio Pro becomes the final stop in burning your finished DVD.

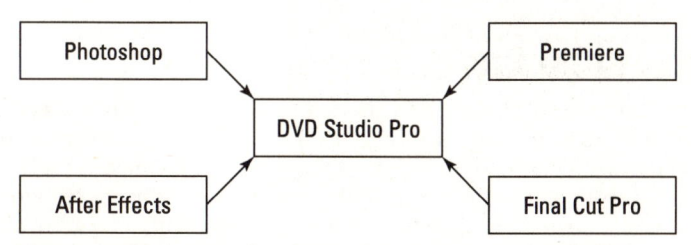

Figure 1-4: An example of typical programs used with DVD Studio Pro to make the DVD authoring cycle complete

DVD Studio Pro building blocks

For most projects created in DVD Studio Pro, you'll use two basic building blocks — tracks and menus. You can think of tracks as containers that hold video and audio. A user reaches the tracks is through the menu by using their remote to make a selection in the menu and play the track.

DVD Studio Pro simplifies the process of setting up your project by visually summarizing project elements as you go so that you can easily click any individual track or menu and change its properties or readjust the relationships between them (see Figure 1-5).

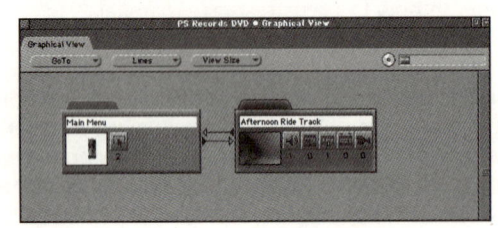

Figure 1-5: A menu and track tile

DVD Studio Pro can go beyond tracks and menus and also create digital slideshows. Digital DVD slideshows enable the audience to experience a series of pictures, with or without audio. Some readers may remember gathering in the living room, sitting in front of a vast, white screen, and jockeying for the nonwireless remote control, and skipping over the embarrassing pictures. A DVD can probably hold several thousand pictures, so break out your digital camera!

Cross-Reference For more information on digital slideshows turn to Chapter 13.

Tracks in DVD Studio Pro

DVD Studio Pro organizes your prerecorded video content into individual tracks, and a project can include up to 99 of these video tracks. Each track in turn can have up to eight *audio streams* associated with it. For example, you may have a DVD project that includes 10 different short films. You want to have each film represented in its original form, but you also want to give the director and certain actors the ability to have their own running commentary on the film. No problem! The video for each film ends up having its own track, and each track has several audio streams — one for the original sound in the movie and separate audio streams for each person who will be commenting. The user can switch between the two audio streams as he watches the film. You can also set up *markers* within a video track to enable the user to jump to a particular scene in the video.

Within each video track, you can have up to nine individual *angles*, sometimes known as *camera* angles. The angle feature essentially enables you to add multiple video streams. And although you can technically shoot a video segment by using nine different cameras, more likely, you'll use the angle feature simply to enable the user to see alternative visual content running over the same audio.

Menus in DVD Studio Pro

In DVD Studio Pro, you can have as many menus as you want, and you can put up to 36 buttons on each menu. The only limits are the storage capacity of the DVD medium and how much time you want to spend creating the graphics.

Still menus

With DVD Studio Pro, the number of menus you can put on a DVD is limitless, as long as you don't exceed the storage capacity of the disc. Each addition of a menu corresponds to a certain decrease in disc space, depending on the size of the files you use with a menu. Most of these menus will be *still menus*, where nothing is moving, except when you click the remote and move between the selections on the screen. As you will discover, the most typical application you will use to create still menus is Photoshop.

> **Cross-Reference**
> For more information on still menus, see Chapter 6.

Motion menus

A good example of a motion menu on commercial DVD releases is where animation runs in the background, while you decide which part of a DVD to experience. The individual menu choices can even be a video clip from a particular scene.

Preparing the source material for use in a Motion menu often involves a technique known as *compositing*, where you take various clips and merge them into one overall video segment. Compositing is done outside of DVD Studio Pro, in a program such as Adobe's After Effects.

With a basic DVD project that features two music videos, you might start out by planning the main menu. Pictures could be taken from the videos to be the menu buttons, where a person selects one of two images, and it becomes highlighted, and when they click the remote the music video starts playing.

Then the client might come back and say that they want something more lively, so you suggest that you could do a motion menu, and you go back and select a few seconds from each music video and make loops out of them. Then, using After Effects, you take the original background image and overlay the two video loops you just created, and you end up with a video segment that has the two separate clips running simultaneously next to each other. These short video loops end up being the buttons that allow you to select either of the entire music videos.

Cross-Reference: More information on Motion menus check out Chapter 6.

Tip: For those who want to try their hand at Motion menus, but who don't have After Effects or Final Cut Pro, compositing can be fudged by using the Motion Control feature in Adobe Premiere. A fully-functional trial version of Premiere is available on the DVD-ROM.

Menus and multilayered image files

DVD menus usually incorporate a background image and images for buttons. The buttons are then linked to video tracks or other menus. DVD Studio Pro is designed to work with multi-layered image files, where you create separate layers in the file for the various parts of the menu (see Figure 1-6).

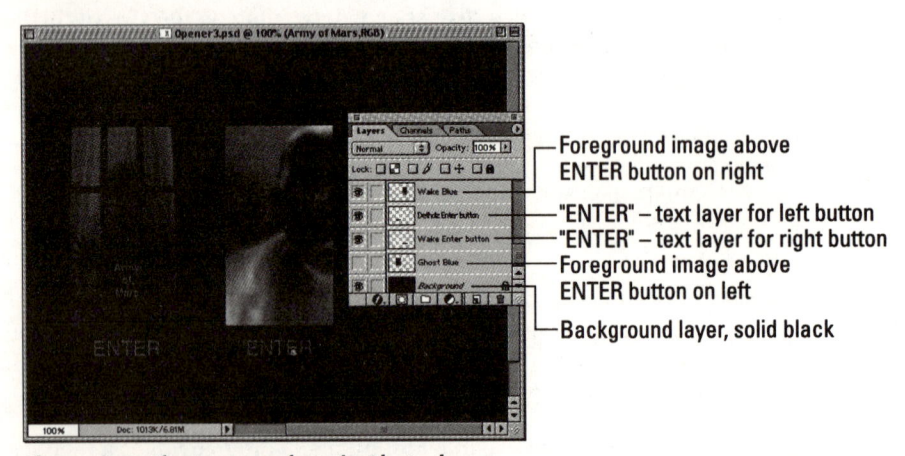

Labels on figure:
- Foreground image above ENTER button on right
- "ENTER" – text layer for left button
- "ENTER" – text layer for right button
- Foreground image above ENTER button on left
- Background layer, solid black

Figure 1-6: The Layers palette in Photoshop with background and button layers

One interesting characteristic of working with menus in DVD Studio Pro is that you import Photoshop files without flattening the layers first. This enables you to import the layers into the menu and work with each layer. If you were saving an image for the Web, first you need to flatten the image (no layers) and then export the image. The previous layers would not be available to you later.

Tip: Macromedia's Fireworks program is an alternative to Photoshop. Fireworks generates multilayered files in the Photoshop .psd format. Using Fireworks may be a route for you if you are interested in cross developing content for DVD and Web delivery. (More on cross-developing content in Chapter 22.)

When you create a menu with DVD Studio Pro, importing a multilayered .psd file enables you to treat the layers as individual units. For example, you can make one layer the background image for a menu and then make a different layer into a button that starts a video segment playing.

> **Note** When you import your Photoshop file into a DVD Studio Pro project, if you need to make a change to the original file, you can go back into Photoshop, edit the image, save it, and then when you go back into DVD Studio Pro, the change will be reflected in the project file.

Working with Project Files

A program such as DVD Studio Pro creates *project files*. A project file includes data about the assets used to create the project, making reference to these files, instead of embedding the assets themselves. Another example of a program that works like this is Adobe Premiere, which has project files. The Premiere project file is not a video file. Instead, it provides a description of what settings are used in the project and how the project elements relate to each other. Storing the data instead of the project results in a reduction of space.

Unlike Premiere, when you preview projects in DVD Studio Pro, the program does not generate *preview files*. Instead, DVD Studio Pro generates previews instantly, so you do not need to be concerned about rendering preview files or keeping track of them. Knowing how project files relate to asset files helps to keep track of both the project file and all asset files used in the project (something especially useful if you have to move from machine to machine or to a new location.)

Starting a project

When you start DVD Studio Pro, it automatically creates a blank project file, ready for importing assets and starting the DVD project (see Figure 1-7). The DVD Studio Pro workspace is explained in Chapter 2.

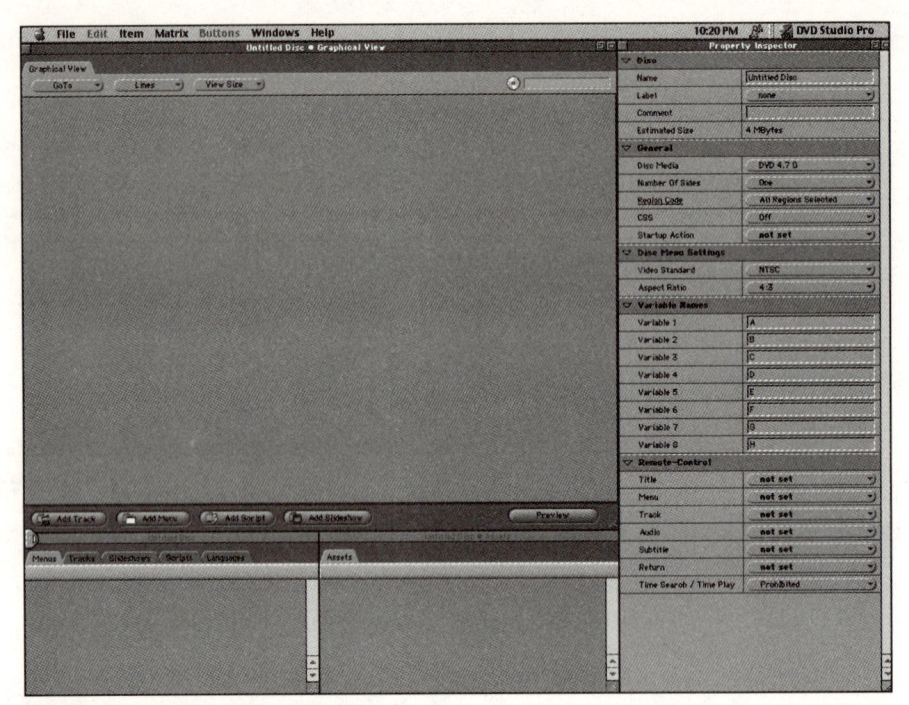

Figure 1-7: Fresh new project file

Saving a project

Saving a DVD Studio project file is a straightforward process. Any assets that have been imported are left in their original location. The only information saved with the project file is the information on the interactive relationships that have been set up between imported assets. The file size of a typical project file is less than 1MB and can theoretically be stored on a floppy disk.

> **Tip** Confucius says, brush your teeth, save regularly, and always back up your project files! Confucius knows that average G4 owners have no floppy disk drive and may not have a CD burner if they aren't running DVD Studio Pro on the 733Mhz tower. But you probably have an Internet connection, and e-mailing the file to yourself is an easy way to make a temporary back-up.

Remember 3.5-inch floppy disks? Now, does anyone out there remember 5¼-inch floppy disks? I wonder if someone out there has managed to set up their Apple IIe with an Ethernet connection could also take the challenge of saving a DVD Studio Pro project file to an antique 5¼-inch floppy disk.

File management

Folders, folders, and more folders! Setting up a folder structure for your project even before you create the project file is worth the effort. Creating folders saves time, money, blood, sweat, and tears. The Mac OS and DVD Studio Pro are fairly smart at finding files when you move the files around. But a time may come when you need to do more than one DVD project or take an entire project to another location. You may start wondering where all those video clips and graphic files really are.

One method that may be helpful is to create an overall project folder and then to create subfolders for video, audio, and graphics (see Figure 1-8).

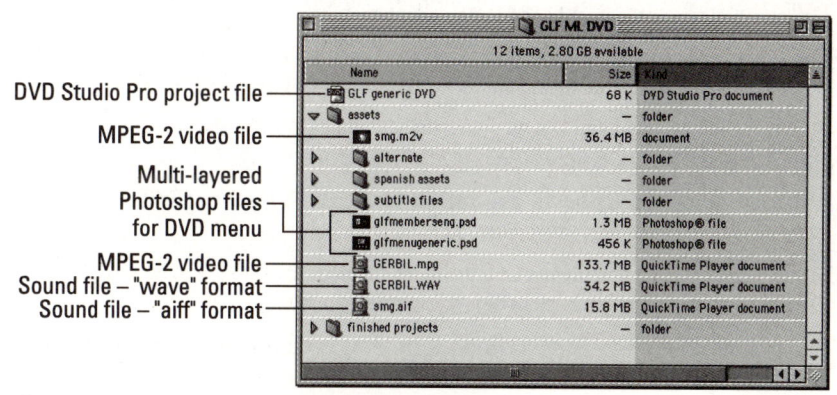

Figure 1-8: A basic approach to file management

Try another approach. Consider setting up separate folders for each section of a DVD project. If you are going to end up with a lot of asset files and want to keep them straight (see Figure 1-9), this approach works well.

Figure 1-9: The extended approach

Do it. Set up a folder structure. Whatever works for you. Ask yourself the question, if your client calls with a last-minute request to change the picture of the vice president that appears in a nested menu buried deep within the DVD, and you were just about to leave to catch a plane to Malaysia to visit a chip factory, how long would it take you to find the file?

Summary

- The storage capacity of a standard, single-sided DVD, also known as the DVD-5 standard, is 4.7 gigabytes.
- DVD Studio Pro uses the standard MPEG-2 DVD video format.
- DVD Studio Pro can generate DVD projects with up to 99 video tracks, and each track in turn can have up to eight audio streams associated with it.
- Still menus are motionless, where individual still images are used as buttons that lead to playing a selection or lead to another menu.
- Motion menus incorporate animation or video loops by using a video clip or composite as a background image.
- Compositing for Motion menus is done outside of DVD Studio Pro by using a third-party application, such as Adobe After Effects.
- DVD Studio Pro generates a project file that contains data about interactivity and how the assets relate to each other, but it does not store the assets themselves within the project file.

✦ ✦ ✦

Inside the DVD Studio Pro Workspace

CHAPTER 2

In This Chapter

Understanding Graphical View

Understanding the Project View tabs

The Assets Container

Using the Property Inspector

Additional ways to view a project

In this chapter, you take a visual tour of the DVD Studio Pro environment and walk through each primary interface element to get better acquainted with the workspace.

The DVD Studio Pro environment is based on a screen called the *workspace* (see Figure 2-1). In the workspace, you assemble, modify, and use all program features.

The workspace is made up of four primary areas. The large window occupying the upper-left portion of the screen is labeled Graphical View. This space is used to visually represent the various elements of the project. Beneath the Graphical View are two smaller windows with tabs. The window at the bottom-left corner is the Project View, where you can view project elements as you might view files in a filing cabinet, with tabs for easy access to different subjects. To the right of the Project View is the Asset window, also known as a *container*. On the right of the workspace is the Property Inspector, where you can enter or modify project settings and properties.

When you are putting a project together, you will be spending a lot of time adjusting settings in the Property Inspector. Just about anything you add to a project will require some kind of adjustment in the Property Inspector, whether it is a menu, a track, or some other project element. And the most common way you select project elements is to click directly on them in the Graphical View window.

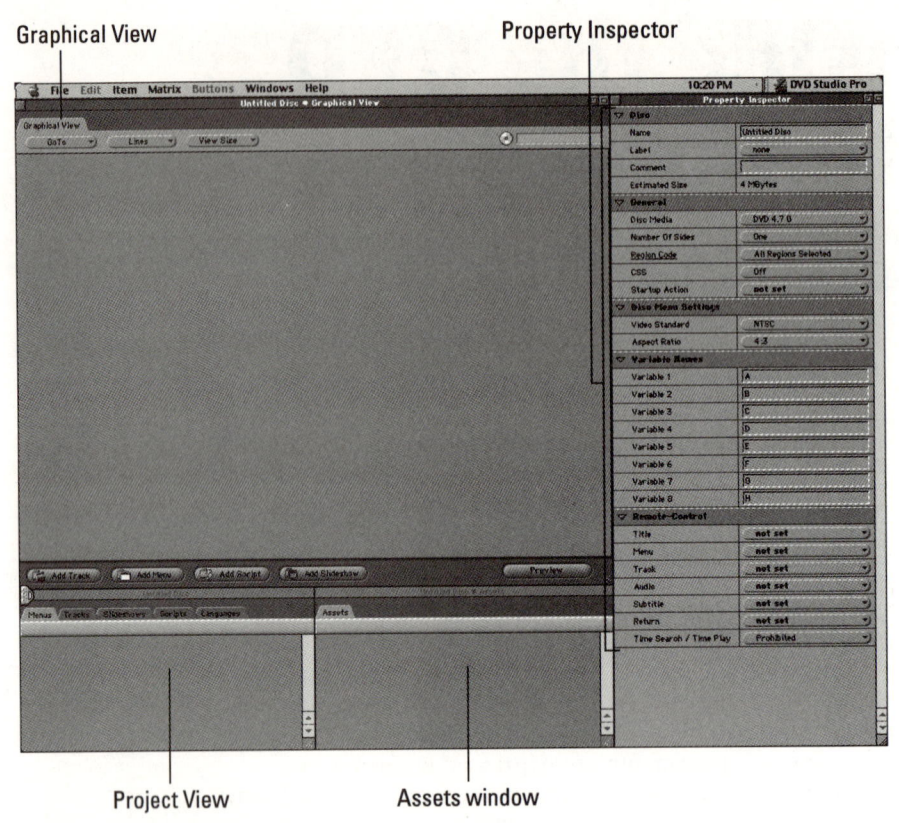

Figure 2-1: The overall DVD Studio Pro workspace

Understanding the Graphical View

The Graphical View enables you to work with the various elements that form your DVD project; the content and the interactive elements are usually represented as tiles. This section briefly introduces the prominent things you will see in the Graphical View window. Typical tasks include the creation of menus and tracks. Figure 2-2 shows a close-up of the graphical view with some sample project elements.

Getting to know Tiles

Tiles are the basic building block of a project, providing a visual summary of project elements. Adding tracks, menus, scripts, or slideshows to a project causes a corresponding tile to appear in the Graphical View area. Tiles in the Graphical View are like icons on your Mac's desktop. You can click them, hold the mouse down, and move them around, or you can double-click them to open and edit the project element the tiles represent. For example, Figure 2-3 illustrates a project that has an untitled track, script, menu, and slideshow.

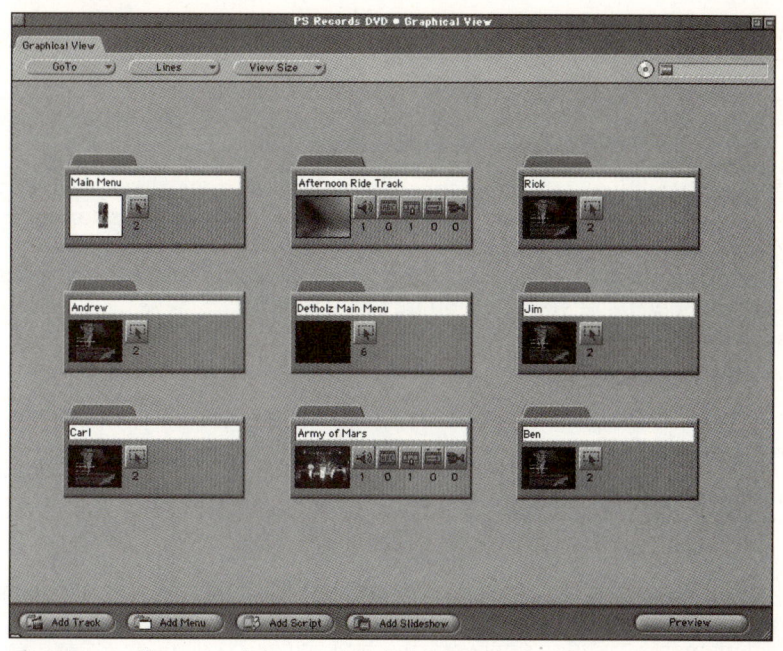

Figure 2-2: Close-up of the Graphical View window with sample project elements

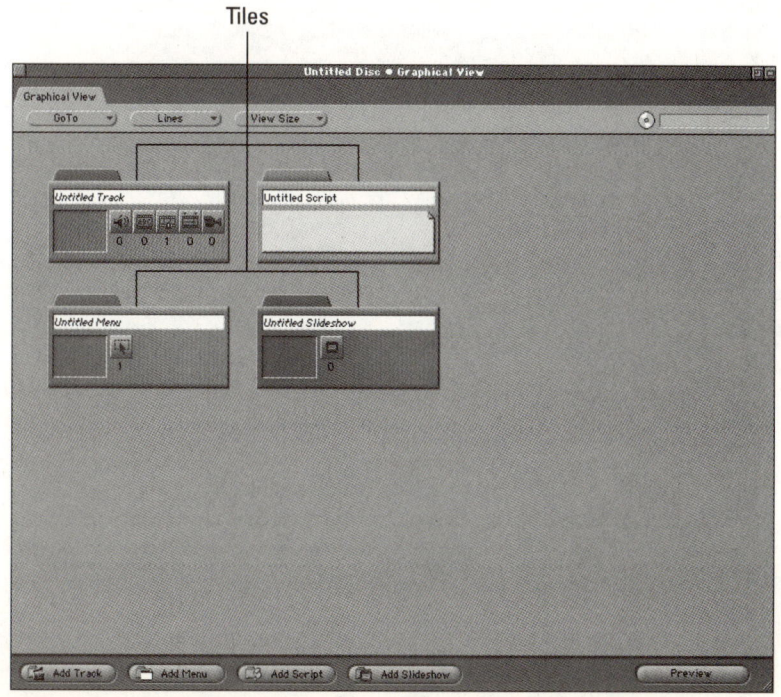

Figure 2-3: Tiles representing track, script, menu, and slideshow

> **Tip** Each type of tile is assigned a unique color to make organization simple.

Exploring the drop-down menus

Beneath the Graphical View tab, you find a row of three buttons, each of which is a drop-down menu. The drop-down menus become active as material is added to the project, helping you to keep track of how the project is organized. The drop-down menus are especially helpful as a project grows in complexity, and when you review how various components of your DVD relate to each other.

Using GoTo to access tiles

The first button is labeled GoTo (see Figure 2-4). As you create project elements, the available elements appear in the GoTo menu as a stacked list. For example, in Figure 2-4, the drop-down menu shows six menus, and then the Army of Mars video track, and there is a line to separate the items by type. You can select GoTo to quickly access the element you want to work with. DVD Studio Pro has a unique feature — it automatically organizes elements in drop-down menus according to type. As you work with DVD projects, you'll find that this feature can make it easier to access project elements by name. This can be especially helpful when you have a lot of project elements and don't necessarily want to search through everything in the Graphical View window.

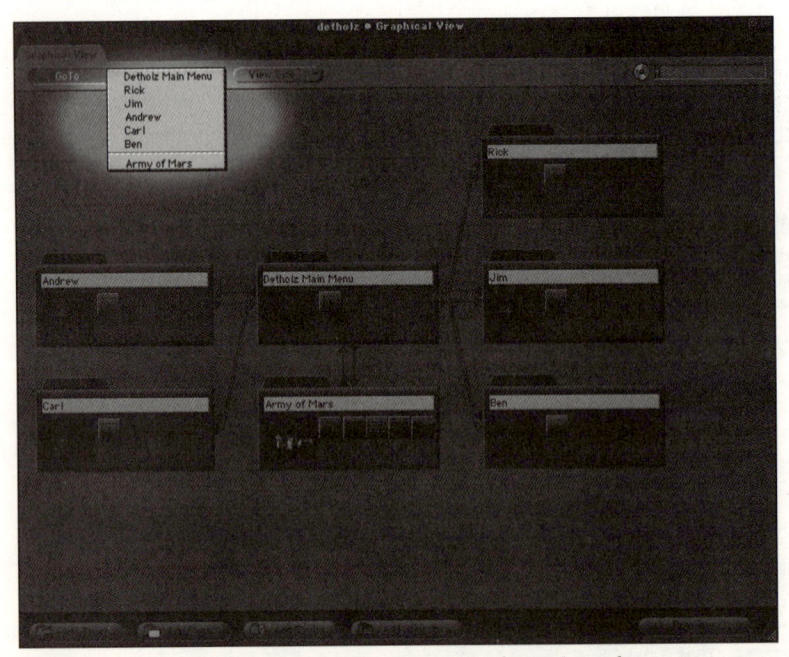

Figure 2-4: GoTo drop-down menu with sample project elements

Reviewing interactivity by using Lines

Accessing the Lines drop-down menu displays an overview of the project, which represents the inter-relationships between project elements by drawing lines between them. The Graphical View window ends up looking similar to a flowchart. (See Figure 2-5.) This tool is useful in reviewing the interactive relationships between menus, buttons, and their associated content (such as video clips). For example, in Figure 2-5, there is a line going from the Detholz Main Menu tile to the Rick Tile. This represents a button in the Detholz Main Menu screen which allows the user to go to the Rick screen.

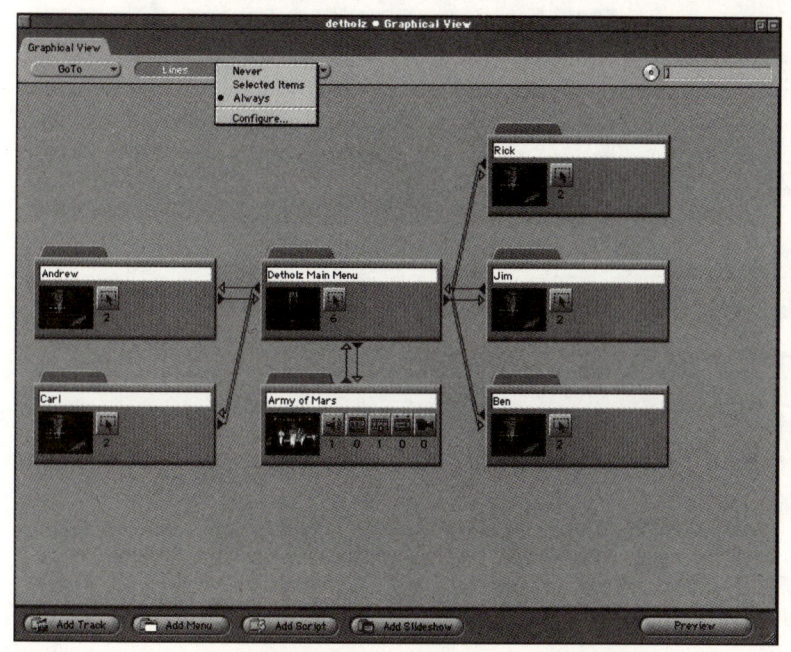

Figure 2-5: Lines drop-down menu, with menu and track tiles

Adjusting the view size of your tiles

The View Size drop-down menu as shown in Figure 2-6 enables you to view tiles as either large or small. Large tiles look like tabbed file folders. Large tiles contain a small preview window for a thumbnail graphic and buttons with numbers indicating the amount of assets being used within that particular tile. Selecting a small view simply shows tiles as rectangular blocks, providing increased space for complicated projects.

Project element buttons

At the bottom of the Graphical View are five project element buttons that provide functionality, allowing you to add elements to a project, as well as preview a project.

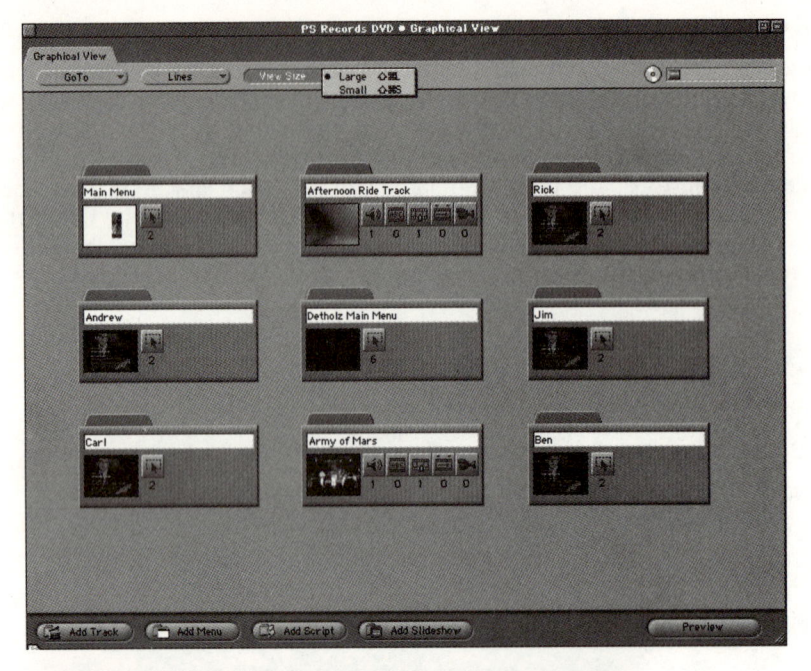

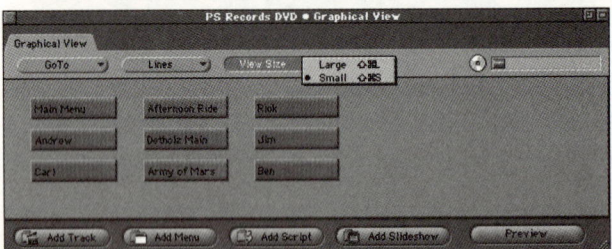

Figure 2-6: View Size drop-down menu

Using the Add Track button

Click the Add Track button to add an empty track tile to the workspace. When they are developed, tracks will contain *assets,* individual video and audio files that become the core of a DVD's material. Use the Add Track button to add up to 99 tracks to your projects.

For example, you first add a track to the project and then incorporate the assets within the newly created track. The track might contain something like a video clip of a home movie. Adding the track allows you to link to the video clip from a menu. After the track is created, you create a menu in your DVD project with a button that plays the track.

The following steps describe how to add a track to the DVD Studio Pro workspace:

1. **Import a video file by selecting File ⇨ Import.** The Import dialog box appears; select the appropriate file and click Add. Then click Import. The file is added to your DVD project, and is accessible in the Assets window.

2. **Click the Add Track button located in the Graphical View window.** An untitled track tile appears as shown in Figure 2-7.

Figure 2-7: Before — Untitled Track tile

3. **Label the track by replacing the text in the Untitled Track text field.** This label represents the track in the Project View and Property Inspector.
4. **Locate the imported video file in the Assets window.**
5. **Drag the file from the Assets window onto the track tile in the Graphical View (see Figure 2-8).** This action makes a connection between the video file and the track tile; the video is added to the track, and it can now be previewed.

Figure 2-8: After — named Track tile with thumbnail video image

6. **Click the Preview button to view the track.**

> **Tip:** To delete a tile from the Graphical View window, select the tile and click the delete key or choose the clear option from the Edit menu.

> **Note:** Click the tab at the top of the tile to select it. For example, if you click in the text field, clicking delete clears only the text, not the tile.

Add Menu button

When you look at a DVD screen, a DVD menu allows you to choose what you are going to watch. Menus consist of backgrounds and buttons that are linked to Tracks; they are the visual, interactive elements of a project that ultimately allow a person to navigate a finished DVD. Backgrounds can be either motion (video) or stills and buttons can be layered on top of these to create engaging, dynamic navigation.

In DVD Studio Pro, adding a menu tile to a project is the first step in developing a DVD menu. To add a menu tile, all you have to do is click the Add Menu button. Then you can follow the overall process of linking a particular graphic to it, to give you the images you have prepared in Photoshop for the background and buttons. For more information on menus, see Chapter 6.

> **Note:** Tracks and menus are also referred to as *containers*, because they contain video, audio, or graphics.

To add a menu to the DVD Studio Pro workspace, perform the following steps:

1. **Import a layered image file, or the file named mainmenu.psd located in the Tutorial/CH2 folder on the DVD, by selecting File ⇨ Import.**
2. **Click the Add Menu button located in the Graphical View window.** An untitled menu tile appears as shown in Figure 2-9.

   Figure 2-9: Before — untitled menu tile

3. **Label the menu by selecting and replacing the text in the Untitled Menu text field.** After naming the menu, the name of the menu appears in the Project View. When the menu tile is selected, the name appears in the Property Inspector.
4. **Locate the imported layered image in the Assets window.**
5. **Drag the file from the Assets window over the menu tile in the Graphical View (see Figure 2-10).**

   Figure 2-10: "After" — named menu tile with thumbnail image

Add Script button

This button brings up an untitled script tile that enables you to add a script to a DVD project (see Figure 2-11). Scripts are used to add sophisticated interactivity to a project. The DVD specification includes its own scripting language (discussed in Chapter 17) that enables the use of additional navigation methods such as randomized playback. Typically, scripting enables the user to customize which menus and buttons they see on-screen or allows them to experience a randomly selected series of video or audio clips.

Figure 2-11: Script tile

Add Slideshow button

This button brings up an untitled slideshow tile that enables you to add a slideshow to a DVD project (see Figure 2-12). If you have explored extra features found on commercial DVDs, you may recognize the term *slideshow*. Slideshows are sequences of still images or video that advance automatically or when the user clicks the remote. The Slideshow incorporates still images, audio, and video in a navigable presentation format.

Figure 2-12: Slideshow tile

Simulating the finished product with the Preview button

Use the Preview button to preview items as the project progresses. After material is added to tiles in the workspace, click the Preview button for a navigable preview of what the finished item looks like and how its various features work. The Preview button simulates how the video, audio, and interactive menus work when the finished project is played on a DVD player.

Estimating available space by using the disc space indicator

This handy indicator, located near the upper-right corner of the Graphical View (Figure 2-13), shows how much DVD disc space was used in the creation of a project and how much space remains for additional content.

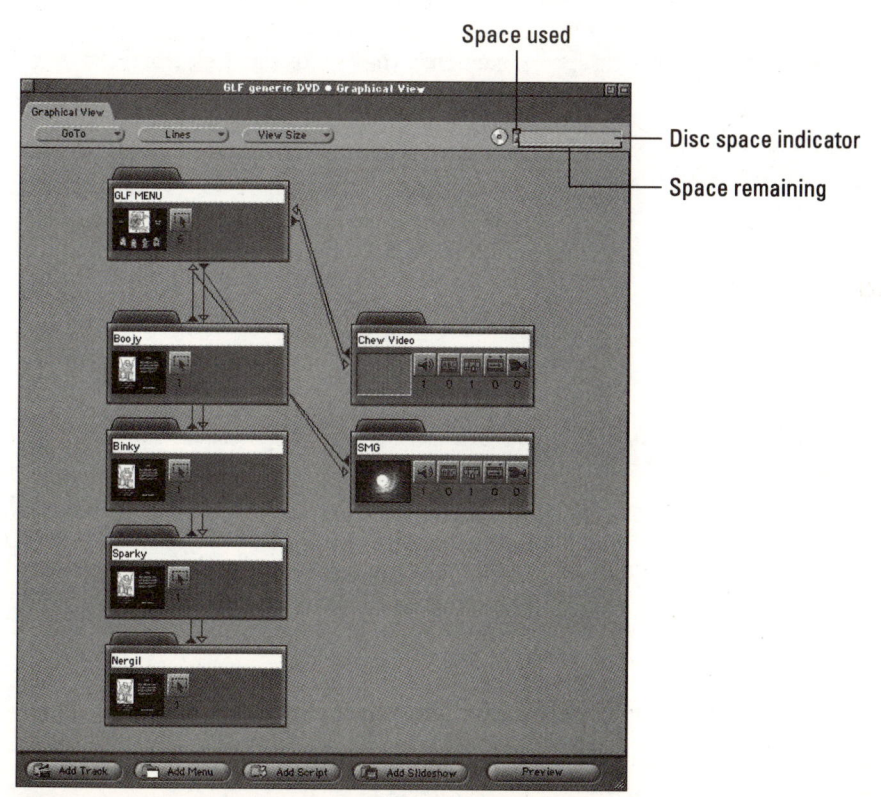

Figure 2-13: The disc space indicator in the Graphical View window, showing space used and space remaining

This feature is particularly useful when trying to fit as much information as possible on a DVD. The bar appears red if you exceed the size set for the finished project.

Understanding the Project View Tabs

The Project View is a place to easily access elements used in the creation of a disc. It provides a convenient text-based representation of project elements, helping with the management of a project (see Figure 2-14).

Tabbed folders provide efficient reference and access to similar elements, providing users with additional means for working with the material. For example, you can identify all the menu tiles in the Graphical View window based on their individual colors, but clicking on a tab in the Project view gives instant access to all project elements at the same time.

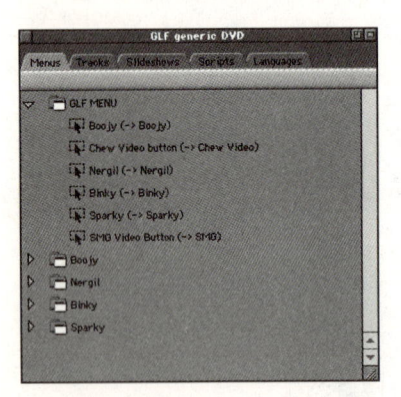

Figure 2-14: The Project View area provides a convenient means of accessing project elements. The Menus tab is selected, showing some of the menus in the sample project.

The Assets Container

Located on the bottom of the Graphical View, the Assets Container allows you to easily access all of your imported video, audio, and graphics files. It's the central location for all materials used in the creation of a project. Before you can use any file in a DVD project, it needs to be imported by using the Import command in the File menu, and then it becomes an asset, and shows up in the Asset window, also known as the Assets Container (see Figure 2-15).

> **Tip** To easily import a file, drag it directly into the Assets window.

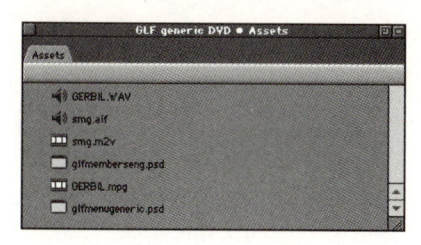

Figure 2-15: The Assets Container, showing graphic and audio assets for a sample project

Using the Property Inspector

Located to the right of the Graphical View Window, the Property Inspector is perhaps the deepest and most complex portion of the DVD Studio Pro workspace. In the Property Inspector, you enter or modify the settings and properties of various project components, such as a track or menu (see Figure 2-16). Properties available in this window are dependent on the type of item selected. For example, clicking on a track tile in the workspace brings up its options in the Property Inspector, but these options are different when a menu tile is selected. If no item is selected, properties for the overall project are displayed, also known as Disc properties.

The Property Inspector contains subpanels and controls that configure and control disc properties, general properties, disc menu settings, variable names, and remote control settings.

Figure 2-16: The Property Inspector

Disc properties

Disc Properties display when you start a new project or click an empty section of the Graphical View workspace. Properties for the overall project are set here. Data entered in the first Disc menu for name, label, and comments are for your reference only and do not appear as a property on the finished disc. The useful Estimated Size indicator estimates the total file size of the completed disc, based on the file sizes of assets added to the project thus far (see Figure 2-17).

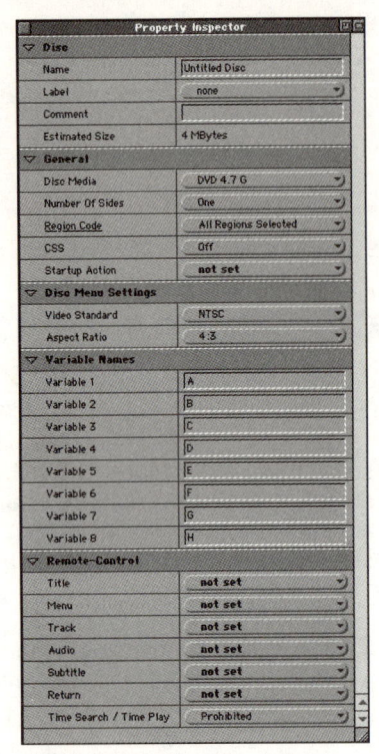

Figure 2-17: Disc properties

General properties

Menu choices include settings for size of disc media, the number of sides for a finished disc, the region codes used in a project, copy-protection or CSS, and Startup Action (see Figure 2-18).

Disc media

Choices include DVDs with capacities of 2.6, 3.95, 4.7, and 8.54 Gigabytes. Note that DVD-5 (4.7 GB) discs are the only type that can be burned by consumer DVD-R recorders at this point in time. Larger projects (such as DVD-9 or 8.54 GB) can be sent to service bureaus capable of duplicating or replicating projects of this size, including those with multiple layers or sides. Choosing disc capacity larger than 4.7 GB in this section would require outputting to a DLT tape of sufficient capacity. DVD-R media will work in most consumer DVD players, but if a project is sent off for manufacturing, the resulting DVD disc can be played in any consumer DVD player, regardless of the storage capacity of the disc.

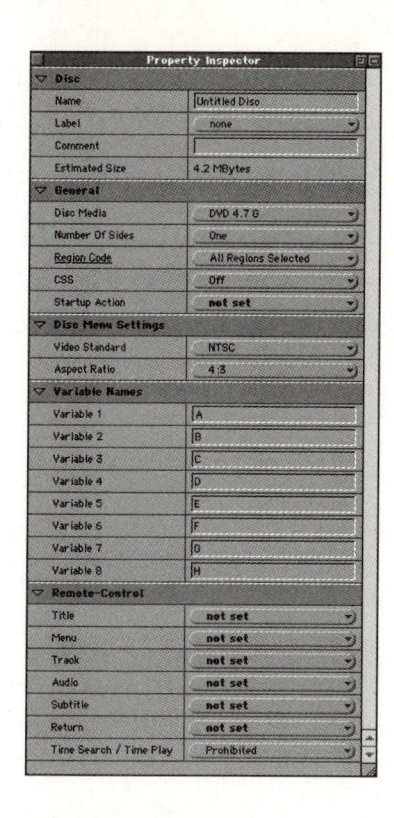

Figure 2-18: The General properties for a project in DVD Studio Pro, with the Disc, General, Disc Menu Settings, Variable Names, and Remote-Control sections.

Number of sides

You can specify a single or double-sided disc. Based on your choice of sides, the disc media selected, and the number of assets used, DVD Studio Pro estimates the size of the project and alerts you if the capacity for the finished disc exceeds the limits.

Region code

Region codes determine what area of the world your disc plays in. Different regions of the world, such as China, have specific codes for players and only discs with matching codes play on them. In Chapter 20, we discuss the significance of regions and how they came about as well as potential resources for finding players and software to work with codes.

Content Scramble System (CSS)

Content Scramble System (CSS) copy-protects a DVD by encrypting data. CSS differs from other methods of copy protection, such as Macrovision, and both are discussed in Chapter 20. The intent of copy protection is to prevent illegal copying or duplication of creative and intellectual property.

Startup action

The startup action is what happens when you insert a disc into a DVD player. To use the Preview mode in DVD Studio Pro, a startup action must be selected first. Setting a startup action is discussed further in Chapter 14.

Disc menu settings

In disc menu settings, you set the Video Standard for menus in the project (NTSC or PAL) and the aspect ratio (4:3 or 16:9) (see Figure 2-19). The type of Video Standard you should use is discussed further in Chapter 11.

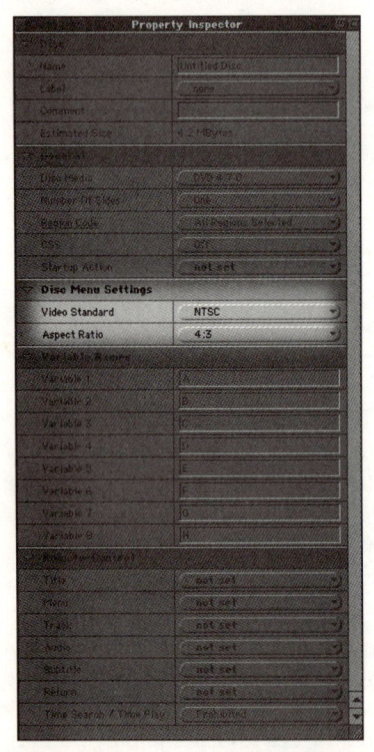

Figure 2-19: Disc menu settings

Variable Names

Global variables store data for scripts and are viewed and named in the Variable Names menu. A total of eight global variables can be used in a project (see Figure 2-20). Variables are discussed further in Chapter 17.

Remote-Control

DVD players come equipped with remote controls, each having basic keys with default functions. This property enables you to assign what you want to happen when a person presses these keys on their remote control unit while viewing the DVD presentation (see Figure 2-21).

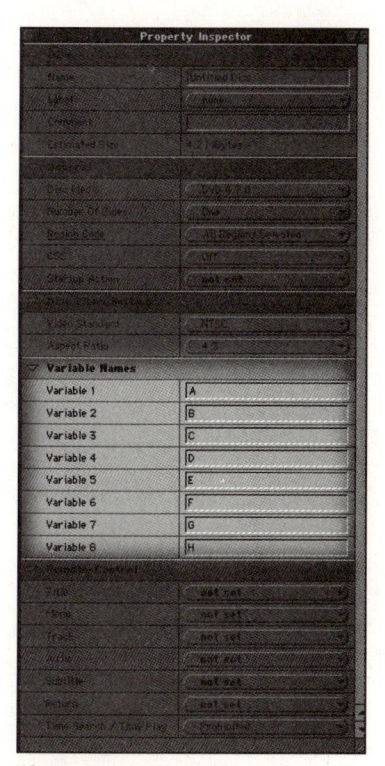

Figure 2-20: Variable Names

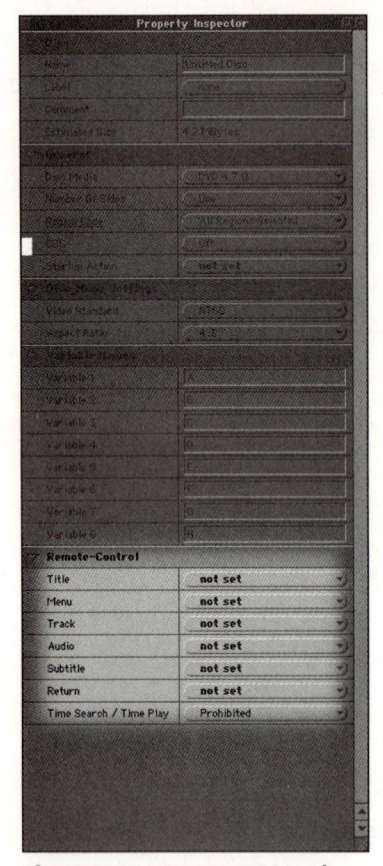

Figure 2-21: Remote-Control

Track properties

When a track tile is selected in the Graphical View window, the track properties appear in the Property Inspector. For more information on adjusting these properties, see Chapter 12.

- ✦ **Access:** Enables you to specify if a Web browser opens when a URL is encountered on a disc.
- ✦ **General:** Menu choices include Pre-Script, Jump when finished, User operations, and Macrovision.
- ✦ **Video:** Includes choices for video assets assigned and the Display Mode for those assets (4:3 or 16:9 — Pan-Scan, Letterbox, and Pan-Scan and Letterbox).
- ✦ **Remote-Control:** DVD players come equipped with remote controls, each having basic keys with default functions. This property allows you to assign different functions to each key.

Menu properties

When a menu tile is selected in the Graphical View window, the menu properties appear in the Property Inspector. For more information on adjusting these properties, see Chapter 6.

- **Access:** Allows you to specify whether a Web browser opens when a URL is encountered on a disc.
- **General:** Menu choices include Pre-Script, Default button, and Return button.
- **Timeout:** Choice of action that occurs when a time limit is exceeded for a menu selection. Allows for automation of a disc, which is especially useful for kiosks or other self-running presentations.
- **Picture:** Allows you to choose an asset which is used for the background of a menu.
- **Button Hilites:** Includes settings for button properties. You specify colors and hilites for Activated or Selected button states.

Tip The maximum number of buttons for each menu is 36.

Script properties

When a script tile is selected in the Graphical View window, the script properties appear in the Property Inspector. There is hardly anything in the script properties; you can name a script, and make a short comment. DVD Studio Pro only allows a certain number of scripting commands, and you can check the script properties and see how many script commands are used, and how many are left. (For more information on scripting, see Chapter 17.)

Slideshow properties

When a script tile is selected in the Graphical View window, the script properties appear in the Property Inspector.

- **Access:** Allows you to specify whether a Web browser opens when a URL is encountered on a disc.
- **General:** Menu choices include Pre-Script and Jump when finished.
- **Languages:** Selections for up to eight separate audio languages used in the Slideshow.
- **Remote-Control:** DVD players come equipped with remote controls, each having basic keys with default functions. This property allows you to assign different functions to each key.

Additional Ways to View Projects

After you understand the basic areas of DVD Studio Pro, you can look at aspects of a project in several different ways. These views include four *editors*, three *matrix views*, and two troubleshooting windows that you become familiar with as you learn the process of creating a DVD project.

Using the editors

The *editors* are windows that are opened when you double-click on a menu, slideshow, or script tile. There is also an editor for *markers* (also known as chapter markers). Markers are created in the midst of a video track to allow the user to jump to or start play at a specific location. Markers are discussed in further detail in Chapter 12.

The following list gives some more information about each kind of editor that is opened when the respective tile is double-clicked.

♦ **Menu Editor:** To open up the Menu Editor, click the Add Menu button in the Graphical View to create a menu, and then double-click the menu tile. (See Figure 2-22) Menus are covered in Chapter 6.

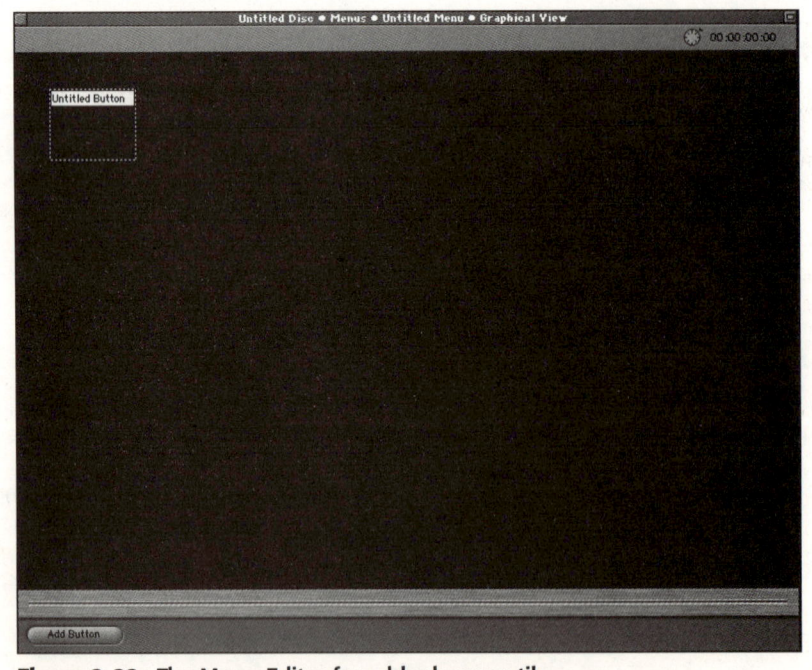

Figure 2-22: The Menu Editor for a blank menu tile

✦ **Slideshow editor:** To open up the Slideshow editor, click the Add Slideshow button in the Graphical View to create a Menu, and then double-click the Slideshow tile. (See Figure 2-23.) Slideshows are covered in Chapter 13.

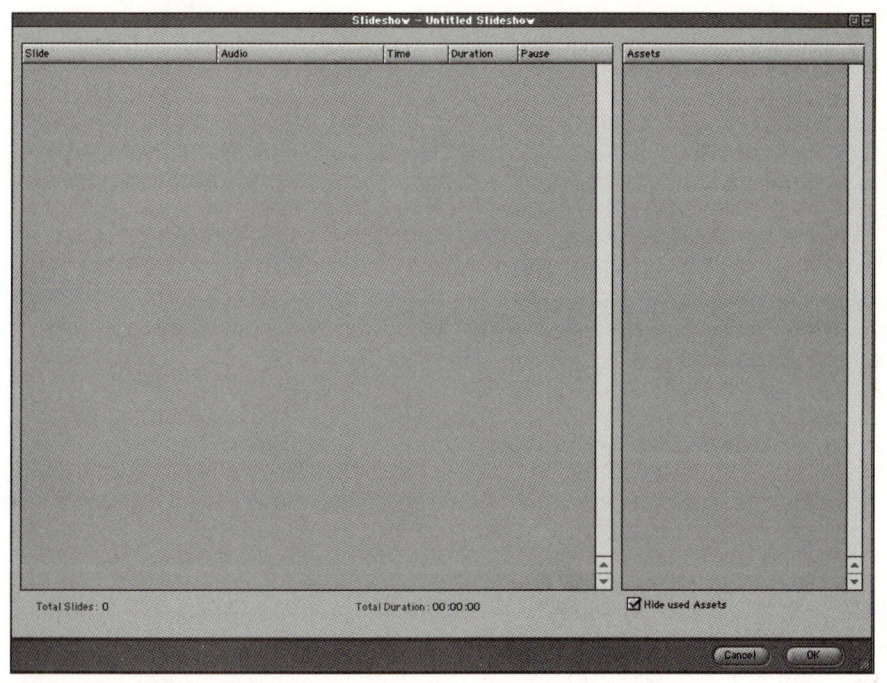

Figure 2-23: The Slideshow editor, ready to create a slideshow

✦ **Script editor:** Open the Script editor by clicking the Add Script button in Graphical View to create a Script and then double-click the Script tile. (See Figure 2-24.) Scripts are covered in Chapter 17.

✦ **Marker Editor:** The Marker Editor is discussed further in Chapter 12.

Figure 2-24: The Script editor, ready to start creating a script

To see what the Marker Editor looks like, follow these steps:

1. **Click the Add Track button in Graphical View to create a track.** A track tile appears. (See Figure 2-25.)

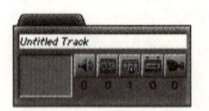

Figure 2-25: Track tile with thumbnail area

2. **Double-click thumbnail area of Track tile.** The Marker Editor appears. (See Figure 2-26.)

The Matrix views

What is the Matrix? *The Matrix* is a great movie and a good DVD. You should go out and buy it. When you watch it, notice how at one point Keanu Reeves says "Whoa!" which is surely a lapse into his former role in Bill and Ted's Excellent Adventure. But within the world of a DVD Studio Pro project, the Matrix views are convenient methods of looking at an overview of a project, and the way project elements relate to each other. Whoa! Excellent!

The Matrix becomes a friend in time of trouble, helping you to solve the mysterious ways of a complex DVD project that has a lot of elements, and a few evil unsolved mysteries running around inside that are trying to escape notice.

Chapter 2 ✦ **Inside the DVD Studio Pro Workspace** 49

Figure 2-26: The Marker Editor, ready to start creating markersThe Matrix makes it easy to:

- ✦ Edit settings in a central location
- ✦ Check settings made in other parts of the program
- ✦ Evaluate the inter-relationships between project elements
- ✦ Replace assets in a central location

The three versions of the Matrix each have a special perspective, allowing you to see your DVD project through different eyes.

Asset Matrix

Have you ever heard of a bank freezing the assets of a patron? What the Asset Matrix (Figure 2-27) does is to give you a freeze-frame cross section of all the assets you have imported into a project, and the way they relate to menus and tracks (containers). It shows video, audio, and subtitle streams across the top, and menus and tracks down the side. There is a series of rows of squares, and the little dots represent connections between project elements. The Asset Matrix is discussed further in Chapter 3.

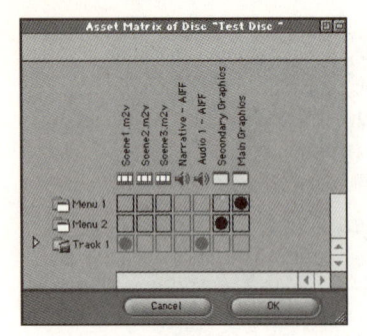

Figure 2-27: The Asset Matrix, with dots representing connections between project elements

Jump Matrix

There is a wonderful moment in *The Matrix* where Keanu Reeves jumps to a beautiful, hovering Huey helicopter from the top of a building. In DVD Studio Pro, when you want to get a sense of where a user can jump within a DVD project, and how they get there, you can consult the Jump Matrix (Figure 2-28). The Jump Matrix shows all possible jump actions along the top, and the jumping off points along the side, such as the overall disc, individual buttons, and any other spot where the joint is jumpin'. There is a series of rows of squares, and the little dots represent connections between project elements. The Jump Matrix is discussed further in Chapter 15.

Figure 2-28: The Jump Matrix, with dots representing connections between project elements

Layer Matrix

In *The Matrix,* the heroes get to press just about every kind of button to battle various kinds of computer-generated evil. In DVD Studio Pro, sooner or later you find excitement in the Layer Matrix, when you are tracing the relationship between buttons and the corresponding Photoshop layers. The Layer Matrix (Figure 2-29) shows Photoshop layers along the top and button states along the side (normal, selected, and activated). There is a series of rows of squares, and the little check marks represent which layers are visible when a particular button is inactive, active, or selected. The Layer Matrix is discussed further in Chapter 5.

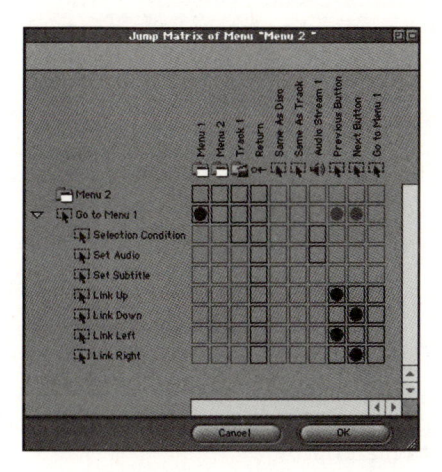

Figure 2-29: The Layer Matrix, with check marks representing which layers are visible when particular buttons are inactive (Normal), selected, or activated

The Troubleshooting Windows

How does one "shoot" trouble? Through a window, of course! Whether it is approaching zombies, aliens with an appetite, or simply a little bug creeping around in your DVD project, Troubleshooting Windows come to the rescue.

The Troubleshooting Windows help you to:

- ✦ Find missing files
- ✦ Achieve a sense of peace and well-being
- ✦ Test your ability to deal with the stress of error messages
- ✦ Get your DVD project running smoothly again so you can invoice your client and have enough money to buy more copies of this book as a collector's item.

You may be laughing, but this is all part of a plan. Laughter is the best medicine for the trouble you will bring down upon yourself if you don't read Chapter 15.

You curiosity is certainly peaked and you want to see the Troubleshooting Windows now, but you have to deal with the sense of anticipation you are feeling. We can't stress how important it is to read Chapter 15 in its entirety BEFORE you start making a DVD.

Summary

- ✦ The Graphical View window is the center of the DVD Studio Pro workspace, with drop-down menus for viewing various aspects of the project. The Graphical View window also has buttons for adding tracks, menus, scripts, and slideshows.

- ✦ The Project View window has tabs that enable you to access project elements as they are added to the project.

- ✦ Video, audio, and graphics files used to make a project in DVD Studio Pro are referred to as assets.

- ✦ The Property Inspector window is where settings or properties are entered or modified for project elements. The window initially shows the Disc properties, but changes according to which type of project element is selected.

- ✦ Disc Properties represent the overall project, including the type of media used and global variables.

- ✦ Tracks are collections of video and audio that play as a whole, containing the assets that become the core of a DVD's material.

- ✦ Menus consist of backgrounds and buttons that are linked to Tracks; they are the visual, interactive elements of a project which ultimately allow a person to navigate a finished DVD.

- ✦ Slideshows are sequences of still images or video with the ability to advance automatically or when the user clicks his or her remote.

- ✦ If you are sitting at a computer and haven't taken a break in the last hour, please get up and drink a glass of water. If the water from the closest faucet isn't tasty, try drinking spring water. Your body thanks you for the resulting relaxation and hydration.

✦ ✦ ✦

Working with Assets

CHAPTER 3

In This Chapter

Understanding compatible file formats

Importing assets

Working with assets

In this chapter, you learn about assets. Assets are the files that make up a DVD project, such as audio or video clips and Photoshop files, which you import into a DVD project. In this chapter you find out more about which file formats are compatible with DVD Studio Pro, how to import the asset files into your project, and various options for viewing the assets after they are imported, including the Asset Matrix. Taking the time to master these skills can help you to develop your DVD project smoothly, because you gain the ability to easily access, modify, and organize the files that make up your DVD.

Think of assets as ingredients. Working with assets is much like putting a good meal together, starting with preparation of ingredients, adding them together, and eventually serving.

In DVD Studio Pro, when you import a file to use it in a project, the program creates a link to that file called an *asset*. Technically, as far as DVD Studio Pro is concerned, an asset is really only a link to a file, not the actual file. This concept becomes clearer as you become familiar with the various ways of viewing and working with assets.

To have assets, the files you import must be in one of the compatible formats.

Understanding Compatible File Formats

The bread and butter of a DVD project are video and audio. The following example is a typical scenario. You hook up your MiniDV camcorder and input a clip through iMovie, and then you open up the same clip in QuickTime Pro and export into

MPEG-2 format. Your video file has the characteristic .M2V file suffix, which indicates that the MPEG-2 file has been generated by QuickTime Pro. You also end up with a separate audio file, also referred to as an audio stream, in AIFF (Audio Interchange File Format), the standard audio file format on the Mac.

> **Cross-Reference** *Encoding* is the process of taking an audio or video clip and converting them for use in a DVD project. Audio encoding is covered in Chapter 10, and video encoding is covered in Chapter 12.

Understanding video compatibility: MPEG and MPEG-2

When you are preparing your content for a DVD project, eventually all video winds up in either MPEG-2 or plain MPEG-1 format, characterized by a filename ending in either MPG, MPEG, or M2V. The MPEG-2 format is part of the DVD standard, with a higher image quality, but it is possible to use basic MPEG-1 video. (Most of the time you will probably end up using MPEG-2 video in a project, but in certain instances plain old MPEG-1 will come in handy, such as when the amount of video you want to put on a DVD is more important than having the highest image quality.)

Within the video file itself, when working with MPEG-2 in NTSC, the best results come from using a frame size of 720 × 480 pixels and a frame rate of 29.97 frames per second; and in PAL, use a frame size of 720 × 576 pixels and a frame rate of 25 frames per second. Furthermore, having an aspect ratio of either 4:3 or 16:9 works best.

	Frame Size (in pixels)	*Frame Rate (in frames/second)*	*Ratio*
NTSC	720 × 480 pixels	29.97 fps	4:3 or 16:9
PAL	720 × 576 pixels	25 fps	4:3 or 16:9

Considering audio compatibility

In contrast to video, with audio file formats in DVD Studio Pro you have a bit more freedom. In DVD Studio Pro, audio must be converted into PCM audio (which includes AIFF, SoundDesigner, and WAVE formats), MPEG-1 audio, or Dolby AC-3. The most commonly used audio format with DVDs is Dolby AC-3, but independent DVD producers may find themselves using the basic AIFF format more often. See Chapter 9 and Chapter 10 for more detail on audio.

Considering graphics compatibility: Photoshop

Preparing graphics to be used as assets in DVD Studio Pro comes down to creating multilayered Photoshop files in the .PSD (Photoshop Document) format. You can also create appropriate Photoshop compatible PSD files in other programs, such as Macromedia's Fireworks. See Chapters 5 and 6 for more information on using Photoshop in the creation of DVD projects.

Understanding file extensions

A file extension is a period followed by a series of letters that represent the *file format*. Generally, with DVD Studio Pro you don't need to be concerned with file extensions. However, if you end up working with assets that are going to and fro between operating systems, having an understanding of their nature in Mac and Windows environments is helpful to avoid confusion.

Working with file extensions in Mac OS

In Mac OS, when you create a file in an application, even though you may not see a *file extension* in the file name, such as .PSD, the file format stays the same. For example, you can create a file in Photoshop and save the file as menu.psd. You can see the file extension. You could change the name to Final without the .PSD at the end, and the file format would stay the same.

Working with file extensions in Windows

Gasp! Mention of Windows, in the very heart of Mac territory! (Penance will now be made to Steve Wozniak and Steve Jobs, with a moment of nostalgic reflection on the venerable Apple IIe computer, and affectionate meditation on a classic video game, Miner 2049'er.)

You may end up in a situation when you work with clients or colleagues who provide source material created on a Windows platform.

If you open a Mac-generated file in Windows, keep in mind that Windows is reliant upon file extensions, and file extensions are *always* three letters. For example, if you use your Mac to save a file in PICT format and send the file to someone with Windows without typing in .PCT after the file name, the Windows user may not be able to open the file.

Working with file extensions and DVD Studio Pro

Most of the time you don't have to think about file extensions. DVD Studio Pro does the thinking for you. On the Mac, files without extensions still have their telltale icons.

When you export MPEG-2 from QuickTime Pro, it creates the audio file and video file with the same name, but a different file extension. If you want to keep the same file name, you have to leave the file extension on one of them. (See Figure 3-1.)

QuickTime files ⎯⎯ Photoshop files

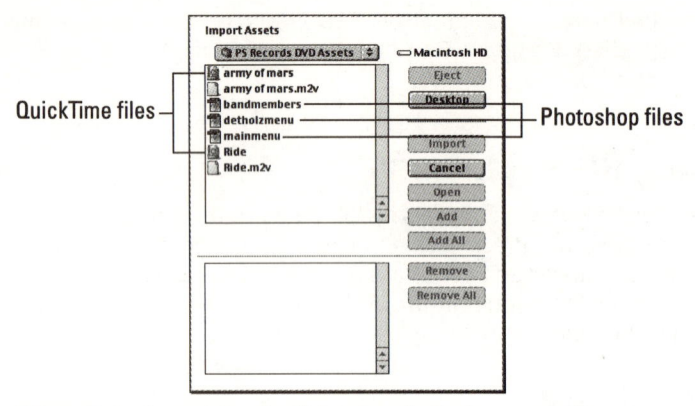

Figure 3-1: Import Assets dialog box featuring some files without file extensions

Realizing that a *single* file format may have *multiple* file extensions is important, because you may be dealing with files from multiple platforms or files generated by different applications and/or people. For example, the AIFF audio format may be represented by .AIF or .AIFF. Always use the appropriate three-letter file extension to avoid confusion as you are importing and managing assets, and to account for the possibility that you may have to go back and forth with a Windows machine.

Importing assets

You can import assets in three ways — by using the file menu option, with the drag-and-drop method, or by creating a new asset through the Item menu. If you have a DVD-ROM drive, insert the DVD-ROM that came with this book for easy access to sample files.

Importing through the File menu

As with all other programs in the known universe, the most common way to import files into a program is through a menu. In this case, it is the file menu.

To import through the file menu, follow these steps:

   1. **Choose File ⇨ Import.** The Import Assets dialog box appears. (See Figure 3-2.)
   2. **Navigate to the location of the folder that contains the assets you want to import.** In Figure 3-2, I chose the PS Records DVD Assets folder.

Chapter 3 ✦ **Working with Assets** 57

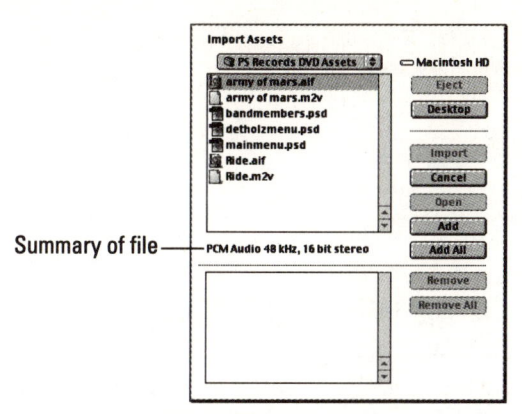

Summary of file ──── PCM Audio 48 kHz, 16 bit stereo

Figure 3-2: The Import Assets dialog box. Notice the fill settings summary that appears between the upper and lower halves of the dialog box.

3. Select the file(s) you want to import and click the Add button.
(See Figure 3-3.)

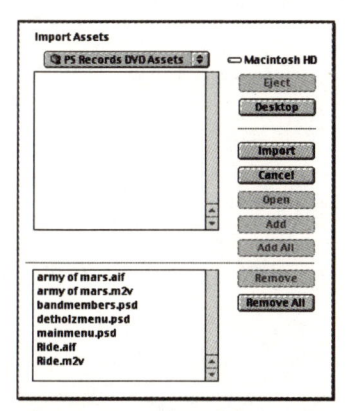

Figure 3-3: Files added to the Import Assets dialog box

4. After adding all of the files you want to import, click the Import button. The files appear in the Assets Container, and you can incorporate them into your project. (See Figure 3-4.)

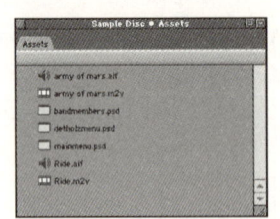

Figure 3-4: The imported files appear in the Assets Container.

Importing with the drag-and-drop method

Depending on your working style on the Mac, you may find it helpful to be able to go directly to a folder containing the files you want to import and to then drag the files into the Assets Container, rather than going through the Import Assets dialog box.

To import files by using the drag-and-drop method:

1. **Locate the folder that contains the files you want to import.**
2. **Choose View ⇨ as List.**
3. **Move the folder window to the lower-right of your screen.** (See Figure 3-5.)

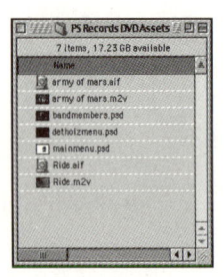

Figure 3-5: The folder containing assets, ready to import

4. **Switch to DVD Studio Pro and choose Window ⇨ Property Inspector to temporarily close the Property Inspector.** Performing this action reveals your window, allowing easy access to the files in the folder.
5. **Select the files you want to use and drag them into the Assets Container.** (See Figure 3-6.) Choose Window ⇨ Property Inspector to bring the Property Inspector back and restore the workspace to its normal state. After the Property Inspector is back, you are now ready to begin putting your project together.

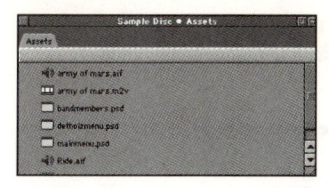

Figure 3-6: After dragging the assets you can organize them.

Importing by creating a new asset

When you click the Assets Container, a New Asset option appears in the Item menu. This option is the equivalent of importing a file, but the dialog box that appears is a scaled-down version of the Import dialog box. (See Figure 3-7.) You select the file you want and click the Open button.

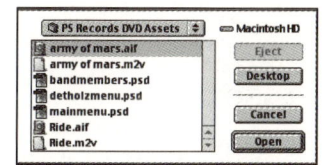

Figure 3-7: You import files by selecting the file you want to import. Click Open.

Now that you know how to import assets, you can begin to work with them.

Organizing and Controlling Your Project Assets

After you have imported assets into your project, the primary ways of working with the assets are the Assets Container, the Asset Matrix, and the Asset Files function in the Item menu.

Understanding the Assets Container

The Assets Container, also known as the Asset View, shows a visual summary of all files that have been imported and are available for your DVD project.

Selecting assets

Click an individual asset in the Assets Container to see information about the asset in the Property Inspector. The information in the Property Inspector is determined by the file format of the Asset. For example, the information on a Photoshop file includes the width, height, and number of layers. (See Figure 3-8.)

Figure 3-8: The Property Inspector reveals the details of a Photoshop asset.

Renaming, labeling, and commenting assets

When you import a file into DVD Studio Pro, the resulting asset is named the same as the original file by default. Within DVD Studio Pro, the asset serves as a link to the file you have imported. An asset is like an *alias* in Mac OS.

> **Note:** An alias is a file that you can create in Mac OS that points to another file. A typical example — an icon on your desktop that allows you to start an application. By creating an alias, you can leave the application in its original folder. Simply double-click the alias, and it points to the original file.

After you select an asset, you can rename it in the Property Inspector. The changes are reflected in the Assets Container. The Asset information area in the Property Inspector also allows you to choose a Label or type in comments for future reference.

To modify an asset:

1. **Click the Assets Container or choose Window ➪ Assets Container.**
2. **Click the asset you want to modify.**
3. **Select the Property Inspector by clicking on it or choosing Window ➪ Property Inspector.**
4. **If you want to change the name of your asset, click in the Name text field and type in a name.** (See Figure 3-9.)
5. **If you want to give the asset a label, select a Label from the Label drop-down menu.**
6. **If you want to add a comment, click in the Comment field and type in a comment. The length of comments is limited to 27 characters.**

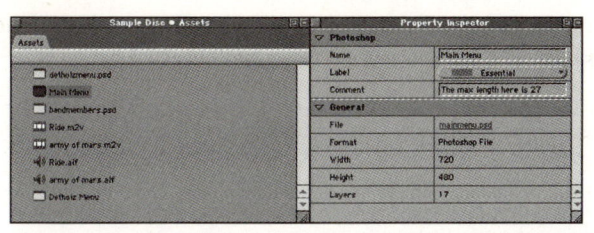

Figure 3-9: The original file name of this modified asset was Main menu.psd. The new name is Main Menu. The change in the asset's name in the Property Inspector is reflected in the Assets Container.

Customizing labels

You can customize the names and colors of the Labels that appear in the Label drop-down menu in DVD Studio Pro by going into the Mac OS and modifying the Label names and colors. Changing the Label names and colors affects the Labels in the rest of DVD Studio Pro, the Mac OS and every other program that uses Labels.

To modify labels:

1. **Switch to the Finder. (Click the menu in the upper-right corner of the screen that shows what program is running and make sure that Finder is selected.)**
2. **Choose Edit ⇨ Preferences and click the Labels tab.**
3. **Change a label name by clicking in a text field next to one of the color icons and typing in the new name.** (See Figure 3-10.)

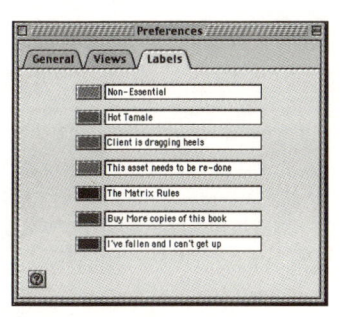

Figure 3-10: Modified Labels in the Mac OS

4. **If you need to kill more time, click on the colored icons next to the Label text fields and choose from a cornucopia of colors to customize the Label colors.**

Sending Assets to the Land of Oz

In computer jargon, an *Easter egg* occurs when an odd function has inadvertently worked its way into the program. This particular Easter egg may come in handy if you want to play a prank on someone who has imported a lot of assets into a DVD Studio Pro project file, but who either hasn't read this book, or hasn't read it carefully.

To send assets temporarily to the Land of Oz:

1. **Click the Assets Container.**
2. **Click your heels together.**
3. **Press the Oz key on your keyboard (also known as the End key). Any assets imported into your project have temporarily disappeared.**
4. **Press the Home key. Doing so brings the assets back to Kansas.**

Note: For the changes to take effect in DVD Studio Pro, you need to close and reopen the project file you are working on.

Now that you are done customizing the label names, the customized labels are available in the Property Inspector. (See Figure 3-11.)

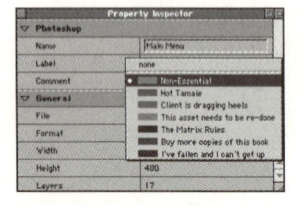

Figure 3-11: Customized Label names available in Label drop-down menu in the Property Inspector

Sorting assets

You can rearrange the order that assets appear in the Assets Container manually or by choosing Item ⇨ Sort Assets. To move an asset manually, simply click it; then drag it to the Assets Container to the desired position. The options for sorting through the Item menu are as follows:

- **Item ➪ Sort Assets ➪ By Usage:** When you have started to incorporate assets into a project, this function places the assets that haven't been incorporated yet at the top of the list.
- **Item ➪ Sort Assets ➪ By Name:** This function sorts the assets alphabetically.
- **Item ➪ Sort Assets ➪ By Type:** This function groups the assets together according to the file format, starting with Photoshop files, video files, and then audio files.

Deleting assets

Click the asset(s) you want to delete in the Assets Container and press the Delete key on the keyboard or choose Edit ➪ Clear. (Naturally, if you accidentally delete an asset because you have been drinking too much Jolt cola, you can click Apple+Z to undo.)

Understanding the Asset Matrix

The Asset Matrix is a convenient way of getting a snapshot view of how the various assets in a project relate to menus and tracks. The Asset Matrix can also serve as a way to easily assign imported assets to menus and tracks.

Viewing assets

To practice viewing assets, follow these steps:

1. **Locate the PS Records DVD folder in the Tutorial section of the DVD-ROM.** Double-click the PS Records DVD file.
2. **After the project file loads into DVD Studio Pro, choose Matrix ➪ Assets of Disc PS Records DVD.** Assets are displayed along the top, and Containers are displayed along the side. The various menus and tracks in the project are shown in Figure 3-12. The black dots represent which assets are assigned to which Containers. The gray dots are an indication that assets are assigned within a particular track. In this example, the Afternoon Ride track is expanded, showing the video (Ride.m2v) and audio assets (Ride.aif) that are assigned to the track.

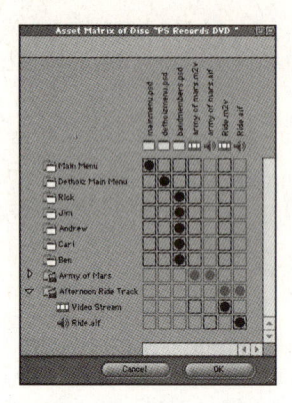

Figure 3-12: The Asset Matrix for the assets of the PS Records DVD

3. **Click the triangle next to the Army of Mars track to expand the track.**
   The black dots represent the video (army of mars.m2v) and audio (army of mars.aif) assets assigned to the track. (See Figure 3-13.)

Figure 3-13: The expanded track excerpt of the Asset Matrix for the assets of the PS Records DVD

4. **Try moving the mouse pointer over the various squares in the Asset Matrix.**
   Vertical and horizontal bars follow the mouse, as a visual aid to help locate the names of assets, menus, and tracks.

5. **Try moving the mouse pointer over the uppermost black dot in the upper left-hand corner of the Asset Matrix.** Notice how a line of text appears in top of the Asset Matrix window, Main menu / mainmenu.psd. (See Figure 3-14.) This summary indicates the connection between the menu and the asset, and is a convenient way of not having to rotate your head 90 degrees to view the asset names in the Matrix. When you move the mouse pointer over a square where an asset has not been assigned, the same kind of summary appears, indicating a potential connection between an asset and a menu or track.

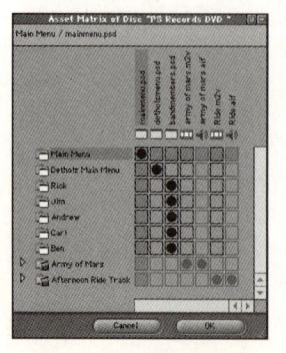

Figure 3-14: The Asset Matrix for the assets of the PS Records DVD indicates the assignment of the asset mainmenu.psd to the Main menu.

Chapter 3 ✦ **Working with Assets** 65

Assigning assets in the Asset Matrix

In an ideal world, a DVD project is planned out completely ahead of time, and all the assets are ready to go when you start importing them. In the real world, you probably need to assign new assets to a menu or track the middle of a project. If you are a visual person, you may find yourself simply dragging and dropping the assets onto their respective Tiles in Graphical View, but the Asset Matrix is an alternative.

To practice assigning assets, perform the following:

1. **Locate the CH3 folder in the Tutorial section of the DVD-ROM.** Open the CH3 folder and double-click the PSR DVDA project file. If you don't have files, the menus and tracks are already created for your convenience; you will be importing and assigning assets.

2. **When the project file loads into DVD Studio Pro, choose File ➪ Import and locate the PS Records DVD Assets folder in the Tutorial section of the DVD-ROM.** Open the PS Records DVD Assets folder, and click the Add All button in the Import Assets dialog box. Click the Import button.

3. **Choose Matrix ➪ Assets of Disc PSR DVDA to bring up the Asset Matrix.** The Asset Matrix shows a blank checkerboard pattern, indicating no assets have been assigned yet.

4. **To make it easier to assign the assets to the appropriate menus and tracks, choose Window ➪ Asset View.** Then choose Item ➪ Sort Assets—>By Type, which places the Photoshop files at the top of the list in the Assets Container.

5. **Next, choose Matrix ➪ Assets of Disc PSR DVDA again.** The various file formats are now grouped together along the top of the Asset Matrix.

6. **Try clicking on the appropriate boxes in the Asset Matrix to assign the Assets to the right menus and tracks.** Don't peek at Figure 3-15 until you finish. It should exactly like your screen.

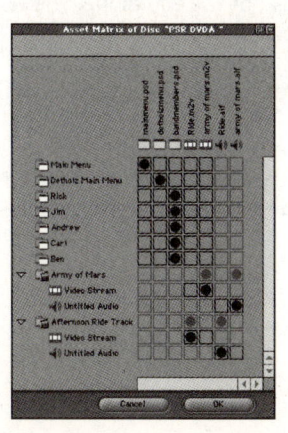

Figure 3-15: The Asset Matrix for assets of PSR DVDA shows the assigned assets.

Understanding the Asset Files window

The Asset Files window enables you to locate missing files. Or you can assign a new file to a particular asset.

The Asset Files window displays information in four columns, with tabs at the top. (See Figure 3-16.) The four categories are the following:

- **Assets:** The list of files used in the project
- **Volume:** The storage location of those files
- **Folder:** The folder location of the files
- **File:** The filename of the file that the asset links to

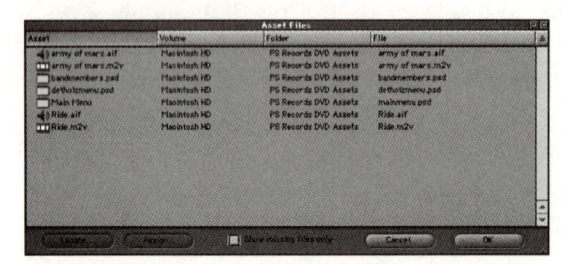

Figure 3-16: The Asset Files window

Clicking one of the rectangular tabs at the top of the Asset Files window sorts the Asset Files list according to that column. If the asset names are the same as the file names, and the files are all on the same hard drive and all in the same folder, clicking these tabs does not have any effect.

Note: An asset is technically only a link to a file and is not the actual file. The Asset Files window is one place where this concept is most apparent.

The sorting feature can be useful if you have several assets, folders storing the assets, or multiple hard drives.

The Locate and Assign buttons at the bottom of the window remain inactive until you select an asset. When an asset is selected, these buttons activate, enabling you to *Locate* or *Assign*.

Locating missing files in the Asset Files window

DVD Studio Pro is excellent at the game of Hide and Seek; it excels at keeping track of asset files. If you move a file anywhere on your hard drive, (even in the Trash), or rename a file, it still keeps track of it for you. But when a project is opened, if for some reason DVD Studio Pro can't find a file that had been imported into a project, it brings up the Asset Files window.

Alas, sometimes it happens. If you ever need to locate a missing file, it is probably because a project was transferred to your computer and one of the asset files was not included, or else the missing file was accidentally deleted. But accidental deletion is not a problem, because you are regularly backing up your entire project.

Right?

Hiding a file from DVD Studio Pro is tricky, but it has been done so that you can experience the joy of locating.

To practice locating a missing file:

1. **Open the PSR DVDC project file in the CH3 folder (in the Tutorial folder) on the DVD-ROM.** The Asset Files window appears.
2. **Click the Show missing files only check box at the bottom of the Asset Files window to deselect it.** This causes the window to display all the asset files in the project. The missing file is indicated in red. (See Figure 3-17.)

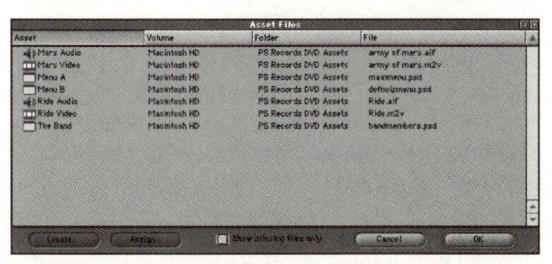

Figure 3-17: The Asset Files window, with the missing file in red

3. **Click the file to select it.** (See Figure 3-18.)

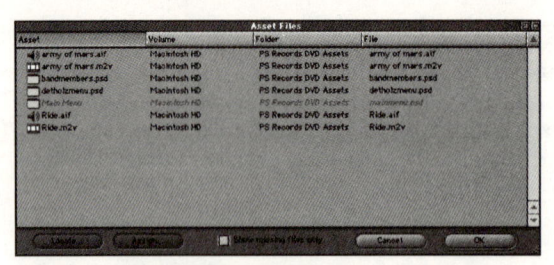

Figure 3-18: With file selected, the Locate and Assign buttons are now active.

4. **Click the Locate button.** (Double-clicking on an asset in the Asset Files window has the same effect as selecting and clicking on the Locate button.) This generates an open dialog box. (See Figure 3-19.) Locate the file by finding and opening the PS Records DVD Assets folder.

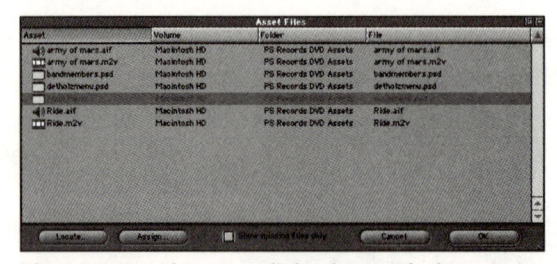

Figure 3-19: The open dialog box with the missing file selected

5. **Click the open button, and the file has now been located!**

Assigning a new file to an asset in the Asset Files window

You may find yourself creating a DVD project complete with menus and tracks, all of which have assets associated with them, but then the client (surprise!) asks to make changes to some of the graphics in the project. One option is to go into the Photoshop files and make the changes yourself, reopen the project, and the changes are reflected in the DVD Studio Project file.

In certain cases, you may need to *assign* a new file to an existing asset. Assigning a new file is helpful when the client wants to redo the file themselves, and they will be e-mailing it to you to incorporate it into your DVD project. In this case, you need to assign the new file.

To assign a new file to an asset, follow these steps:

1. **Open the PS Records DVD project file, which is located in the Tutorial section of the DVD-ROM.**
2. **Choose Item ⇨ Asset Files. Select the mainmenu.psd asset.**
3. **Click the Assign button at the bottom of the Asset Files window.** A dialog box appears providing a list of files to select. In this case, navigate to the CH3 folder in the Tutorial section of the DVD-ROM, open the CH3 folder, and select the newmainmenu.psd file; then click the open button. (See Figure 3-20.)

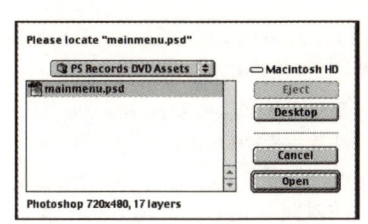

Figure 3-20: The Asset Files window, with the newly assigned replacement file. Same asset, new file.

Assigning files to assets using the Property Inspector

An alternative to using the Asset Files Window to assign a new file to an asset is to use the Property Inspector.

To assign a file to an asset by using the Property Inspector, follow these steps:

1. **Choose Window ⇨ Asset View to bring up the Assets Container (Asset View).**
2. **Select the asset you want to assign a new file.**
3. **Choose Window ⇨ Property Inspector.** You see the Property Inspector's summary of the asset. In the General section of the summary, in the File line, notice the blue link that looks like a hyperlink on a Web page. Click the link to bring up the Asset Files (See Figure 3-21.) dialog box.

Assign button   New replacement file

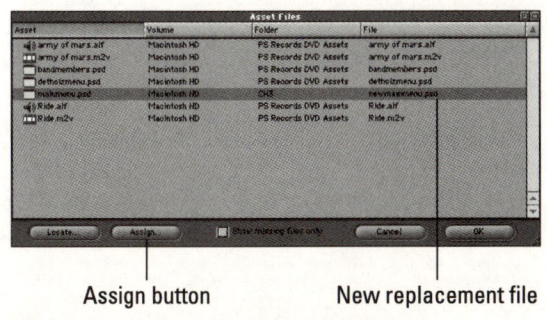

Figure 3-21: The Asset Files dialog box

Summary

In Part I, Working with DVD Studio Pro, you acquainted yourself with the workspace and the various ways to look at your projects and their elements. In this chapter, you have worked with assets — the resources that make up the heart of your DVD Project.

- ✦ Within DVD Studio Pro, an asset is only a link to the file you have imported; it is like an *alias* in the Mac OS.
- ✦ After an asset is imported, the asset's name can be changed, while the original filename stays the same. You rename the asset in the Property Inspector.
- ✦ DVD Studio Pro uses the MPEG-2 and MPEG-1 video formats exclusively. MPEG-2 is the ideal choice, but MPEG-1 work well, too.
- ✦ DVD Studio Pro incorporates either PCM audio, (which includes AIFF, SoundDesigner, and WAVE formats), MPEG-1 audio, or Dolby AC-3.
- ✦ Assets can be imported through the File ➪ Import option, the drag-and-drop method, or by creating a new asset through the Item menu.
- ✦ The Assets Container, also known as the Asset View, shows a visual summary of all imported and available files to use in your project.
- ✦ Clicking on an individual asset in the Asset Container provides information about the asset in the Property Inspector.
- ✦ You can customize the names and colors of the Labels that appear in the Label drop-down menu in DVD Studio Pro. The Label drop-down menu appears in the Property Inspector when you have selected project elements, such as assets.
- ✦ You can rearrange the order in which assets appear in the Assets Container manually or by using the Item ➪ Sort Assets function.
- ✦ Delete assets in the Assets Container by clicking on the asset(s) you want to delete and press either the delete key on the keyboard or select Edit ➪ Clear.
- ✦ The Asset Matrix provides a snapshot view of how the various assets in a project relate to menus and tracks as well as serving easily connect imported assets to menus and tracks.
- ✦ The Asset Files window enables you to locate missing files or to assign a new file to a particular asset.
- ✦ After an asset is selected in the Asset View, you can assign a new file to an asset by selecting Item ➪ Asset Files or by clicking on the blue File link in the Property Inspector.

Now that you are familiar with the DVD Studio Pro interface, and the various ways you can work with individual project elements and assets, you will move on to taking a look at the bigger picture. In Part II, you discover how to create an interactive experience. Chapter 4 starts you off by introducing you to DVD project planning — an essential part of creating a positive and enjoyable user experience.

✦ ✦ ✦

Creating Interactivity

PART

II

In This Part

Chapter 4
Planning and Prototyping

Chapter 5
Working with Buttons

Chapter 6
Working with Menus

Chapter 7
Using Subtitles

Chapter 8
Exploring Interface Design

Planning and Prototyping

CHAPTER 4

In This Chapter

Considering the Audience

Storyboarding

Prototyping DVD Interactivity

Many issues affect the planning of a DVD project and this chapter helps you to recognize what the issues are and how to manage those issues. You are also exposed to several methods of prototyping DVD interactivity.

Some people may enjoy building a project as they go, without thinking too deeply about it, or perhaps just thinking intuitively. An intuitive free-form approach can serve arguably well, especially where the project has the freedom to grow and evolve according to inspiration that comes along. At the very least, it can be wise to build flexibility into any project, to leave room for ideas that don't occur to you until you are deep into the workings of the DVD.

But there will be situations that require planning ahead of time, perhaps where the project is the result of a team effort, where each person has ideas to contribute, and the various ideas need to be resolved into a cohesive whole. Or maybe a single client has very particular ideas of what they wish to do, and a prototype is a good tool to give the client an idea of how things will look. There is also a sub-paragraph of the Statutes of Murphy's Law, (the case of DVD Author vs. Father Time), which states that time flies faster than you think it will. No matter how you approach the organization of a DVD project, it can be helpful to at least chart out what you want to accomplish, and give yourself a timeline with goals of when you must achieve certain elements.

Considering the Audience

It is wonderful to contemplate that no matter what kind of DVD project you create, there is an audience out there for it. What's more, DVDs will be a novelty for a long time coming.

A professional, well-designed DVD with interesting content will stand out, even after DVD burners become commonplace.

Consider the Internet. Today, a lot of people have the capability to make their own Web pages; but not all of them are enjoyable or make sense. As Carl Sagan might have said, there are "billions and billions" of Web pages out there, but only a handful have been created after careful consideration of the audience.

The interface you create to navigate a DVD is one of the central facets of a DVD project, and in the chapter about Interface Design, Chapter 8, there is a discussion of several principles to consider, partly drawn from lessons learned by the developers and designers of Web pages. The purpose of this chapter is more general; the intent is to stimulate thought about the overall planning of the DVD project.

Considering enjoyability

The concept of *enjoyability* is a complement to the concept of *usability*. Usability, as jargon and principle, has been popularized by Jakob Nielsen, a highly respected Internet professional and speaker who has created a well-visited Web site at `www.useit.com`. The principle of usability in the context of the Internet urges Web designers to place a particular emphasis on ease of use, but it can be taken to the extreme, where any dynamic element that is thought to interfere with functionality is abandoned. For an interactive experience which is primarily informational, the principle of usability makes particular sense on its own, but in the context of entertainment, whether it is on the Web or on a DVD, the concept of enjoyability should be considered as well.

As more and more people have gone online, the aggregate perceptual expectations of Web surfers have shifted as well, as more and more individuals come to the Web from a mainstream lifestyle which includes a higher amount of television and pre-recorded watching. Ultimately their decisions of what television channel to watch or which movie to rent and whether to watch the trailers are based not on what is most functional, but what is most entertaining. Similarly, when a modern surfer goes online, they may be seeking information, but they are also there to enjoy the Web. In looking for fun, they find that they have an endless variety of Web sites to enjoy, and they are bound to head for the ones that are most enjoyable.

Therefore the enjoyability of a Web site can be defined as the combination of usability and entertainment value, relative to the user. The more a site is designed to be a research tool, the more expectations of enjoyability are likely to eclipsed by a desire for usability. A person searching for bibliographic references does not care if they are animated. The reverse is not true though; a person who is looking for fun is always going to appreciate usability.

Asking the questions

When you are considering the audience for your DVD project, you may want to ask yourself one of the following questions, from the standpoint of usability and entertainment value.

Who is the intended audience for the project?

When a company is considering how to market a product, they often think of the "audience" for their product in terms of *demographics*. The word demographics is defined in *Webster's Collegiate Dictionary* as "the statistical characteristics of human populations (as age or income) used especially to identify markets." Even if you aren't necessarily selling something to a person, it is helpful to consider an audience from the perspective of their age, income level, likes and dislikes, etc.

And you don't necessarily have to get into the field of statistics to better understand your audience. A simple way to gather information is to think of people you know or with whom you are acquainted, who might be representative of the audience you want to reach with a DVD project. You can learn a lot by asking them questions; running ideas by them, to get their reaction. If you don't have some kind of prototype to share with them, you can still ask questions like "what if you had the choice of switching audio streams," or something like "which do you think would be more appealing, having the videos run all together, or separated with different buttons leading to each one?" Doing this kind of research at a planning stage can be helpful in guiding the shape of a project, especially if you are experimenting with something new, and are looking for confirmation of the direction you're going in, such as the design, or the way one menu leads to another.

One of the best ways to get peoples' reaction is to give them something to look at or try out. For example, if you have the freedom of developing several interface design ideas, you can share them with people and ask them to choose the one which they like the best. You could set up the menus of a DVD and ask someone to try clicking through the choices to make sure they make sense. Even if they make sense to you, the way things work might not be apparent to others. DVD-R media may be an ideal way to run your project by a few people, asking them to take it home and try it in their player.

Caution

If you are planning to share a prototype or finished DVD project with someone, keep in mind that DVD-R media is not 100% compatible with all DVD players. It will run in most newer players and DVD-ROM drives, but the only way to know for sure is to test it ahead of time, or to ask what model a person has, and consult compatibility lists such as the one Apple maintains at `www.apple.com/dvd/compatibility`.

When you have taken the time to identify some of the reactions from people who are representative of your audience, you can use this information to help you plan your project. Consider the question of whether to include a particular DVD feature or content from the standpoint of whether it would be received well by your audience.

Many companies put together a formal version of this kind of research which is called a *focus* group. A specific group of people is assembled, usually from outside of the company, people who fit the intended audience for a project. They are gathered together, and asked to react to the results of a particular project. The thing a focus group is asked to try could be a certain food, a series of songs, a series of designs for a Web page, anything.

Whether you are considering the audience informally or formally, it doesn't need to be a one-time thing. Getting into the habit of letting people react to your DVD projects as they develop will help you to develop a more professional, enjoyable, and functional end result. This kind of research can ultimately be a source of inspiration from the very start of a project, as you ask yourself what kinds of things the people you're shooting for might enjoy, what might be relevant to them.

Is it a "captive audience" or a "free audience"?

A captive audience for a DVD project would be one where the project is being distributed to all the members of a group, where they are expected or obligated to experience or utilize the DVD. With a captive audience, you may have more freedom to include content or make references that wouldn't make sense to, or wouldn't be understood by, a more general audience. With a free audience, you may not be able to pinpoint demographics, and may need to keep things as generic and general as possible. Other differences between a captive and free audience may include consideration of advertising in the DVD itself, where with a free audience you might want to take advantage of every reasonable opportunity to emphasize a company logo and contact information either in the DVD or on an accompanying DVD-ROM presentation. (See Chapter 21.) With a captive audience, you may wish to emphasize the appeal of the content over "internal advertising," to limit the negative association some people may develop in being forced to experience repeated messages or images.

What will the audience expect when they play the DVD?

In general it is safe to assume that potential user's expectations center on functionality and enjoyment. Users want the DVD to work and to enjoy it. What is not safe to assume is that the user will necessarily understand "how" everything should work, or that what makes sense to the DVD author necessarily makes sense to the user.

It can be helpful to assess a project as it develops from this standpoint, allowing people who have nothing to do with the planning of the project, to "try it out," especially if there is any degree of complexity in the project. Do they know where to find the information? Do they know how to use the Web links?

Many people in Web and DVD design are asking the question, "What is the user experience like?" The user experience for a DVD, simply put, is how someone experiences your DVD. That is, what kind of impressions they form, do they enjoy it, do they get frustrated, does it meet their expectations, and so on.

Author's note: One of my favorite DVD interfaces is the one for the movie *Requiem for a Dream*. The movie is often fairly dark, but there is a humorous side, which has been carried over very well into the DVD interface. A friend told me it took them a while to figure out how to actually start the movie after they inserted the disc — the way you're supposed to use the interface was not immediately apparent. When I inserted the disc, it was similar for me. I don't want to give it away, but the DVD interface is in the context of humor. Whether or not the difficulty was intentional, it is an example of users not immediately understanding. Generally you want to stay away from this as much as possible. *Requiem for a Dream* was all in fun, but sometimes a bad experience can be very frustrating. There was another time I remember trying out the Special Features section of a DVD, and I wanted to change which language the soundtrack it would be in, but the way it worked just didn't make any sense at all, and I gave up on it. It's probably good to for DVD authors to remember that if we are building a DVD around someone else's creative content, that the DVD interface can have a definite impact on the impression people form about the content of the DVD. Probably the best thing to aim for would be that a DVD interface looks good, sounds good, and allows you to get to where you are going without frustration.

If the content already exists in other formats, or will be available in multiple formats, how will the DVD relate?

The process of *re-purposing* content is taking something you have already created and using it in a different way. For example, if you have a short film which you put on a DVD, you can re-purpose the content by encoding the video a different way and hosting it on a Web site so that people can see the movie on the Internet. *Cross-purposing* content is the process of planning to deliver content through a variety of mediums. For example, a company might launch a new line of instructional videos, and to launch the new product, they decide to distribute a sample course to as wide an audience as possible. The resulting video is delivered on every available format, via CD-ROM, DVD, VHS tape, on television, and on the Internet.

If you are "re-purposing" content which is already available to the audience in another format such as VHS tape, or "cross-purposing" content, planning for delivery in a variety of mediums, it is helpful to consider the relative capabilities of each medium. For example, in order to make the DVD stand out as an attractive option, you may wish to take advantage of as many features as possible of the DVD medium, such as multiple tracks or languages. Alternatively, in a case where you don't want a person who doesn't have a DVD player to miss anything, you may wish to build the DVD project in a way that is as close as possible to what a person would experience with VHS or CD-ROM, so that the same message or experience is getting across. Thinking about these kinds of issues can be helpful if the DVD project is being planned or produced parallel to other related projects, so that overall goals can be met, taking the relative capabilities of each medium into account.

After you have considered the audience, or in conjunction with your research, you may want to *storyboard* a project. Storyboarding allows you take some of the ideas you have and put them together in simple visual form, so you can try ideas out without having to fully develop them.

Storyboarding a DVD Project

Storyboarding is the process of putting together a visual summary of a project, representing a series of scenes or screens with images or pictures, so that you can look back at them and get a better sense of how the project may turn out.

Storyboarding can help the DVD author to get a better sense of how various scenes in a video segment will work together. Storyboarding is often used in video and film production at the early stages of a project to help the producers and directors get a sense of how particular visual sequences flow. You can get as complex and detailed as you like with storyboarding, but generally it amounts to making some kind of rough sketch on a series of sheets of paper, where each sketch represents a particular "shot" or "scene", and some kind of brief description is written beneath each sketch. To make things look nicer, a page may be divided into sections to represent separate screens, as in the sample of the very simple storyboard in Figure 4-1.

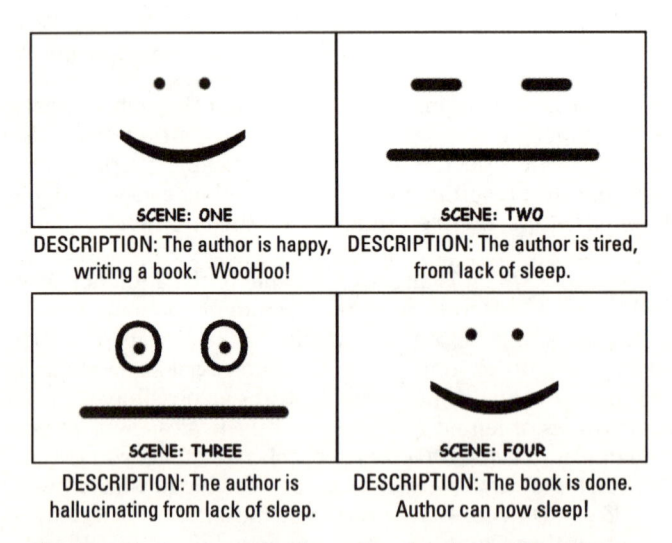

SCENE: ONE	SCENE: TWO
DESCRIPTION: The author is happy, writing a book. WooHoo!	DESCRIPTION: The author is tired, from lack of sleep.
SCENE: THREE	SCENE: FOUR
DESCRIPTION: The author is hallucinating from lack of sleep.	DESCRIPTION: The book is done. Author can now sleep!

Figure 4-1: An example of an ultra-simple storyboard page created in Microsoft Word

On the DVD-ROM in the Tutorial/CH4 folder, there is a simple sample of a storyboard document, a Microsoft Word file storyboard.doc, very similar to the one represented in Figure 4-1, which you can customize for your own uses. It is blank, and you may wish to simply print it out and hand draw and write the sketches and descriptions, or you may also find it convenient to type in simple descriptions ahead of time and expand the description section.

If you want to use digital images in a Microsoft Word-based storyboard, all you need to do is have the images ready to insert in the Word document. Word will import a wide number of image formats, such as PICT, PCT, JPEG or GIF.

For example, if you were planning a short video segment that was filmed in different rooms of a house, you could take a digital camera and take a picture of each room. When you transfer the pictures to your computer, you could insert them in a storyboard document to represent the flow of scenes.

Assuming that you have the images that you want to include in your storyboard, you can place them in the blank Microsoft Word document I have provided on the DVD-ROM. To add the files to your document, follow these steps:

1. **Open Microsoft Word and locate the storyboard.doc file in the Tutorial/CH4 section of the DVD-ROM.**
2. **Open the storyboard.doc file and click in the upper left table cell.**
3. **Choose Insert ⇨ Picture ⇨ From File.**
4. **Locate the file you want to use for a frame storyboard.** If you do not have an image to work with, you can use the walking.pct file in the CH4 folder. (You may have to delete some of the spaces in the table cell if the picture file you inserted makes the table cell larger than you wish.)
5. **Click Insert.**
6. **Repeat the insert operation for each of the cells, using subsequent images that represent the sequence of events.** In my situation, I used the train1.pct, train2.pct, and light.pct files from the CH4 folder. You might want to experiment with typing in a short description for each screen shot. When you are done, the file should look something like Figure 4-2.

While storyboarding can be helpful in planning the video segments in a DVD a project, another process to consider would be prototyping the interactivity, making screen shots link to one another, to create a simulation of what the DVD might be.

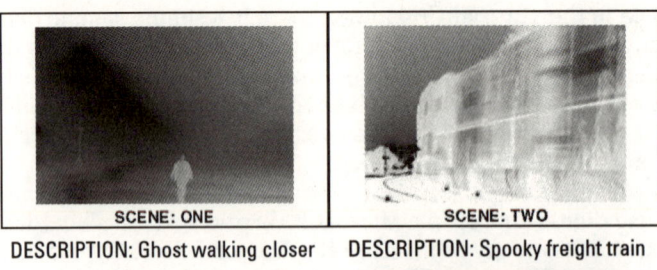

SCENE: ONE	SCENE: TWO
DESCRIPTION: Ghost walking closer	DESCRIPTION: Spooky freight train
SCENE: THREE	SCENE: FOUR
DESCRIPTION: If you lived here you might be home by now.	DESCRIPTION: Egads! Is it an alien ship? No! It's a lamppost!

Figure 4-2: A storyboard page for a video sequence

Prototyping DVD Interactivity

DVD Studio Pro makes putting a project together easy, but if you are creating a project for a client, you might want to consider *prototyping*, that is, a simulation of how the DVD might work. The simulation is made of screen shots from video segments and images representing the menus that you are developing for your project.

The upside of prototyping is that it gives the client some instant gratification; they get to see something of how things will work. Depending on how you do the prototype, it could provide them with an easy way to share the project via e-mail and in the case of a Microsoft Word document, you could request to have feedback and commentary be incorporated right into the prototype itself. The downside of prototyping is that the time for feedback can extend the timeframe required to get the project done, and you may open yourself to more feedback than you bargained for.

Ultimately the question depends on how important it is to share an idea of the DVD with someone who couldn't sit down with you and look at a preview you could generate in DVD Studio Pro. A prototype could also be a natural stage in the project plan, where you have your resources gathered, and create DVD menu images in Photoshop, and then combine the menu images with screen shots from the video and share them with the client before going any further. One option would be just to print the images out, or attach them to an e-mail, but if you have the time and inclination, a simple prototype would make you look more impressive!

Considering a DVD-R Prototype

If you have a DVD burner and DVD-R media to spare, you may just prefer to go all the way and make the whole project in DVD Studio Pro, burn a DVD-R, and hand it to the client.

Caution: If you are planning to share a prototype or finished DVD project with someone, keep in mind that DVD-R media is not 100% compatible with all DVD players. It will run in most newer players and DVD-ROM drives, but the only way to know for sure is to test it ahead of time, or to ask what model a person has, and consult compatibility lists such as the one Apple maintains at www.apple.com/dvd/compatibility.

Apple's SuperDrive, and the various standalone FireWire DVD-R burners that will soon be available, do not necessarily generate a DVD-R which would be compatible with all consumer DVD players. This might be a factor in considering whether or not to prototype with a "rough" DVD-R, where there is a risk that if the client takes the DVD-R home it might not work if they have an incompatible player. You should know that DVDs have a better chance of being compatible with computer DVD players than consumer DVD players. When you want a 100% compatible DVD, you want a manufactured DVD, sent to at a manufacturing plant via DVD-RAM, DLT, or DVD-R. The process of manufacturing a finished DVD is commonly referred to as replication.

Note: DVD-R stands for a recordable DVD. It is the DVD equivalent of a CD-R. Pioneer makes a DVD-R burner (The DVR-S201) in the $3000-$4000 range which was essentially the only option until the recent advent of Apple's SuperDrive and the various emerging standalone DVD-R drives. At present, the media for Pioneer's DVR-S201 is NOT compatible with Apple's SuperDrive, and vice versa.

Apple maintains a list of consumer DVD players that are compatible with DVD-Rs burned in the SuperDrive at www.apple.com/dvd/compatibility/.

So if you want to do a prototype, but have limited time and budget, you may just want to go with a good old Microsoft Word prototype.

Prototyping with Microsoft Word

A Microsoft Word document is a simple way to convey a rough approximation of the look and feel of a DVD project. Fortunately or unfortunately as the case may be, a vast percentage of colleagues and clients have Microsoft Word 97 on their computers, and e-mailing them a Word document allows your clients to click through the document, print the prototype out, show it around at meetings, and make commentary right in the course of the document, save it, and send it back to you.

Wait, did you catch the phrase *click through*? Yes, it is actually possible to have a hyperlink in a Word document lead from one page to another. You can insert screen shots in the document with no links, but giving someone even a slight feeling of interactivity in a Word document could bring a prototype one step closer to representing a DVD.

At present, you can only place links to follow the screen shot. Currently there is no way to create hot spots in the image itself that would be compatible with Microsoft Word 97, the version of Word that most people in the Windows world are using right now.

Author's Note: Authors are known to not be omnipotent, so if someone can come up with a way to do this, we'll put it up on www.dvdspa.com. If you don't see it up there already, feel free to send an e-mail in and let us know how, and we'll give you full credit!

The technique of linking from one spot to another in a Word document is achieved by placing bookmarks in the appropriate spot, and then inserting hyperlinks on another page linked to the bookmark. The bookmark is a destination; it gives the hyperlink a place to go to. For Web-savvy DVD authors, the bookmark is kind of like an anchor link, where you click on a link at the top of a Web page and it leads to a spot on the same page, rather than a different Web page altogether.

To make hyperlinks that lead from one page to another in a Word document, follow these steps:

1. **Take a deep breath.**
2. **Open Microsoft Word and create a new document, and type in the words, "go to page two" or something similar.** Place the cursor next to the word "go" in the line you just typed in, and choose Insert ⇨ Bookmark; when you get the Bookmark dialog box, type in PAGEONEBOOKMARK (or anything else you want to). Then click the Add button. You now have a bookmark on page one.
3. **Place your curson next to the first word in your bookmark.** In my case, I placed the cursor to the left of the word *go*.
4. **Choose Insert ⇨ Bookmark.** The Bookmark dialog appears.
5. **Type the name you want to call the bookmark.** I named my bookmark PAGEONEBOOKMARK.
6. **Click the Add button.** You now have a bookmark on page one. A finished example, simplelink.doc, can be found in the Tutorial/CH4 folder of the DVD-ROM.

Note: To view bookmarks, Choose Tools ⇨ Options, then click the View tab, and select the Bookmarks check box. Bookmarks display as square brackets ([...]) on the screen. The brackets do not print.

7. **Place the cursor after the line you typed in on the first page.** Now press the return key several times, and choose Insert ⇨ Break ⇨ Page Break ⇨ OK. This brings you to a second page.

8. **On page two, type the words "go to page one".** Then insert a bookmark next to the line you just typed in, and call it PAGETWOBOOKMARK.

9. **On page two, select the text "go to page one", and Choose Insert ⇨ Hyperlink.** This brings up the Hyperlink dialog. Click on the lower Browse button, to use the "Named location in file" function. This brings you to the Bookmark dialog, where you can choose the place you want your hyperlink to lead to. Select PAGEONEBOOKMARK and click OK, and then click OK again in the remaining Edit Hyperlink dialog . You now have an active hyperlink on page two, which leads to page one.

10. **On page one, select the text "go to page two," and this time, when you create the hyperlink, select PAGETWOBOOKMARK.**

The result of this exercise is that you are able to click on one or the other links, and it gets you to the next page. After you become familiar with hyperlinking in a Word document, you can use it to create a prototype that will give a client a partial sense of what it would be like to click between the various screens in a DVD.

Figure 4-3 shows what appears on each page in a sample prototype which utilizes the hyperlinking feature you just learned. On each page, there are text hyperlinks above the screen shot images, which simulate how the buttons in the DVD would work. The prototype is called mainmenu.doc, and is available in the Tutorial/CH4 folder on the DVD-ROM.

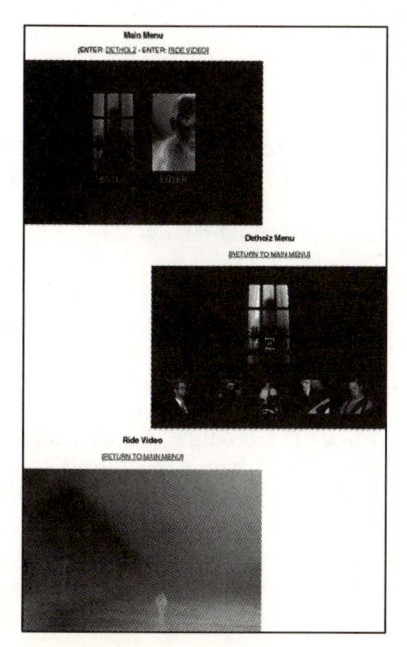

Figure 4-3: A composite of what appears in mainmenu.doc, a simple interactive DVD prototype created in Microsoft Word. On the main menu page, there is an ENTER DETHOLZ hyperlink link above the screen shot of the main menu. This represents an ENTER button in the main menu of the DVD. So when you click on the ENTER DETHOLZ hyperlink in the Word document, it brings you to the page where the Detholz Menu is represented, as if you had activated the button in the DVD.

When preparing images for use in Microsoft Word for an interactive DVD prototype, I recommend that screen shots be re-formatted in Photoshop. I suggest you resize each image to something like a six-inch width. Use the Save As function to save the file in a 16-bit color depth PICT format to keep the file size down.

For the mainmenu.doc example, page breaks were inserted before the title of each page, and the bookmarks were inserted directly next to the titles to ensure that clicking on a link brought you to the top of each page. The links were placed above the images to make the interactive experience smoother by ensuring that the viewer always see the links, regardless of the size of the Microsoft Word window. If you placed the links following an image, a person may come to a page and see part of the screen shot and not see the subsequent links, and would have to know to scroll to get to those links.

There are several advantages to using Word for prototyping:

- Word is so common, most people know how to already use it
- Word is transportable and found on most systems
- Word allows potential reviewers to insert comments

Compared with prototyping in Word, prototyping with HTML would be the "next step up." Using HTML would allow you to have hyperlinks in the visual space of the screen shots, rather than having hyperlinks above the images, so the simulation of a DVD would be one step closer to the real thing.

Prototyping with HTML in Dreamweaver

Even if you have no experience in HTML, it might be worth taking the time to learn. If you create a prototype in HTML and put it on the Web, it's very easy to just give a link to someone, and make changes to the prototype which the person can come back and check, just as if they were looking at an updated Web page.

Macromedia's Dreamweaver is the leading visual HTML editor, and if you're interested in prototyping DVD interactivity with HTML, Dreamweaver should be considered seriously. Dreamweaver can save you a lot of time. It can take care of the HTML code for you, allowing you to concentrate on making the best DVD prototype.

Dreamweaver allows you to create HTML documents (Web pages) without necessarily having to edit the actual HTML code. It represents things visually, allowing you to add elements to an HTML document one by one, as it generates the code in the background for you. It is definitely a good idea to know the basics of HTML when working with a program like Dreamweaver, but they've certainly gone a long way in making it as easy as possible to make a Web page.

Understanding basic HTML

HTML stands for Hypertext Markup Language, and is the basic building block for the Internet. When you look at a Web page through Netscape or Internet Explorer or America Online, you are using a Web browser to view the page. In a nutshell, the Internet is basically a big network of computers, so when you type in something like `www.gerbilfront.com`, it is an electronic address on a network. When you get to a particular Web page, notice that in the address or location area of the browser, the address changes slightly as you click on different links, and usually the last part of the address ends in .html. When you type an address such as `www.psrecords.net/wake/index.html`, the browser retrieves the index.html document over the network. The HTML document contains code, something close to the English language, instructing your browser to display a page, telling the browser which text to display, what color, where to get the images, and so on.

> **Note** When naming HTML documents, you can use a file extension of either `.htm` or `.html`. Programs used to make Web pages sometimes differ in the way they save HTML documents, but preferences can usually be set in the program to save with either file extension. Whichever way you choose to go, you do need to be exact when linking to an HTML document. For example, if you name a document `example.html`, a link may not work if you use example.htm for the address, omitting the *l* at the end. And when you are creating "index.html" files, under certain conditions you have to use the html extension, htm may not work. (Index.html files are used to shorten Internet addresses. If you had the address `www.psrecords.net/examples/home.html`, if you wanted to shorten the address to `www.psrecords.net/examples`, you would change the name of home.html to index.html in the examples directory on the Web server; the browser knows to look for the index.html file when it hits the address `www.psrecords.net/examples`.)

Most people use some kind of HTML program like Dreamweaver, but HTML documents could be created in a text editor as simple as SimpleText. The following HTML code in corresponds to the Web page located at `www.psrecords.net/storyboard.htm`:

```
<html>
<head>
<title>Macworld&reg; DVD Studio Pro Bible</title>
</head>
<body bgcolor="#FFFFFF" text="#000000">
<center>
<img src="storyboard.gif" width="360" height="255">
</center>
</body>
</html>
```

When you look at a page through the Web browser, the HTML instructions are acted upon and the page is put together, but you never see the code (see Figure 4-4).

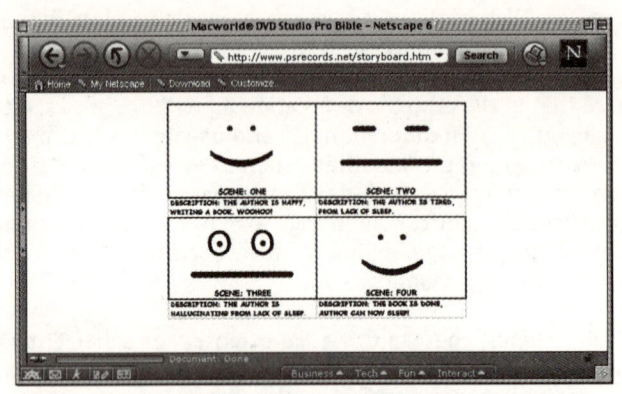

Figure 4-4: The finished result of the HTML code, as seen through the Mac version of Netscape 6.0.1

The browser identifies the HTML code (also known as tags) by the brackets <> which appear on either side of the code. If you look at the example code, and compare it with the actual page, it makes a certain amount of sense. You may also notice that the HTML tags are repeated. For example, the <title> tag appears before and after the text of the title, and the tag which appears after has a slash </title>. The slash (/)tells the browser that this is the end of the <title> tag, so then the browser can move on to the next piece of code. The result of the <title> tag is that the text "Macworld" DVD Studio Pro(tm) Bible" appears at the top of the Web page.

Another common HTML tag is <body bgcolor>, which sets the background color of a Web page. The tag uses a special hexadecimal format to represent the color you want. You can choose the color for your Web page by consulting a Dreamweaver color chart and allow it to generate the code for you. (See Figure 4-5.)

The point of this section has been to introduce you to the basics of HTML, which is the basic code of the Internet, and the code which Dreamweaver is designed to help you generate. Dreamweaver can save you a lot of time in generating HTML for you, but you can't completely ignore HTML code itself. There are occasions where you may have to go into the code and adjust something directly. So it is good to learn a little HTML, which will enable you to use a tool like Dreamweaver more effectively, whether you are making an HTML-based DVD prototype, or a Web page to promote your DVD project.

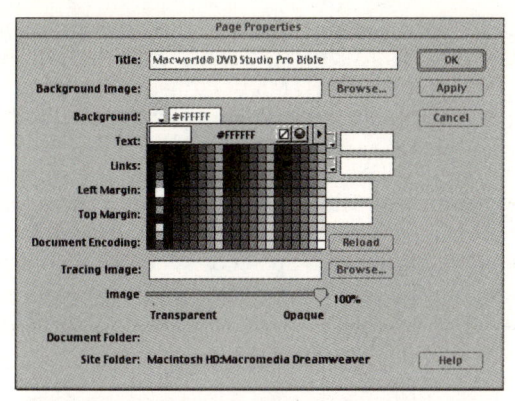

Figure 4-5: Select a background color in the Dreamweaver Page Properties dialog box, using the pop-up Color Picker to generate the appropriate hexadecimal value (#FFFFFF represents white).

Understanding Dreamweaver

There isn't enough room here to fully equip you with the skills needed to make a Web page in Dreamweaver, but the tutorials that come with the program will take you a long way, and there are several Web sites such as `www.webmonkey.com` that provide several helpful tutorials on HTML and related topics.

If you are new to HTML and/or Dreamweaver, but would like to learn it, a free, fully functional 30-day trial version of Dreamweaver is available on the DVD-ROM. You can also download a number of related programs and files directly from Macromedia's Web site at `www.macromedia.com`.

In general, the process of creating a Web page in Dreamweaver is somewhat similar to creating a DVD project. DVD Studio Pro is a DVD *authoring environment*, a place where you put a DVD project together. Similarly, Dreamweaver is an HTML authoring environment, where you assemble your graphics, text and other files that you want to incorporate in an HTML-based Web page.

Dreamweaver allows you combine various resources together to craft a page or series of pages. (These resources may include graphics, text, animation, video or audio, anything that a Web browser such as Netscape can be made to understand.) With HTML documents, a hyperlink leads from one page to another. This is kind of like DVD menus, where you have buttons that link one screen to another.

With Dreamweaver, you create the HTML documents on your own computer and preview them to get a sense of what your Web pages will look like when they are on the Internet. Then, when you are done, upload the HTML documents, images, and other files to a host computer, which is specially designed to be connected to the Internet all the time.

In order to put Web pages on the Internet, you will need a space to put them. The same company that sells you your Internet service can also set you up with space on a host computer for the Internet. Often, your dial-up account comes with some free space for putting up a Web page, but the address you are given may not be desirable, and may be limited to being an extension of the Internet Provider's Web address, something like `www.internetprovider.com/~yourusername`.

Nowadays most people go to a Web site such as `www.register.com`, and simply register their own domain name. A domain name is essentially what appears between `www` and the domain `extension` such as `.com`, `.net`, or `.org`. For example, in the case of `www.hungryminds.com`, `hungryminds` is the domain name. Often you can set up Web hosting and a domain name through the same company. The possibilities are endless, but if you're not sure where to start, visit `www.earthlink.net`, a national provider which has just about everything.

> **Note** If you create an HTML prototype, you can also gather the HTML documents and images together, and place them on a CD-ROM or some kind of magnetic media such as ZIP or floppy disk, and share them with a person that way. You don't have to put them up on the Web. As long as you give the person all the correct files and that person has a browser program such as Netscape Navigator or Internet Explorer, he or she should be able to view the HTML documents you provide in much the same way as if they had been on the Web; the only real difference is that he or she will either need to double-click one of the HTML documents to start up the browser program, or open the HTML file through the File menu in their browser program.

In order to do an HTML prototype in Dreamweaver, then, there are some skills to master, but if you haven't acquired them already, knowing how to do a Web page can come in handy.

Making a DVD Prototype with HTML

Making a simple DVD Prototype with HTML is relatively straightforward; you can get as complex as you like, but essentially what HTML allows you to do is put up approximations of DVD menu screens on the Web.

Getting Ready to Rumble

The general process of utilizing HTML to simulate DVD screen click-throughs is via the hotspot feature of HTML, also known as image maps. Basically, you take the multi-layered Photoshop file you are using to prepare menus for your DVD, select the layers you want to appear on the page, re-size it as desired, save it in an appropriate Web format (`.gif` or `.jpeg`), then bring it into Dreamweaver and add the hotspots over the buttons in the image, and then create separate HTML documents for the various screens.

> **Note** If you are making an HTML prototype and using a Photoshop file that is also used by a DVD Studio Pro project file, you should make another copy of the Photoshop file. This is especially true if there's any chance you will be re-sizing the image. Work with a separate copy, to avoid the possibility of changing the original and affecting the DVD project.

Preparing Images in Photoshop

To prepare images for an HTML prototype, it is necessary to work in an image editing program like Photoshop. You can use a Photoshop file that has already been prepared for a DVD menu.

If you do not have any images to work with, the CH4 folder in the Tutorial section of the DVD-ROM has a folder called HTML, which includes two Photoshop .psd files, basicmenu.psd, and bandmembers.psd, which can be used to create a simple HTML prototype in Dreamweaver.

To prepare an image for your prototype, follow these steps:

1. **Create a folder on your computer's hard drive and name it something that will make it easy to recognize later as your prototype.**

2. **Create two subfolders inside of the folder you created in Step 1 and name them so that you can easily recognize the types of files you want to put in them.** For example, I named my subfolders images and resources and placed them inside prototype. The purpose of these two folders is to organize the different elements you will use within your HTML document.

> **Tip** It is common practice when working with HTML documents that you name both files and folders all lower case. This is not strictly necessary but can help to avoid confusion, with situations such as when you are conveying Web addresses to people.

3. **Move the images you would like to use in your HTML prototype into one of your subfolders that you've designated to hold images.** If you do not have any images to use yet, you can use a couple of images I've provided on the DVD-ROM. The files are located in Tutorial/CH4/HTML/ folder and are named basicmenu.psd and bandmembers.psd.

> **Tip** DVD Studio Pro is one of the only programs where you import .psd files directly; in most other cases, a Photoshop format file (.psd) serves as a resource to save an image in a particular format, such as .jpg, .gif or .tiff. When working with HTML, you may wish to create a resource folder when working with .psd files, to keep them separate from the final files that are actually used in the HTML document, that may be stored in an images folder.

4. **Open up a Photoshop file that you have already created for a DVD menu.** If you don't have one to work with you can use the basicmenu.psd file in Photoshop; there are five layers corresponding to each of the buttons, and a Background layer. (See Figure 4-6.) The image has already been resized to a width of 540 pixels, roughly three-quarters of the original size, to save on the ultimate download time for the graphic when on the Web. Depending on the ultimate audience, you may wish to leave the image at its original size if they have a high-speed connection. The higher bandwidth available, the higher quality of an image you can save, since on the Web, higher quality generally means a larger file size.

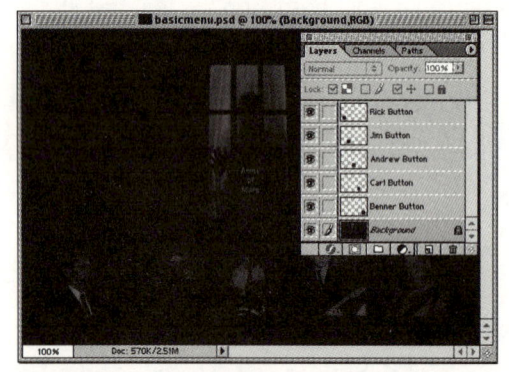

Figure 4-6: An example of a basic multi-layered Photoshop file originally created for use in a DVD menu, with the Layers palette superimposed, showing the layers for the buttons across the bottom

5. **Choose File ⇨ Save for Web.** This brings up with Save For Web dialog box, which allows you to save a version of your image for the Web and optimize compression settings.

6. **On the right side of the dialog box, in the settings area, choose JPEG for an optimized file format, and Medium quality. (See Figure 4-7.)**

> **Note** JPEG format and .JPG are the same thing.

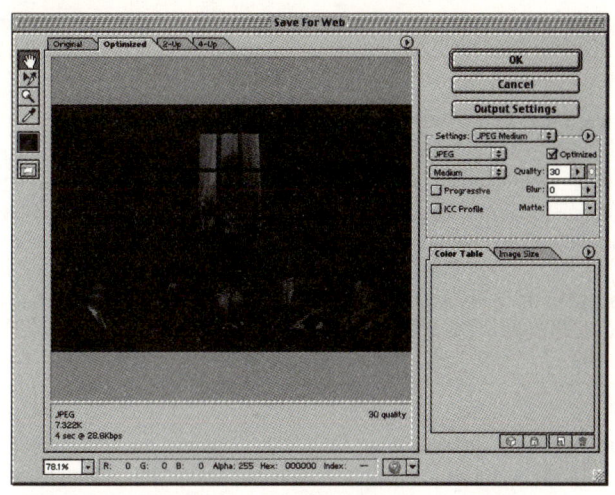

Figure 4-7: The Save For Web dialog box provides the opportunity to experiment with different optimization settings for you to obtain a balance between the best picture and the best size of the file for downloading.

7. **Click OK, and save the jpeg file in the folder where you want images to be stored, using the Save Optimized As dialog box (see Figure 4-8).**

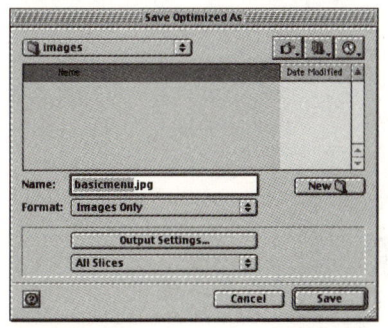

Figure 4-8: Saving the optimized image with the Save Optimized As dialog box

You now have an image (basicmenu.jpg) in your prototype folder, which can be used in the HTML prototype.

Creating the HTML documents in Dreamweaver

To create an HTML prototype, you need to create a series of HTML documents in Dreamweaver, set the properties of the page, insert the images you just saved, and create hotspots to link the pages together. This tutorial is not meant to be in-depth, but is meant to expose you to the possibilities.

In the following example, brackets are placed around specific file names which are suggested if you are working with the example files from the DVD-ROM.

To create a prototype:

1. **Open up Dreamweaver and choose File⇨New to create a new HTML document.** Choose Modify⇨Page Properties and set the title background color as you wish; in this case the background color is set to black. Save the document in the prototype folder, naming it in a way that allows you to remember which page it is, something like basicmenu.html.

   Note: If you want the background color of your screen image to exactly match the background color of a Web page, you may want to learn more about color palettes as they relate to the Web, so that you can modify an image in Photoshop so that the background color in Photoshop can exactly match a Web page background color (See Figure 4-9.) Sometimes if you don't match these backgrounds exactly, you will be able to see a border around the original image. One way around this is to intentionally set the background color to something like white, so that the "screen" will stand out.

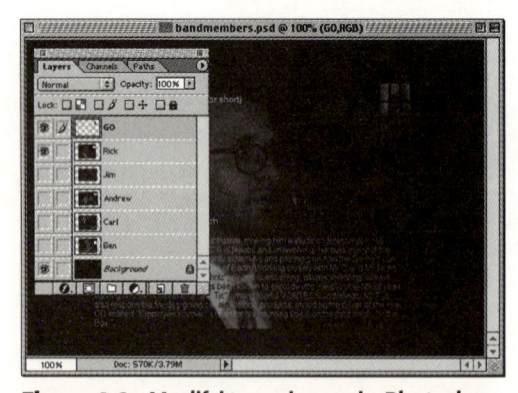

Figure 4-9: Modifying an image in Photoshop

2. **Place the cursor at the top of the document and choose Insert ⇨ Image.** Select the image you saved in the images folder in the prototype folder basicmenu.jpg, and click the Open button. Save the file but leave it open in Dreamweaver.

3. **Create a new document, adjust the color and title if you like.** This second document would represent a sub-menu on the DVD, a secondary screen that links from the main menu. Save the document, then insert an image, and save again. (Save as rick.html, insert the rick.jpg image, and save.)

4. **Choose the Window menu and select your original document.** [basicmenu.html] Using the Window menu allows you to get back to the other document you have open, which represents the DVD's main menu. Click on the image you imported into the document [basicmenu.jpg]. Choose Window ⇨ Properties, to make sure the Properties window is displaying, and expand the Properties window to display the "extended" set of buttons (by clicking in the lower right hand corner of the Palette window). Select the Rectangular Hotspot Tool. (See Figure 4-10.)

Figure 4-10: The Rectangular Hotspot Tool is used to capture the main details of the figure.

5. **Using the Rectangular Hotspot Tool, click and drag a rectangular hotspot over the area you want to make into a button.** [The lower left hand photo in the example menu image (Rick)]. In the Properties window, click on the small folder icon in the link area to connect the hotspot to the HTML document you created [rick.html] (see Figure 4-11).

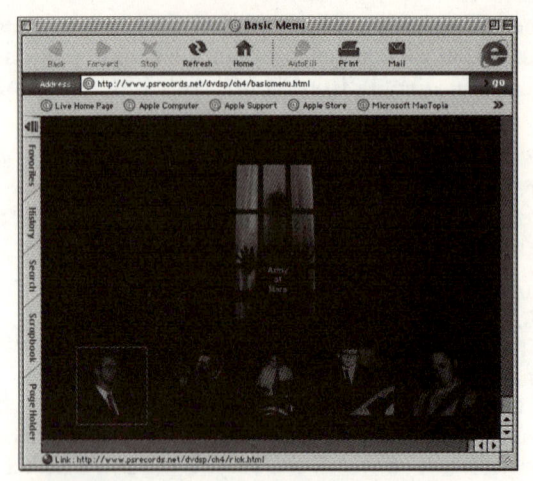

Figure 4-11: Overall image selected in Dreamweaver, hotspot created (indicated by small squares at edges of hotspot), and hotspot linked to a sample HTML document (rick.html)

6. **In the HTML document which represents your DVD sub-menu, create a hotspot over the button which links back to the main HTML document representing the main DVD menu.** [basicmenu.html]You now have a simple interactivity prototype. If you like, practice further by creating a page for each of the band members and links back and forth from the basic menu page. The fully flushed out prototype is in the finished files folder in the CH4 folder in the Tutorial section of the DVD-ROM, and it is also live at `www.psrecords.net/dvdsp/ch4/basicmenu.html` (see Figure 4-12).

Figure 4-12: Live sample prototype at `www.psrecords.net/dvdsp/ch4/basicmenu.html`, showing Rick link selected, with link indicator at bottom of window pointing to rick.html

Congratulations! You have seen how Dreamweaver and HTML can be used to create a prototype of your DVD project that can easily be shared with clients and colleagues. It is an example of how the Photoshop skills which are used in DVD Studio Pro can also be used for developing the project in the larger sense, giving you the ability to post a design and get feedback. This technique maximizes the resources you create in a Photoshop file by allowing you to use it outside of DVD Studio Pro, in a way that can be helpful for improving the product as you develop it. The technique also gives you a way of developing additional credibility by putting a professional-looking prototype up that can give a person a way to easily navigate an approximation of your intended interface.

In this chapter, you have taken a look at various techniques you can employ in planning projects. Considering the audience can help you develop a more effective or more entertaining DVD project, one that is best suited to the needs and expectations of those who will be experiencing it. The process of charting out potential scenes in a video segment through storyboarding may prove useful. And whether you use a DVD-R, HTML, or Microsoft Word, a prototype of your DVD project can be a useful tool, allowing you to present design and interactivity ideas to clients, colleagues, or representative members of the audience, giving them a taste of how the DVD might turn out while the project is still being developed.

Summary

✦ The usability of an interactive experience amounts to the combination of functionality and ease of use.

✦ The enjoyability of a DVD can be defined as the combination of usability and entertainment value, relative to the user.

✦ The word *demographics* is defined in *Webster's Collegiate Dictionary* as "the statistical characteristics of human populations (as age or income) used especially to identify markets."

✦ As you plan a project, asking questions about audience expectations and demographics can be helpful in helping you to identify and clarify what your approach will be in shaping the content for that audience.

✦ Storyboarding can be a useful tool as you plan a DVD; it can help you to get a better sense of how various scenes and screens work together.

✦ Using the bookmark function, Microsoft Word documents can be customized with hyperlinks that lead from one page to another within a document, and this can be used as the basis for a simple DVD interactivity prototype which may be a convenient way to share screen shots with clients or colleagues.

✦ HTML stands for Hypertext Markup Language, and is the basic building block for the Internet.

✦ The Internet is basically a big network of computers. When you "hit" an address such as `www.psrecords.net/wake/index.html`, what the browser is doing is going over the network and retrieving the index.html document.

✦ When you look at a page through the Web browser, the HTML instructions are acted upon and the page is put together.

✦ A program such as Dreamweaver can be used to create HTML documents that contain images from DVD menus, in order to prototype DVD interactivity. The interactivity is simulated using the feature of HTML known as "hotspots" or image maps, where you can have a link from an image on one HTML page lead to another separate page, thereby simulating the successive screens of a DVD.

✦ The Save For Web Dialog box in Adobe Photoshop is a great way to optimize images for use in HTML documents. It enables you to strike a balance between getting the smallest possible file size with the highest possible image quality.

The next chapter, on Buttons, takes you another step along the path of interactivity, demonstrating how to create the clickable elements of a DVD project. Buttons form the basis for giving a user choices in a DVD, allowing the user to choose which content to experience.

✦ ✦ ✦

Working with Buttons

CHAPTER 5

In This Chapter

Creating buttons in Photoshop

Creating buttons in ImageReady

Setting button properties

Buttons are literally the first link to creating a navigable DVD project. Like a television remote control, buttons in a DVD are essential for getting from one channel or screen to another. Buttons enable the user to view content and to interact with what's on the screen. Borrowing from the simplest features of Web pages and interactive CD-ROMs, buttons should be practical and engaging at the same time, enticing us to venture deeper into a disc and giving us a clear path for making a selection.

DVD Studio Pro uses buttons in menus to link video assets or other menus together. By clicking on a button, the user can view a slideshow, select a video track from a list of thumbnail photos, move between menus, or make choices about which disc properties to change. Although buttons are obviously essential to building a DVD project, what is not so obvious is how to create them in a way that allows for layers or properties, such as *roll-overs* (buttons that change when they are selected or the cursor is passed over them), to be imported and worked with easily. In this book, we focus primarily on Adobe's Photoshop and ImageReady programs to create our buttons. Although other image editing programs may be used, they must be able to generate compatible Photoshop (`.PSD`) files.

In this chapter, we discuss how to create buttons efficiently. A general knowledge of Photoshop is assumed, but don't worry if you are not familiar with the program. This chapter leads you through some of the basics of the program. DVD Studio Pro also includes some built-in features for making button selections in motion menus (which are set up differently than still menus) that we discuss at the end of this chapter.

Note: When talking about Photoshop in this book, we assume that a person has either Version 5.5 or 6. A trial version of Photoshop 6 and its somewhat simpler counterpart Photoshop Elements are available on the DVD-ROM that accompanies this book. Photoshop Elements is another option for a DVD author is to consider, a less expensive alternative to the full version of Photoshop. which can generate multi-layered .PSD files. A third option, which is mentioned in a couple different places in the book, is to use Macromedia's Fireworks, which can also generate multilayered .PSD files.

Creating Buttons in Photoshop

DVD Studio Pro requires that all custom buttons are created in the Photoshop format, as part of a multi-layered image file, with the extension PSD. This format allows for the creation of buttons with additional layers, such as those buttons with layers for selected and activated states. (The one exception to this rule is with overlays, where you only need one layer. Overlays are discussed later in this chapter.) Using multi-layered image files for DVD menus makes the designing of menus simpler because all backgrounds and buttons can be kept in a single file. By importing that file, you have easy access to all the button states and optional background images in one manageable location. As you work through a project, you really begin to appreciate the flexibility and ease this affords. First, focus on creating your designs within Photoshop and customize them to your satisfaction. At the same time, create multiple button images for each state, testing them as you would in the final DVD project by turning layers on and off.

Note: Every menu must contain at least one button that links to a track, slideshow, menu, or other disc element, otherwise navigation on a disc is not possible. You can create as many menus as you want although a maximum of 36 buttons can be used per menu.

Preparing layers for use in a still menu

You can set up a new Photoshop document to create a set of buttons for use in a still menu. To do so, follow these steps:

1. **Choose File ⇨ New. Name the file and set the image size to 720 × 540 pixels (Figure 5-1) with a Resolution of 72 pixels/inch for NTSC video (the video standard used in the United States and Japan).** Doing so sets the stage for graphics that we later resize to a slightly smaller dimension.

    Remember, too, that the RGB color model must be NTSC compliant to avoid display problems. To make sure that your RGB color model is NTSC compliant, follow these steps:

1. **Choose Edit ⇨ Color Settings then select advanced mode.** Under Working Spaces for RGB select SMPTE-C. Save this as SMPTE-C Video and enter the comment NTSC safe video color gamut. Select OK ⇨ OK.

2. **Select View ⇨ Gamut Warning.** This command tells Photoshop to gray out colors that are not "legal" for video display from the color picker.

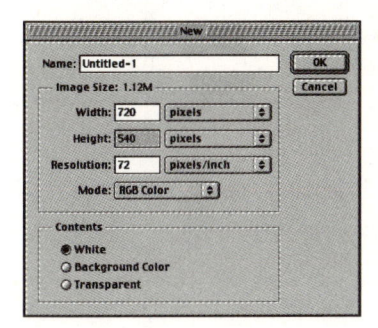

Figure 5-1: To create your menu, set the initial image size to 720 ×540 pixels.

2. **Click the new layer icon (Figure 5-2) in the lower-right corner of the Layers palette or choose Layer ⇨ New ⇨ Layer from the menu bar at the top.** Name the layer Button 1 by selecting Layer ⇨ Layer Properties.

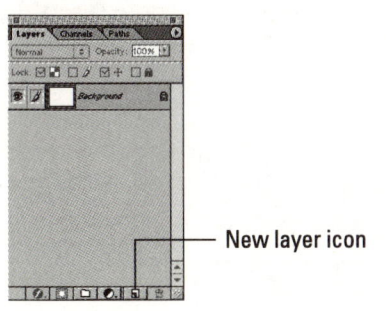

Figure 5-2: The new layer icon is located in the lower-right corner of the Layers palette (next to the trash can icon).

— New layer icon

Tip

Always be consistent when naming layers. By labeling buttons and background elements according to their function, you can avoid confusion later. Establish a logical naming system and stick with it!

3. **Draw a shape for the button by using the shape, pen, or paint tools located in the Toolbox to the right.** Oval shapes or rectangles work well, as shown in Figure 5-3. Add text or other elements as desired. If you use text, remember to merge the text layer to the button it belongs with by choosing Layer ⇨ Merge

Down when the text is above the desired layer. Also be sure to flatten any effects layers, as Photoshop effects layers do not work in DVD Studio Pro. (In Chapter 8, we discuss some design considerations when creating buttons and backgrounds.)

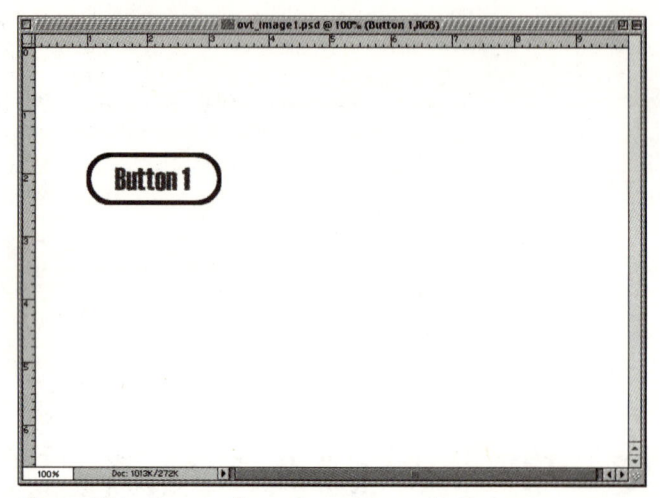

Figure 5-3: Draw a shape for your new button.

Tip

Try experimenting with layer effects for your buttons. Choose Layer ⇨ Layer Style (Figure 5-4) to select an effect. Bevel and Emboss are popular choices for buttons.

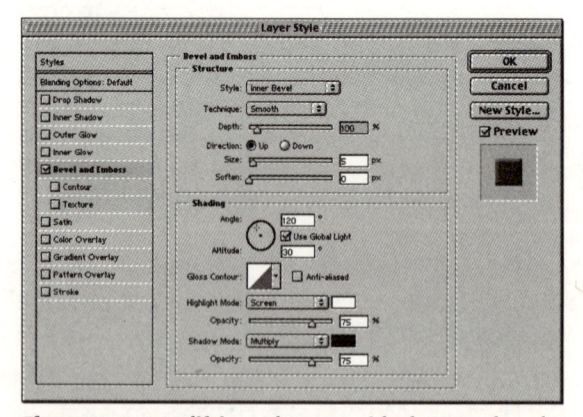

Figure 5-4: Modifying a button with the Bevel and Emboss layer effects

4. **Duplicate the current layer by choosing Layer ⇨ Duplicate Layer from the menu bar.** Name the layer Button 1 — Selected. You use this layer when the button is selected (before it is activated) by clicking enter on the remote. If you want a completely different look for the selected button state, you may create a blank layer instead as described in Step 2.

5. **Modify the duplicate layer by selecting different colors, by stretching the proportions, by adding and subtracting elements, or by adding a drop shadow.** Use your imagination. A good place to start looking for examples is on the Internet where button state changes called roll-overs are frequently used. Unfortunately, the DVD spec has no room for Flash effects or animations; the only way you can incorporate a Flash animation into a DVD is to export it as a video file. Chapter 22 discusses considerations when cross-developing for the Web and CD-ROM.

6. **Repeat Step 4 to create another layer, this time name it Button 1 — Activated.** You see this layer when the button is selected and enter is clicked on the remote control. Activated buttons typically employ a color change or a depressed look.

7. **You have now completed one set of buttons.** Repeat these steps to create as many buttons as your project requires. A nice background wouldn't hurt either!

8. **After your buttons with all three states have been created (normal, selected, and activated) and your background image has been added, it's time to resize your image (Figure 5-5).** Choose Image ⇨ Image Size and set the Pixel Dimensions to 720 × 480. Make sure that the Constrain Proportions box is unchecked at the bottom; otherwise you are unable to alter the dimensions. The graphic looks stretched and wider than normal. This problem is fixed when the graphics are imported into DVD Studio Pro. Later in this chapter, we discuss why this change in proportions is necessary to create graphics for video.

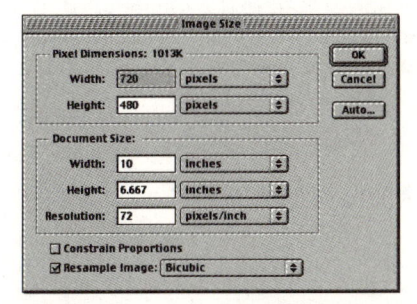

Figure 5-5: Resize your finished menu to work with DVD Studio Pro.

9. **Choose File ➪ Save As and save the file as a PSD file.** Now we are ready to see what the graphics look like in DVD Studio Pro!

10. **Open DVD Studio Pro. Choose File ➪ Import and select the PSD file you just saved from its current location. Click Add and then Import to bring the layered file into the Assets Container.**

11. **Click the Add menu button at the bottom of Graphical View. Drag your PSD file from the Assets Container and place it on top of the new menu tile.**

12. **With the menu tile selected, click the pull-down menu for Layers under Picture in the Property Inspector (at first it says No Layers Selected) as shown in Figure 5-6.** Click all the button image names including the background to select them.

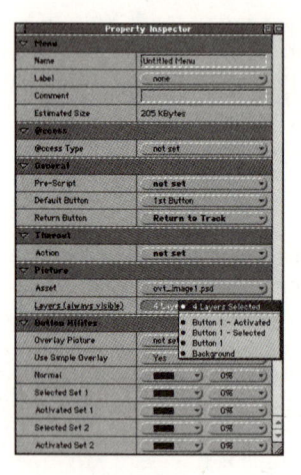

Figure 5-6: Select the image layers that you need for your menu.

13. **Double-click the thumbnail window in the menu tile that is displaying a small preview of your image.** A screen called the Menu Editor opens to present you with an image preview similar to what you see in Figure 5-7. Here, you make button settings and assign assets.

14. **Choose Item ➪ Preview Menu or press ⌘+P.** This preview is the window that uses the Apple DVD player software to test a project.

Tip

If your buttons or graphics contain horizontal lines, make sure that their height is greater than 1 pixel. Lines that are only 1 pixel high appear to flicker on a TV screen.

Note

In previous versions of Photoshop (5.5 and earlier), you were limited to a maximum of 99 layers. In Photoshop 6, you can use hundreds of layers per file although DVD Studio Pro recommends that you do not exceed 100 layers.

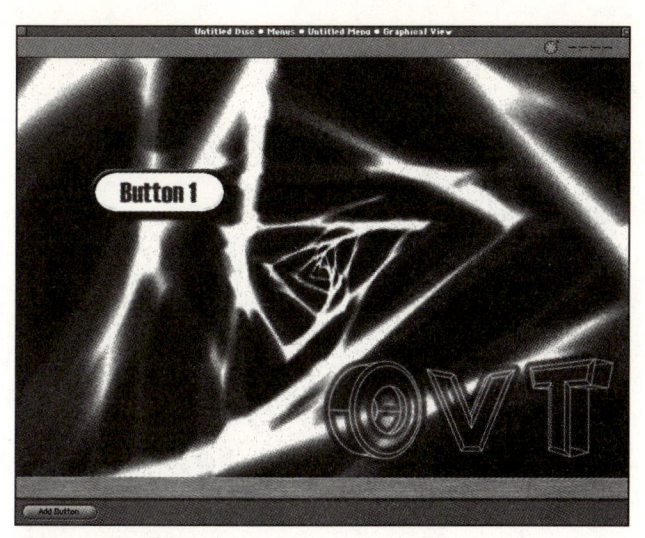

Figure 5-7: This is how the menu graphics appear in the Menu Editor.
Background image courtesy of OVT Visuals

In the next few sections, some general techniques that may be helpful for creating button images in Photoshop are discussed. Photoshop's versatility enables you to create virtually anything you can imagine. The tools available to you in this application are rich with features that you may never need to use, but it's nice to know that you have them. Regardless of experience, you'll always have something new to learn in Photoshop, and you can often learn better ways of accomplishing tasks that you have been doing the hard way!

Matching colors in Photoshop

You frequently use the Eyedropper tool in Photoshop. This tool is helpful for browsing potential color choices and for selecting a color range for replacement as well as matching color palettes from other documents or areas of the same graphic. Most important for our purposes, the Eyedropper is used to maintain accuracy and consistency when working with various layer elements. Also, you can keep tabs on all your graphics' color palettes that you use in creating a complete DVD project.

The following steps describe how to select a color with the Eyedropper tool and what values are important to make note of when creating a color palette for your project.

1. Select the Eyedropper tool (Figure 5-8) from the Toolbox at the left of the screen.

Figure 5-8: The Eyedropper tool is located in the Toolbox.

2. **Click a color that you want to sample.** In the Color palette like the palette shown in Figure 5-9, the RGB values are displayed for that color. Write these values down if you want to remember them for future images. At this point, you may want to open a photo, video still, or other document to sample a color. Doing so is useful if you want to match a particular color from a video for a matching menu or button element.

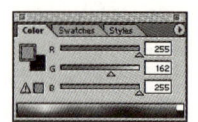

Figure 5-9: The Eyedropper tool is used to select colors from the Color palette or from another image.

3. **Adjustments can also be made to the Foreground color by using the sliders in the Color palette.** Move each of the color sliders to see the effect in the Foreground color box in the Color controls box in the lower-left corner beneath the Toolbox. You can change the background color to the foreground color by clicking the arrows in this box.

Using Photoshop shapes and styles

The latest version of Photoshop offers new options that make complex tasks easier. One such addition is the expanded number of shapes available for use. Another improvement is the simplicity with which you can add complex styles to a shape or layer. When creating buttons these tools are particularly useful, because they allow for consistent, yet unique shapes for standard button designs.

Shapes

Drawing difficult shapes is now as easy as pressing a button and dragging your shape. The six basic shape tools are outlined in Figure 5-10.

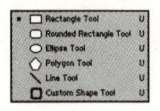

Figure 5-10: The shape tools are easily located in the Toolbox in addition to the Options bar.

Rectangle tool
The most basic shape you can create is now made even easier, without resorting to the Marquee tool. Holding down the Shift key and dragging creates a square.

Rounded Rectangle tool
This tool, as pointed out in Figure 5-11, is similar to the one found in ImageReady. You have the ability to choose a Radius value that determines the amount of curve on the corners (think of feathering without the soft edges). The bigger the Radius value, the more rounded your corners appear. This shape is particularly useful for creating standard button designs.

Radius value

Figure 5-11: The Radius value, which affects the roundness of corners, is specified in the Options bar.

Ellipse tool
An elliptical shape can be drawn with this tool. By holding down the Shift key and dragging, you can create a circle.

Polygon tool
This unique tool enables you to create polygons — symmetrical shapes with straight sides (Figure 5-12). Whether triangles, squares, hexagons, or octagons, you can specify as many sides as you want by entering a value in the Sides box. Another way to select the number of sides for a shape is by pressing the bracket keys ([and]) to cycle through the numbers and increase or decrease the value.

Number of sides for a shape

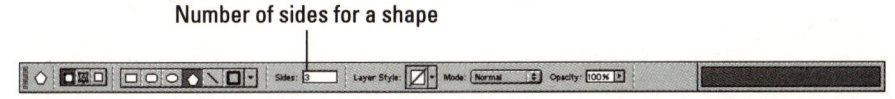

Figure 5-12: You can specify the number of sides for a shape with the Polygon tool.

Line tool

The Line tool can only be called a shape when you add weight to it. By entering a pixel value in the Weight box, you determine the thickness of the line. You can quickly draw elongated rectangles.

Custom shape tool

Selecting the custom shape tool (Figure 5-13) enables you to choose from a symbol library of preset shapes (14 shapes come standard with Photoshop). Drawing a star, heart, or even a foot has never been easier. You can also create your own shapes, but if that sounds to complicated, look into companies that sell stock shapes that you can import into Photoshop.

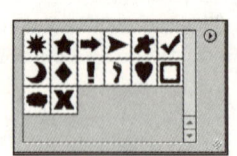

Figure 5-13: You can choose from several shapes with the Custom shape tool, or you can create your own.

Additional settings are available for each of these shape tools by pressing the arrow to the right of the shapes in the Options bar (Figure 5-14). Also, shapes can interact with other shapes by, for example, overlapping or intersecting to create new shapes. These compound shapes or paths can be made from existing or new shapes by selecting one of the four choices in the upper-left corner of the Options bar. The choices are Add to shape area, Subtract shape area, Intersect shape areas, and Exclude shape areas. Try playing around with the various options to see what each one does. These options can be very useful for creating complex shapes.

The options that display are different for each shape

Figure 5-14: The Compound shape options enable you to make complex forms by intersecting shapes with each other.

The following steps demonstrate how to create and modify a shape by using the Shape tools in Photoshop:

1. **Select a color for the foreground of your new shape by clicking the Set Foreground Color box at the bottom of the Toolbox.** After you have selected a color by using the Eyedropper tool, click OK.
2. **Select the Shape tool you want from the Toolbox or from the Options bar at top.**

3. **You can modify the properties of the shape you have chosen in one of two ways: Click the arrow to the right of the shapes in the Options bar to specify Geometry options, such as Fixed Size for exact dimensions or Snap to Pixels for alignment (Figure 5-15).** In the Options bar, you can also change the number of sides for Polygons, the Radius for Rounded Rectangles, or the Weight for Lines.

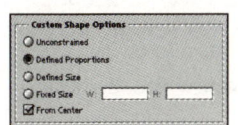

Figure 5-15: Modify your shapes by changing the available options in the Options bar.

4. **Draw the shape by clicking and dragging in the Image window.**
5. **If you want to combine shapes, select a Compound path option from the upper-left corner of the Options window and draw another shape that intersects or merges with the first shape you drew.**
6. **To modify options such as color or opacity for the shapes, click the Layer clipping path thumbnail in the Layers palette.** Numerous options are available — Bevel and Emboss, Drop Shadow, and various color and gradient overlays (to name a few).

Styles

Styles are automated layer effects that enable you to add complex colors, shadows, edges, and more to an image layer. Interesting textures and overlays are just a button click away. You can either create your own style or select from a library of presets like the presets found in Figure 5-16.

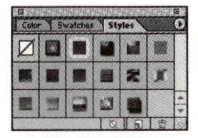

Figure 5-16: Some of the standard Photoshop Styles

The following describes how to add a style to a button image.

1. **Select the layer you want to affect and click the Styles tab to display the Styles palette.**
2. **Click the thumbnail for the style you want to apply.** You can choose from approximately 16 preset layer styles. The effect is automatically applied to your image layer.

Using text in Photoshop

With the arrival of Photoshop 6, we have greatly improved text tools at our disposal that makes working in the program more intuitive and fun. You may have noticed the elimination of the text box, which makes placement and visualization of text on the canvas much easier. The improved text wrapping and resizing of text fields, per-character adjustments and rasterization options are something else to be happy about. The most enjoyable text improvement is the text-warping feature. With the warping capability, text can be bent and twisted to fit any design imaginable.

Warping a line of text in Photoshop 6 is simple. Just follow these steps:

1. **Select the text that you want to warp by highlighting it.**
2. **Click the Create Warped Text button (Figure 5-17).**

Figure 5-17: Click the Create Warped Text button in the Options bar to alter your text.

3. **Choose a style from the drop-down menu (Figure 5-18).** Choices include Arc (Lower and Upper), Arch, Bulge, Shell (Lower and Upper), Flag, Wave, Fish, Rise, Fisheye, Inflate, Squeeze, and Twist. Within each of these styles are settings that affect Bend, Horizontal Distortion and Vertical Distortion.

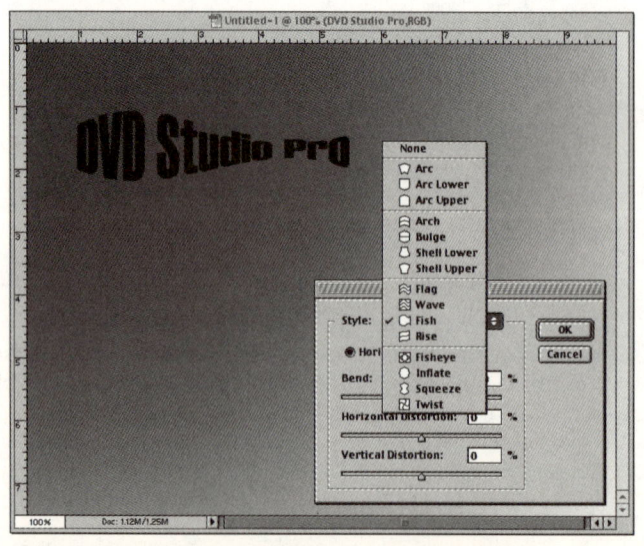

Figure 5-18: Choose a style for your warped text.

4. **If you want to apply various filters to your text you first need to rasterize it, which creates an ordinary uneditable (from the text standpoint) image layer.** To do this, click once on the text layer in the Layers palette and then choose Layer ➪ Rasterize ➪ Type. You may now treat the layer as you do any other image layer.

Creating Buttons in ImageReady

ImageReady is the companion program to Photoshop and is included with every copy of Photoshop sold. Typically, it is used to create graphics for the Web. In the latest version of Photoshop, more tools are migrating from ImageReady and becoming a part of Photoshop's expanded capabilities, further eliminating its reliance on its smaller companion.

Even though Photoshop may accomplish many of the same tasks, ImageReady can still be a very useful application. Its smaller size requires less memory to run and as a result it is good for running in tandem with memory hungry programs. Its slicing tools and its animation functions are very good and are also quite useful particularly for creating Web pages.

In this section, we discuss how to use ImageReady to create buttons for DVDs, CD-ROMs, and the Web — something you might consider if you have thought about cross developing for several platforms.

> **Note** A trial version of ImageReady can be found on the DVD-ROM that accompanies this book.

The following describes how to set up a new ImageReady document to create a set of buttons for use in a still menu.

1. **Choose File ➪ New to create a new ImageReady document.** Name the file and set the image size to 720 × 540 pixels for NTSC. This sets the stage for graphics that we later resize to a slightly smaller dimension.

2. **Click the new layer icon in the lower-right corner of the Layers palette or choose Layer ➪ New ➪ Layer from the menu bar at the top.** Double-click the new layer and name it Button 1.

3. **Select the Rounded Rectangle Marquee Tool from the Toolbox at the left.** Draw a button and fill in with the Paint Bucket Tool by using the Eyedropper to choose a color.

4. **Select the Type Tool to add text to the button.** Make sure to choose a different color for the text first; otherwise the text is the same foreground color as the button, and you won't be able to see it. Also, try making the text size large so that it can be seen. Position your text.

5. **Select the text and button layers one at a time and Choose Layer ➪ Duplicate Layer to make a copy of t each on top of the originals.** Name the new graphic and text layers Button 1 — Selected. Move the layers together in the Layers palette window.

6. **Change the color of the new text layer and the color of the new button graphic if desired.**

7. **Repeat Steps 5 and 6 for another button state by choosing Layer ➪ Duplicate Layer to make another set of copies.** Name this Layer Button 1 — Activated. Change the color of the new text layer and button graphic to something different than the first two.

8. **Merge the text layer from Button 1 to its button image by choosing Layer ➪ Merge Down when the layers are stacked on top of one another.** Repeat this step for the other button states (selected and activated buttons).

9. **You have now completed one set of buttons.** Repeat these steps to create as many buttons as you want, making certain to keep normal, selected, and activated buttons states on their respective layers. Remember, when you create your first button layer, you must create all of your buttons for that button state at the same time. Normal state buttons are on one layer while activated and selected states have additional layers. You can keep the background white or fill it with another color or graphic.

The Layers palette should look something like Figure 5-19 when you have completed a set of buttons.

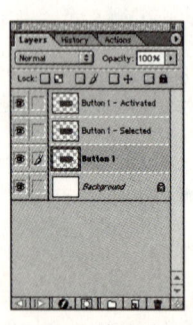

Figure 5-19: The Layers palette with a completed set of buttons.

10. **After your text buttons with all three states have been created (normal, selected, and activated) and your background image has been added, you need to resize your image.** Choose Image ➪ Image Size and set the Pixel Dimensions to 720 × 480 pixels. Make sure that the Constrain Proportions box

is unchecked at the bottom; otherwise you are unable to alter the dimensions. The graphic looks stretched and wider than normal. This problem is fixed when the graphics are imported into DVD Studio Pro.

> **Note** When working with graphics for video, it is important to make certain that colors are safe for display on television sets. In ImageReady, you can instantly make your colors safe for inclusion in NTSC video by using the NTSC Colors filter. To activate the filter, choose Filter ⇨ Video ⇨ NTSC Colors. Use the filter once you have finished creating all of your graphics — before you export the file.

Understanding Square versus Non-Square Pixels

A problem you may encounter when working with video in an image-editing program is the relationship between screen resolution and pixel aspect ratio. Photo programs and image editors, such as Photoshop, work with square pixels that you can easily see by zooming in close on a photograph. When looked at from a distance, those square boxes appear to compose a picture (much like our individual molecules and atoms combine to create us). Examples of this phenomenon can be seen in the art of such painters as Georges Seurat who developed a technique known as pointillism. A variation of this technique is also used in print work. For example, looking at a photograph in a newspaper with a magnifying glass reveals hundreds or thousands of black dots (for a discussion of print issues and the potential problems they may cause, turn to Chapter 13).

Video comes in many formats. The format experiencing the most success lately is DV or *digital video*. (A detailed discussion about understanding digital video in included in Chapter 11.) The DV video format has a screen resolution of 720×480 pixels for NTSC, which is the same resolution that DVD Studio Pro uses to create projects. The pixel aspect ratio for DV is .9:1 as compared to the 1:1 ratio used in Photoshop. When creating square pixel graphics to be combined with non-square formats such as DV, special preparations must be made. When specifying an image size for a Photoshop file, create it at a resolution of 720×540 pixels. Later, after your graphics are ready for export, resize the image to the DV format of 720×480 pixels. By doing this, you eliminate a stretching effect that ordinarily takes place.

Software DVD players can effectively display non-square video at square ratios although computer monitors can sometimes be deceiving in the way they display that same video. The horizontal pixel spacing on a monitor makes an image appear slightly stretched or wider than normal. When the same video is viewed on an external video monitor this apparent distortion no longer exists. Therefore, previewing material on a separate video monitor whenever possible is helpful and highly recommended — particularly video that has graphic elements added which were created in an image editor, such as Photoshop or an effects program, such as After Effects.

Considering Workflow Possibilities

When you begin a project, consider the ways in which you may end up using or distributing the finished product. Should it only be delivered on DVD? Should you decide to create something that just as easily may find its way to the Web? Consider these and other possibilities carefully — you may save yourself a lot of work in the long run. By planning out your project before you start creating the necessary elements, you are in a better position to cross-develop for various formats and distribution options. (For an in-depth discussion of cross-developing for DVD, CD-ROM, and the Internet, refer to Chapter 22.)

First, we want to recognize that quality does vary depending on the medium. The current state of video on the Web cannot compare to that delivered through our televisions, DVDs, and our VCRs. As technology improves many of these limitations should be overcome (broadband Internet connections are the first steps in this direction), but until these limitations are overcome, we must recognize that different mediums call for different methods.

Designing for video is a relatively easy task. Beginning with the editing process, you design and create exactly what you want to see. Add in layers and effects, throw on some titles, render it out to a single file, and then prepare it for several methods of delivery in an electronic manner similar to turning a few knobs, checking the playback, and pressing record. The Web is a different beast altogether. Certainly, products such as Macromedia's Flash and Director are an advancement for the WYSIWYG inclined (What You See Is What You Get), but the image quality limitations imposed by the Web can be difficult to overcome. Add to that the complexity of true interactivity and designing single pages for maximum usage of concurrent graphics and text, and anyone can feel a bit overwhelmed.

Multimedia on the Web is all about getting the most out of as little as possible. DVD, on the other hand, has only minor bit-rate concerns to contend with, while DVD-ROM content is only limited by the speed of the recipient's computer and the amount of data that can fit on a disc medium holding several gigabytes of material. If only the Web had it this easy! In the not-too-distant future, the Internet may be at this point, in which case a few gigabytes would be nothing compared to the virtually endless stream of content available from the entire world. Watch out, DVD!

Adobe has created an entire suite of professional media development products. Most people recognize the name Adobe for their flagship product Photoshop, but equally valuable are their tools for developing Web content, design, and video production. Adobe's GoLive and LiveMotion are solid choices for integrating with Photoshop and ImageReady. GoLive allows for the precise visual placement of elements, providing many easy-to-use tools to accomplish the task. LiveMotion can create advanced motion effects and animations for the Web.

Macromedia is also a leader in the field of Web and interactive design. Macromedia's products, such as Dreamweaver and Fireworks, are generally considered the ones to beat in the Web arena. Their proven history of interactive design and style has cemented them as the standard Web development software among many professionals. Even so, Adobe's offerings may be a good choice for professionals seeking tight integration with other Adobe products they may already use.

For the purposes of this chapter, think about ways to create elements in ImageReady and Photoshop that allow the graphics to be used in Web products as well as alternative interactive formats like CD-ROM and DVD-ROM. As discussed earlier in this chapter, images created for eventual output to DVD must comply with certain dimensions relating to pixel ratios for video (refer to the section in the chapter on square versus non-square pixels). When creating buttons and menus in Photoshop, you begin by defining a new canvas with a Width of 720 pixels and a Height of 540 pixels. At this stage, consider what you may need for Web output as well. If the buttons and backgrounds you create are to be used in a Web page, make certain to save an alternate version of the document, prior to resizing down to 720 × 480 pixels for video. Choose File ➪ Save As and retain your layers by saving to PSD format. Later, you can go back to this document and save out individual layers that are optimized for the Web and CD-ROM/DVD-ROM, which do not have the same pixel ratio considerations.

Of course, you must address other considerations when creating for DVD versus the Web. With DVD, you are somewhat limited in what you can design for an interactive interface. (For an in-depth discussion of interface design, turn to Chapter 8). Buttons are generally of standard size and proportion on a DVD (although this does not always have to be the case) and with the exception of remote-control options available, the structure and layering options have a hard time competing with what

Web page layout can do. To simultaneously prepare graphics for the Web and video, think about creating space on a menu or Button design for slicing images into tables by using ImageReady or even the new slice tools available in Photoshop 6 (see Figure 5-20).

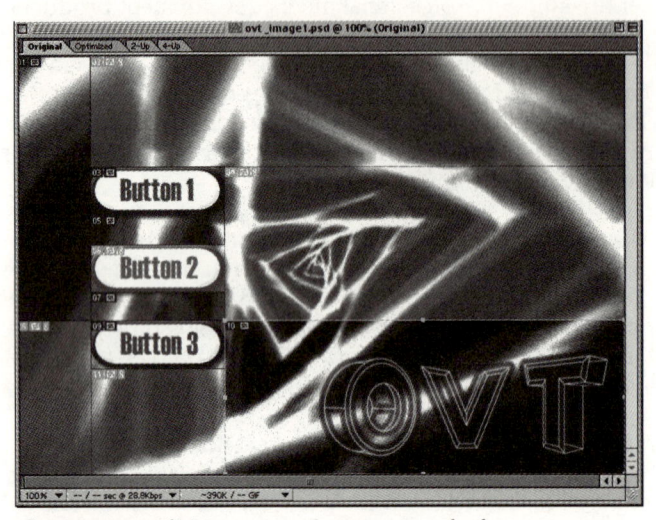

Figure 5-20: Slicing a menu in ImageReady that was originally created for a DVD project
Background image courtesy of OVT Visuals

Another consideration when designing for the Web and DVD is to choose colors that are Web-safe. Not only does this make your choices for button and menu elements a little easier (sometimes too many color choices can lead you into endless debates about what shade of forest green best suits the spirit of your project) but also it ensures that your finished project looks its best no matter what kind of monitor it is viewed on.

The following steps describe how to select a Web-safe color:

1. **Open the button or menu document you are using for your DVD project.**
2. **Click the Foreground color square at the bottom of the Toolbox on the left.**
   Doing so brings up the Color Picker dialog box. In this box, you are able to select a color that is used by various drawing and painting tools.
3. **Check the Only Web Colors box of the Color Picker dialog box (Figure 5-21).**
   When this box is checked, only Web-safe colors are displayed for you to choose from.

Chapter 5 ✦ **Working with Buttons** 117

Make sure Only Web Colors is checked

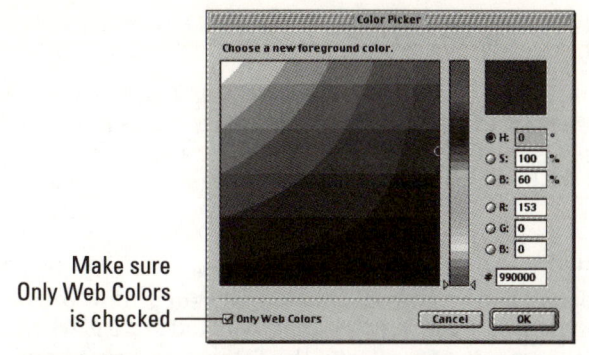

Figure 5-21: Make sure that the Only Web Colors box is checked in the Color Picker to select colors that display accurately on all monitors.

4. **Select a color by clicking it.** Notice the hexadecimal code for the color chosen (Figure 5-22). This code is unique and is used to specify absolute color values for the Web. Write this code down and use it to fill in this box for other layers that you want to assign the same color.

Hexadecimal color code

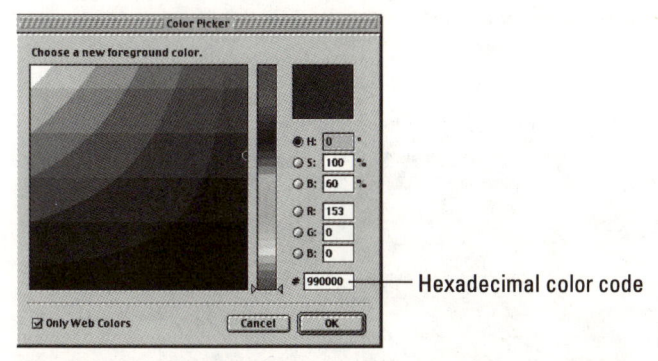

Figure 5-22: The hexadecimal code is useful for establishing consistency when designing for the Web.

Note

Remember that when working with graphics for video it is important to make certain that colors are safe for display on television sets. In ImageReady and Photoshop, you can instantly make your colors safe for inclusion in NTSC video by using the NTSC Colors filter. To activate the filter, choose Filter ➪ Video ➪ NTSC Colors. Use the filter after you have finished creating all of your graphics — before you export the file.

Setting Button Properties

Any disc element can be linked to a button, whether it's a Track, Slideshow, or another menu. Buttons can also be used to link actions, such as jumping to a Script. This linking action is set in the Property Inspector under such item settings as Jump When Activated. In the next chapter, we discuss how to create menus with button links.

Every button you create has three possible states that must be specified. These states are normal, selected, and activated. The normal state is how the button appears in a menu when it is not yet selected. The selected state is when a button has been chosen with the remote control, and the activated state is how the button appears after it is clicked by the user when he or she presses enter on the remote control. As discussed earlier in this chapter, each button must be created with these three layers in Photoshop before being imported into DVD Studio Pro.

The following steps describe how to assign states to a button and preview the results.

1. **Open the Menu Editor by clicking the menu tile's thumbnail.**
2. **Select the button you want to work with in the Menu Editor (Figure 5-23) by clicking it.** You should have a button selection around your button. If you do not have a selection box already, click Add Button to create one and stretch it to encompass your button.

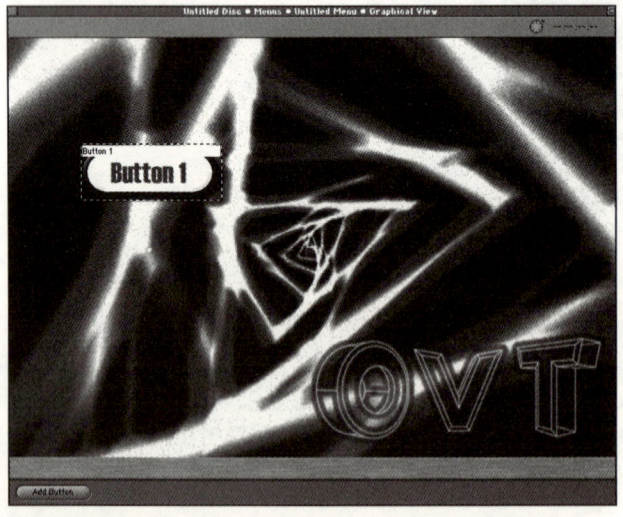

Figure 5-23: Choose a button you want to assign a Display property.
Background image courtesy of OVT Visuals

3. In the Display section of the Property Inspector (Figure 5-24), choose the button layer you want to assign as the Normal state from the drop-down menu (if you are naming your buttons consistently it may be named something like Button 1).

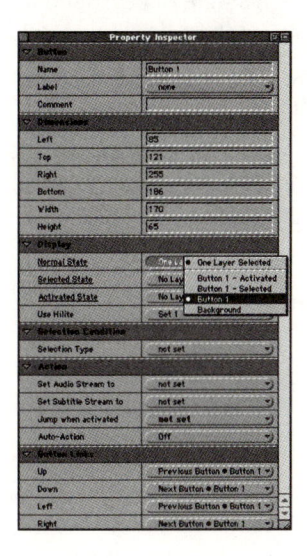

Figure 5-24: Choose the button layer that you want to assign for a particular state — in this case, the Normal state is specified.

4. Then choose the button layer you want to assign as the Activated state from the drop-down menu.
5. Close the Menu Editor.
6. Select the Menu tile in the Workspace and click the Preview button.
7. Use the keys in the Preview window to test your buttons.

> **Tip:** You can test your buttons while you work in the Menu Editor by choosing Show Normal State, Show Selected State, or Show Activated State from the Buttons menu at top.

Creating overlay images in Photoshop

The following steps demonstrate one way to create an overlay image in Photoshop:

1. **Create your button designs exactly as you ordinarily would on a 720 × 540 pixel, white canvas.** You may also create your overlay from button designs that you have already created. In this case, open a .PSD file with your button layers.

2. **Discard any background graphics and keep only one set of button states (most likely your normal state buttons).** You should have a white background with your button layers on top.

3. **After you finish creating your buttons, flatten all of your layers.** Choose Layer ➪ Flatten Image. You should now have a single Background layer with your images. Make sure that you save a copy of your original Menu file.

4. **Set the mode to Grayscale by choosing Image ➪ Mode ➪ Grayscale.** Make sure that the mode is set to 8 bits/Channel (Figure 5-25).

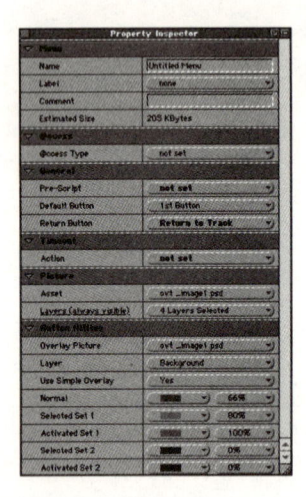

Figure 5-25: Your overlay image must be set as a grayscale image before saving.

5. **Choose Image ➪ Image Size and resize your image to 720 × 480 pixels.**

6. **Choose File ➪ Save As and save the overlay as a PICT file.** Your overlay is now complete.

Note: With overlay graphics, the darkest portions are the most opaque, and the lightest portions are the clearest. Design with this in mind — typically, your buttons are darker than your text. But rules don't apply, so feel free to experiment to achieve the best results.

Working with Button Hilites

When working with Motion menus (menus with video as the background instead of a still image), ordinary buttons using Photoshop layers do not always suffice. (Refer to Exploring Motion Menus in the next chapter.) Sometimes, you may want to create Motion menus with simple buttons that merely consist of selecting a rectangular portion of the background video clip. Other times, you may want to create buttons with unique shapes and borders for your Motion menus. In either case, you can use

Button Hilites for creating button selections. Button Hilites add a glow to the perimeter of a button instead of the usual type of action typically assigned when working with still menus. Each button state can be assigned a unique hilite color and transparency.

The following steps demonstrate how to create Button Hilites in a Motion menu.

1. **To create Button Hilites, first you need to create an overlay for the Motion menu.** Follow the steps in the previous section on how to create an overlay images and then move on to Step 2.

2. **Open the Menu Editor for the menu you want to edit.**

3. **Make certain that the Show Selected Hilites is selected in the Buttons menu. Click the background in the Menu Editor.** You should not have buttons selected at this point.

4. **In the Button Hilites section of the Property Inspector (Figure 5-26), choose a color for Selected Set 1 from the drop-down menu and set the transparency to a percentage.** When your buttons are selected, they are covered with the color that you have chosen.

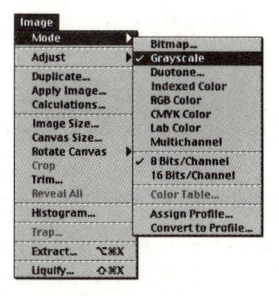

Figure 5-26: Select a color and transparency value for each Button Hilite.

5. **In the Button Hilites section of the Property Inspector, select a color for Activated Set 1 and set its transparency to a percentage.**

6. **Choose Show Activated Hilites from the Buttons menu to preview how these buttons look.** All your buttons should be covered with the last color you chose. When your button is activated, it should display the hilite you just selected.

Note

You can adjust the color and transparency settings to whatever is appropriate for your project. The numbers used in the steps are merely suggestions and can be deviated from. In many cases, experimenting with settings brings unpredictable yet pleasing results. Don't be afraid to try new things and remember — you know better than anyone what is best for your project. Just be sure to have fun with it!

> **Tip** You can merge your Normal button state with the background layer in Photoshop. By doing this, you can eliminate the need for an extra layer.

Using buttons as interactive markers

In addition to being used in menus as a way of linking to other menus, tracks, or slideshows, buttons can also be placed on top of tracks as *interactive markers*. Using buttons as interactive markers enables the user to choose various actions as they view a video stream, in much the same way as they function for standard menus.

However, using buttons as interactive markers can be a bit tricky. Before you do anything else you need to create an overlay image in Photoshop that you can add to the video. Overlay images determine the position and shape of the buttons and the button's hilites. (Refer to the section Creating an overlay images in Photoshop, earlier in this chapter.)

Once your overlay image has been created, you are ready to add it to a video stream in a Track and create an interactive marker.

1. **Add a Track to the Workspace by clicking Add Track at the bottom of Graphical View.**
2. **Assign a video Track by dragging it from the Assets Container over the Track tile, or choose an asset from the drop-down in the Video section of the Property Inspector.**
3. **Open the Marker Editor by double-clicking the Track tile's thumbnail.**
4. **Under Overlay Picture in the Button Hilites section of the Property Inspector, choose the overlay that you have created and imported from the drop-down menu.**
5. **Choose a color for the Normal state of the button and set the transparency to a high percentage, such as 80 percent.** Setting a high transparency ensures that your buttons display well even over very bright or busy video clips. Of course, you can set the transparency to a lower percentage if it is appropriate for your project.
6. **Choose a different color for Selected Set 1 and set the transparency to another percentage, such as 66 percent.**
7. **Choose another color for the Activated Set 1 (different than the previous two color selections) and set the transparency to a high percentage, such as 100 percent.** Now it is time to draw selection boxes around your button images.

8. **Click Add Button at the bottom of the window to create an Untitled Button that you can move around the screen and position around your button selections, or simply click and drag the cursor in the window to automatically draw a selection around your buttons.** When you draw the outline, the hilite automatically matches the shape of the button images you drew for the overlay image. Be precise with placement of button selection boxes to avoid overlaps or other unforeseen errors. The hilites that outline the buttons should now be visible.

9. **If the hilites are not visible, choose Show Selected Hilites or Show Activated Hilites from the Buttons menu at top.** Also, click the buttons in the window to test the Normal and Selected states.

10. **Click the buttons and name them in the Button section of the Property Inspector.**

11. **When a button is selected, look under the Action section in the Property Inspector to set the properties for each button.** Is there a particular track, marker, or menu you would like the button to take you to? The Action section is where you may set that function. Choose another disc element from the drop-down menu next to Jump when activated. By clicking that button, a user is linked to the asset you have specified in this menu.

Understanding the Matrix Views

Matrix Views enable you to see the connections between the elements in your project. They also enable you to quickly reassign assets or edit *settings* in a single location. Out of the three possible Matrix Views, Jump Matrix and Layer Matrix are relevant to buttons. The Jump Matrix gives you the option of assigning links to buttons while the Layer Matrix lets you assign Photoshop layers for the various button states. At the top of the Jump Matrix view, you see the available assets and along the left side you see the containers that include your menus and buttons. In the Layer Matrix, the Photoshop layers are on top and the button states are to the left. In the next chapter, we discuss how to use these Matrix Views to assign assets to buttons and menus.

Summary

- Buttons are created by using Photoshop layers to delineate normal, selected, and activated states.
- DVD Studio Pro uses video with non-square pixels, which can make working in an image editing or graphics program a little tricky.

- ✦ The latest version of Photoshop, Version 6, offers many useful tools, such as shapes and text warping, for creating unique button designs for menus.
- ✦ ImageReady is another useful program for creating buttons with cross-development in mind — although you use it primarily for creating and optimizing Web graphics.
- ✦ Buttons can also be added to video tracks as interactive markers utilizing overlays created in Photoshop.
- ✦ Understanding what you may need for future projects is useful to know before beginning a new project so that you can determine the proper workflow solution.

In the next chapter, we discuss how to integrate buttons into menus and how to link them to other disc elements. Aside from the preparation of assets for your project, menus are most likely your first real step in the creation of a DVD project. By working through the tutorials and referring back to this chapter on buttons, you should be able to gain a better understanding of how menus work and what options they provide to you.

✦ ✦ ✦

Working with Menus

CHAPTER 6

In This Chapter

Working with layered image files

Exploring motion menus

Enabling menus

When you put a DVD into a player, the first item you see is a menu (if you do not count an FBI warning or an animated company logo). Menus are essential to making a disc truly interactive. Without menus, you would only have collections of video tracks linked in a linear fashion or by pressing the Next Track key on a remote control. Menus can be very simple, incorporating a button or two and a static background, or they can be very dynamic, using looped video and layers of activity over which buttons can be placed. Look at the last DVD you bought that is distributed by a major studio. Chances are that the disc includes extensive use of motion graphics and audio in its menus. Studios pay a lot of money these days to outdo each other in the arena of DVD menus and special features. Even though complex images may hold your interest, simple menus can also be very effective if they are designed with style and usability in mind.

In this chapter, we discuss the many uses for menus and describe how to create them. Preparation of video content for inclusion in menus is also touched on in a section devoted to Apple's professional video editing software, Final Cut Pro. Also, Animations have become a popular way to introduce and link menus and are discussed briefly at the end of this chapter. Of course, before you can begin creating your menus, you must have a suitable background and buttons created. Again, Photoshop is the tool to use for creating and preparing many of your assets. As you work through this chapter, refer back to Chapter 5 for suggestions and instructions on how to create buttons, the elements that ultimately give menus their functionality.

Working with Layered Image Files

Layered image files are produced with image editing applications like Photoshop. Essentially, these files contain individual graphic layers, which can be edited and modified without affecting the other layers. This allows a designer to apply a style or change an attribute of a specific graphic element, while leaving the remaining layers unaffected. Also, layered image files provide access to separate layers in any order necessary, providing the ability to show particular layers (or graphics) by turning certain layers on and off. Creating menus for a DVD requires the use of different graphic layers to display buttons or other elements to indicate a particular state or user action. Still menus (where the background is not animated) make extended use of layered image files, since each button requires several layers to show different states (normal, selected, and activated).

To work with layered image files, you first need to be aware of the necessary elements and tools required to create them. Many new DVD authors start out by creating still menus, as still menus are easy to create and provide the backbone for the majority of projects. Using Photoshop to create still menus is ideal, because Photoshop produces the required .PSD format for DVD Studio Pro. (You can also use PICT files for backgrounds, but considering that you need to create buttons in multilayered .PSD format anyway, PICT can be impractical.) DVD Studio Pro takes advantage of Photoshop's advanced layer capabilities that effectively allow you to import a single file with perfectly aligned interface elements. Even if you have never worked extensively with Photoshop layers, the principles discussed here are not difficult to master.

Menus are essentially composed of two elements — backgrounds and foregrounds. Backgrounds are for the still pictures or motion video clips, and foregrounds are for the buttons that sit on top of the background frames. Creating backgrounds for DVD is often the easy part. Getting the backgrounds to successfully mesh with the foreground button layers is a bit more difficult. Buttons undoubtedly provide the most challenges for first-time users. (Refer to Chapter 5 for more detail about how to create buttons.) Apart from design considerations, such as the placement and layout of interface elements, technical limitations, such as the problems imposed by video safe colors, must also be dealt with.

Working with backgrounds

Start your menu project by creating a suitable background for the eventual placement of buttons. Consider your designs carefully. A bit of previsualization and planning are almost always beneficial. Are your background designs too busy or do they have hard to distinguish shapes and lines? What about your choice of color? Is it a good match with the buttons you want to use?

If you are creating background graphics from scratch, make sure that the graphics comply with DVD standards as required by DVD Studio Pro. You should create all still graphics at the 720 × 540 pixel resolution that are later resized to 720 × 480 pixels before importing into DVD Studio Pro. (Refer to Chapter 5 for an explanation of why this is necessary.) Later in this chapter, we discuss the considerations for motion menus.

When creating still menus, consider creating your normal button states (those buttons which have not been selected or activated by the user) as part of the background. Although doing this isn't necessary, it can save you an additional step in the setting up of button layers. By creating your normal button states as part of the background, you only need to create layers for the selected and activated states of your buttons. This should prove less of a hassle for you in the long run, although you may want to create backgrounds that are completely independent of buttons to be able to turn layers on and off as you desire.

You can easily create backgrounds in Photoshop by using the tools available with the program.

1. **Open the Photoshop application and create a new file by choosing File ➪ New.** Set the size of the document size to 720 × 540 pixels.

2. **With your new document open in Photoshop, select the background layer then choose a color and use the Paint Bucket tool to fill in your background.**

3. **You may also select the Gradient tool (as shown in Figure 6-1) from the Toolbox (it is located in the same menu as the Paint Bucket, which is accessed by clicking on the Paint Bucket tool, holding down the button and selecting the Gradient tool).** In the Options menu at top you can choose a particular type of gradient. For our purposes, select the Foreground to Background gradient. You have many options to choose from, so experiment if you are so inclined. Drag a line in the direction you want to apply the gradient, usually starting at one corner or edge and dragging to the opposite.

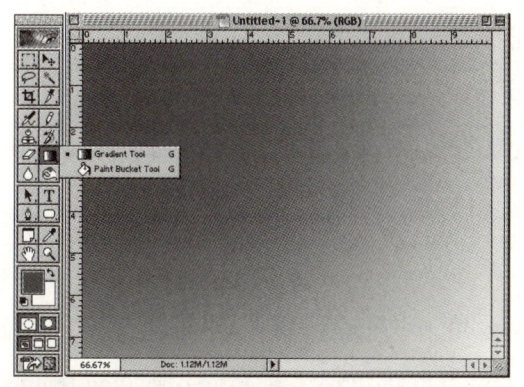

Figure 6-1: Use the Gradient tool to create a more interesting background.

4. **Choose the Type tool and select a font type, size, and color for your text from the Options menu at the top.** Fill in your titles and other text elements.

5. **Open other graphic files to cut and paste image selections to fill in space or create your own designs by using the Shape and various Paint tools.**

6. **After you have completed your background design, flatten your layers by merging them to the layers beneath them. (See Figure 6-2.)** Choose Layer ⇨ Merge Down or Layer ⇨ Merge Visible. The reason that you are flattening these layers is because they all form the background. When you have a layer which represents an interactive element, it needs to remain a separate layer.

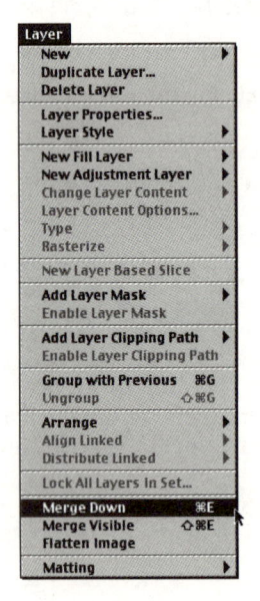

Figure 6-2: Merge your layers to create a single background image.

7. **You have completed a basic background image.** Now you can add your button layers. (Refer to Chapter 5 for more suggestions on creating buttons.)

Another method for creating images for your backgrounds is to take a still from a video clip. Still images captured from video are very effective for maintaining consistency within a project and they also mesh well with foreground elements. For example, export an image from Final Cut Pro by using QuickTime to create a background still.

To capture a still from a video clip with Final Cut Pro, follow these steps:

1. **With a video clip open in Final Cut Pro, place the playhead (Figure 6-3) at the frame in a clip that you want to export.**

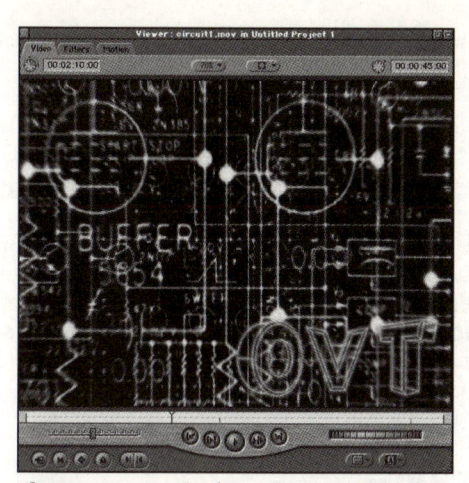

Figure 6-3: In Final Cut Pro, position the playhead where you want to grab a still frame from a video clip.
Background image courtesy of OVT Visuals

2. **Choose File ➪ Export ➪ QuickTime and select Still Image as shown in Figure 6-4 from the Format drop-down menu.**

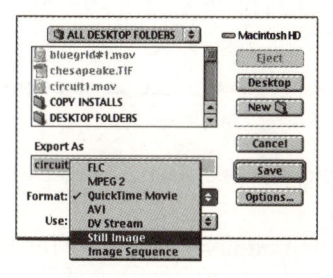

Figure 6-4: Select Still Image from the QuickTime export options in Final Cut Pro.

3. **For the Use option, select Picture, 29.97 fps.** Selecting this option ensures that you export an *interlaced* video image.

> **Note** Interlaced refers to the recombination of horizontal video lines that are separated into odd and even fields to create a complete image on a television or on a monitor.

4. **Save the file to your desktop or choose another location, such as an images folder.**

5. **Open the image in Photoshop and choose Select ➪ All to select the entire image.**

6. **Paste the image into a background layer for a menu with finished buttons by choosing Edit ➪ Paste.** Make certain that you have resized your button document to 720 × 480 pixels before pasting the video still into the background.

Note Remember that video stills are most likely to have dimensions of 720 × 480 pixels. This presents a problem because to create your buttons, you must begin with an image document whose dimensions are initially set to 720 × 540 pixels and later resized. (Refer to Chapter 5 for more details.) Create your buttons in a separate document, then after you are finished resizing them you can copy the video still and paste it into the background layer.

Working with foregrounds

We refer to *foreground elements* as anything that is placed over a background. Essentially, all foreground elements are buttons, as buttons are all that we add to our menu backgrounds. When working with button layers in Photoshop it can be confusing depending on the number of buttons you are working with. Always be organized! Create a naming system and stick with it. Naming buttons according to function, such as their state (normal, selected, and activated) is a good method.

To create your buttons in Photoshop, start with a background image and build from there. (In Chapter 5, we discuss how to work with and how to create button images in more detail.)

The following is a basic example of combining foreground button elements with your background still image:

1. **With your background layer open, click the Create a new layer button in Photoshop's Layers palette.**

2. **Create your Normal state buttons on this first layer.** Use all of the Photoshop tools at your disposal to create interesting buttons; however, do not forget to merge your text layers to your button graphics.

3. **Continue adding new layers and button graphics for the remaining button states (Selected and Activated).**

4. **Save the PSD document and import it into DVD Studio Pro for use as a menu.**

Note Photoshop effects layers do not work in DVD Studio Pro. Make certain to flatten all layer effects before importing into the program. When you flatten a layer, do not accidentally flatten the entire file, otherwise it does not work in the your DVD project.

> **Tip:** To make the task of organizing layers easier when working in DVD Studio Pro, create dividers by using empty layers named with a hyphen. When you import the PSD file into DVD Studio Pro the empty layers should act as dividers. The best way to accomplish this is by grouping all of your buttons together according to their names (all Button 1 states together, all Button 2 states together, and so on).

Exploring Motion Menus

What makes a menu a *motion* menu is the fact that a motion menu uses a video clip for the background, over which you may place a layer that contains your buttons. Buttons for motion menus use hilites to designate a selected state. Before creating a Motion menu, make sure you have all of the necessary elements. First, you need a video clip, which you may or may not loop. Next, depending on the type of buttons you choose to create, you may need to have an overlay image for your buttons that was created in Photoshop.

The following steps describe how to assemble a motion menu.

1. **Click Add Menu from the bottom of the Graphical View.** A new Menu tile appears in the Workspace.

2. **In the Picture section of the Property Inspector, choose the video clip asset that you want to use for your background.** Of course, you want to make sure that all of your menu elements have been imported before attempting to create your menu.

3. **In the Audio section of the Property Inspector, choose an audio asset for your motion menu.** You do not have to use audio, but it may make your menus more interesting.

4. **Launch the Menu Editor by double-clicking the Menu tile's thumbnail.** Your video clip opens in the window with an Untitled button already created.

5. **Drag the Untitled button selection box over to your first button.** Your buttons are actually part of the background video clip. The background video clip can be created and composited (layered with video) over a motion video clip by using a program like Final Cut Pro or After Effects. For example, you can create separate boxes of video within another clip; these boxes represent video thumbnails of your button selections.

6. **Drag the corners of the box to approximate the size of your button.** At the bottom of the Menu Editor, click Add Button to create additional button selection boxes for your remaining buttons.

7. **Name each of your buttons by clicking the button and filling in the Name field in the Menu section of the Property Inspector.**

8. **Set the Hilites for the activated and selected states of the buttons by clicking the background to deselect any buttons.** Make sure that the Show Selected Hilites option is checked in the Buttons menu at the top. This preview enables you to see if your hilites are working correctly.

9. **In the Button Hilites section of the Property Inspector, choose a color for Selected Set 1 and set the transparency to a percentage.** You may need to experiment to determine the best level of transparency for your buttons. Your buttons should now be covered with the color you selected, and this is also how your buttons should look when they are selected by a user.

10. **In the Button Hilites section of the Property Inspector, choose a different color for Activated Set 1 and set the transparency to a percentage.** Select the Show Activated Hilites option located in the Buttons menu at the top of the screen. Your buttons should now be covered with the color you just selected, and this is also how your buttons should look when they are activated by a user.

Note When working with motion menus, remember that all video assets must be in the MPEG-2 video format before importing. One solution is to create your video backgrounds in Final Cut Pro or After Effects, and then use the new QuickTime MPEG-2 encoder included with your copy of DVD Studio Pro, to produce a DVD-compliant file.

Preparing video with Final Cut Pro

Anyone producing DVD projects with DVD Studio Pro probably has experience in or at least interest in the art of video editing. No matter what format they may be intended for, creating a video project requires a good degree of editing along the way.

However, professional video editing systems have long been beyond the reach of most consumers and even many smaller professional studios. Over the last few years, a major revolution has taken place in the marketplace, providing affordable video editing solutions for consumers. This phenomenon is largely due to the faster processing speeds of today's computers. The general acceptance of digital video formats allows users to bypass the usual complexities encountered with traditional video editing methods, especially when trying to input and edit video without a significant loss in quality.

Final Cut Pro is Apple's flagship video editing software, targeted at the serious hobbyist and professional user. This software is a very powerful and scalable solution for editing everything from home movies to high definition video and film. In addition to the standard video editing capabilities included with the program are effects and transitions for compositing video and producing smooth edits. As an added bonus, you can also use Adobe's After Effects plug-ins from within Final Cut Pro. Final Cut Pro makes creating motion video menus simpler with these integrated effects options.

Chapter 6 ✦ **Working with Menus**   133

To begin, let's bring a video clip into Final Cut Pro and cut it into a loop to be used in a motion menu. Using the MPEG-2 QuickTime encoder, we can also prepare the clip for importing.

1. **Open a video clip by selecting it from your current project's Browser or import a clip by choosing File ⇨ Import ⇨ File (see Figure 6-5).**

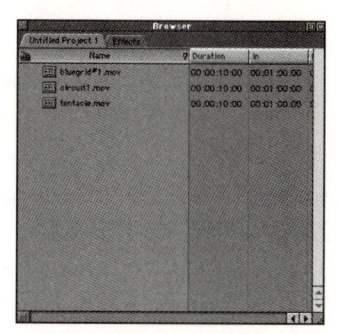

Figure 6-5: Imported video clips can be found in the Final Cut Pro Browser.

2. **In the Viewer, mark in and out points to designate a portion of the clip for creating a *subclip*.** (See Figure 6-6.) Subclips are portions of a larger clip and can exist without destroying the original. Subclips are useful for creating several loops from one piece of material, to test which loops work the best.

3. **Choose Modify ⇨ Make Subclip, shown in Figure 6-6, to create the subclip.**

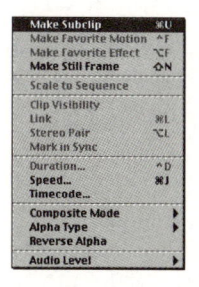

Figure 6-6: Mark in and out points on your master clip, then choose Make Subclip to create a subclip.

4. **To export the subclip, select it from the Browser and choose File ⇨ Export ⇨ QuickTime.**

5. **Name the clip and select MPEG2 from the Format drop-down menu.** You can specify the bit-rate (as shown in Figure 6-7) and other settings for the MPEG2 compression by clicking Options. (Encoding options for MPEG-2 are discussed in detail in Chapter 12.)

Part II ✦ Creating Interactivity

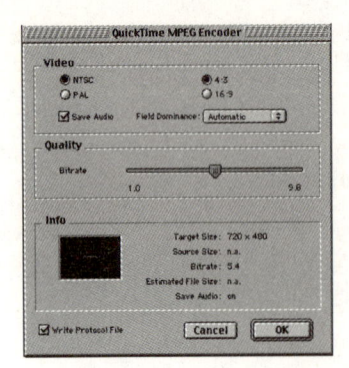

Figure 6-7: QuickTime's MPEG-2 encoding options

6. **Save the clip to the desktop or other location.** An MPEG-2 video clip and audio file should appear where you have designated.

7. **Import the video clip into DVD Studio Pro for use in a motion menu.** Delete the audio file by dragging it to the trash, unless you want to use it in a motion menu along with the video.

> **Tip**
>
> Portions of video images are often cut off when viewed on a TV or external video monitor due to the *overscan* (an area outside of the screen view) not being shown. To make sure that you are selecting a section of video whose elements can be seen and distinguished on a TV, use the clip with the Title Safe boundary overlay showing. Choose View ➪ Title Safe to display the boundaries in the Viewer.

Looping video

Video used in a Motion menu can be looped to create interesting backgrounds with minimum effort and space. Most users have probably made a menu selection before your video starts looping, but repeating the video ensures a smoother user experience.

To loop video clips, follow these easy steps:

1. **Add your video background to your menu as described earlier in the chapter by using the Property Inspector, or simply drag it from the Assets Container over the Menu tile.**

2. **Click the Menu tile to launch the Menu Editor.**

3. **In the Timeout section of the Property Inspector, choose On for the Loop option as shown in Figure 6-8.**

Figure 6-8: Set the Loop option in the Timeout section of the Property Inspector to On.

4. **Close the Menu Editor and preview your loop.** Is it long enough? Is the motion smooth? Unfortunately, a slight lag between the end of a loop and its beginning always occurs. This problem is a limitation of the DVD spec and not the authoring program, yet it may be improved in the future when we see faster access times for discs and players.

> **Tip**
> To eliminate the lag at the end of a loop, try creating longer versions of your looped video file. Take the same loop and repeat it several times in your video editor. Then export it to MPEG-2 and import it into DVD Studio Pro for use in a Motion menu. This method is especially helpful for very short loops, which would experience frequent lags between plays.

Considering animations

Many powerful 3D applications are available today that can yield impressive results for even a beginning animator — with some practice of course. By using 3D, you can create environments for motion menus that really draw the user into the experience. Professionally produced discs, such as special edition *Terminator 2: Judgment Day* or *The Abyss* sets, are excellent examples of what is possible.

Another advantage of using computer-generated (CG) images for DVD is the compression factor. CG images look very clean when encoded for MPEG-2 video, because any analog interference or static is completely absent. *Toy Story 2* and *A Bug's Life* are examples of movies whose digital origins contributed to the wonderfully crisp and clean DVDs.

The learning curve is fairly high with any 3D application you choose, although programs are becoming more user-friendly with each new version. LightWave, Maya, Universe, and 3D Studio Max are a few of the high-end and complex options available to date (not to mention expensive). Bryce and Amorphium Pro are a couple of the more consumer, less-technically oriented (though relatively powerful) options out there. Whichever 3D package you choose to create animations, the first requirements are an imagination and an original idea. Of course, being artistically inclined wouldn't hurt either!

Enabling Menus

Now it is time to take the menu background and buttons that you have created and apply them to a project. This involves setting up your menu properties and linking assets to your buttons. Before we jump into the actual process of enabling menus, you should have an understanding of the options available to you in the menu's Property Inspector.

Using the Property Inspector with menus

In the Property Inspector, you make many of your button and general menu settings. Going over the different sections and their functions is helpful at this point. Remember that you can open or close sections of the Property Inspector by clicking the arrows to the left of the headings. When the arrows are down-turned, a tree of options becomes available. They can be closed at any time to avoid confusion when working with different sections, but you really don't need to close them unless you do not have adequate space for displaying all the options on your monitor. For the most part, expanding the different sections in the Property Inspector should not be an issue if you are viewing the program with the computer's display set to at least 1,024 × 768 pixels.

Menu

The Menu section of the Property Inspector shown in Figure 6-9 presents the basic settings for the menu with which you are working.

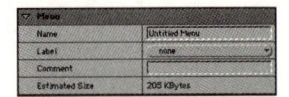

Figure 6-9: The Menu section of the Property Inspector presents you with a menu's basic settings.

Name

The Name section of the Property Inspector is where you can give the menu a name. You may also name the menu directly on the Menu tile and bypass this box.

Label

The Label options enable you to assign a level of importance to the menu in the same way you can label Macintosh system folders and files. The Label option is strictly an organizational tool and does not affect the functioning of the program.

Comment

A brief note about the menu you are working with may be written in the Comment box. Perhaps you want to have a reference for where material originated or a brief description about the Menu that you cannot fit in the Name box. The Comment box is a good location for those notes. (Comments do not appear on the finished disc.)

Estimated Size

The Estimated Size refers to the total size of the DVD project you are working with. The Estimated Size is an indicator of how much memory has been used in the creation of the disc so far, providing you with information to make an educated guess about how much additional material you can add. The number displayed here is not always 100 percent accurate, and sometimes the calculated size has difficulty refreshing when you add new material to a menu. If the number displayed appears to be inaccurate, try deselecting the menu and clicking the Workspace or select a Track tile to refresh the Estimated Size number.

@ccess

This option allows you to add a URL that is automatically opened when the menu is encountered. (See Figure 6-10.) Users watching a DVD on their computers should see a browser window pop up when they arrive at the Menu and there is a value entered here. The default setting in the pull-down is none, but clicking the arrow and selecting the URL presents you with spaces where you can enter a name for the link (@ccess Name) and the actual URL (@ccess URL, such as www.apple.com).

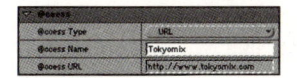

Figure 6-10: You can specify a URL to activate in the @cess section of the Property Inspector.

General

The General section (see Figure 6-11) gives you options pertaining to a few primary menu functions. These options are described here.

Figure 6-11: The General section gives you a few simple options relating to default Menu functions.

Pre-Script

The Pre-Script option gives you the ability to assign a script that runs before the menu starts. Scripts are written instructions that perform a specific command, such as randomly generating a number to change the result of a menu selection. (Turn to Chapter 17 to find out more about scripts.)

Default button

The Default button is the button that is selected first when a menu opens. Any button can be specified as the default. Choose from a list of all the available buttons by clicking the arrow for a pull-down menu with all the possible choices.

Return button

The Return button option enables you to select where you want users to arrive when they press the return key on their remote controls while in the Menu. For example, if you hit the Menu button while watching a video track, you may want to return to that same video track. You can specify if you want the button to return users to the track from which they came, or you can choose to send the users to a different menu.

Timeout

Have you ever been so mesmerized by a Menu's animation, graphics, or music when, before you have realized it, you were sent to another menu or to the beginning of a movie without pressing any buttons? The Timeout option (Figure 6-12) enables you to set an Action that automatically sends a user to a different track, menu, slideshow, or other disc element. The default action is not set. When you click the arrow for the drop-down menu and select a destination such as a video track, a box appears that enables you to set a time Duration in seconds before the menu automatically redirects the user.

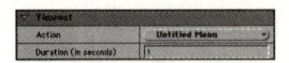

Figure 6-12: The Timeout section allows you to specify a time limit for a video loop in a menu.

Picture

The Picture section (Figure 6-13) is where you select which graphics and layers of your Photoshop files can be seen on the screen. After you select the PSD file that has your Menu layers, you can select which particular layers are visible. If you named your button layers and backgrounds correctly, determining the choices that you need should be easy. When an asset is assigned, the option, Layers (always visible) becomes available. It is in this section where you do the actual assigning of layers. You can select the individual layers from the drop-down menu, or you can click the link to open a window with all of your layers listed. Check an item in this window to select it as always visible. You can keep this window open if you want and minimize it as needed, which can be a good way to see which buttons are in use and which are left over without having to go back to the drop-down menu in the Property Inspector.

Figure 6-13: The Picture section of the Property Inspector enables you to assign graphic files for the menu.

> **Tip**
> When the Layers (always visible) window opens in the Workspace, you can select all or some of the layers by clicking in a space, holding down the Shift key, and clicking selections above or below to automatically select several layers at once. You may also drag the cursor to select several layers at once by clicking and holding down the mouse button while dragging over the layer options.

Audio

When you are working with motion menus, the Audio section enables you to specify an audio track to play with the motion menu. The Audio section only becomes available when you have selected a video asset in the Picture section of the Property Inspector.

> **Note:** To include audio with a DVD menu, it is necessary to have a video asset that you can select in the Pictures section of the Property Inspector. But this video file does not necessarily have to display a moving image. If you want to have a still image in a menu and have audio running at the same time, you can generate a video file where every frame is the same image, and have an audio file serve as the soundtrack. QuickTime Pro makes it easy to do this kind of thing. For more information, look in the Case Studies area of the DVD-ROM for the Still Image Audio Menu.

Button Hilites

In the Button Hilites section (Figure 6-14), you make settings pertaining to which graphics, colors, and transparencies are used for your Hilites.

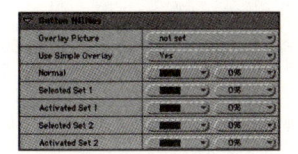

Figure 6-14: The Button Hilites section gives you the option to set button Hilite properties.

Overlay Picture

An overlay picture is the graphic that defines the outlines of button shapes for highlighting when the buttons are added to a motion menu or video track. Use this option to select the graphic file that should act as the overlay image. For more information, see Chapter 5.

Use Simple Overlay

This option enables you to choose from two values, Yes and No. Yes, the default value, gives you a limited set of choices for assigning colors and transparencies to button layers. Most users probably never need to use the No option, with which you can specify four color and transparency values for each button state.

Normal

This section is where you specify which layer is to be used for all buttons in the normal state. When a button has not been selected or activated, it is highlighted by the color you choose from the drop-down menu to the right. Make certain to indicate a unique color (a color that is not used by the other button states, unless you do not want the states to be differentiated from each other) and adjust the transparency as desired.

Selected Set 1

In this section, you specify which layer is to be used for all buttons in the selected state. When selected with a remote control, these buttons are highlighted by the color you choose from the drop-down to the right. Make certain to indicate a unique color (a color that is not used by the other button states, unless you do not want the states to be differentiated from each other) and adjust the transparency as desired.

Activated Set 1
This option enables you to specify which layer is to be used for all buttons in the activated state. When the Enter key is pressed or OK button is clicked on the remote control, these buttons are highlighted by the color you choose from the drop-down menu to the right. Make certain to indicate a unique color (a color that is not used by the other button states, unless you do not want the states to be differentiated from each other) and adjust the transparency as desired.

Selected Set 2
With this option, you can assign unique properties to an additional set of buttons, as you did for Selected Set 1.

Activated Set 2
With this option, you can assign unique properties to an additional set of buttons, as they were for Activated Set 1.

Linking assets to menu buttons

Any asset can be linked to through a menu by creating buttons and specifying where these buttons should take you when activated. You can nest menus by linking from one menu to the next.

The following steps describe how to link assets to buttons. For our examples, we assume that you are starting by linking menus to tracks; however, you can link to slideshows and other menus just as easily, following these steps.

1. **Click the Add Menu button to add a Menu tile to the Workspace.**
2. **Drag your Menu PSD file containing all of your button and background layers over the Menu tile.** This may also be done with the Property Inspector by choosing the PSD file from the Asset option in the Picture section (Figure 6-15).

Figure 6-15: Choose a layered PSD file to assign for your Menu graphics.

3. **Double-click the Menu tile's thumbnail photo to launch the Menu Editor.**

Chapter 6 ✦ **Working with Menus** 141

Tip

By clicking on your Menu PSD file in the Assets Container, the Property Inspector (see Figure 6-16) displays the properties for that Photoshop file. In addition to simply providing you with information on the file's graphics, such as width, height, and number of layers, you are also able to enter a new name for the file or change its label, and add comments. You can also assign a different file to replace the current one by clicking the blue hyperlink under File in the General section of the Property Inspector. You are then presented with a choice to select a file from any location on your computer.

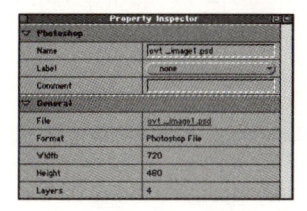

Figure 6-16: The Property Inspector can display information for a particular asset, such as a PSD file.

4. **In the Picture section of the Property Inspector, select the layers that you want to use by checking all of them in the drop-down menu to the right.** An alternative method to checking all of the layers is by clicking on the underlined link labeled Layers (always visible) to open a small window where you can click and drag to select the layers you want to use. (See Figure 6-17.) The layers should now appear in the Menu Editor window. Notice that the dotted box named Untitled Button automatically appears when you open the Menu Editor. You can use these boxes to select your buttons and to designate the active area surrounding them.

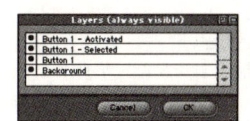

Figure 6-17: An optional window can be opened from the Picture section of the Property Inspector and used to select active layers. (It is opened by clicking on the Pictures link in the Property Inspector as if it were a hyperlink.)

5. **In the Menu Editor, clicking the box turns your cursor into a hand, indicating that you can pick up the box and move it around the window.** Drag the box over to your first button. (See Figure 6-18.)

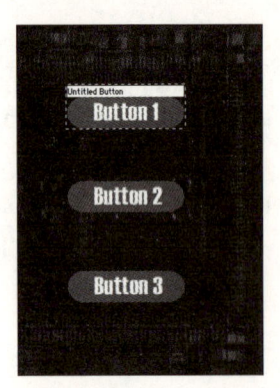

Figure 6-18: A new Untitled Button is used to select an active button area in your Menu.

Background image courtesy of OVT Visuals

> **Note**
> All menus must have at least one button to function. A maximum of 36 buttons can be used in one menu. For most DVD authors this is not an important limitation. Remember that you can create as many menus as you need, so if 36 buttons are not enough just link to another menu with more buttons.

6. **Drag a corner of the box to resize it.** Make the box a little larger than your button graphic and position it accurately around it.
7. **Name your button by clicking on it and filling in the Name field of the Property Inspector.** (See Figure 6-19.)

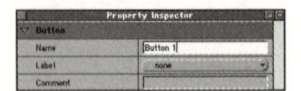

Figure 6-19: Name your new button in the Property Inspector.

8. **Create another button by clicking the Add Button at the bottom of the Menu Editor.** Name the button and repeat Steps 5 through 7 for your second button.
9. **Repeat these steps to create as many button selections as you need.**
10. **Click your first button. Notice the options that are available to you in the Property Inspector.**

    - **Button section.** This section enables you to give the button a name, a label, and comments just as the other Property Inspector allows you to.
    - **Dimensions section.** This section enables you to input exact sizes and positioning for your buttons. The Width and Height are the actual pixel dimensions for the button selection box, while the Left, Top, Right, and Bottom correspond to the position of your selection box on the page.

For you perfectionists, this is a good place to check consistency of the selection boxes you are drawing including their relative alignment and adjust accordingly.

- **Display area.** This area is where you assign the particular button layers and states associated with a single button (Normal, Selected, and Activated) as well as which set of Hilites are to be used (Set 1 or Set 2).

- **Selection Condition.** This option gives you the choice of Selection Type you wish to assign a button, which Hilites the button depending on whether a Track, Audio Stream, or Subtitle Stream was playing before the Menu appeared.

- **Action section.** This section gives you choices for setting the default Audio Stream, Subtitle Stream, or Auto-Action assigned to the button. It also lets you assign where the button is linked to with the Jump When Activated option — the most important action of all.

- **Button Links.** This area specifies which button is selected when a user presses the arrow keys on their remote control. The default is Previous Button and Next Button whose order is determined by the placement of buttons in the Project View. (See Figure 6-20.)

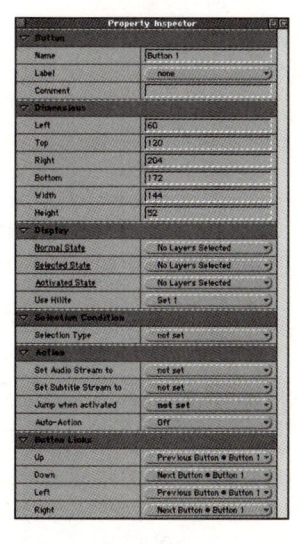

Figure 6-20: The Property Inspector enables you to make many choices for each of your buttons, including their relative positions and where they should jump to when activated.

11. **Click your first button to select it. In the Display section of the Property Inspector, select the button layer that you want assigned as the Normal State.**

12. **Continue assigning button layers for the other Selected and Activated States.** If you named your button layers according to their functions as suggested, you should find them very easily. (See Figure 6-21.)

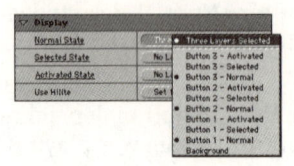

Figure 6-21: Select which states you want assigned to each button.

13. **To link your buttons to another disc element, such as a track, select your button and make a choice from the drop-down menu for Jump when activated under the Action section.** A list of all available assets that you can link to is here. Select a track that should be linked to from that button.

14. **Repeat the linking of assets for the rest of your buttons.** (See Figure 6-22.)

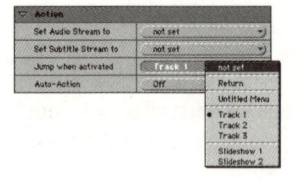

Figure 6-22: Link your menu buttons to other disc elements, such as a track.

After you have created your menu and completed assigning disc elements to each button, systematically check the menu. Make sure to test it immediately after creation to ensure less confusion later when trying to determine a source of failure.

Testing completed menus

Testing menus can be a bit tricky if you have a lot of buttons and links. If you have a lot of buttons and links, you should conduct some serious testing to ensure that you have not forgotten to link a button correctly or that all of your Hilites and layers are in their proper place. Also, it is often helpful to have another person check your menus (and the finished disc for that matter), since someone unfamiliar to the project can often detect problems that you may not have noticed while editing your own work.

After you have created a menu, you can test it by using the Preview function. To do so, follow these steps:

1. **First, close the Menu Editor and select the menu tile that you want to preview.**

2. **Click the Preview button in the lower-right corner of the Graphical View.**
   You see a screen that displays the menu you created. How your menu appears here is exactly how it appears and functions on a finished disc, minus the black border, of course.

3. **Use the navigational arrows at the bottom of the Preview window to select a different button in your menu and click OK to test the links.**

Tip

Develop a systematic approach and check off every item that you have already tested. Did you specify a Timeout action? If so, make sure that you wait long enough to see if it functions properly. Consider all of the actions that you have created and test them and your standard button links and Hilites.

Using Lines to display connections

The Lines feature in the Graphical View can also be useful for testing the links in a menu. (See Figure 6-23.) By clicking Lines and then selecting the configure option, you can configure lines to display connections between disc elements (visually presented as tiles) and the menu whose buttons they are connected to. (See Figure 6-24.) If you have created a few menus, this is a useful way to instantly see where they are pointing. An arrow on the line indicates the direction of the link (if the line is coming from a menu the arrow is usually headed away).

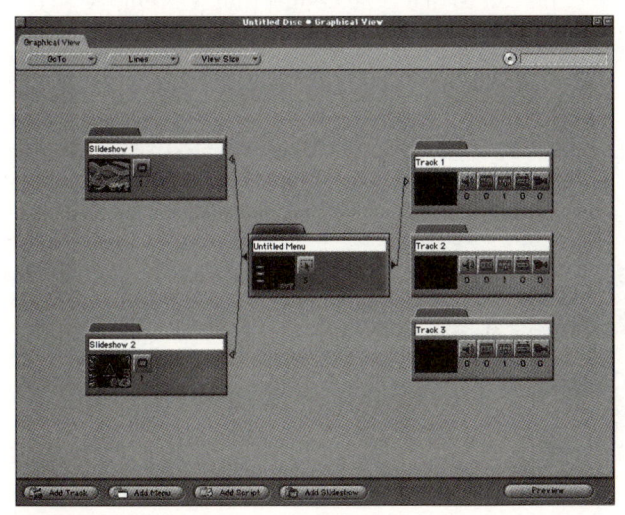

Figure 6-23: Lines are an easy way to view connections between menus and other disc elements.

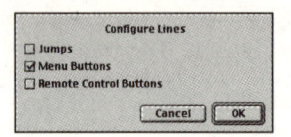

Figure 6-24: Open the Configure Lines dialog box to display connections with menu buttons.

Matrix Views

The Matrix Views in DVD Studio Pro provide you with a means of setting and previewing connections between the various elements in your DVD project. Although you do not need to use the Matrix Views, these views may be a useful tool for easily checking and modifying your project in a central location. For the purposes of working with menus, focus on the Jump Matrix and the Layer Matrix views.

Jump Matrix

When you select Jump Matrix from the Matrix Views menu, a screen, which looks like a giant Connect Four game, opens up over the Workspace. (See Figure 6-25.) This chart shows the Actions available for you to use along the top of the view and all of your menus, buttons, tracks, and other containers along the left side. Clicking an arrow to the left of an item on the left side of the view expands that item or container, like a file tree, showing all available elements inside. Assigning an action involves simply clicking in a box that corresponds to the intersection of the items that you want to link. A dot is placed to mark the connection. To remove a connection, just click the dot again and it disappears. The menu you were working with can be checked by working your way down the left of the view, making sure that every item you want is connected to the appropriate action. If they are not connected correctly, it is only a matter of re-assigning items by placing a dot. After you are comfortable working in the Matrix Views, you may decide to use this tool instead of working with the traditional Graphical View and Property Inspector.

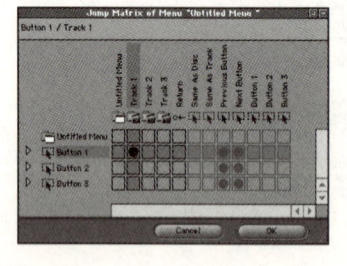

Figure 6-25: The Jump Matrix lets you check your menu connections and make changes in a single location.

Layer Matrix

In the case of menus, the Layer Matrix is used in the same way as the Jump Matrix, though instead of assigning actions it allows you to assign Photoshop layers to menus and buttons. All of the possible layers in a Photoshop file are listed along the top of the view while the button states are listed along the left side. (See Figure 6-26.) As stated for the Jump Matrix, assets are assigned by clicking in the intersecting spaces. Doing so places a dot to signify a connection between the items. By viewing a menu in this way you can often spot mistakes that are otherwise difficult to see.

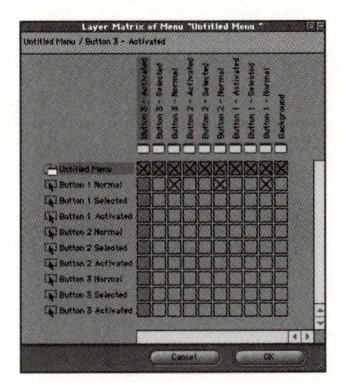

Figure 6-26: The Layer Matrix lets you check your Photoshop layer assignments to make sure that all of your menu buttons and backgrounds are correctly placed.

Summary

- ✦ Stills and video can be used to create backgrounds for a menu.
- ✦ Menus are created from the combination of background and foreground elements.
- ✦ Photoshop layers enable us to create the interactions betweens button states.
- ✦ Final Cut Pro is useful for creating video loops for inclusion in Motion menus.
- ✦ Looping video in Motion menus can produce lags between where a clip ends and where it begins; however, this can be circumvented by creating longer clips with combinations of the same loop.
- ✦ Matrix Views and Lines can be useful for testing a Menu's connections.

In the chapters that follow we discuss interface design (Chapter 8) and working with video (Chapter 10) and audio (Chapter 12). In the next chapter, we discuss how to create subtitles for a project using the Subtitle Editor.

✦ ✦ ✦

Using Subtitles

CHAPTER 7

♦ ♦ ♦ ♦

In This Chapter

Understanding the Subtitle Editor

Working with key frames

Using and selecting fonts

Working with external subtitles

♦ ♦ ♦ ♦

Have you ever wondered how subtitles are created and synced to the dialog in a movie? If you are a fan of foreign films, or even if your first exposure to the filmmaking of other cultures was *Crouching Tiger, Hidden Dragon*, you are certainly familiar with the role of subtitles. Adding subtitles to film involves a complicated syncing process that eventually results in burning the text into a finished release print that is sent to theatres. In the case of DVDs, subtitles can play an expanded role, providing the viewer with a potentially unlimited set of language choices to accompany their favorite movie, regardless of the country in which it originated. In addition, DVDs enable you to switch between the various language streams and turn subtitles on and off, unlike videotape or laserdiscs that burn the subtitles into the picture.

Aside from use in straight foreign language translations, subtitle tracks in DVDs can also be used for purposes ranging from text for karaoke, written commentary to describe a scene, overlays for menus (refer to Chapter 6), or as an aid for the hearing impaired. Previously (without a dedicated subtitling program), most subtitling tasks for video were accomplished through a combination of programs, such as Adobe Photoshop, to create and line up text, and hit or miss methods of manually matching frames to dialog in a video editor. DVD Studio Pro provides you with a new, convenient interface called the Subtitle Editor, which is used to create your own subtitles quickly and accurately. What seemed like a difficult and laborious task is now a matter of learning a simple interface. If you are an aspiring kung fu director hoping to create the next digital movie craze, learning how to add subtitles to your video could be the first step.

Although DVD Studio Pro's subtitle feature is most often used to produce text for video, it also allows for some relatively sophisticated image overlays. The subtitle streams on a DVD are composed of individual pixels, similar to how a computer displays images on a monitor. The ability to overlay images makes working with any language or character set possible, as well as providing a way to create menus on top of video clips, in the case of motion menus. (Refer to Chapter 6 for details about motion menus.) You may also use the subtitle feature to create simple animations or add motion to graphics and text. Colors and transparency are the only significant areas that have serious limitations, imposed by the nature of the DVD player itself.

In this chapter, we discuss how to use DVD Studio Pro's Subtitle Editor to create subtitles for a video. We also discuss working with fonts, as well as providing alternative methods for using subtitles that can be generated by other programs or service bureaus. In addition, we show you how to use the subtitle feature as a method of generating multiple layers in a DVD project. Don't be afraid; launch into the Subtitle Editor right away — feel free to use the video files included in the tutorial folder on the DVD-ROM included with this book if you do not have a kung fu movie in progress.

Working with the Subtitle Editor

The Subtitle Editor is actually a separate program from DVD Studio Pro. You can find it in the DVD Studio Pro folder on your hard drive after you install DVD Studio Pro. Launching a separate application is inconvenient, but doing so does make a certain amount of sense because assets must be prepared by the user *before* importing them into DVD Studio Pro. As you may have noticed, DVD Studio Pro can take up a significant amount of memory and has quite a few windows going already, so it's probably just as well that the Subtitle Editor is on its own. The other advantage of the fact that the Subtitle Editor is a separate program is that it can run alongside virtually any video editor or playback environment you may want to work in while subtitling.

Selecting project settings

After you click the Subtitle Editor icon, the program launches and presents you with a window called Project Settings. Before you advance to the actual Subtitle Editor, you must first make a few selections and settings for your overall subtitle project (see Figure 7-1).

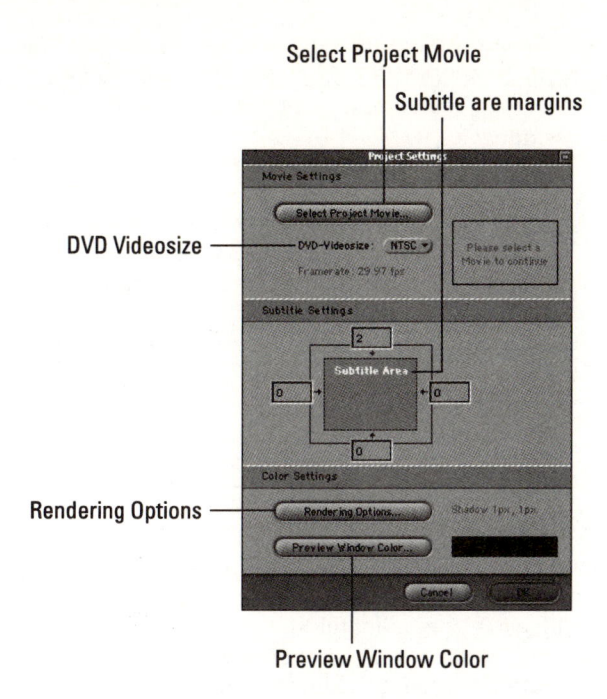

Figure 7-1: The Project Settings window is where the overall settings for the Subtitle Editor are made.

The Project Movie

When you are creating a subtitle to go along with a video segment, the Subtitle Editor requires the use of a *Project Movie*, which must be in QuickTime format. The Project Movie can be the exact same movie for which you are creating the subtitles, and even saved in a smaller screen size, as long as it's in QuickTime format. Essentially, the Project Movie is an aid to help you accurately place (or *key frame*) your subtitles over time; it can be video, video + audio, or audio only. A typical way to prepare video for a DVD that contains subtitles is to generate one video file in MPEG-2 format for the final DVD and another video file at a screen size of 320 × 240 pixels, 30 frames per second (fps), for the Project Movie to be used in the Subtitle Editor.

When you start up the Subtitle Editor, you must first click the Select Project Movie button in the Movie Settings section of the Project Settings window to choose a QuickTime movie file to which you want to add your subtitles.

After you click the button, choose your movie file from its location on your computer and click Open or simply double-click the appropriate file name to add it to your project. After your movie is chosen, it should display as a thumbnail photo to the right in the Movie Settings window.

You may also choose an audio file as long as it is in the QuickTime .MOV format. Using an audio file may be preferable for some people who simply want to translate dialog in a video. Using audio as a Project Movie is particularly useful for narration that may be added to slideshows or, for example, as narration in travelogue or sales videos. Ultimately whether you choose video or audio files to sync your subtitles to doesn't matter, because the resulting subtitles are exported from the Subtitle Editor independent of the Project Movie.

DVD-Videosize

DVD-Videosize (or video standard) refers to the picture size and frame rate of your source video. The DVD-Videosize is set by selecting NTSC (the default) or PAL from the drop-down menu in the Project Settings window. Choosing NTSC displays a frame rate of 29.97 fps, and PAL displays a frame rate of 25 fps. When creating subtitles, the frame rate is important because you must sync your subtitles to individual frames. Therefore, your DVD-Videosize should match the standard of the video to which you are adding the subtitles — otherwise your subtitles won't appear at the proper times or frames. If your subtitles are intended for a video created in NTSC, you should not choose PAL and vice versa.

> **Note** The Subtitle Editor does not convert video standards, but if you are working in NTSC, you may create subtitles for an eventual PAL version of the video by choosing an NTSC source file, setting the DVD-Videosize to PAL, and working with the Subtitle Editor as usual. The Subtitle Editor automatically limits the frame rate to 25 fps for PAL, even though your NTSC source originated at 29.97 fps. The use of varying frame rates for your Project Movie is possible because subtitles are created independently of the video, and the Project Movie is simply a means to assist in syncing text to video. However, this method can be risky, particularly for a long video program, as conversion of frame rates may differ slightly, and the shift over time may become noticeable.

Subtitle Settings

Your next step is to set the margins that define the area in which your subtitles should appear in the frame. This is done in the Subtitle Settings section of the Project Settings window. Type a value in the appropriate boxes, which corresponds to the number of pixels from the edge of the frame that your margins should be set. These values set boundaries for your text, ensuring that your text is aligned correctly and remains visible at all times. For instance, you might try setting margins of 25 pixels for left, right, and bottom values, with 400 pixels for the top as a standard setting when working with NTSC video.

Color Settings

In the Color Settings section of the Project Settings window, you are presented with two buttons, one labeled Rendering Options and the other labeled Preview Window Color. Both buttons give you options for setting the color palette you want to work with in the Subtitle Editor.

Clicking the Rendering Options button opens the Rendering Options window, which enables you to select what your subtitles should look like when they are rendered for output (and subsequently ready for importing into DVD Studio Pro). First, you must specify whether you want a Shadow or a Border for your text. If you choose the Shadow option, you are able to select a color for the shadow from the Color drop-down box. You may also set positive and negative values for Offset X and Offset Y, which determine the depth and position of the shadow in the horizontal and vertical planes, respectively. A high Offset X value gives the appearance of a dimensional shadow cast far behind the foreground text, which is the same size, although moved further up (for negative values) or down (for positive values) in the frame, while the Offset Y value shifts the position of the text further to the right (for positive values) or left (for negative values). (See Figure 7-2.)

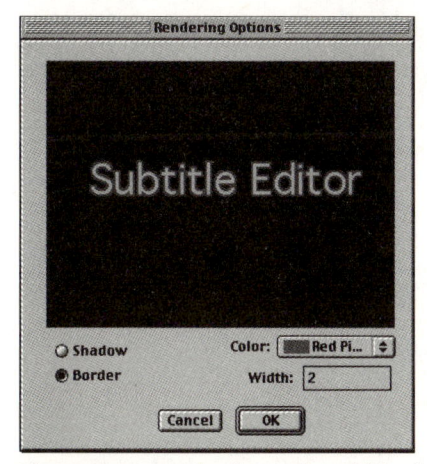

Figure 7-2: A red color and a width of 2 pixels are values selected for the border of subtitle text in the Rendering Options window.

If you choose the Border option you are able to set the color and width of the subtitle's text border. Select a color for the text's border from the drop-down menu and specify a value for the width in the box. The maximum value for a border's width is 20 pixels. (See Figure 7-3.)

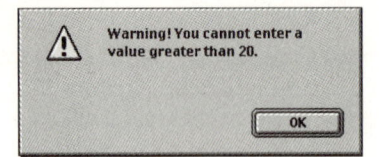

Figure 7-3: The maximum border width for subtitle text is 20 pixels.

Clicking the Preview Window Color button opens a color picker window, which enables you to select a special background color for the preview window in the Subtitle Editor. The color you select only pertains to working in the Subtitle Editor and does not appear in the rendered subtitle stream. If you are working with colored text (with a border or shadow), you may want to consider changing your background color from the default black to a color that complements your text. The color of your background is mostly irrelevant, although you may find working with a particular color generally more soothing than others. (See Figure 7-4.)

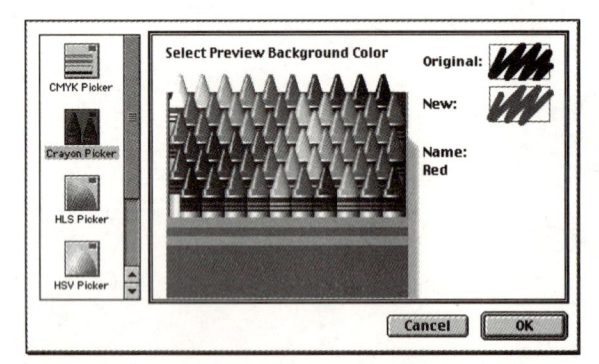

Figure 7-4: By clicking the Preview Window Color button, you may select a background color for the Subtitle Editor interface using one of the color picker windows.

As soon as you have completed making your initial project settings, click OK in the Project Settings window. The main Subtitle Editor window opens and covers the desktop. If you want to go back and make changes to your project settings as described previously, you may do so at any time from within the program by choosing Project Settings from the File menu.

Preferences

Before exploring the many windows in the Subtitle Editor, choose File–>Preferences to take a look at the Preferences. The Preferences window is shown in Figure 7-5.

Figure 7-5: The Preferences window enables you to specify some settings for the Subtitle Editor including the Startup Action, which occurs when the program is first launched.

The Preferences window is where you adjust the default settings for the Subtitle Editor. The first option you find in the Preferences window is the Startup Action, which determines what happens when you click the Subtitle Editor icon to launch the program. The default action is to create a new project, although you may also choose to open an existing project or simply do nothing. Below the Startup Action settings is a checkbox labeled Ask before deleting an Entry. Make certain that this box is checked, or you could delete a subtitle without intending to do so. Clicking the Color Menu Settings button launches a dialog box that enables you to change the names and color value for any of the sixteen colors permitted in your subtitle project. (See Figure 7-6.) Use the color picker windows to choose a new color or stay with the default values. When you are finished working with the color settings, click OK to close the box.

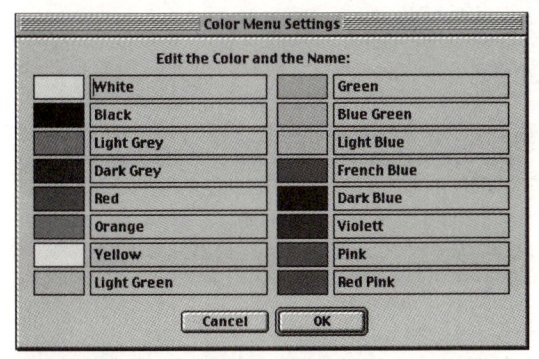

Figure 7-6: The Color Menu Settings window enables you to edit the colors and names of the colors used in the subtitle project.

Note: In case you were wondering, 16 is the maximum number of colors allowed for subtitle overlays in a DVD project, whether they are text or images. However, using more than four colors at any one time is not advisable (and in many cases not even possible). Subtitle text in particular should remain limited to two colors. These restrictions are due to the amount of memory available on DVD players, which have limited capacity for displaying multiple subtitle colors simultaneously.

Exploring the Subtitle Editor's windows

Now you can begin exploring the Subtitle Editor interface. Note the three main windows that appear when you launch the program: working left to right and down, they are the Preview, Marker, and Subtitles windows. (See Figure 7-7.) The windows can be hidden or shown at anytime by selecting them from the Windows menu at the top.

Figure 7-7: The Subtitle Editor is composed of three windows that are used to create subtitles and to navigate within your project movie.

Chapter 7 ✦ **Using Subtitles** 157

> **Considering Memory and Bit rate Limitations**
>
> If you choose to use more colors than is recommended, do not be surprised if your subtitles display incorrectly when you attempt to play them. Always keep memory considerations in mind when creating a DVD project. Even though it may be possible to include a large number of subtitle tracks on a DVD, remember that you are limited by the final bit rate of your project, which is the amount of data processed by the player at any given moment. The more elements you add to a scene on a DVD, the higher your bit rate needs to be to accommodate them. As a result of requiring a higher bit rate for your project, you lose additional space on a disc and are therefore limiting the amount of time available on your disc. A 4.7 GB DVD-R, for example, can rapidly approach its limits when you are using a great number of subtitles, alternate audio streams, and multiple angles.

The Preview window

The Preview window displays your Project Movie along with a picture of the subtitles as you create them. (See Figure 7-8.) This window lets you navigate through your Project Movie to locate key frames for your subtitles. Use the scroll bar at the bottom of the window to scrub or move through your Project Movie, just as you would an ordinary QuickTime movie.

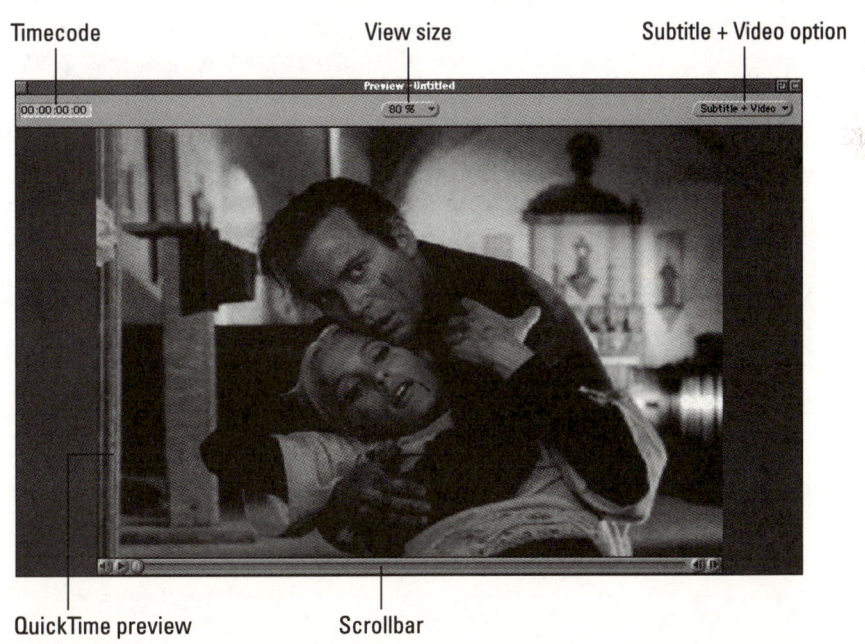

Figure 7-8: The Preview window displays your movie and subtitles as you work.

Three elements are at the top of the Preview window. On the far-left corner, a time code box indicates the frame you are on in your project movie. Click the Play button or move the indicator in the QuickTime Preview window to see how this value changes and make note of how frames are counted. This counter is invaluable when you are locating and placing the position of a key frame, and you should refer to it often. By typing a time value in the box, you jump instantly to that point in the movie.

> **Note** Don't worry if you are unfamiliar with time code. Essentially, the only difference between reading time code and a regular clock is that time code adds another value on the far-right by breaking down the seconds into the number of frames video plays per second. For NTSC, every 30 frames equals 1 second, and for PAL every 25 frames equals 1 second.

To the right of the time code box is a percentage value that indicates the size of the Preview window picture for your project movie. You can change this to a smaller or larger value (in increments of 10%) at any time, although the default value of 80% should work adequately for most people because it is exactly the right size to fit all three windows on the screen at a resolution of 1024 × 768 pixels, without overlapping or covering other windows.

You may choose to make the Preview window smaller and expand your Subtitle window — this allows you to work with more subtitles expanded or available at a time. If you want to run other applications along side the Subtitle Editor, you may want to make the window smaller to save space, or if you have an extraordinarily large screen size or dual monitor configuration, you may choose to view your preview window at the maximum value of 100%.

In the far-right corner of the Preview window is a drop-down menu that is used to select whether you want the subtitles you create to be displayed along with the video (Subtitle + Video) or with only the subtitles visible (Subtitle) while editing. Working with the video on or off is simply a matter of preference. If you are using an audio track and no video as your project movie, this drop-down menu is grayed out and not available as a viewing option.

The Marker window

The Marker window provides a convenient means of keeping track of scenes in a movie as well as providing a quick method for navigation. (See Figure 7-9.) For example, if your movie is very long, you may set markers at various strategic points throughout the movie and name them to describe the action or other specifics of that scene.

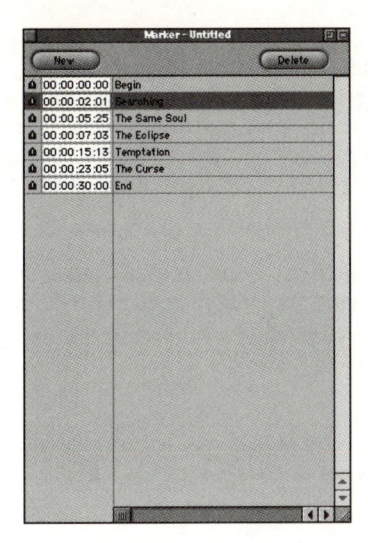

Figure 7-9: The Marker window helps you to keep track of points within your movie.

When you want to return to a particular marker, you can simply click the marker you want in the Marker window, rather than scrolling haphazardly through the entire timeline until you find it. You may also select a particular marker from the Markers drop-down menu, located in the Markers menu at top. To the left of each marker is a time code box that displays the exact location of the marker in the movie. You may enter a different value in this time code box to easily reposition a marker. One of the nicest and most useful features of markers is that they are automatically arranged chronologically in the Marker window, regardless of when you have created them. Also, in and out points of markers are independent of each other, so overlapping markers is not a concern. Organization is extremely simple.

Creating a marker

To create a marker, scroll to a point in the movie where you want to place it. After you have located the exact frame you want, click the New button to create a marker, which is placed in the correct chronological order in the Marker window. If this is your first marker, it is created between the Begin and End markers, which are created by default and delineate the beginning and ending points of your movie. Double-click the untitled box in the New Marker window and name your marker. Click a marker in the window to be instantly whisked away to its location in the movie. If you want to erase a marker, select it in the Marker window and click Delete.

The Markers menu at the top provides you with another way to create and navigate with markers. You can create a marker by scrolling to a point in the movie as before, only this time when you click the New Marker option you are presented with a dialog box listing the Time (you can change the time code as desired) and a box for the Name of the marker. (See Figure 7-10.) Selecting the Edit option in the Markers menu when you are located on a marker is another way to change the time code and name for a marker.

In most cases, the edit option may be a little impractical, because you can always edit the markers by clicking them in the Marker window. However, the edit option may be useful if you wanted to get rid of the Marker window to make room on your desktop.

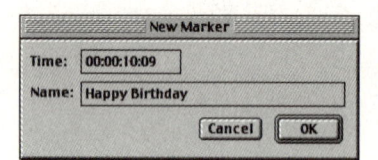

Figure 7-10: The New Marker dialog box enables you to specify a Time and Name for a marker that you create.

Using the Find feature
Find is a useful feature of the Markers menu if you are working on a long movie with a lot of markers. Click Find to launch a dialog box and fill in the Name field to search for a word in the marker's name to find it. The find function works even if you enter incomplete words.

Using the Delete option
The Delete option works as you can guess by first selecting a marker and then choosing Delete to erase it.

Navigating with the Markers drop-down menu
The Markers drop-down menu is simply another method for navigating between your markers by selecting a marker name to be taken to.

The Subtitle window
The Subtitle window is where you do all of the creation and typing of your subtitles. Essentially, the Subtitle window is a word processor, where you type in the text that is key framed to your movie. Subtitles are created and typed into boxes in the Subtitle window. (See Figure 7-11.)

Figure 7-11: What do you want? is typed in the Subtitle window.

Along the top of the window are options that determine the look, size, and position of your text. Also, clicking the New button creates a new subtitle as well as buttons to *split* or *join* current subtitles. Split enables you to split a subtitle into two parts and can only be used when you are placed between the in and out points of a subtitle. Join combines two subtitles that are in direct consecutive order to each other. When you create a new subtitle by clicking the New button or by choosing Subtitle ⇨ New Subtitle from the menu at the top, you see a black box in the middle of the Subtitle window where you do all of your typing and two time code boxes on the left side of the window where you specify the duration of your subtitles.

Creating Subtitles with Key Frames

To place subtitles at specific times in a movie, we must create subtitles with *key frames*, which are in and out points that determine when a particular subtitle should display and when it should disappear. The following sections describe how to accomplish the creation of subtitles by using key frames and other features within the Subtitle Editor — the subtitling program included with your purchase of DVD Studio Pro.

Creating subtitles with a Project Movie

Before you can begin creating subtitles, you must have a QuickTime file to sync with the subtitles. You may choose an audio or video file for this purpose. If you are using a video file, the exact dimensions do not matter. You can use 640 × 480 pixels, 320 × 240 pixels, 160 × 120 pixels, or any size video that you prefer. (The Subtitle Editor automatically scales the project movie you select to fit into the standard 720 × 480 pixel dimensions for DVD.) Using a low-quality movie as a reference is sometimes preferable to conserve hard drive space, which can be of concern in longer movies. Essentially, movies used to sync with subtitles are irrelevant to the final product, as the subtitles are exported and imported independent of any movie file.

The only important consideration is that the movie file (if it is video) matches the frame rate of the video standard you are using for eventual output. If your subtitles are intended for NTSC video, you must have a project movie with 30 frames per second to use as a reference (PAL requires at least 25 frames per second). If you decide to use a low-quality version of your program as a reference video, make sure that the frame rate has not been changed to a lower value, such as 15 frames per second, otherwise when you play back your DVD, you may notice that the subtitles are not timed correctly with your video. After you have a project movie to use as a reference, you are ready to start creating your subtitles.

Creating subtitles is fairly straightforward when you know the procedure. The following steps guide you through the creation of subtitles using the Subtitle Editor:

1. **Look in the DVD Studio Pro folder on your hard drive and locate the Subtitle Editor program.** Click the program icon to launch the program. When you launch the Subtitle Editor program, you should see the Project Settings window. The Project Settings window is where you make the general settings for your project.

2. **Under the Movie Settings section, click the button labeled Select Project Movie.** Select a video or audio file that you want to use as a reference to place your subtitles on top of from its location on your computer.

   Note: If you want a video file to practice with, look in the CH7 folder in the Tutorial section of the DVD-ROM that accompanies this book.

   As soon as you have selected the file you want, click Open or Convert. Convert is the option that appears if the file you choose is not ready to be used in the Subtitle Editor. In that case, the Subtitle Editor converts it for you. Remember, the Subtitle Editor works with QuickTime files only.

3. **Next, choose your DVD-Videosize in the Movie Settings section.** If you are using an NTSC video standard for your movie, choose NTSC (29.97 frames per second). If your movie is in the PAL video standard, choose PAL (25 frames per second). Choose wisely! The subtitles you create are key framed to the appropriate frame rates. If frame rates do not match the intended video standard, subtitles appear out of sync on the finished DVD.

4. **In the Subtitle Settings area of the Project Settings window, set margins to define the subtitle area where the subtitles should appear.** Subtitles may be placed anywhere in the frame. (See Figure 7-12.) Let's start with a typical setting for NTSC that places subtitles at the bottom of a frame. Click in the boxes and set the top margin to 400 pixels and set the remaining left, right, and bottom margins to 25 pixels.

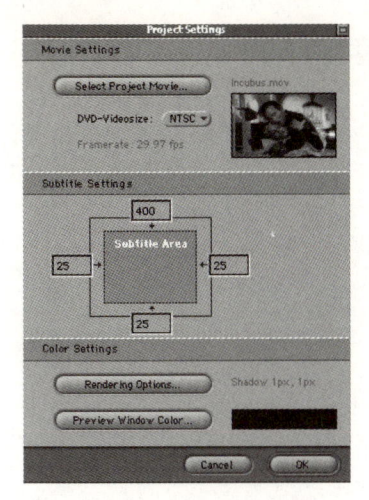

Figure 7-12: Set your margins to appropriate dimensions for the placement of subtitles within a video frame, such as these basic settings for NTSC (top, 400 pixels; left, right, and bottom, 25 pixels).

5. **In the Color Settings area of the Project Settings window, you have the option to make changes for the way colors appear in your rendered subtitles and in the main Subtitle Editor interface.** Click the Rendering Options button to open a window where you can make choices for your subtitle text.

6. **In the Rendering Options window, select either a Shadow (which adds depth to the text) or a Border (which adds an outline to the text) to make your text stand out more when placed on top of video.** This should make reading your subtitles easier for the viewer. (For example, select Border and specify a width of 2 pixels by typing it in the text box.)

7. **Next, you may select a color for your Border by clicking the drop-down box to the right in the Rendering Options window.** Choose a color that highlights your text for easy reading. (See Figure 7-13.) For example, select Black from the drop-down list. Highlighting your text with a black border should make it easier to read when the text falls on white (or generally brighter) scenes in your movie. How many times have you missed reading lines of subtitles in a movie because they were shown against a white wall or other bright object in a movie? We are trying to avoid this problem by using borders and shadows.

Figure 7-13: Select an option, such as Black, for the color of your subtitle text's border or shadow.

8. **If you prefer to view your subtitles over a background other than black while editing, click the Preview Window Color button in the Project Settings window.** Choose a color by using any one of the color picker tools in the window. The color that you select is for preview purposes only, and it does not appear in the rendered subtitles that you export from the Subtitle Editor. Most people stick with the default black, but there may be times when you want to see what your text looks like placed over other colors.

9. **After all of your project settings have been made, click the OK button at the bottom of the Project Settings window to proceed to the Subtitle Editor interface.** After the Subtitle Editor opens you are presented with three windows.

10. **In the QuickTime Preview window, scroll to a point in the movie where you want to place your first subtitle.** Navigating in the Preview window should be familiar if you have used video editors such as Final Cut Pro or even if you have only used the ordinary QuickTime player. If there is a specific point in the movie which you have the time code for, enter it in the box in the upper-left corner, and the movie should automatically advance to that point.

11. **Click the New button in the Subtitle window to create a new subtitle stream.** You may also choose New Subtitle from the Subtitles menu at the top.

If you want to add additional subtitles continue clicking the new button. (See Figure 7-14.)

Note: A maximum of 32 subtitle or graphic overlays may be added to a DVD project. Even so, a maximum of 16 languages can be set for a DVD project. If you want to use more than 16 subtitle language streams (however unlikely this may be), the remaining subtitles are unlabeled and cannot be assigned as languages.

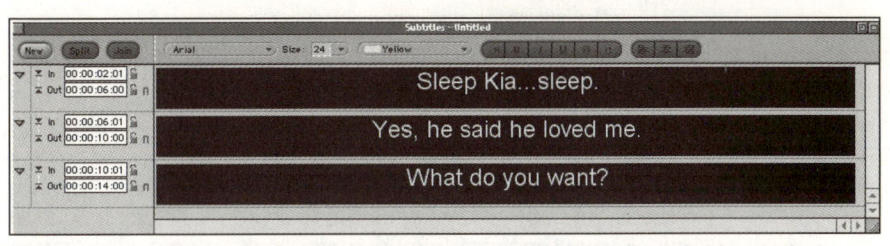

Figure 7-14: You can work with more than one subtitle in the Subtitle window.

12. **Double-click the black line in your new subtitle or click the triangle on the far left to open the subtitle for editing.**
13. **Begin typing your text for the subtitle.** The position of the flashing cursor indicates whether your text is justified to the left, right, or centered. The majority of subtitles are centered, so make sure to click the center button in the bar at the top of the Subtitle window. (See Figure 7-15.)

Figure 7-15: The bar at the top of the Subtitle window offers options for font choices and alignment of text.

Figure 7-16: In addition to typing values into the time code boxes, in and out points can be locked to ensure that subtitles always appear at a certain point.

14. **Now set the duration of your subtitle by using one of two methods.**

    - **Method 1:** Type new in and out points into the time code boxes in the Subtitle window to the left of your subtitle text. Assuming that your in point has already been set by creating a new subtitle, scroll through your movie to find the time when you want the subtitle to go away. Look at the time code box in the upper-left corner of the Preview window and make note of the value. Type this value into the out time code box in the Subtitle window. A duration has now been set for the subtitle.

    - **Method 2:** Scroll to the in and out points and then select In and Out, respectively, from the Subtitle menu. An additional lock feature is available to the right of these time code boxes. By clicking the lock icon or by selecting Lock from the Subtitle menu, the subtitle is set to always appear at that exact point regardless of any other changes made to other subtitles or time codes. (See Figure 7-16.) If you are certain that this value should not change click the lock icon. Clicking it again or choosing Unlock from the Subtitle menu unlocks it. You can also move the subtitle to different times by selecting Move from the Subtitle menu and then choosing a time code value that is Absolute or Relative. *Absolute* values correspond directly to specific time codes in the movie. (An absolute value of 11:21 is 11 seconds and 21 frames from the beginning of the movie.) (See Figure 7-17.) *relative* values are the duration from the point you start. (If you start at 5:05 in the movie and select a relative value of 11:21, you end up at 16:26.)

    Figure 7-17: An absolute value of 00:00:11:21 corresponds to 11 seconds and 21 frames from the beginning of the movie.

    Note
    If users are viewing your DVD project with the subtitle feature activated, and they press fast forward or rewind on their DVD players, they may experience a lag before the subtitle for that scene appears. This aberration occurs if your subtitles for a scene have been set for a long interval. Your player is looking for key frames to display the subtitles, but when you land in the middle of a scene by fast forwarding or rewinding, the player does not automatically look back to see where a key frame may have occurred. Instead, the player begins the subtitles at the next key frame it encounters while moving forward, which creates the apparent lag. To remedy this problem, repeat your subtitles for long scenes at short intervals, such as every five seconds.

15. **When you are finished creating all the subtitles for your movie, it is time to compile the subtitles into a stream that can be read by DVD Studio Pro.** Choose Compile Project from the File menu and select a location to save the .SPU file.

Chapter 7 ✦ Using Subtitles   167

The creation of subtitles is now complete! You next step is to import the subtitle into a DVD Studio Pro project.

Importing subtitles

To import your subtitles into DVD Studio Pro, follow these steps:

1. **Open DVD Studio Pro and create a new Track tile by clicking the Add Track button located at the bottom of the Graphical View.**
2. **Choose File ➪ Import and select the subtitle file you have created from its location on you computer.** Click Add and then Import to bring the file into DVD Studio Pro.
3. **Click the subtitle icon on the Track tile to open a Subtitle Streams window.** This window displays any available subtitle assets when you have assigned them to this track. (See Figure 7-18.)

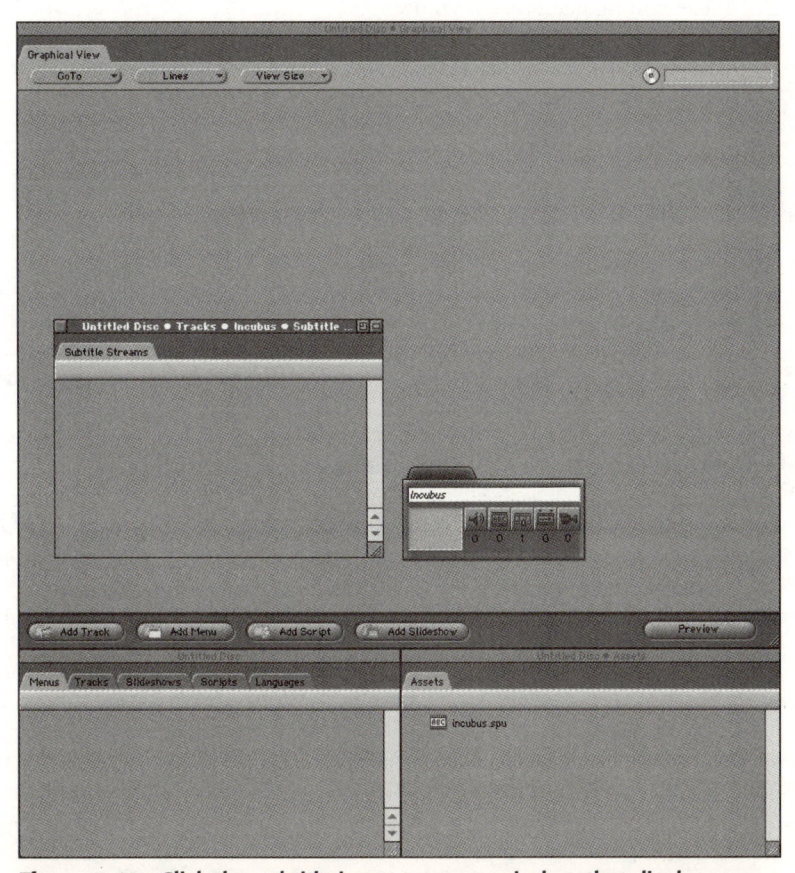

Figure 7-18: Click the subtitle icon to open a window that displays your available subtitle streams when you have asigned them to a track.

4. **Choose Item ⇨ New Subtitle to create a new subtitle assignment in the Subtitle Streams window.** Name the subtitle in the Property Inspector. (See Figure 7-19.) You probably want to name this subtitle the same as the one you are importing.

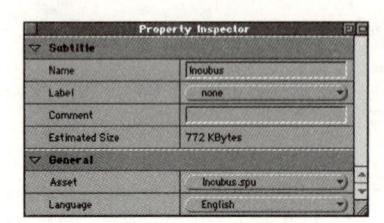

Figure 7-19: Adjust settings for your subtitle in the Property Inspector window.

5. **In the Property Inspector, select your subtitle from the Asset drop-down menu to assign it to the subtitle you have just created in the Subtitle Streams window.** Also, select the language that corresponds to the subtitle from the Language drop-down menu.

Your subtitle has now been added to the DVD Studio Pro video track! In the next section, we discuss how to position your subtitle text correctly within a video frame.

Positioning subtitle text in a frame

You can use a few options to position text in the Subtitle Editor for positioning text. The first option you encounter is in the Project Settings window. In this window, you are given the option of creating margins to define the subtitle area. The values that you place in all four boxes correspond to pixels from the edge of the frame. Only one setting can be made for the entire subtitle project, so choose carefully. A common setting that places text close to the bottom of a frame (the way most foreign films are subtitled) is top=400 pixels, left, right, and bottom=25 pixels for the NTSC video standard and top=450, left, right, and bottom=35 for PAL.

Depending on the video, you may want to work with different values. Feel free to experiment with the margins. Just remember not to go too close to the edges, as doing so may cause some text to fall outside the video safe boundaries. One example where you may want to experiment with different values for your margins is when you are working with letterboxed videos. Depending on the aspect ratios being used, you may want to center the text better within the horizontal black bar at the bottom of the frame. If you are looking to produce precisely placed text for a special video, try working out the required pixel margins by first experimenting with still frames in Photoshop. After you know the pixel dimensions you require, return to the Subtitle Editor and try them out.

Text may also be aligned left, right, or center by clicking on the options in the bar at the top of the Subtitle window (these appear similar to equivalent options in a word processor). Also, breaking lines of text manually can be accomplished by pressing return when typing your subtitles. This method is not very precise for spacing, but if you want to separate lines of text for some reason, it does work. Consider using this method (pressing return to create spaces or paragraphs) combined with the left, right, and center alignment options for placing text next to a speaker in a movie. I have seen examples of text aligned in relation to the speaker in some animated DVD titles, where who was speaking wasn't apparent, and this method of placing subtitles made following the story considerably easier. Another way to help avoid confusion when multiple speakers are present is to use a hyphen at the beginning of each line of dialog. This provides a clue to the reader that the text belongs to different people.

> **Tip** Aesthetically, limiting subtitle text to two lines is preferable. Reading dialogue is much easier for a viewer and reduces the chance of covering video in a scene.

Working with Fonts

When working with fonts for video, it is particularly important to be aware of issues that may affect the quality and readability of the text. Because almost any video program, particularly one with subtitles, makes use of text at some point, you should be concerned with the creation of type elements and make them better for the viewer if possible.

Selecting fonts and styles for subtitles

In the Subtitle Editor, size, color, and styles of fonts can be chosen in two ways. One way is to select them from the bar at the top of the Subtitle window by clicking the drop-down menus and buttons corresponding to the change you want to take place. You can also select the same options from the Text menu at the top. One of the great features of subtitles on a DVD is that any font or character set may be used. All fonts that you have on your system are available for use. This feature is particularly useful for creating text with non-western character sets, such as Japanese or Chinese.

Considering fonts and the user experience

Choose a sans serif font (fonts without "feet") for your subtitles and avoid the use of serif fonts (fonts with "feet") if possible, or you may notice a flickering effect around your letters on a television screen. This undesirable effect is caused by the interlacing of lines on a television screen, producing considerable difficulty when trying to display horizontal lines composed of a single pixel width. Sans serif fonts (such as Arial) are generally easier to read, as well. Also, make sure to set the size to at least 24 for maximum readability. You can set the size smaller, but again, you run the risk of flickering due to small lines, not to mention eyestrain for viewers.

One of the most common problems with subtitles results when white-colored text is placed over white-colored video, causing the text to become unreadable. To prevent this problem from happening, place a border or shadow around your text by clicking the corresponding buttons at the top of the Subtitle window or by selecting the option from the Text menu at top. Black is commonly used as a border, along with yellow.

In the next section, we discuss what to do with subtitles that were not generated in the Subtitle Editor.

Importing External Subtitles

Although the Subtitle Editor provided by DVD Studio Pro is a great application for creating subtitles, it is not always the best option for every project. In addition to the Subtitle Editor, a couple other methods for working with subtitles are compliant with DVD Studio Pro projects.

Acquiring subtitles created by another Subtitle Editor

Although the Subtitle Editor included with DVD Studio Pro is extremely useful for creating subtitles, you may find that another program works better for you or simply adds more options. Another possibility is that you have old subtitles created by some other means, which you want to use in the Subtitle Editor. In either case, you can import subtitles by choosing Import ➪ Subtitles from the File menu in the Subtitle Editor. Subtitles may be imported from a text file such as one generated by an EDL (Edit Decision List) from a video editing application. If you choose to import subtitles from a text file, make sure that there are three fields for each subtitle, which specify the subtitle's start time code, end time code, and the subtitle text. For those familiar with HTML, by inserting the <p> paragraph tag, you can create line breaks in the subtitle text.

Obtaining subtitles from a service bureau

If you choose to have your subtitles done by a professional subtitling company you need to be aware of some considerations. Of course, with the exception of foreign language translations, why would you need anyone else to do your subtitles when you have a perfectly capable Subtitle Editor included with DVD Studio Pro? Very long programs requiring a great deal of transcription and translation may suggest this option as a possibility. When you approach a company about their subtitling services, make sure that they can produce subtitles with the .SPU extension or at least can produce plain text files as described in the previous section. Make sure that you give them as many notes and specific requests as you can about the placement and style of your text to avoid more work and confusion for yourself when you go to import them into DVD Studio Pro or the Subtitle Editor.

Summary

- ✦ The Subtitle Editor allows you to create text for projects involving multiple languages, narration, or graphic overlays.
- ✦ The Subtitle Editor is included with DVD Studio Pro as a separate program that makes the creation of subtitles easy to manage.
- ✦ The Subtitle Editor is primarily composed of three windows (Preview, Marker, and Subtitle) where you create and work with your subtitles.
- ✦ Key frames, or in and out points marked in your timeline, are used to introduce subtitles in your movie.
- ✦ The position of subtitles in a video frame is important for the readability of text.
- ✦ Font style, size, and color are a few of the most important factors that contribute to how your text is viewed and understood.
- ✦ Subtitles can be created in other programs or created by subtitling services and then imported into DVD Studio Pro.

In the next chapter, we explore the concepts and principles pertaining to interface design. Through the discussion of interface design, we hope to illuminate some factors that can help you to create a more enjoyable and functional user experience.

✦ ✦ ✦

CHAPTER 8

Exploring Interface Design

When you purchase a DVD videodisc from a store, you are probably buying it because you like the movie, music, or other element of content on the disc. What you may not realize until you put the disc into your DVD player is that good interface design contributes greatly to a satisfying viewing experience. Without an interface that engages the viewer, or is at least logically planned, people might not be so eager to buy these shiny, round things we call DVDs.

At the onset of a DVD project, it is extremely important to consider the flow and arrangement of material on a disc as well as the manner in which users will access the content. Interface design is the key to unlocking a disc's potential. A disc's design leads the user through a world of choices, hopefully reflecting the spirit and atmosphere of the video and audio content in a seamless manner. This chapter explores how you can use interface design to improve your project's appeal and usability.

Understanding the Principles of Interface Design

The design of an electronic interface determines our experience of the material being presented. How many times have you checked your e-mail or visited a Web site in the past 24 hours? Most likely, in that time, you interacted with a lot of different material on many different levels and invariably came in contact with several unique interfaces. Of course, the Web has its own set of design rules, which most sites (for better or worse) adhere to. Still, the variations in experience are enough to give us a sense of what is right and what is wrong with interface designs. From our experience, two of the most important elements of any interface design are the enjoyability and the usability it affords the user.

In This Chapter

Investigating design principles for DVD interfaces

Evaluating the enjoyability and usability of an interface

Critiquing DVD interfaces

Considering the technical limitations of DVD design

Looking at third-party applications for design solutions

Enjoyability

The user's capacity to enjoy or feel involved with an interface often determines the success or failure of an interface design. If you are unable to gain a user's interest, apart from the usability factor, then you have already lost the war. Thinking in terms of what users want from their interactive experience is important if you want to sell them on your idea — without a good incentive to use an interface, users may choose not to return as often to your disc. Begin a DVD project by thinking about what you might want from an interface and you should soon see that providing an interface with at least a few considerations for enjoyability is the beginning of a good design.

Rewards

It is often nice when a user is rewarded for doing something right. This could include scanning through a 3D animated environment with the remote-control arrows and being pleasantly surprised by finding a door or other item that is highlighted — an indication that more material or experiences await the user. A transitional animation between menus or disc elements is also an effective way to engage a user's attention. For example, by clicking on a door, the user may be whisked though the door and out the other side where the next menu or video clip awaits. *The Abyss* DVD is comprised of such an environment. When users click an object, such as a door, they are led to another menu or video clip with a satisfying animation or sound effect. This reinforces participation with the users, which is highly desirable if you want them to continue using or searching through a disc. Good interactivity is about providing incentives for a user. In Web terminology, this may be considered creating *sticky* pages, which hold a user's attention, keeping them on your site (or, in the case of DVD, a disc) for longer periods of time.

Unfortunately, DVDs and other menu-driven media platforms are largely considered to be interactive in a passive sense, meaning they fail to involve a user in the same manner as video games, which provide a level of true immersive interactivity. While this is true in most cases, the use of appropriate graphics and high quality animations may enable a user to experience a great degree of immersion with DVDs — which is different but equal to many games, such as the ever-popular *Myst* computer game series.

Truly successful interfaces go beyond the interface itself and enable users to feel as of they are interacting with real objects or environments. In most cases, this involves choosing objects or backgrounds that pull viewers into the mood or aesthetic of the movie or video featured on the disc. For example, if your movie is about a dentist gone mad, you might want to create an interface with an open mouth graphic or animation and use the teeth or dental instruments as buttons to activate other menus or video content. If it is a nature video, you might want to add three-dimensional trees, flowers, or animals as buttons that are set against a subtly moving background.

Easter eggs

Giving the user something to search for, while not detracting from the usability, is a popular gimmick (if that is the correct word) to increase interest in a disc. *Easter eggs* are a popular way to add something unique apart from the main content, while encouraging users to thoroughly explore a disc — especially complex projects with a number of nested menus. Essentially, Easter eggs on a DVD are bonus items, such as video clips, photos, or other elements, that are hidden somewhere on the disc. You can find them in menus by clicking on hard-to-find buttons, which are highlighted by paging left, right, up, or down on a menu where you otherwise would not try to find a button. *The Matrix* DVD provides easter eggs through a metaphor of a white rabbit and pills, which, when displayed briefly as overlays on the screen or in menus, link the viewer to behind-the-scenes video content. In the *Terminator 2: Judgment Day* Ultimate Edition DVD, viewers can find an Easter egg by punching in a special numerical code on a menu screen, unlocking an extended version of the movie. Easter eggs, or giving users something to search for, can also be a fun way to add material that would otherwise not make its way onto a disc.

Usability

One of the most important elements to remember when designing an interface is to establish clear rules for navigating the disc. You may decide to make the options very clear, such as a button that says, "click here" or an arrow that says "next page." Or, you may let users explore the environment and present logical puzzles for them to solve. Whether you choose to challenge users or not, consistency in navigation is crucial to being able to effectively use an interface. Try to give users clues that enable them to make informed decisions about which button to choose.

Providing users with practical feedback is also important. For instance, if they are on a page searching a scene index, they need to know which scenes they are viewing with either numbers, letters, or names to identify them. If users cannot find the information they want within a short period of time, most users become frustrated and may even give up. Remember, you should be able to challenge users without testing their patience. For the most part, being straightforward about an interface is the best method. Avoid confusion at all costs, particularly if the disc is dense with a lot of material and menus to navigate.

Suggestions for Designing DVD Interfaces

Consider using still frames or short clips from your main video program as backgrounds for menu interfaces. This is an easy way to obtain material that meshes with the other content on a disc. After extracting the sequence or picture you want to use, apply effects or composite other elements onto it to make an interesting background. For example, you may have a picture of a skyline that you have taken

Interface Design Considerations By Alan Martin

Bio: Alan Martin has been designing video games since 1989. Some of his titles include "Taz-Mania" (SNES), "Superman" and "NHL All-Star Hockey '95" (Genesis). The co-creator and designer of the "Mutant League" sports series (and ensuing animated series), Alan's most recent works include "Looney Tunes Space Race" for the Dreamcast and Infogrames' upcoming mascot game. His full resume of games can be found at www.alansgames.com.

When designing an interface for a DVD (or for any interactive product), there are two main areas of consideration — ease of navigation and aesthetics. Depending on who the target audience is, one of these areas may get a heavier emphasis than the other. (In my experience designing video games, there is usually a healthy balance between the two; in designing Web site interfaces the emphasis is usually on the navigation, as bandwidth issues still prevent many Web users from viewing high-end graphics.)

A successfully designed interface should be first and foremost simple to navigate. The main interface is usually the first (and most frequent) controllable area that the user interacts with, so it is important that the selections are clear and the control is intuitive. Having a beautifully designed interface that accurately reflects the style and theme of the project is ideal, but if the user can't access a feature he's looking for, the design is a failure. (That is unless a feature is specifically designed as something to be searched for – like an Easter Egg.)

Another important aspect to consider in the design of an interface is any "standards" that a particular medium utilizes. While this may not yet be as important an issue for DVDs yet, Web sites and individual video game console makers have quickly established standards (like left-side or upper frame Web navigation). In some cases with game consoles, these standards have become mandatory for approval on the platform (the A button advances screens for N64 games, the X button for PS2 games).

The aesthetics of an interface is what can make it stand out from the rest (and sometimes, in the case of DVD movies, even stand out from the actual content!). A visually well-designed interface should reflect the style and thematic elements of the content. Truly great interfaces are so intertwined with the feeling of the product that the user doesn't even feel like he's "left" the content when accessing the sub-screens.

For video games, this is something you really shoot for in the design. In one action game I designed, the Pause Menu and game options all appeared on a PDA type of device that the player character actually held in his hands. (You could even see his fingers around the device.) As the character took "damage" (and the PDA was theoretically jostled around), the player would lose access to some of the maps and item displays on this PDA. (An added treat would have had drops of blood falling onto the PDA screen as the character was near death.) In the near future, game consoles will be able to achieve examples like this, as well as having the ability to include DVD-quality movies within the PDA's mini-screen during game play.

With the DVD platform, load times must also be taken into consideration when designing aesthetics. While it is ideal to use interfaces as linking devices to the exciting (and large) content that DVDs can provide, it's frustrating for the user to wait endlessly between menu screens while "unnecessary" content loads. (Unnecessary not to the designer, but to the end-user.)

Chapter 8 ✦ **Exploring Interface Design** 177

The latest in the video game home consoles, the PlayStation2 (PS2) and the X-Box (due for release in November 2001), both have the ability to not only play video games but music CDs and DVDs. And while they won't quite compare to the features of high-end DVD players, the X-Box plans to have full DVD video and audio support, including HDTV. The GameCube (also due in November 2001) will not have the ability to play DVDs; however, Matsushita plans to release a combo DVD-player / GameCube at a higher price point sometime in 2002 in the US. (The combo unit's price should be comparable to the PS2 and X-Box.)

In the next few years, as game developers learn to master the "next generation" consoles and all their capabilities, designers will find new and exciting ways of blending the DVD medium into the interactive gaming experience. One of the areas with the most potential is the ability to truly tell a story and immerse the user in the interactive experience – an area that is frequently lacking from most video games today.

from a video. (In our example, shown in Figure 8-1, the skyline is of Chicago.) Export the still from your video editor to an application, such as After Effects, or use the built-in effects provided by Apple's Final Cut Pro (which is also capable of using After Effects plug-ins). Apply various video filters and export to a graphic file.

Figure 8-1: Using Final Cut Pro and After Effects plug-ins, such as the glow effect provided by Digieffects Delirium, produces interesting images for interface elements or backgrounds.

Interface Design: Creating An Experience
By Justin Kuzmanich

Bio: Justin Kuzmanich, Creative Director of ad2, Inc. reports that many of the Web sites and ROMs created at ad2 deal directly with films. When designing an interface for a film project, the design must allow the user to seamlessly interact with the experience. After the user is comfortable within the interface, we can focus on delivering an experience that extends the message of the film.

Movies provide a temporary form of escape. They take us on a journey, transporting the audience into the subject matter. You sit with others in a dark room totally engrossed in the events the director has put on the screen. We as designers, like the director, have the power to take the users on a journey. The first step, is making sure that the user has a solid interface to navigate through the experience. By building an effective interface we can create experiences where the user not only becomes the controller, but also the creator. This happens when real time feedback from the user, by way of the interface, is incorporated into the experience. When this happens, the user feels as if he or she is helping to effect and potentially create the experience. The demand from users for a more active role in experiences will give birth to new forms of entertainment where observers will become participants. It's going to be fun over the next few years discovering and creating new interfaces that allow users to have a deeper and more personalized level of control.

What are the qualities that lead to an effective interface?

- It should be compelling. Users should be drawn in, totally engrossed in the subject matter.
- It should be as invisible as possible to the user. The interface should allow for smooth transitions from subject to subject. If users are comfortable within the construct of the experience they are more apt to be engrossed in the subject rather than aware of its interface.
- It should be communicative. Users need to know where they are within the experience at all times. Disoriented users tend to leave quickly
- It should provide a high level of feedback. The more we show the user that they are affecting the experience the closer they will be drawn to it.
- It should reflect the emotional tone of the experience. The way the interface looks, sounds, and feels should hint back at the overall experience, even adding to it in a subtle way.
- Lastly, it should be personal. If we can let the users see their input reflected within the interface, they will feel a deeper sense of involvement with the experience. This involvement can lead to a sense of ownership or membership. If personalization can be incorporated into the interface, the user can in effect become a part of the experience.

To create exciting and compelling interface designs, you not only need the passion but you also need the right tools. One of the tools we use at ad2 to create effective interfaces is Flash from Macromedia. Using Flash, we create experiences, but with filmatic qualities by weaving sound, imagery, and motion to tell a story, much like a director would with the

Chapter 8 ✦ Exploring Interface Design 179

added bonus of interactivity currently not possible in film. When it was first released, developers quickly embraced the potential of this new technology because of its ability to create exciting animations and sounds in file sizes small enough to be shared across the Internet. Today, we see Flash being used for interface design across a wide variety of platforms in addition to the Web, including CD and DVD-ROM, wireless devices, and gaming platforms. The future is anyone's guess, but if you have the right passion and the right tools, you'll be ready to meet the challenge head on.

For still menus, you can use Adobe Photoshop to accurately edit and enhance your stills (see Figure 8-2 for a example of what can be done with an image). If you had a still of a skyline, you could remove the sky and, perhaps, add a glow to the buildings. Composite your button elements onto the picture along with any other material or text and you have an interface, ready to use. You may also try extracting a portion of the picture to use as an interesting background element. Creating montages like this can produce interesting results.

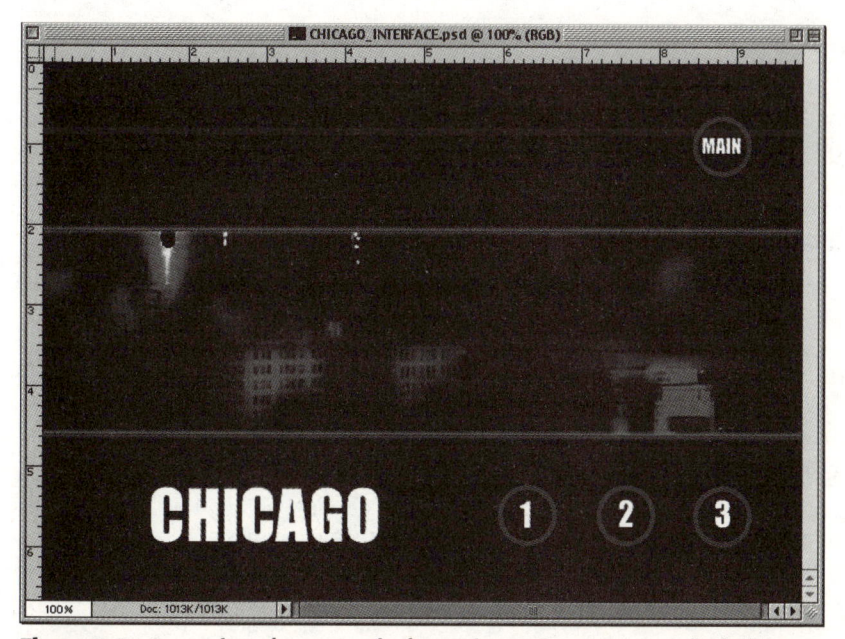

Figure 8-2: Removing elements of a larger image is one way to find elements for DVD interfaces.

Try using real-world objects or backgrounds that might be associated with the content you are going to view on the disc. For instance, if the movie takes place in a coffee shop, try simulating that environment as your background. By adding even a few subtle elements, such as steam rising from a cup of coffee or a neon sign flashing

Designing for DVD By Dan Fenster

Bio: Dan Fenster has been designing still and motion graphics as a freelancer since 1996 and recommends it to anyone who loves puzzles and people as much as he does (`hiwidesign@aol.com`).

As with any task, in order to do it well, you need to be familiar with its parameters. DVD menu design and production can be addressed from both a mechanical standpoint and a creative one.

Creating menus and animations for use on a DVD definitely uses both sides of the brain. One side generates the idea behind the information delivery, and what it looks like (the "style" of the disc's interface), while the other keeps tabs on where all the little bits go so that they can all be "authored" successfully, in my case, by another individual. Communication between the designer and author needs to be very clear, as the designer is basically making the puzzle pieces that the author is going to have to put together.

This kind of work method requires me to be a realistic artist. (Is there really such a thing?) I really have to know what the limits to a given job are. This balance of ideals and hard numbers is probably the reason why it is so hard to find designers and authors who are capable of doing this scope of work well. The designer should be at once an artist and producer, constantly looking at the client's needs and the DVD's appropriate representation.

When I start to consider doing a job for a client, the producer that I will be working with has usually handed me a menu flow. This document outlines the menu's contents and navigation for that particular job. This step is extremely helpful, and not always possible to produce in the timely fashion that they are needed, so good communication with the producer is of great value. After all, what good are all of my efforts if what I design is not what the client wanted, or if the time I spent creating my design elements went way over budget.

After digesting the shows contents, be it a musical, movie or episodic DVD, I start thinking about what people who go out and buy these discs are likely to want to see in the menu's interface. How is the feature styled? How does my client feel about the look of the menus? This phase of the DVD interface design is very important, so I carefully choose fonts and colors that not only are NTSC safe and friendly, but also will match the feature's mood and not betray its significance.

This is where the focus begins to shift over from the nebulous, to the more technical details that greatly concern the author of the disc. If a client's budget allows, I will animate transitions from one menu to another. These are created as small-animated clips that are generally 20 frames to several seconds in length. These clips must match the in and out frames of the surrounding menus, and/or animations. This is no small feat considering the desirability of seamless joins between these elements, and their need to loop. I always think these details out far in advance before I unleash the digital powers of Adobe products onto them.

The design I did on behalf of Crush Digital for Whitney Houston's Greatest Hits DVD, a job which required me to composite green screen footage of the performer, animate transitions and integrate the album's design concept, ended in success, by homogenizing various pictures, text, and moving video into the final product.

Starting with only pencil and paper, I worked out logical and attractive proportions between Whitney's image and the text elements that were to be the menu choices. Timings for animating these design elements on and off, while ruminating over the actual look and execution of the piece was an enjoyable challenge. All of these variables were considered to some degree before a computer was even turned on.

Of course there were interlacing and compression issues with the footage, design and color issues to be worked out with the client, but thanks to good planning and open communication with the producer and author, the disc was finished successfully and won a DVDA award to boot! A word of advice here: never be shy in asking for exactly what you need for the job to turn out well. If I didn't voice my concerns regarding the digital transfers or an unsuitable idea here or there, these would have been my problems to correct. Speak up if you see a troublesome aspect of your job.

I've been designing for broadcast since 1989. Without a doubt, it is the extent of my experience with video and design that enables me to consider many of these factors at one time. It is this experience that I constantly use to quickly and accurately make decisions about how a DVD will be designed and created within a given budget. Also as a freelance designer, it requires having a good, raw understanding of how to apply years of accumulated knowledge into one DVD product that should be easily usable by the viewer.

After all, they are the ones who will ultimately be irritated by the ill-applied efforts of a menu designer that ruined their Saturday night entertainment because they couldn't get the feature and all its extras to play.

outside the window, you can create the basis for your navigation. From there, you may choose to highlight an item, like a jukebox or cash register, as a button to provide links to items on a disc. You could complete the entire picture in After Effects, or start by designing a 3D environment in an application, such as Electric Image or LightWave. Editing a background image in Photoshop and then using After Effects to add elements, such as steam, through the use of plug-ins, is another popular route. Whatever your level of expertise or skill, there is usually a logical solution to your design problem close at hand. Of course, some may produce more impressive results than others. However, through experimentation, often you can arrive at conclusions that set your project apart and make it truly unique.

Evaluating DVD Interface Designs

Commercial DVDs produced by major studios are one of the best places to get inspiration, because they represent a wide variety of quality interfaces and indicate what the industry expectations are for production values. As you can see by looking at even a few titles, standard conventions are employed to design most commercial DVD interfaces. The more popular a title, whether judged by box office receipts or cult status, the better the interface. *Terminator 2: Judgment Day* (the Ultimate Edition version) has one of the best DVD interfaces around with animations, graphics, sound effects and layers of complex interactive options that are all top notch.

> ### Designing Interfaces for DVD versus DVD-ROM
>
> The major difference between creating interfaces for DVD and DVD-ROM is that DVD-ROM provides the potential for a virtually limitless degree of interactivity. With DVD-ROM, anything that you can think of or create for a disc-based game or the Web is possible. This is in contrast to DVD videodiscs, which may only contain certain types of material and relatively limited interactive options. Some of the features you might see on a DVD-ROM and not on a DVD videodisc are windows that simultaneously display a screenplay and trivia while the movie plays back in another resizable window. Also, complex games can be created that even link to Web sites to share information. If you can design it, it can be put on a DVD-ROM. Popular interactive design software, such as Macromedia Director, can be used to create high-impact experiences for users. Once you know the type of interface you want to create, choose the programs that can help you turn your idea into reality.

Although complex interfaces are wonderful to explore, some of the best interfaces are relatively simple and don't necessarily include 3D animation or layers of complex scripts. Often the best interface designs contain no more than a clever concept or an interesting link to the material on a disc. Flashy interfaces can be great, but they aren't always the most appropriate way to access material. An example of a great interface concept is the *Requiem for a Dream* DVD. The interface is a convincing mock infomercial video screen, with flashy "buy now" style graphics and clever sound design, such as an announcer's voice and phones ringing in the background to complete the effect. It's so convincing that, upon first playing the disc, you might be fooled into thinking that you accidentally hit the TV button on your remote control.

When DVDs were first produced a few years ago, it was common to see still menus with barely any special navigation or animation features. Usually, DVD interfaces were (and still are, in some cases) comprised of a simple static background image with a few text buttons and highlights. Today, it is not uncommon to see even the most low budget film treated to at least some animation or specially designed DVD interfaces. Hopefully this is a sign that studios are trying to do more for consumers by increasing interest with quality products. More likely, it is due to the increase in knowledgeable disc authors and designers having access to better tools. Also, consumer expectations have increased over time, requiring production companies to work even harder at producing well designed, feature packed discs. Whatever the reason for high quality interfaces, it raises the bar of excellence for those just getting into the field of DVD design. This should actually encourage new designers, because it indicates a growing interest in design and an appreciation for their art form.

Technical Considerations When Designing for DVD

When designing for DVD, remember that your medium is television, and design accordingly. Television introduces a variety of special considerations, such as aspect ratios, title, and action safe areas, as well as color and resolution issues not encountered in designing for personal computers. (Chapter 11 discusses many of these issues in detail.)

One important consideration is the use of screen space. Remember to design primarily within the title safe area of the screen. Of course, background graphics can venture all the way out to the edge if they do not contain more important information, such as buttons. Keep your buttons within these limits and you should be fine. Also, if you are creating graphics with thin lines, such as borders or text, remember to make them at least 2 pixels high to avoid the flickering that occurs with interlaced video.

Finally, make sure that the colors you use are safe for either the NTSC or PAL video standard. When creating video in After Effects, determining safe colors is easy. Select Effect ⇨ Video ⇨ Broadcast Colors and adjust the luminance and saturation, or choose Key Out Unsafe from the pull-down menu (See Figure 8-3).

Figure 8-3: After Effects enables you to select colors that are safe for broadcast video.

Design Considerations and Tips for Designing Graphics for Television By Jason Zada

Bio: Jason Zada is the co-founder and creative director of evolution | bureau, a convergence marketing and design studio in San Francisco. His work has been featured in Communication Arts' Interactive Design Annual, Print Magazine's Interactive Design Annual and numerous Internet books and magazines, and he has received A.V. Multimedia Magazine's Top 100 Media Producer Award for 1998 and 2000.

Digital design has been going through nothing short of a revolution with rapidly advancing software and hardware setting expectations higher and deadlines tighter. And with that revolution, designers are encountering a new set of challenges. As televisions and the computer world seem to be converging, there is a huge distinction between designing for screen or for television.

Pixel Shape

One of the first and not-so-obvious distinctions you will run into is the square pixel vs. non-square. On the computer monitor, you have equal-sided pixels. On a television monitor, however, you work with non-square pixels – the vertical and horizontal measurements are not usually equal. The effect, in converting from a computer to television screen, could make the design look slightly squeezed. The problem is easily solved by designing your menus at 720x540, and before exporting the menu to your DVD program or into After Effects, use the Image Size in Photoshop to resize the image to 720x480 (standard DV format). The menu or graphic will appear compressed in Photoshop, but will look fine when you bring the image into After Effects, Final Cut Pro and output it through FireWire to video.

At evolution | bureau, we use a great tool, a Photoshop plug-in called Echo Fire, from Synthetic Aperture, that allows you to preview your work on a video monitor via a FireWire cable. You will, of course, need a DV-to-Analog converter (i.e. the Sony DV-Analog Converter or a DV camera with the S-VIDEO or Composite outputs connecting to a video monitor) to view the image. While working in Photoshop, you can easily do a File ➪ Output to Echo Fire and view your image on a video monitor. This is useful in checking a variety of differences from screen to video, such as color variances, line width issues and buzzing and bleeding.

NTSC and Color Reproduction

Color reproduction on television versus a computer monitor will be another trap you could fall into. We try to never use 100% white on a video monitor, especially for type. We usually keep our whites at around 80-90%. Otherwise, you will find that the white is so strong that it sometimes causes text and other items to start "bleeding." For designers that enjoy designing with bright, fluorescent colors, be forewarned, your video monitor does not like them. And the lower the quality of the television, the worse the colors will look. The key is to keep outputting to a video monitor as much as possible to try and find a workable solution. We generally output to a monitor that can handle S-VIDEO and Composite since they will look noticeably different. In S-Video, your image will look crisp and clean, with colors appearing a bit more vibrant. In Composite, the image will be blurred slightly and the colors will become noticeably more saturated.

Eliminating the Buzz

Small thin lines are one element that will drive your video monitor crazy. The lines will "buzz" and jitter, making them uncomfortable to watch. To easily solve the problem of the lines bleeding and buzzing, use lines on-screen no less then four pixels wide. Using small type is another taboo. The smaller the type, the worse your type will render on television. Without proper kerning of each character, allowing enough definition and distinction between each one, your type will also bleed into itself on-screen. One way to avoid this problem is to use Photoshop 6.0's Layer Style option by double-clicking on any given layer. There are two features that will save you time and energy: the "Drop Shadow" and "Stroke" Layer Styles. When you are trying to set type on a moving image, such as DV footage, a good practice is to pull the type away from the background. This will give your type distinction and will make it more legible on a television or video monitor. Using even a slight "Drop Shadow" will allow you to place any color type on any color background and have some visual separation making the type appear more legible. Using the "Stroke" command and setting your stroke color to either a lighter color if your type is darker and a darker color if your type is lighter will add to its legibility. If your type is a light gray, set the stroke color to the same gray, and use the Color palette to make it slightly darker. We usually set the stroke to one pixel or higher depending on your type and the image against which you will be compositing.

Designing for television is sometimes a tricky beast, but if you are aware of the issues and problems upfront, you will save yourself a lot of headaches during the process. Checking your final menus, graphics or composites on as many different televisions as you are able will also help you understand the different issues that are presented in each situation. A good reality check is to look at your image on a 13-inch television that you can buy at a Goodwill or garage sale. A percentage of the world still is viewing DVD's, commercials, etc. on lower quality screens. If you can make your graphics look good in an imperfect environment, they should look excellent anywhere else.

Case Study: evolution | bureau DVD Project

Evolution | bureau, the convergence and design studio I co-founded, wanted to create a self-promotional collateral piece that could be used to showcase our work for prospective and current clients. We opted to showcase our portfolio and reel on DVD, since that presented our forward thinking process in the most innovative and engaging way. Since our studio has worked on a variety of projects in the past year, this feat was easier said then done. Our challenge was to create a DVD that showed our whole body of work including Flash, HTML-based Web sites, DVD interfaces, video pilots, printed collateral, and more.

A highly overlooked step in DVD production process is the understanding and creation of an information architecture. When creating anything interactive, you have a responsibility to the audience to make your product as usable as possible with as little of a learning curve as possible. We started our DVD project by outlining all of the information we would be presenting on the disc and organizing the pieces into "buckets," or containers, that would hold that information. We decided to break down our DVD into three main categories: Portfolio, Play Reel, and About. This architecture allowed us to put a wide variety of work into our Portfolio, which would easily be the largest section on the disc. Within the Portfolio section we broke things into two sub-sections, Video Work and Interactive Work.

Continued

Continued

To start actual production on the disc, we organized the work into folders, and imported all of the non-video materials into a Photoshop template measuring approximately 720x540 pixels. We output the screen shots through FireWire to check their image on an NTSC monitor. One of the main issues that we had to deal with is the title-safe area of the NTSC output. If we did a full-bleed image in 720x540, the image was cropped on the sides and the top and bottom. The goal was to find a tolerable solution to showcase the image as large as possible without the image cropping at all. By dragging out guides in Photoshop we made a 720x540 template with title-safe guides that helped us place each image perfectly.

After all of the assets were imported and resized into the Photoshop template and all of the video and Flash assets were organized into folders, we moved to the next step of converting our assets to video and compressing them in the MPEG-2 format. We used Final Cut Pro as our primary tool during the entire DVD. Final Cut Pro reads virtually any file and allows you to drag it to the timeline and output to video. We decided to present our Interactive Portfolio with two different options, the first as a slide show with music, the second as a highly interactive, user-controlled slide show. We decided to do the "slide show with music" option in Final Cut Pro, since we could use transitions and dissolves that would sync with music. Even our Flash items were brought into Final Cut Pro to be outputted to video via the MPEG-2 format.

Once every asset was output to MPEG-2, it was time to create a general feeling and design for the disc. We opted for a highly engaging opening, complete with special effects sequences, after-effects sequences and images from throughout the disc. Once the viewer arrived at the Main Menu, we wanted the menus to have motion and sound to help set a tone and a mood. Another objective for the menus was an easy one: make the text large and legible and easy to navigate. Although every project calls for a different plan of attack, we wanted to make the experience of evolution | bureau, quick, elegant and painless.

Creating a unique and engaging experience was our primary goal as well as presenting evolution | bureau as a top convergence marketing and design studio. Since we completed the DVD in under three weeks, with beautiful motion menus, transitions and a high quality showcase of our work, we feel that we have exceeded our goals. The project is a success.

Looking at Third-Party Applications for Compositing, Animation, and Special Effects

If you intend to create motion menus for your DVD project, or even if you just want to liven up an otherwise uninteresting still image, you need an application that can work with layers of video, animation, or customizable effects. Adobe Photoshop is the standard choice for most still menu designs, while Adobe After Effects is the obvious choice for video and animation. Apart from Adobe's strong offerings, Macromedia has a few programs that, while not made specifically for video or film, permit you to easily cross-purpose designs for use in either video or Web applications.

After Effects

Ubiquitous in its use as a tool for video, Adobe After Effects has long been considered the de facto standard by which all other compositing, animation, and special effects applications are measured. With the release of its latest software version, 5.0, After Effects is once again at the top of its class. Aside from a solid video editor like Final Cut Pro, After Effects is, perhaps, the only other video application you may ever need. Of course, After Effects integrates seamlessly with other applications you may use, such as Photoshop and Premiere, using many of the same tools and similar interface elements (see Figure 8-4).

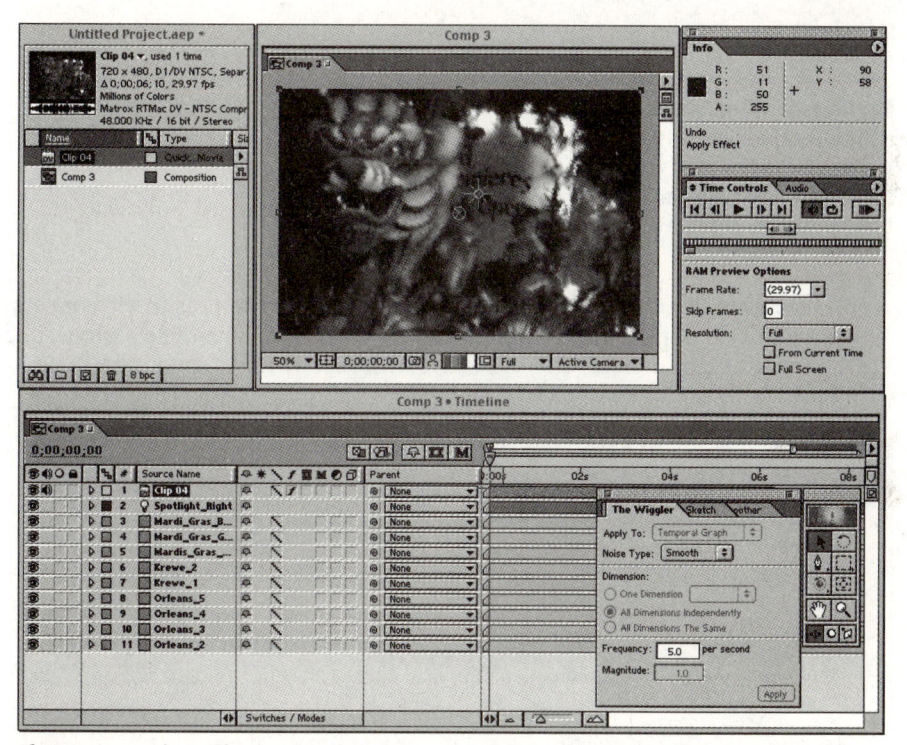

Figure 8-4: After Effects 5.0 offers many capabilities necessary for creating great composites and animations.

After Effects is great for compositing video elements in a motion menu. It also provides unique lighting and 3D capabilities to two-dimensional graphics. In fact, its 3D capability is probably the most exciting addition to version 5.0 (see Figure 8-5).

Figure 8-5: After Effects 5.0 comes equipped with the capability to manipulate 2-dimensional graphics in a 3-dimensional space, while adding lights and other effects, such as this image with two spotlights applied to separate 3D layers.

A unique aspect of After Effects' 3D feature is the use of cameras to view layers (see Figure 8-6). Every aspect of the camera can be customized, just like a physical camera, enabling complex views and matching with real world cinematography.

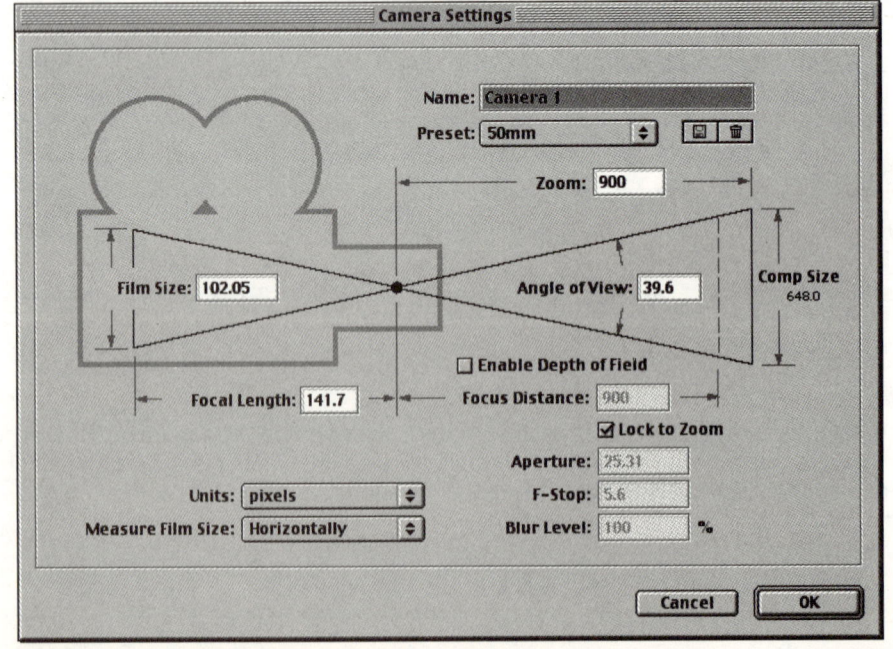

Figure 8-6: Various cameras with unique settings can be used to view a composition with 3-dimensional layers.

Apart from its use in combining layers of video and graphics, After Effects is also known for its special effects, which includes blur, distort, key, paint, and numerous other customizable options (see Figure 8-7).

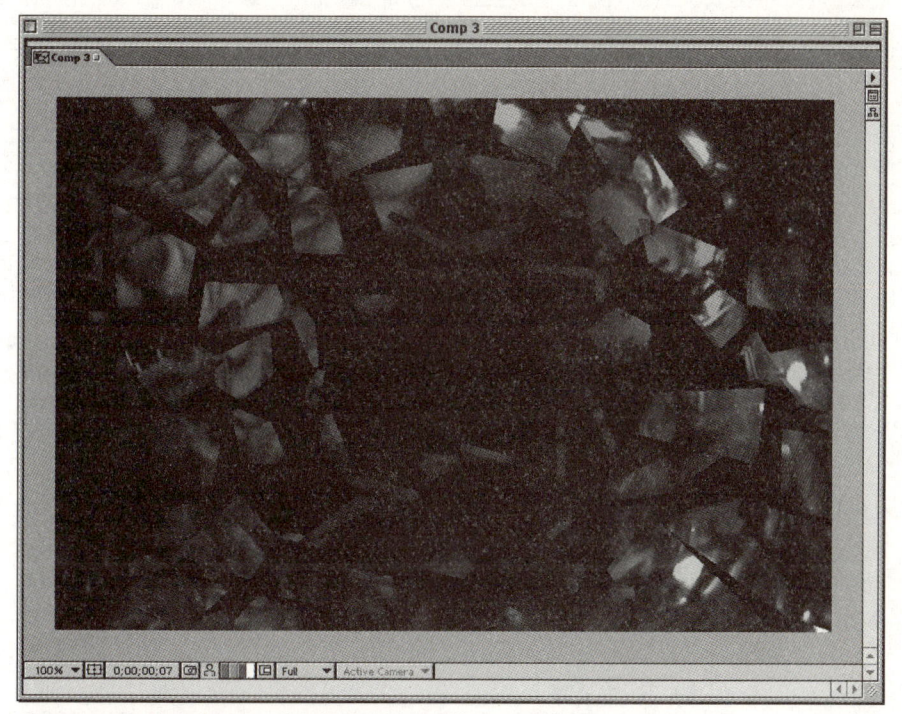

Figure 8-7: The shatter plug-in for After Effects can blow apart a video or graphic layer into different shapes — which is also an interesting way to create a transition between menus and tracks.

CineLook and CineMotion

DigiEffects has produced several After Effects plug-ins that can enhance and stylize the look of your video. Among these plug-ins are CineLook and CineMotion, which produce effects typically associated with film — as it would ordinarily appear (or not so ordinarily) when transferred to video. By applying these film-look effects to your video, you may produce some interesting backgrounds for motion menus that reflect the aesthetic of a particular movie or video. For instance, if you were creating menus for a film about the Wild West in the late 1800s, you might consider adding a scratchy, sepia tone effect to a background video layer, thus evoking a turn-of-the-century feel.

Motion Menu Backgrounds By Brian Dressel

Bio: Brian Dressel has entertained audiences around the world as a video performance artist with OVT Visuals, produced and directed numerous music videos, acted as director of photography and motion control operator for a wide variety of broadcast spots, produced multiple award winning programs and is an accomplished graphic designer.

When thinking about good design work on motion menus, the first DVD releases that come to mind are *Stir Of Echoes*, and *The Matrix*. Both are easily accessible at your local video store and represent some great characteristics in motion menu backgrounds.

The Matrix uses scenes from the movie as a background for the menu overlays. The scenes are composited with binary elements to stylize the footage, differentiating it from the movie. *Stir Of Echoes* uses scenes from the movie, in deep, beautiful montages in its backgrounds. Both of these motion menus represent editing and compositing with very simple menus over the clips.

Motion menu backgrounds are easy to create and enhance using Adobe After Effects. Most of the plug-ins discussed are available as downloadable demos from their respective sources. There are many ways to effect your footage, giving it an otherworldly or enhanced look. You can add a filter, or combination of filters to your clips.

Some of the my favorite plug-ins are:

- **DigiEffects CineLook / FilmDamage.** Allows you to add grain, correct color and add film artifacts like dust, scratches, stains, hair and more.
- **DigiEffects Delirium Schematic Grids.** Creates cool cyber grid effects, when used with transfer controls add a 'tech' element to any footage.
- **DigiEffects Delirium Glower.** Creates a glow around an image with interesting compositing and color effects.
- **Final Effects Light Blast.** Gives the 'flashlight beam through fog' effect.
- **Boris Multitone Mix and Tint-Tritone.** Allows precise, stylized coloring of your clips.
- **Boris Continuum Looper.** These plug-ins are very handy when you want to loop a sequence. It offers many controls for blending the end frame back to the start frame of your clip to give a seamless loop. Based on the tracking speed of current DVD player technology, it may be wise to create longer pre-looped clips. Add a fade in/out at the start/end of your looping clip. This may help avoid the video 'hiccup' associated with a marker change.

Experiment to see what works for your particular footage. Try keyframing the effects to create transitional elements for menu changes.

Utilizing transfer modes can also give your clips an enhanced look. Duplicate the clip in the timeline, apply one of the transfer modes to the top clip to get interesting results. Try using 'Multiply,' 'Hard Light,' 'Soft Light,' 'Screen,' or 'Add' to start. Experiment with variations in opacity levels of both layers, add a 'Gaussian Blur' to the bottom layer, or 'Levels' to adjust contrast.

It is also fun to composite various clips with each other. This can get quite complicated, depending on the clips. Animated titles, stills, graphic elements can all be added to make the clip more interesting. If the DVD is about technology, try compositing a clip of rapidly changing binary numbers, DigiEffects Delirium Schematic Grids, or some 3D wireframes over the clip. If your DVD has the theme of water, composite some looping water caustics over the clips. You get the idea. There are no limitations to what you can do to enhance your clips, giving them new life.

There are many effects to be had simply by playing with the presets, some of which may be nearly perfect "as is" for your video (see Figure 8-8). Choosing among various professional film stock presets may initially produce video with too much grain or contrast, but a few simple adjustments can take care of any flaws. Also, CineLook offers many powerful options for manipulating color (chromamatch), grain (stockmatch), curves and gamma for adjusting the luminance, along with time integration (timematch) for a standard 3:2 pull down effect (refer to Chapter 11 for a discussion of 3:2 pull down and the telecine process).

Figure 8-8: DigiEffects CineLook plug-in provides a great degree of control when attempting to simulate the look of film.

FilmDamage is a special module that comes with CineLook, which allows you to add the authentic look of dirt, scratches, and even hair to your video (see Figure 8-9). This simulates the appearance of film that has been mistreated or well worn.

(Remember those old driver's education movies?) If you have ever wanted your video to look like film that was left in the attic for the past 50 years, then this is the effect for you. Graphic designers may find that using FilmDamage can enhance their title sequences by making the picture appear flawed in unusual ways, such as with scratches and jitters, as in the introduction to the movie *Seven*, and music video directors may use it to switch between media styles and formats, similar to those achieved by such movie makers as Oliver Stone.

Figure 8-9: FilmDamage can make new video look like old, weathered, or severely abused film.

CineMotion is the companion plug-in to CineLook, offering improved controls for achieving the motion associated with telecined film. (Refer to Chapter 11 for a discussion of telecine, 3:2 pull down, and other considerations when transferring film to video.) Although CineLook offers a 3:2 pull down effect, CineMotion gives you better control and less strobe and flickering, which has always been a problem for many film simulators. After Effects offers a 3:2 pull down effect as well, but CineMotion makes adjusting the effect intuitive, accurate, and ultimately painless (see Figure 8-10 for some of the parameters available).

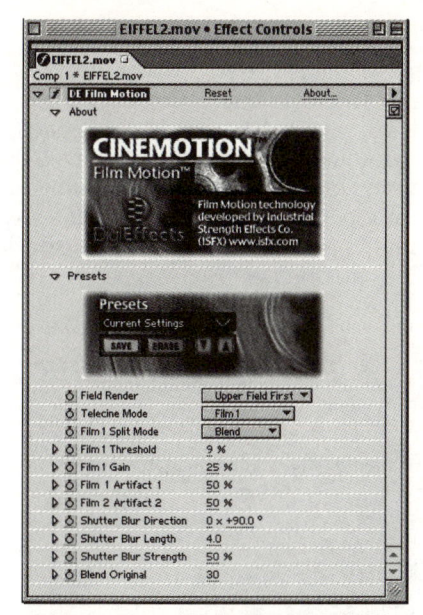

Figure 8-10: DigiEffects' CineMotion, the companion plug-in to CineLook, produces fluid film motion for video.

Plug-ins can produce some very convincing film effects. If you have always wanted to work with film but your budget won't allow it, take a look at the plug-ins offered by DigiEffects (also available as AVX plug-ins for Avid and other non-linear editing systems). Although they are not a substitute for film, they might be all that you need for a low to moderate budget project. The effects are flexible and it's a no-hassle way to get that elusive film look.

However, like many plug-in effects, CineLook and CineMotion can take an incredibly long time to render. An ICE hardware board is one option (albeit expensive) that can speed up the rendering process. Try to budget your rendering time with these plug-ins and use effects judiciously, giving the client what he wants but also giving him a completed project on time. It's easy to get carried away with trying out new parameters. If you have the choice between using Final Cut Pro or After Effects to render your effects, choose After Effects, because Final Cut Pro can take even longer to render. In addition to better rendering times, After Effects has an improved preview window for adjusting settings prior to rendering.

Delirium

Another plug-in offered by DigiEffects that may enhance your video or motion menu is called Delirium. With this plug-in you can create interesting backgrounds by adding special effects such as bubbles, snow, rain, or lens flairs (see Figure 8-11). Consider a few of the other effects you may produce: fire, sparks, fog, clouds, fireworks, fairy dust, schematic grids, framing gradients, visual harmonizer, video malfunction, day for night, retinal bloom, flow motion, grayscaler, sketchist, solarize, smoke, and more. Together, these effects should provide you with enough special effects to create truly exciting textures for DVD motion menus.

Figure 8-11: Adding electrical arcs, lens flares, rain, and fog are a few of the many effects you can produce with DigiEffects' Delirium plug-in for After Effects.

Tip Try using Delirium filters on static graphics to add motion to boring pages, or create animations that act as transitions between menu pages or video tracks (see Figure 8-12). For instance, you can use the bubbles filter to create a fun way of moving from one menu to another by placing it over the current menu and creating a separate video track that can act as an animation to the next screen.

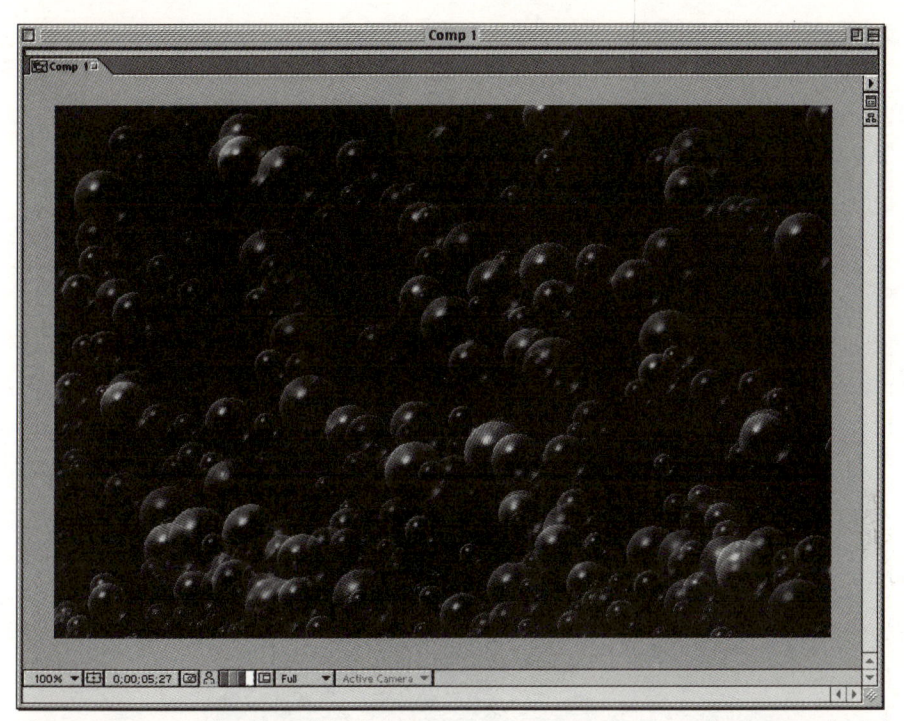

Figure 8-12: Particle effects like bubbles or snow can make menu backgrounds and transitions more interesting.

Flash

Although not typically thought of as a tool for video, Macromedia's Flash is a great application for creating animations that are scalable for video as well as the Web (see Figure 8-13 for an example). In version 5.0, enhancements, such as the Pen tool, new color controls, and improved panels (to name a few), make working in Flash even easier and more enjoyable than before. However, it is Flash's capability to export QuickTime movies that is of interest to video professionals. Exporting to QuickTime means that high quality, vector animations can be created for output to video, without the need for another program such as After Effects. Of course, after a QuickTime file has been created, working with it in other programs — such as After Effects and Final Cut Pro, where elements may be composited and edited — is easy.

Figure 8-13: Macromedia's Flash 5 is a good tool for creating animations that can be exported to QuickTime for use in video applications.

You can, for instance, create an interface for a DVD menu directly in Flash, complete with animations, such as elements that scale (grow smaller or larger) and buttons that glow (see Figure 8-14). Of course, for broadcast video purposes not intended for the Web, you must create a movie without interactivity. If you create a Flash movie with both video and the Internet in mind, you may be able to use it for a Web interface as well. After you have finished creating the interface, simply export it to the QuickTime 4 format.

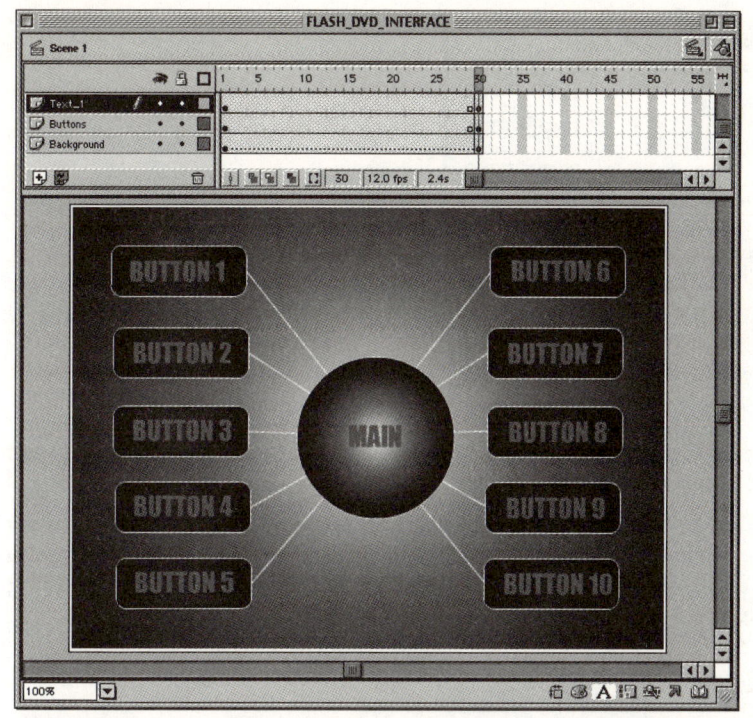

Figure 8-14: Macromedia's Flash 5 can be used to create interfaces for DVD menus as well as for the Web, although the same interactivity does not apply.

To export a Flash movie to QuickTime for video, follow these steps:

1. **After you have created your Flash movie, choose File ➪ Export Movie.**
2. **In the dialog box that appears, choose a location for your new file, give it a name, and choose QuickTime video from the Format drop-down list (see Figure 8-15).** Click Save after you are finished.

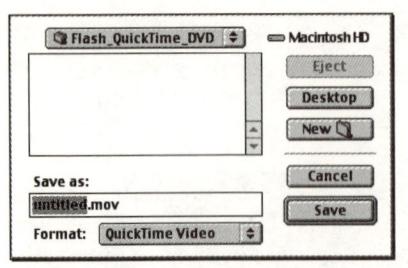

Figure 8-15: To export a Flash movie to QuickTime, you must first name the file and choose a format option from the drop-down menu.

The Export QuickTime dialog box opens and enables you to specify dimensions for your project.

3. **Enter a value for the Width and the Height of your video (720 width × 480 height matches DVD video); select a color format; choose a compression type, quality setting; and decide whether you want sound included if you plan on using audio that was imported into Flash (see Figure 8-16).**

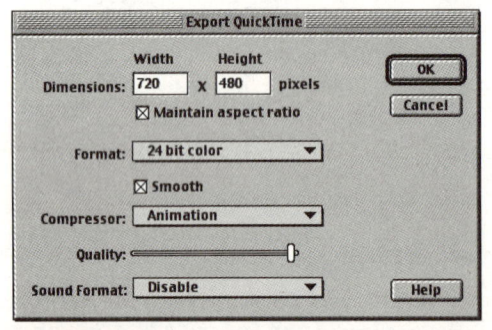

Figure 8-16: The QuickTime export settings in Flash let you specify elements, such as the size, color format, cormpressor, quality, and sound.

Note

The Export QuickTime dialog box offers a Smooth feature, which applies anti-aliasing to the elements in your movie, creating smoother edges that blend better with each other. Depending on the complexity of the movie, this feature may add an undesirable halo effect to some images placed against color backgrounds. You can experiment with selecting this feature, but you may find it better to leave it unselected, especially if you plan on cutting and compositing images from the movie in After Effects.

4. **After you have made all of your export settings, click OK.**

Your file is now in QuickTime video format and ready for editing, compositing, or encoding into MPEG-2.

Freehand

Among the improvements in the latest version of Freehand, Macromedia's illustration tool, is better integration with Flash, making the transition from one application to another even more seamless than before. The tools and interface elements also match other Macromedia products better in version 10, which makes learning the applications easier — something that Adobe has known for some time.

Tip: For those not familiar with Freehand, it may be worth your while to try out a demo of the application or at least consider how your workflow and ideas may benefit from working with it.

Using the extensive set of tools in Freehand, you can create complex vector art (just like in Adobe Illustrator), which can be used for print graphics, Web designs, or even video (see Figure 8-17). For instance, you can create a vector-based logo animation in Freehand that can be imported into Flash for use on a Web site and exported (from Flash) for use in an introductory animation on a DVD project (see Figure 8-18).

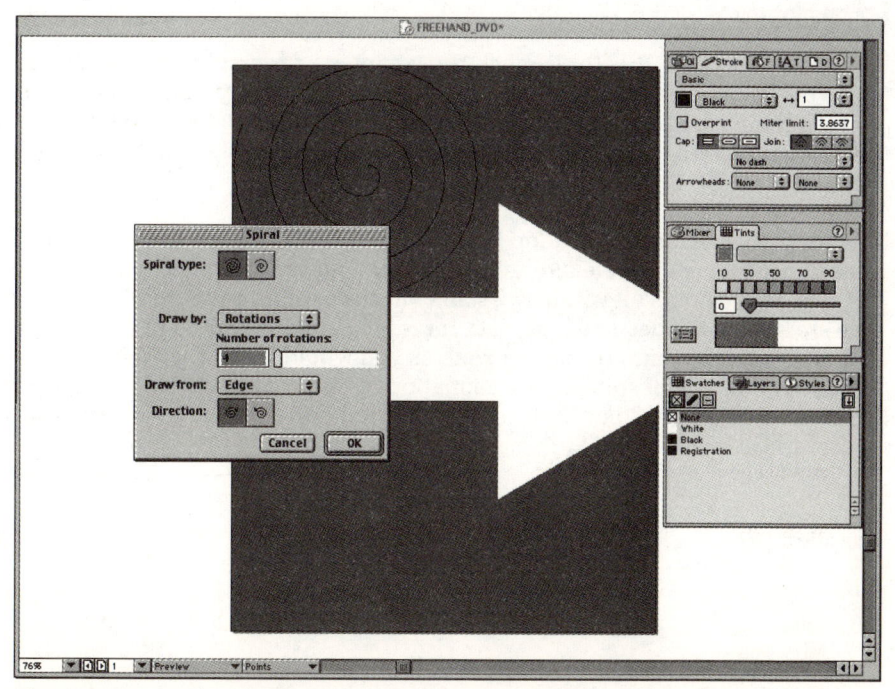

Figure 8-17: Freehand offers a large variety of tools for designing complex images.

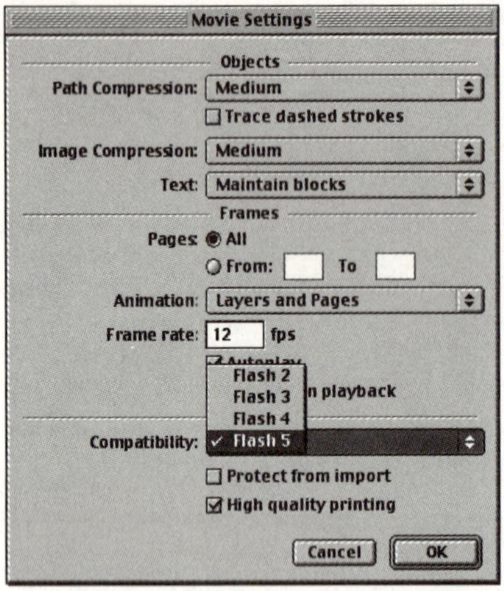

Figure 8-18: Using Freehand, animations composed of detailed vector art can be created, and the movies can be exported to Flash.

Director

While Macromedia's Director is much too complex a program to speak of in detail in this book, it is certainly worth mentioning that it is widely regarded as the standard for complex, disc-based interactive design. With version 8.5, Director expands its capabilities with interactive 3D options, which game designers, in particular, should find interesting.

Director is the perfect application for creating DVD-ROM content (see Figure 8-19). You can produce incredible interactive productions by placing video, graphics, text, and 3D elements all within the same interface. Also, because DVD-ROMs offer large storage capacities and high data rates, you are not limited in the amount or quality of material that you can present, as you would be on the Web (see Figure 8-20 for an example of some of the animation options available for your project). Even Microsoft PowerPoint files can be imported into Director if you are designing presentations for DVD-ROM.

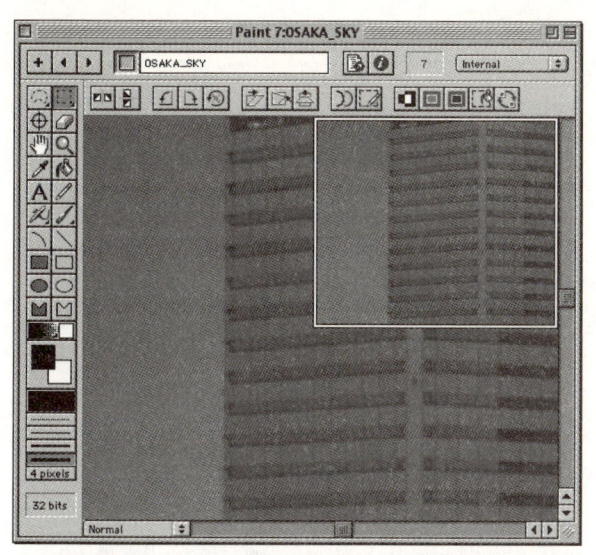

Figure 8-19: Macromedia's Director offers many ways to work with graphics and video for complex, interactive presentations, including image editing and paint controls.

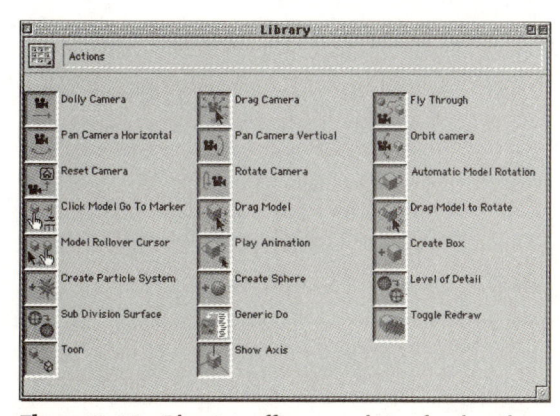

Figure 8-20: Director offers a variety of animation and 3D tools for creating interactivity, including the use of cameras for animating views of objects.

Summary

- A successful interface enlists principles of enjoyability and usability — the two main factors that affect a user's experience
- Designing for television requires special considerations, such as screen size, color, and resolution.
- DVD-ROM offers the possibility for an incredible amount of interactivity, while DVD video has only limited interactive capabilities.
- Adobe's After Effects is the standard tool for compositing and special effects, providing an incredibly flexible set of tools to work with motion graphics.
- DigiEffects CineLook, CineMotion, and Delirium are unique After Effects plug-ins that can add interesting textures to motion menus.
- Macromedia's Flash 5 includes QuickTime output, which makes it a great animation tool for video and the Web.
- Director can produce complex interactive interfaces (including 3D) for DVD-ROM and CD-ROM, as well as the Web.

Now that we have concluded our section on creating interactivity through interface design, we move on to the topic of using and understanding assets related to a DVD Studio Pro project. In the next chapter, we talk about understanding DVD audio.

✦ ✦ ✦

Using Assets in a Project

PART

III

In This Part

Chapter 9
Understanding DVD Audio

Chapter 10
Working with DVD Audio

Chapter 11
Understanding DVD Video

Chapter 12
Working with DVD Video

Chapter 13
Working with Slideshows

Understanding DVD Audio

The advances in entertainment brought by the arrival of DVD were not limited to video picture quality or interactivity. Audio was just as important a consideration when assembling the DVD specifications. After all, what would films and video be without sound? Because we have come to expect a high degree of quality from our entertainment experiences, the best audio formats available were chosen for inclusion on DVDs. PCM, AC-3, DTS, and even SDDS are some of the possible audio formats that you can use. Remember that while all of the formats are high quality options, each has its own special benefits and limitations. Also, noting which formats are available for use in DVD Studio Pro is important.

If you have a home theatre setup that includes a receiver and speakers capable of 5.1 channels of audio (achieved with a maximum of six speakers placed around the room), then you are already familiar with the advantages of surround sound listening. If you are unfamiliar with the format's benefits, this chapter may be enough to convince you that your home listening experience could use some improvement. As you read through the descriptions of the various audio formats, remember that the content delivered through your speakers is more important than how it is delivered. However, just about any soundtrack can benefit from proper use of multichannel audio. Think about the potential that audio gives you to make your DVD projects even more enjoyable for your audience.

CHAPTER 9

◆ ◆ ◆ ◆

In This Chapter

Discussing the benefits of various audio formats for DVD

Exploring the capabilities of multi-channel sound

Understanding how to mix for 5.1 channels of audio

Monitioring 5.1 audio channels

◆ ◆ ◆ ◆

Investigating Audio Formats

DVDs are capable of holding several different audio formats, each with its own pros and cons. While audio purists may argue the case for one format over another, the decision to produce a project with a particular audio format often comes down to the choice of budget, system compatibility, and space limitations. The current audio technologies are the result of many years of research and development, and all of them are able to produce extremely high quality output. Therefore, it is often a matter of how your audio is recorded and mixed that determines the success of your project.

The following audio formats are options currently available to producers of home video. Use these descriptions as rough guidelines when choosing a format that is right for your next project. Choose wisely, because your decision may determine whether your disc wins the next festival award for best sound design!

Linear PCM

The most common and high quality audio format available for DVD is without a doubt linear PCM (or simply PCM), which stands for Pulse Code Modulation. Essentially, PCM is pure digital audio sampled without additional compression. In fact, you are probably already familiar with this type of audio because it is essentially the same audio format used in audio CDs, with the exception that PCM for DVDs is capable of higher sampling rates (96 kHz versus 44.1 kHz for audio CDs). Although PCM can produce audio at higher frequencies, it can be scaled to meet the demands of your project — 16, 20, or 24 bits can be used. When considering your options for sample frequencies, 48 kHz and 96 kHz are the best choices. (See Figure 9-1 for an example of working with audio at 48 kHz.) It is important to remember, however, that some players may not accept 96 kHz audio and will instead downsample the audio to 48 kHz. Because players are required to adjust for these higher values, choosing a higher frequency should not be an issue, rather, concern yourself with recording and mixing at the best possible quality to begin with (see Figure 9-2). That way, our audio options for a DVD project are limited only by considerations of space and efficiency. A maximum of eight audio channels may be used, although using several channels for a single track limits the total available bandwidth for other project elements, such as video, multiple angles, subtitles, and alternate audio streams.

> **Note:** DV cameras record PCM audio at 48 kHz, unlike most audio CDs that record at 44.1 kHz. This can cause problems when trying to mix audio formats. Consult the documentation for your audio editing application to determine the best method for dealing with audio produced with different sample rates.

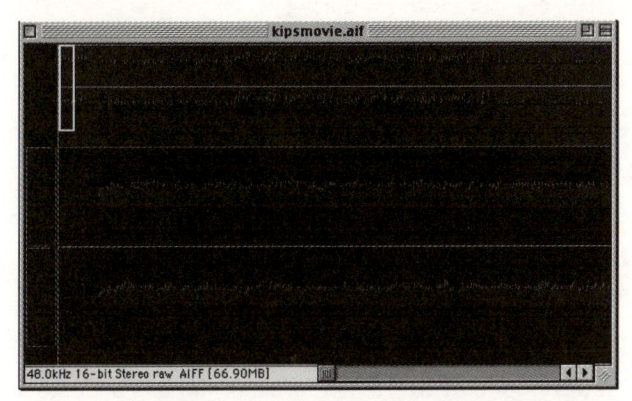

Figure 9-1: Working with PCM audio for a video project, such as an AIFF file, is easy in an application like Peak DV, which is currently included with Final Cut Pro 2.

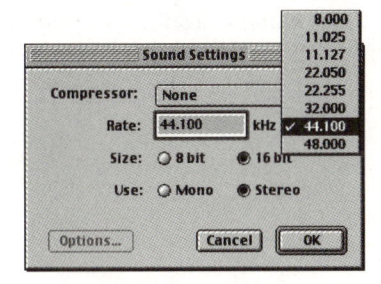

Figure 9-2: QuickTime Pro is especially useful for working with PCM audio, because it allows you to output to different PCM audio formats with various sample rates.

DVD Studio Pro is capable of importing three PCM audio formats. These formats include AIFF, WAV, and Sound Designer.

AIFF

AIFF (Audio Interchange File Format) was first developed by Apple as the standard audio format for Macs. These files do not contain data compression and yield high quality files that are very large. Files created in this format usually end with the .AIF extension.

When using the QuickTime MPEG-2 encoder to produce video, the audio track included with the movie is saved out as an AIFF file. You can import these files directly into DVD Studio Pro and they should sync up correctly with the video files with which they were encoded (see Figure 9-3).

Figure 9-3: The encoding of a QuickTime file with audio into MPEG-2 produces an MPEG-2 video file (`.M2V`) as well as a corresponding AIFF audio file (`.AIF`).

WAV

WAV is an audio file format developed by IBM and Microsoft and has become the standard format for PC audio. WAVs are probably the most common audio file format, although it may not have some of the flexibility of other formats.

Sound Designer II

Developed by Digidesign for its sound editing application of the same name, the Sound Designer II format (SDII or SD2) is one of the popular output/input options for professional audio software. Sound Designer II supports multiple channels and data sizes. Sound Designer II's primary advantage is how it makes it easier for you to work with files when editing or mixing (see Figure 9-4).

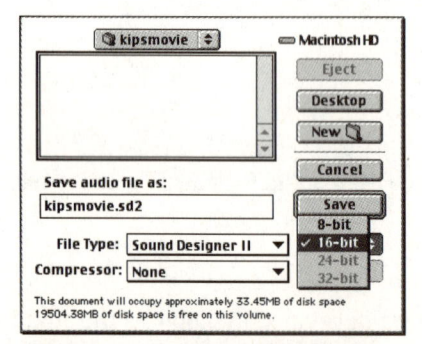

Figure 9-4: Peak DV, included with your purchase of Final Cut Pro 2, is one of many audio applications that enable you to export in the Sound Designer II format.

Viewing audio file information with QuickTime Pro

QuickTime Pro enables you to output and convert your audio to different sound formats, many of which are supported by the DVD specification (the standard by which DVD audio and video elements are judged suitable for inclusion on a disc). It also enables you to check the information for your files.

To look at an audio file's information, follow these steps:

1. **Open the QuickTime Pro player by double-clicking its icon, found in the QuickTime folder on your hard drive (if it's not there already, you may want to make an alias for the QuickTime application on your desktop in order to make it more easily accessible).**

2. **Choose File ⇨ Open Movie to select an audio file from its location on your computer.** Select the file and click Open to load the file.

3. **Select Movie ⇨ Get Movie Properties to open a properties window with information about your file.**

4. **Select Sound Track from the Movie drop-down list on the left.** This action gives you more viewing options for your audio file in the Annotations drop-down menu to the right (see Figure 9-5).

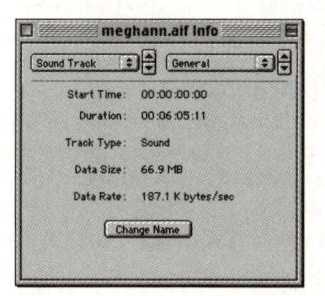

Figure 9-5: QuickTime Pro enables you to view general information about your audio files, such as duration, data size, and data rate.

5. **Click the drop-down list labeled annotations on the right in the window to select another viewing option.** For example, choosing General gives you information about the data size and data rate of your audio file, as well as the start time and duration. Selecting Format shows you the sample rate, sample size, number of channels, and the compression used, if any (see Figure 9-6).

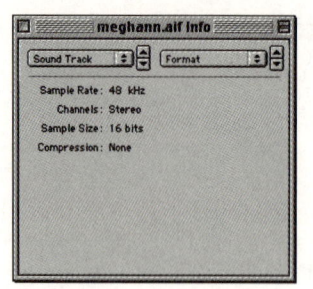

Figure 9-6: QuickTime Pro lets you view additional information about your audio files, such as sample rate and number of channels.

AC-3/Dolby Digital

The most sought after audio format for DVD is AC-3, which is also known as Dolby Digital (we use the terms interchangeably). If you have a home theatre system capable of five or more audio channels, this is the format for you. AC-3/Dolby Digital is a compressed format capable of delivering 5.1 discrete (separately stored and unmixed) channels of audio to left, right, center, left rear, right rear, and subwoofer (.1) configurations. It was developed using Dolby's vast experience with understanding the way we hear sounds. Although DVDs do not require AC-3/Dolby Digital to be included, most discs do contain at least one and sometimes several audio streams using this format. In fact, every DVD player is required to have AC-3/Dolby Digital capabilities; therefore, it is the format of choice for multichannel audio.

The .1 or LFE (Low Frequency Effects) channel is optional and only supplies additional bass information. While it is not required, it can significantly enhance the listening experience if you have a separate subwoofer in your surround sound setup. Explosions, engine roars, and thunder are only a few of the sounds that can benefit from the use of this extra channel. If you do not have a subwoofer, most receivers are capable of redirecting some of the extra bass output to the other speakers. Your playback device carries out this "downmixing," and no additional adjustments are required.

Although AC-3/Dolby Digital is a compressed format, the compression is not exactly what you might equate with the relatively low-quality methods used by graphics and streaming media applications. With AC-3/Dolby Digital, data that is measured as being outside the range of human hearing, and thus unnecessary, is thrown out. What remains is well defined, clear, digital sound, with sample rates beyond what can be produced with conventional audio CDs. A process know as *masking* is used to eliminate sounds at low levels when high level sounds at close frequencies occur at the same time, because the characteristics of human hearing do not enable us to hear these frequencies anyway. Throwing away this essentially useless data saves a considerable amount of space, which is crucial for today's storage requirements. In fact, the space saving algorithms developed for AC-3/Dolby Digital were first developed for theatrical film soundtracks, which also have a limited space for audio data on film prints.

Currently, the maximum sample rate for AC-3/Dolby Digital on DVDs is 48 kHz. This is beyond the 44.1 kHz rate for audio CDs, although it is not quite at the level of 96 kHz PCM audio. The maximum bit-rate is 448 Kbps, a far cry from PCM's 6,144 Kbps. This difference in technical quality is related to the amount of compression required for AC-3/Dolby Digital. To accommodate all 5.1 channels of audio, a lower overall bit rate must be used. Actually, it is not always advised to use the maximum bit-rate, as room must be made for additional disc assets. Instead, 384 Kbps is suggested for full 5.1 channels and 192 for stereo-only streams. This should allow adequate space for video and other disc assets in a typical project. With AC-3/Dolby Digital audio, multichannel surround soundtracks can be created at lower bit-rates than are required for a single track of PCM audio.

Included with your purchase of DVD Studio Pro is an application called A.Pack. With this utility, AC-3/Dolby Digital audio may be encoded after separate audio files have been properly mixed (see Figure 9-7).

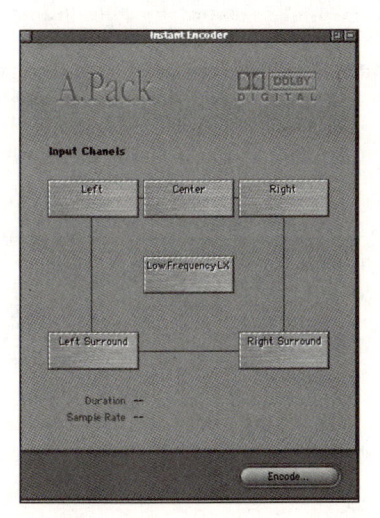

Figure 9-7: The A.Pack application enables you to encode 5.1 AC-3 audio (six channels) for use in a DVD project.

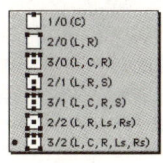

Figure 9-8: Using a pictoral grid to demonstrate the placement of speakers in a surround sound setup, the Audio Coding Mode in the A.Pack application enables you to specify the desired output, setup, and speaker configuration for the encoded AC-3 file.

Surround Sound by Nika Aldrich

Nika Aldrich, recording engineer and in-house Dolby and Surround expert, Sweetwater Sound (www.sweetwater.com), believes that the hottest buzzword in the professional audio industry, the film industry, and the home hi-fi industry right now is "surround." The concept alone conjures the recollection of sound effects flying overhead and an encapsulating scene in a movie in which the audience "could swear that they were really there!"

It also, however, is probably one of the most confusing topics, with terms like "Dolby Digital, Pro Logic, DTS, SDDS, SACD, THX" thrown around unscrupulously. The following is a basic overview with the intent of providing enough background in surround work that the tools available in this software become more of a creative asset than a feared impediment.

Surround work is done by mixing multiple sources of material to "stems" — each stem representing one speaker. Often a piece of software such as Pro Tools is used to pan each audio track to the various stems, which are then sent to their respective speakers. Pro Tools is a professional recording application used by many recording studios. The surround panning can either be done with the software alone or it can be done with joysticks that are connected to the computer and used in conjunction with the software. The process involves focusing on each individual track, panning it around the room to determine where it needs to go, and then checking all tracks together to ensure a satisfactory overall effect.

The computer must have enough audio outputs to represent each of the stems (speakers). A dedicated sound card with multiple outputs, such as a Digi001 from Digidesign is used for this application. This hardware uses a PCI slot and adds eight discrete outputs from the computer so that each of the speakers have its own signal feed. As the panning is done in software, the software routes the appropriate amount of audio signal to each of the respective outputs. When this has been completed the signal sent specifically to each speaker (stem) is rendered or "bounced" to the hard drive, culminating in a separate audio file for each speaker and the subwoofer.

The final step in this process is that the separate files representing the stems are encoded into a file type that can then be put on a DVD. After the file is encoded, a decoder such as a surround-capable home stereo receiver is needed to properly decode the signal. The DVD player sends the digital data to the receiver, which decodes the information and sends the stems to their appropriate speakers.

Several different encoding algorithms can be used depending on the application. Dolby Pro Logic is an older format that utilizes a pair of front speakers, a center speaker, and a signal that is sent to the rear speakers in mono. This is a four-channel surround format that is currently being used on broadcast television and on VCR tapes that have surround material. DTS and Dolby Digital/AC3 are 5.1 systems. This means that they support five discrete speakers (front left, front center, front right, rear left, rear right) and a subwoofer that gets its own signal. These are both six-channel formats and they both require digital technology — they cannot be used on VCR tapes, analog television, or analog cable TV. They can only be utilized in digital formats such as DVD, HDTV, Satellite, and digital cable.

Dolby Digital and DTS are two different versions of the same technology by two competing companies, and each company has its own way of encoding the information for the best

Chapter 9 ✦ Understanding DVD Audio

sound quality. Dolby Digital/AC3 is by far the most used digital surround encoding algorithm available. Every DVD that has a 5.1 surround mix has at least a Dolby Digital/AC3 version of the soundtrack on it; though it may also have a DTS version of the same surround mix on it. All discs will have a regular two channel (stereo) mix as well.

A basic surround mixing system is as simple to assemble as a sound card with enough outputs for each speaker to have its own, the proper number of speakers, and a piece of software that can provide the surround panning ability. DVD Studio Pro then provides the encoding ability, and the rest is up to the producer's creativity.

Working in surround for the first time is an awe-inspiring experience. Effectively using it will provide the audience with a better understanding of the overall image that the producer is trying to convey. Surround work is easy, fun, and very satisfying. Perhaps the best thing about it is that there really are no rules. A person is only limited in what he wants to do by his own imagination.

In addition to choosing a speaker configuration for AC-3/Dolby Digital output (see Figure 9-8), you may specify a bit rate for your project file. All bit rates are constant, not variable, and remain unchanged for the duration of the audio file. Make certain that you set the value high enough if you are using the maximum 5.1 channels, but not so high that the data rate for your disc project is sacrificed. A value often used is 448 Kbps, the default value for the A.Pack application (see Figure 9-9).

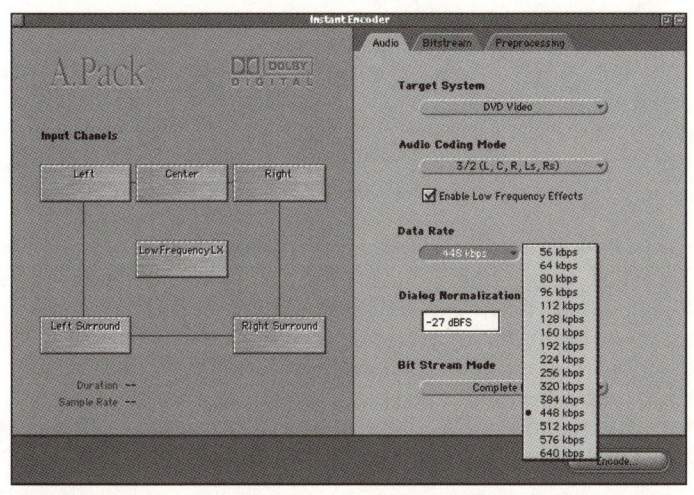

Figure 9-9: The data rate for encoding your AC-3 project can be specified in the A.Pack application, although for general purposes, the default value of 448 Kbps works fine.

MPEG audio

MPEG audio quality is similar to AC-3/Dolby Digital, with the exception that MPEG audio is capable of using 8 total channels of audio — 7.1 versus AC-3/Dolby Digital's 5.1. Also, the maximum bit rate for MPEG-2 is 912 Kbps, which is higher than AC-3/Dolby Digital's 448 Kbps.

The only video standard capable of carrying the MPEG audio format is PAL (Phase Alternating Line, a format used in areas of the world like Great Britain). This limits the appeal of the format and diminishes its use as an audio standard. Even in countries with PAL programming, disc producers often look to AC-3/Dolby Digital for their multichannel requirements. MPEG audio is a rarely encountered audio format.

DTS

DTS (Digital Theatre Systems) is a 5.1 channel format (similar to AC-3/Dolby Digital) that audiophiles can appreciate. It was first developed for movie theatres (*Jurassic Park* was the first movie to officially use the format). Due to its higher than average bit rate and, what some consider, improved compression schemes, DTS brings multichannel audio closer to a pure, crisp sound. Still, it does employ compression and therefore is not quite at the quality level of PCM audio. In fact, most people have a hard, if not impossible time, differentiating between DTS and AC-3/Dolby Digital (or PCM for that matter). Also, not all players are required to carry DTS capabilities as they are required to handle AC-3/Dolby Digital. This limits its appeal, and as a result, only a select number of discs are currently released with DTS audio.

Audiophiles are always eager to argue the benefits of one format over another. This is the case with multichannel audio such as Dolby Digital and DTS. These two formats essentially operate in a similar manner — each is capable of playing back 5.1 discrete audio channels in the same configuration and both employ data reduction methods to meet the storage requirements of film and video. It is the compression methods that ultimately set these two formats apart.

To begin with, Dolby Digital uses a compression ratio between 10:1 and 12:1 while DTS uses only about 4:1. It would appear that DTS has a quality advantage since it is closer to approximating uncompressed audio. Most people would not notice the difference; however, if you are trying to preserve as much signal as possible, DTS is considered the best choice for 5.1 audio. The problem for disc producers lies in the large file sizes that are created with DTS, versus the relatively low file sizes for Dolby Digital. Because the perceived difference in quality is negligible, producers often opt for the format with the lower space requirements and reduced data rates.

When played side by side, DTS may sound better than Dolby Digital; however, some argue that this is only because DTS boosts the signal a little as compared to Dolby Digital. This has led to discussions that Dolby Digital reproduces sound more accurately, while DTS embellishes the sound by raising the overall audio levels. Depending on who you ask, you may get different responses to this question, although it appears that neither accusation is really correct, because the two formats deal with audio in their own special ways. In the end, when you look at the technical specs, DTS does appear to have a slight advantage over Dolby Digital. The question is, does a listener notice or even care? If the answer is yes, then you may decide to follow the course Stephen Spielberg chose and select DTS for your next project!

SDDS

SDDS (Sony Dynamic Digital Sound) is capable of 5.1 and 7.1 channels of surround audio. Similar to AC-3/Dolby Digital and DTS, SDDS works at a sample frequency of 48 kHz, and the maximum bit-rate is 1,280 Kbps — slightly less than DTS. SDDS uses a special data reduction scheme based on ATRAC (Adaptive Transform Acoustic Coding) compression, a method developed for mini-discs. Also like the other multi channel formats, it was first developed for movie theatres, and is trying to find its introduction into high-end home video devices. Special equipment is required for SDDS, and an expanded speaker configuration is needed if you want to make use of the maximum eight channels. However, because SDDS is primarily used in movie theatres, it could be a while before you see it available for your home system.

A form of data reduction originally developed for mini-disc applications, ATRAC (Adaptive Transform Acoustic Coding) can compress PCM audio into approximately one-fifth of its original size while maintaining high quality. Using principles of psychoacoustics, including perceptual coding and the masking effect mentioned for AC-3/Dolby Digital compression, ATRAC significantly reduces the amount of data in an audio file, while retaining excellent quality. The idea behind this form of compression is to remove sound or data that is beyond the range of human perception, rather than include large quantities of useless information.

DVD-Audio

With the advent of new audio technologies such as DVD-Audio and SACD (Super Audio Compact Disc), the future of high-quality listening formats has yet to be determined. Arguably, the format with the greatest potential is DVD-Audio. This new format was recently introduced and special players are hitting the market. Ordinary DVD players cannot play back DVD-Audio unless specifically noted (typically, only a feature on a few high-end players). Of course, as with any new format, the players are initially expensive and the discs are few and hard to find. Also, some wonder whether most people are going to seek out higher-quality audio formats, as the majority of consumers are content with what they have now. Besides, most people could not tell the difference between listening to AC-3/Dolby Digital audio on a DVD-Video disc and listening to 5.1 audio on a DVD-Audio disc.

Therefore, at this point in time, DVD-Audio is targeted at the minority of audiophiles with systems and ears to notice the difference. However, for content creators, audio technicians, and musicians, the idea of preserving as much audio fidelity as possible makes DVD-Audio an attractive proposition. Players that cater to all segments by having the ability to play both DVD-Video and DVD-Audio may make purchasing decisions easier. Also, DVD-Audio discs can hold multimedia content similar to DVD-Video discs, which many people may find unnecessary because DVD-Video already does both very well.

DVD-Audio can play back 24 bit audio data at a sample frequency of 192 kHz for two channels and 96 kHz for full six-channel surround sound. As an added benefit, different channels in the same track can be set to varying sample rates. For example, the front channels can play back at 96 kHz while the rear channels are at 48 kHz. The maximum data rate for DVD-Audio is 9.6 Mbps. To overcome data rate limitations (you could not use six simultaneous audio channels at 96 kHz otherwise) a lossless data compression method called MLP (Meridian Lossless Packing) can be used. Because no data is actually sacrificed with this compression method, there is no reason not to use it. Data can be reduced to about half of the original size, allowing six channels of 96 kHz audio to fit on a disc. Together with DVD's increased storage capacity over traditional CDs and MLP's data reduction schemes, an enormous amount of audio information can be recorded for DVD-Audio. However, it still remains to be seen whether record labels are going to support this format.

Now that we've told you what audio formats are available for inclusion on a DVD, it is time to take a look at what is involved with mixing for multichannel audio, particularly AC-3/Dolby Digital.

Mixing for AC-3/Dolby Digital

Mixing for 5.1 channels of audio has its advantages and disadvantages. On the one hand, it provides the producer with an expanded palette of aural material to enhance a project's appeal. On the other hand, it also introduces new technical challenges that can be difficult to circumvent.

One consideration is whether you want to create a real sounding environment, an artificial space, or a combination of both. Real environments, such as the recording of a concert, do not need a great deal of panning or effects. But environments that are unrealistic can use any number of effects or mixing tricks to simulate new perspectives for a listener. Where do you want to place the listener? Six channels of audio give you a lot of options that may become overwhelming if you do not plan your project out first. Also, remember that you must work with six separate (discrete) audio files to create true 5.1 surround sound. After you are finished mixing, you can import these files into the A.Pack application (discussed in Chapter 10) to encode your final AC-3/Dolby Digital file for inclusion in your DVD Studio Pro project.

Carefully consider where you want to place sound track elements in your mix. For example, important dialogue is often mixed in the center speaker. When mixing for the center channel, you may either place the audio strictly in the center speaker, pan it equally to the left and right (for a "phantom" center effect) or spread it across all three front channels (left, right, and center) to add a sense of depth. Spoken elements can also originate from any one of the other speakers in your setup depending on where you want to place the listener in the surround space.

Tip   Use reverb carefully when mixing for surround sound. This is especially important for dialogue and voice elements in center channels, which may otherwise seem flat without the reverb applied to the other channels. To avoid having a flat center channel, try mixing its reverb in with the left and right channels. Reverb and echo effects can be nice to simulate larger environments; however, they must be carefully balanced.

The L.F.E. channel is often considered unnecessary to a surround mix, because it can complicate how other bass frequencies are processed. Also, downmixed audio for systems that do not have a sixth channel or are simply playing back in stereo cannot read the extra channel. If your production does not need the extra low frequency effect provided by the ".1" channel, then it is best left out of the main mix. Depending on the listener's setup, bass frequencies on other channels are usually more than enough.

However, if you decide to add the extra channel, use it carefully. When monitoring your low frequency bass channel while mixing, you may want to try setting the gain to +10 dB (decibel) relative to the other speakers in your setup. This ensures accurate monitoring, since +10 dB gain is usually added to the channel when it is played back. Also, make sure that you only include information for the L.F.E. that would otherwise be lacking from your other channels. This should help receivers manage data better whether they are capable of the extra channel or not.

Remember, there are no strict rules when it comes to mixing for surround sound. It is still a relatively new format with many variations whose results depend on the medium and genres you are mixing for. Do not be afraid to experiment with your audio to achieve the most interesting effects. It is also a good idea to keep a log of all your mixes; otherwise, you may find it difficult when you need to go back to fix a problem or remix portions of a project.

Before you begin to mix for six channels of audio, you should properly set up your studio to monitor 5.1 surround sound.

Monitoring 5.1 Channels of Audio

When monitoring six channels of audio, there are some basic factors to consider, especially when setting up the position of speakers in the listening environment. Ideally, the center speaker would be the same distance from the listener as all the others. However, this is not practical in real world situations because the TV and front speakers are usually set against a wall. Even more likely, the listener is sitting on a couch, which further complicates the ideal listening position.

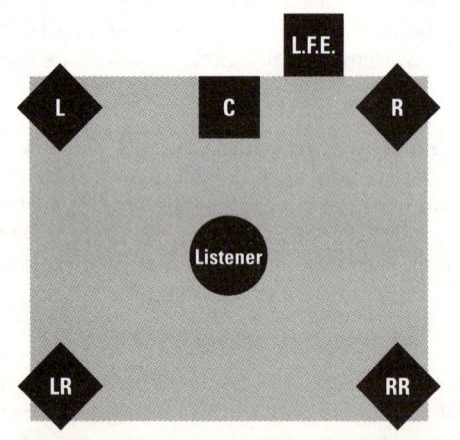

Figure 9-10: An optional setup for monitoring 5.1 audio for a common home video environment, with tight 45 degree angles on the speakers — not ideal, but it is acceptable.

The angle of the speakers is another consideration that experts tend to argue about. Speakers are often set up at a 45-degree angle or less to point toward the listener when the listener is in the center of the sound field (see Figure 9-10). Of course, this also depends on the layout of the room and the position of speakers in your setup. Whatever your setup, remember that you want to simulate the environment where your listeners are going to be hearing the audio. This is probably the most important consideration. After all, most people watching a DVD are not going to be in a studio environment with perfectly aligned speakers. The variations in setups should be accounted for when mixing. Try different setups to find what works best for you and your listeners (see Figure 9-11).

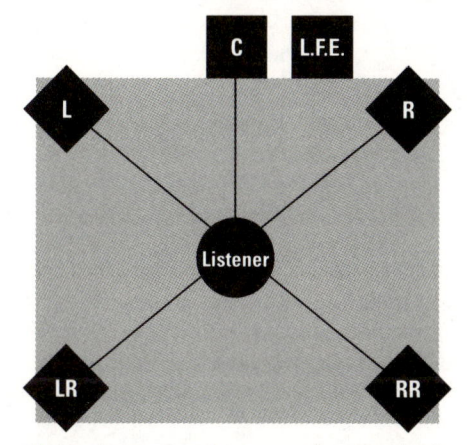

Figure 9-11: Another setup for monitoring 5.1 audio for home video environments, which sounds a little better due to the equal distances of the speakers from the listener.

In addition, the subwoofer (L.F.E. ".1" channel) can be set up in front or back depending on your room acoustics. Some people like to put the subwoofer behind a couch in their home theatre to absorb the extra bass output. However, subwoofers are often located in between the center and the right or left front speakers when monitoring for mixing (see Figure 9-12). Even in a home theatre setup, it is recommended to place the subwoofer in front. Once again, arrange your speakers in a way that sounds best to you. If you are still unsure of the best location for your speakers, your local audio dealer can recommend a preferred setup for the particular speakers you may have.

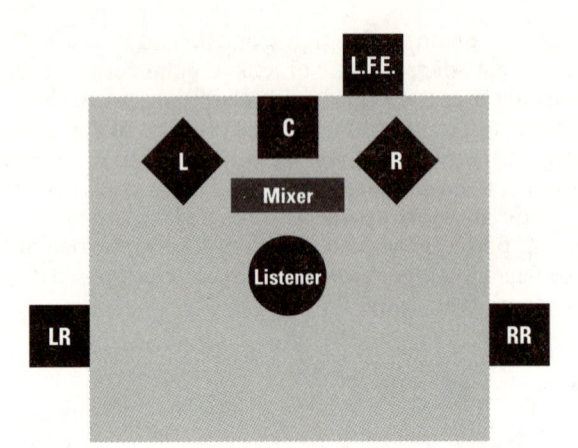

Figure 9-12: This 5.1 audio monitoring setup is even closer to the ideal listening environment for home theatres; however, each setup varies depending on the layout of the room and the speakers being used — make certain to adjust accordingly.

When deciding on what type of speakers to use for monitoring, identical, self-powered speakers are the best. Having the same speakers in front that you are using in back ensures that your sound levels remain consistent. The only way to be certain that what you are hearing is correct is if you use calibrated speakers that are exactly the same.

To test the final mix of your DVD's audio, try burning various mixes to CD or DVD-R and play the discs back in a variety of sound systems. If your friends have different AC-3/Dolby Digital or DTS setups, try visiting as many as possible and make note of any differences you encounter when testing. Take note of the position and height of speakers in the room in relation to where you are sitting, as well as environmental factors such as high ceilings or thick carpeting. At the same time, test your mix on 2-channel systems to ensure that the automatic downmixing of your 5.1 channel mixes sound all right. Many 5.1 channel receivers give you the option to switch back and forth to stereo-only output.

Summary

- ✦ DVD Studio Pro is capable of working with various PCM audio formats, including AIFF, WAV, and Sound Designer II files.
- ✦ PCM is the highest quality audio option available for DVD. It is uncompressed and delivers the highest bit-rates, as well as large file sizes.

✦ DVDs can use multichannel audio formats such as AC-3 and DTS.

✦ AC-3/Dolby Digital is an audio format capable of producing 5.1 surround sound channels with relatively small file sizes and data rates for DVD.

✦ DVD-Audio is the latest high quality sound format with terrific audio specifications but questionable appeal at this point in time.

✦ Mixing for 5.1 audio is a prospect that can be confusing and rewarding at the same time.

✦ When monitoring 5.1 audio while mixing, it is important to remember what type of environment and speaker setup a listener is going to use.

In the next chapter, we discuss how to work with DVD audio using DVD Studio Pro along with third-party applications. We also discuss the A.Pack encoding utility in depth. While working through the examples in the following pages, refer back to this chapter if you have any questions about formats or mixing.

✦ ✦ ✦

Working with DVD Audio

CHAPTER 10

In This Chapter

Understanding audio formats compatible with DVD

Working with audio in other applications

Using A.Pack to encode audio for Dolby Digital

Working with audio files in DVD Studio Pro

Producing projects in DVD Studio Pro with multiple audio streams

---

In the last chapter, we discussed the nature of audio used for DVDs. In this chapter, we discuss the next logical step — working with audio files to create content for DVD projects. Good audio — audio that is produced to please the listener — elevates an otherwise neutral experience into something truly enjoyable, or it takes a good DVD and makes it excellent. Audio is easy to take for granted, but it forms an important part of a DVD. Can you imagine your favorite movie without the soundtrack? Your experience would have been completely different. Professional sound designers and audio engineers work tirelessly to achieve the best effects, in the hopes that the listener will be completely absorbed by the world the filmmakers are trying to create. Designing audio for your own DVD project should be no different. With a little additional work, you can add significantly to the project.

Audio professionals employ a variety of tools and techniques to achieve their results. Some of these methods can be expensive or simply too complex or involved for general DVD projects, but they are worth taking a look at. People traditionally find ways of bringing high-quality video and audio capabilities into the consumer and pro-consumer level. DVD authoring is an example of this, and some audio technologies that relate to DVD seem to be heading in a similar direction. You may want to start with standard audio and then start experimenting as you master the various facets of DVD production. Ultimately, the potential for using higher-quality audio formats with DVD Studio Pro is similar to the situation you have with video. You can do anything within the confines of the DVD standard, but you have to prepare the source material before you bring it into the program.

One popular audio innovation that is receiving a great deal of use these days is AC-3 (also called Dolby Digital), the standard audio format commonly employed in the creation of DVDs.

This format produces a convincing effect that gives the listener a sense of being surrounded by audio. Techniques for producing AC-3/Dolby Digital audio were once a difficult prospect (and still can be, depending upon the complexity of your project), although with DVD Studio Pro, Apple has generously included software to help you produce your own AC-3/Dolby Digital files for use on a DVD.

Note AC-3 and Dolby Digital are the same audio format and the terms may be used interchangeably.

Evaluating Compatible Audio Formats

As discussed in Chapter 9, you can use three main audio formats to create DVDs:

- ✦ PCM (Pulse Code Modulation)
- ✦ AC-3/Dolby Digital
- ✦ MPEG audio

Oddly enough, MPEG audio is not widely supported and thus PCM and AC-3/Dolby Digital are your only sure choices. Of course, the format you will probably experience the most is a PCM audio format called AIFF (Audio Interchange File Format) with the suffix .AIF. This is the audio format created by default with the included QuickTime MPEG encoder. This format is adequate for most standard DVD audio requirements. WAV is another PCM format allowed by the DVD standard. But if you have the mixing capabilities, AC-3/Dolby Digital is an excellent format and provides interesting six-channel effects, versus the two channels you are probably used to.

Preparing Audio in Third-Party Applications

As with video, before audio is brought into DVD Studio Pro, the audio track must first be prepared in other applications. In many cases, if you are not using multiple audio tracks, the audio is dealt with as a part of the overall video clip in a program, such as Final Cut Pro or Adobe Premiere. However, if you have isolated audio tracks (such as commentary as an extra audio track to go along with an existing clip), or if you are running audio clips beneath a slideshow, you need to edit and save them in the proper format before you can use them in the DVD project. Depending on how the audio tracks were recorded, in some cases, you may be able to open them up in QuickTime Pro and encode these extra tracks by exporting them in MPEG-2 format.

But a time may come when having a separate program to edit and to format audio will come in handy. The choice of which application to use depends on the capabilities you need to produce your audio projects. Many audio solutions are available today, from pure and simple waveform editing applications (waveform editing is the process of looking at a sound clip that has a visual representation of the sound waves as a guide), to programs with advanced sampling and mixing features.

Choose a program (or several programs) that meet the needs of the audio you are trying to produce. Also, if your project is exceedingly complex, you may want to consider paying for some studio time, especially when it comes to mixing audio for six channels.

Considering multitrack recording

A typical scenario when working with audio for a DVD project is that you are simply using the audio that was recorded with the original video. Depending on the camera, in most cases this audio is stereo, two-channel audio (left and right). If you want to work with audio in a standalone application, and perhaps add additional segments, such as director commentary, or take a finished music clip and adjust it, a program capable of two channels may be adequate.

Another scenario is that you wish to record an initial track, such as dialogue (or in the case of a typical music album, perhaps the drums), and then record additional tracks, such as voices and sound effects (or musicians). You then mix these multiple tracks down into the standard stereo (two-channel) audio file. This is an example of *multitrack recording*. The power of multitrack recording is that you can record successive tracks and then adjust their individual levels and placement (such as panning a voice or something left or right), before the final audio clip is generated.

Depending on how many simultaneous tracks of audio you intend to work with, you may need to get a faster hard drive than comes standard with a Mac, such as an ultra-wide SCSI drive and controller card, running at a speed 7,200 or 10,000 rpm, where 10,000 rpm is the ideal.

Bias — Peak or Deck

Peak from Bias is a great, straightforward application for working with standard two-channel audio. Peak has powerful signal processing effects and includes support for a variety of output formats (including the ones you should find most useful — AIFF, Wave, and Sound Designer II). A special version of Peak is called Peak DV, which is actually bundled with Final Cut Pro 2 (and is only available with that program).

Deck is another excellent program from Bias that enables you to go beyond two channels and have multitrack capability. Deck also gives you the ability to edit audio for QuickTime movies while synchronizing the audio with frame accuracy; this feature is helpful if you are working with a DVD project in which you have more than one audio stream for a person to choose from. Just this feature alone may make it worth your while to check out the program.

Both Peak and Deck are in the $500 range; however, both programs have upgradeable LE versions with less features that are closer to $100. Also, both programs are available in trial versions from the Web site (www.bias-inc.com) and on the DVD-ROM that accompanies this book.

Adobe — Premiere

Using Adobe's Premiere to work with audio, on its own or in conjunction with video, is possible. Premiere is a powerful program, primarily used for editing video as well as generating specialized video files at smaller sizes for CD-ROM and Web delivery. The latest version includes some helpful features, including support for exporting video directly into Web-based streaming media formats. In terms of audio, version 6.0 includes a professional-level mixing tool, the Audio Mixer, that enables you to have traditional controls such as gain, fade, and pan adjustment for up to 99 individual audio tracks. The Audio Mixer works directly with Premiere's Monitor window, so you can make audio adjustments while you are watching the synchronized clip. A trial version of Premiere 6 is available for download on the Adobe Web site (www.adobe.com), and on the DVD-ROM that accompanies this book.

Digidesign — Pro Tools

Digidesign's Pro Tools 5.1 is easily the standard in high-end digital audio production (www.digidesign.com). The program's multichannel mixing and automation capabilities are some of its best features. DVD creators may be interested to know that with the right equipment, the higher-end version of Pro Tools (24 MIX and MIXplus) have the ability to work directly in 5.1 surround sound. This is an important capability if you want to add AC-3 (Dolby Digital) audio to your production. After you have mixed your surround audio in Pro Tools, you can then bring it into the A.Pack application included with DVD Studio Pro and produce AC-3 streams for importing into a DVD project. Of course, Pro Tools is a more expensive option and requires some hardware to work properly. However, if professional level audio, equivalent to that used in many major studios, is what you seek, you can't go wrong with this program. The free version of ProTools is available on the DVD-ROM that accompanies this book.

Emagic — Logic Audio

Another option for AC-3/Dolby Digital and general audio is Emagic's Logic Audio (www.emagic.de), another audio application on a par with ProTools that has widespread popularity. Logic Audio 4.5, 4.7, and above offer surround sound support when used in conjunction with a sound card such as Emagic's AW8. Emagic also offers an AW8 Surround kit that includes a special version of Logic and the card for around $500. These options can be a viable lower-cost alternative for home Dolby Digital. In addition to possessing features equivalent to those mentioned for other third-party applications, Logic Audio also features support for ASIO — the hardware sound system that lets you monitor true AC-3/Dolby Digital streams on your computer. Logic Audio has advanced and comprehensive capabilities of working with MIDI, such as sequencers, keyboards, and samplers.

> **Note**
> MIDI essentially enables you to record musical instructions rather than the actual audio so that when you play back a MIDI sequence, the computer triggers the various devices you have, such as keyboards, or samplers with sound loops, in effect being an electronic maestro. The advantage of sequencing music is that you can go back and make minor adjustments to individual notes and sounds, and the timing/tempo, with relative ease, instead of having to re-record a part.

In addition to its top of the line series (Logic Platinum being the highest), Emagic also offers an application called micrologic AV, an inexpensive program that includes the capability of creating soundtracks for QuickTime movies.

Motu — Digital Performer

Another multitrack recorder and mixer that you may consider using is Motu's Digital Performer (www.motu.com). As with many of these other applications, Digital Performer offers powerful multichannel capabilities, effects, EQ, and MIDI sequencing. It also offers automation tools, which put it on a par with more expensive mixing applications. QuickTime is also supported for simultaneous playback of audio and video while you edit.

Experimenting with Dolby Digital

If you're interested in getting into Dolby Digital, aside from the hardware and software to record it, you will need to have a Dolby-capable speaker system, to be able to accurately monitor the sound. Although ideal systems with higher quality speakers can get quite expensive, you can get a consumer Dolby stereo system with speakers for as little as $500 from the home theater offerings of a retailer, such as Best Buy (www.bestbuy.com). The idea is that you have computer hardware that allows simultaneous output of as many as six channels, and you hook it up to the monitor system.

If you're looking at the low end of the spectrum, you can theoretically put a system for creating Dolby Digital audio together for not much more than a $1,000, if you put together Emagic's Logic Surround kit (which requires a Mac with an open PCI slot) and a consumer-level Dolby home theater system. To record the audio in the first place, you may need to get a faster hard drive than comes with a typical Mac. The stock Macs typically come with a type of hard drive known as IDE, running at 5400 rpm. It is generally recommended when recording audio (or video for that matter), to get an ultra-wide SCSI drive, 10000 rpm is ideal for multitrack recording. In most cases, you need to get a special controller card, but hard drives and controller cards can be had in all their variety from suppliers such as MacWarehouse (www.macwarehouse.com) or MacMall (www.macmall.com).

If you need assistance in evaluating the various system options, a supplier such as Sweetwater Sound can be helpful; it can put together entire systems from computer to speakers and carries a wide range of audio software including Pro Tools and Emagic (Logic) products. To get a better understanding of how Dolby Digital and other forms of surround sound work, there are some great books out there, including *5.1 Surround Sound* by Tomlinson Holman (Focal Press, 1999), and *Spatial Audio* by Francis Rumsey (Focal Press, 2001). For more information, check out the books area of www.dvdspa.com.

> **Tip:** For more information on books, suppliers, and other resources, visit www.dvdspa.com.

In the next section, we discuss how to prepare audio by using the A.Pack application that comes with DVD Studio Pro. The assumption with A.Pack is that you have a AC-3/Dolby Digital file that has previously been recorded and mixed properly, ready for encoding.

Investigating the A.Pack Interface

Included with your purchase of DVD Studio Pro is an application called A.Pack. This program enables you to encode audio files into the AC-3/Dolby Digital format commonly used for DVDs. AC-3/Dolby Digital produces 5.1 channels of surround audio, which greatly increases the number of options available for placing sounds and producing dramatic audio effects.

The Instant Encoder window

When you launch the A.Pack application for the first time, you are presented with the Instant Encoder window (see Figure 10-1). This window is where you spend most of your time working in the program.

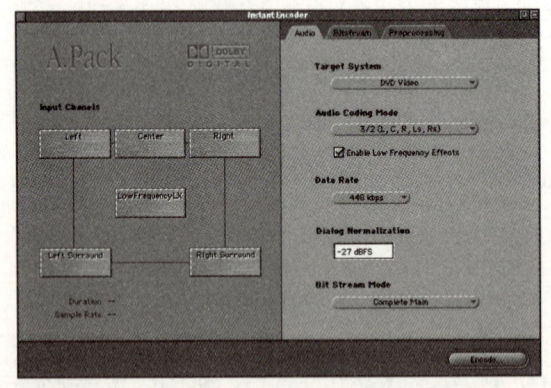

Figure 10-1: Preferences for a project include a startup action that takes place when the application is launched, and a selection of output systems for an AC-3 monitor.

More options become available when you click the tabs on the right side of the Instant Encoder window. Also, you may change the startup action (the action that occurs when you launch the application) from the default Instant Encoder to either

New Batch List or Open Batch List by choosing Edit ⇨ Preferences and selecting the desired option from the drop-down list in the Preferences window (see Figure 10-2).

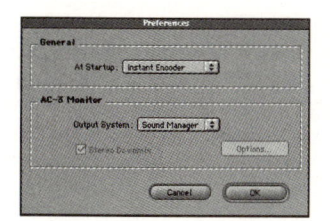

Figure 10-2: The Instant Encoder window is the first window that appears when you launch the A.Pack program and is used to access more parameters for your project.

The Preferences window also enables you to specify what device you want to use to monitor audio in the program. Choices for AC-3 monitors include the default Sound Manager and the optional ASIO. Sound Manager is the standard Mac application utilized for sound, and ASIO is a hardware-based sound system that you may purchase to listen to true 5.1 audio on your computer. (For more information on ASIO, see www.dvdspa.com.)

Using input channels

The Instant Encoder window is composed of several parts. The area on the left side of the window shows the available input channels and is where channels are assigned to physical points in the audio setup. By clicking a box (Left, Center, Right, Left Surround, Right Surround, or Low Frequency LX), the Select Input Channel dialog box opens, asking you to choose the audio file you want to assign (see Figure 10-3). If you choose a file with stereo channels, the program asks you which channel (Left or Right) you want to assign from that file. If the file you are working with has more than two channels, you are given a choice of alternate channels labeled sound 1 through sound 6, depending on the file that you are using.

At the bottom of the input channels window are two indicators. The first indicates the duration of the files you are working with and automatically defaults to the duration of the longest file you are using. Ideally, you should only be working with audio files that are the same length. Although, if you choose files of differing lengths, A.Pack adds silence to the tracks to make them match the duration of the longest file. The second indicator at the bottom of the Input Channels window is for the sample rate. You'll most likely work with 48 kHz audio files, although 32 and 44.1 kHz are also acceptable. However, you must remember that only a single sample rate may be used for all of your audio files.

Figure 10-3: The Select Input Channel dialog box enables you to specify which channel you want assigned from a stereo audio file.

Selections that you make from the tabbed audio window on the right Instant Encoder window may also affect the display of input channels.

Using audio settings

On the right side of the Instant Encoder window are tabbed windows with more project parameters. The first of these windows contains audio options, including settings for Target System, Audio Coding Mode, Data Rate, Dialog Normalization, and Bit Stream Mode. These options affect the larger overall settings for your AC-3/Dolby Digital project.

Target System

This option determines the system that your encoded audio is intended for. Choices include DVD Video, DVD Audio, and Generic AC-3. Video encoded for use with DVD Studio Pro requires the DVD Video setting, although the other options may prove useful in the future if applications for the DVD Audio format attain popularity.

Audio Coding Mode

Use the Audio Coding Mode settings to determine which channels should be encoded in the audio stream that you are producing. If you are only using simple audio and do not require all six audio channels, you can cut down on the amount of bandwidth required by selecting a different number of channels. You have seven choices for the audio encoding mode, ranging from center only to a full 5.1 audio setup (see Figure 10-4).

By clicking the Enable Low Frequency Effects button in this section, you can activate the .1 channel used for very low bass effects, such as rumbles and earthquakes, most often heard in Hollywood movies.

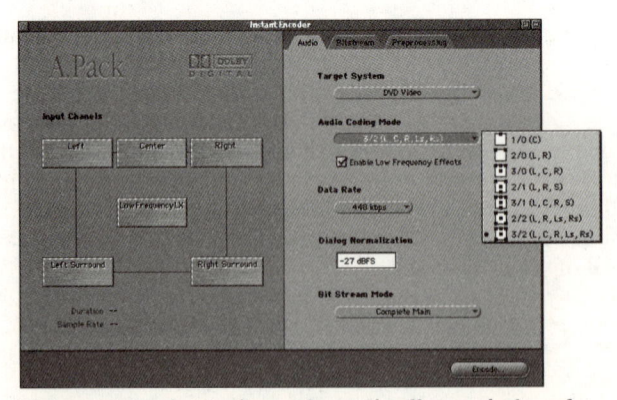

Figure 10-4: The Audio settings tab offers a choice of audio coding modes for your project.

Data Rate

The data rate affects the quality of the encoded stream. By choosing a higher value for the data rate, you are producing better quality audio — although, you are limited by the number of channels and the target system you want to use. For example, after choosing an audio coding mode, the data rate options that are no longer available (based on that choice) are grayed out in the Data Rate drop-down menu (see Figure 10-5).

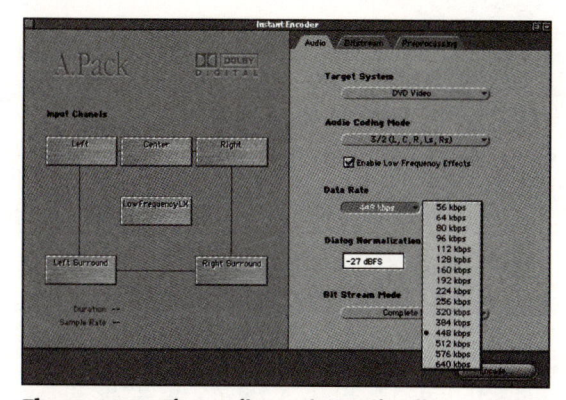

Figure 10-5: The audio settings tab offers a choice of data rate for your AC-3 project.

Dialog Normalization

This setting refers to the average volume of dialog present in your audio files and can be adjusted to ensure consistent levels among all of your channels. *Normalization* is a process that looks for the loudest point in your audio and adjusts the overall levels to set this point as the maximum volume within a given range. The default value in A.Pack is –27 dBFS, and unless you have a specific range in mind, leaving it at this setting is best. The highest setting allowed is –1 dBFS and the lowest is –31 dBFS.

Bit Stream Mode

Bit stream refers to the information passed to the player by the audio signal. The type of audio defined by the information in the bit stream falls into eight categories: Complete Main, Music and Effects, Visually Impaired, Hearing Impaired, Dialog, Commentary, Emergency, and Voice Over / Karaoke. By choosing one of these options, you are telling the decoder what purpose the audio is serving and providing the decoder with additional clues about how to process the audio, ensuring the best experience for a viewer. The default setting is Complete Main, which is an all-purpose setting that is good for most audio (see Figure 10-6).

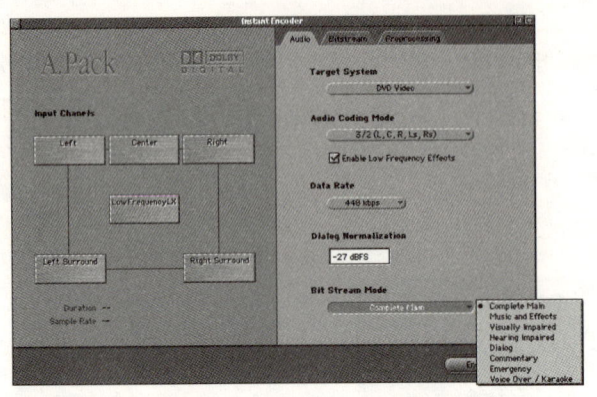

Figure 10-6: The Audio Settings tab offers a choice of bit stream mode that can be interpreted by some decoding systems as instructions for the purpose of the audio.

Using bitstream settings

When you click the Bitstream tab in the Instant Encoder window, additional settings appear for adjusting information that is read by a player. After the initial settings are created by default, they work for most projects and do not require additional changes. However, you may alter these settings if you have a good idea of the technical implications for each one. Otherwise, we suggest leaving these settings at the default values. In addition, certain options only become available based on the overall audio settings that you have made. For example, if you have chosen 2/0 (L, R) audio coding mode, then the center downmix and surround downmix options are grayed out to indicate that they are not relevant to the mode you have chosen.

Center Downmix

The Center Downmix settings indicate the level that downmixed audio should have if the playback system does not support audio with center channels. Values include –3.0 dB, –4.5 dB, and –6.0 dB. The default value is -3.0 dB and is generally the best setting for most audio.

Surround Downmix

The Surround Downmix settings indicate the level that downmixed audio should have if the playback system does not support audio with surround channels. Values include -3.0 dB, -6.0 dB, and -infinity dB. The default value is –3.0 dB and is generally the best setting for most audio.

Dolby Surround Mode

The Dolby Surround Mode setting indicates whether audio signal should include Dolby Pro Logic information for systems with only left and right stereo channels. Choices include Not Indicated, NOT Dolby Surround Encoded (apparently emphasized, so you know exactly what you are choosing), and Dolby Surround Encoded. The default is Not Indicated (see Figure 10-7).

Beneath the Dolby Surround Mode drop-down menu are two checkboxes. These boxes indicate whether Copyright Exists for the audio material and if the Content is Original. This additional information is simply added to a stream that can be read by some players and is not necessary — although including it for any legal purposes that may arise regarding the misuse of copyrighted material is a good idea. Both boxes are checked by default.

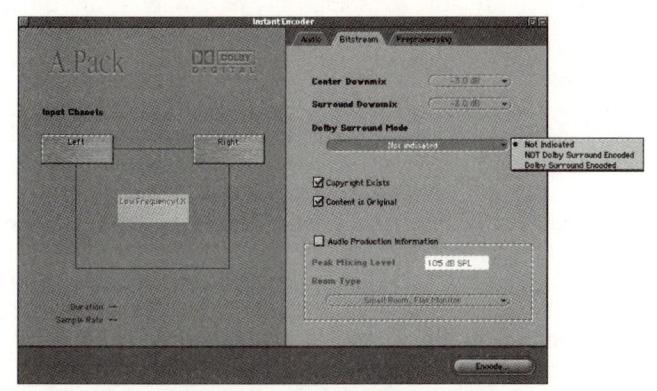

Figure 10-7: Dolby Surround Mode determines how audio with only right and left channels should be handled in regards to systems supporting Dolby Pro Logic.

Peak Mixing Level and Room Type

When you click the Audio Production Information box at the bottom of the Instant Encoder window, additional options become available. Audio production information settings place information in the stream to indicate how audio was mixed. This information can be read by some players to adjust settings for optimum playback of encoded material. The default value for Peak Mixing Level is 105 dB SPL and indicates the peak sound pressure level present in the environment that this mix was mastered. The default value is fine for most productions; otherwise the range may be changed from 80 dB to 11 dB by audio pros who know what they are doing.

The Room Type setting indicates information about the studio environment where the audio was mixed. The default value is Small Room, Flat Monitor, but other settings include Not Indicated, and Large Room, X Curve Monitor (see Figure 10-8).

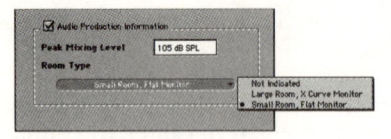

Figure 10-8: Optional audio production information passes data to a player about the room type (or studio environment) used to mix the audio.

Using preprocessing settings

Special settings that were made prior to encoding can be found under the Preprocessing tab in the Instant Encoder window. Again, default settings should be fine for most productions, but you are able to change any parameters that do not meet your specific needs.

Compression

By selecting a compression mode from those available in the preprocessing section, a preset dynamic range can be applied to your AC-3/Dolby Digital audio (see Figure 10-9). By specifying these settings prior to encoding, audio performance can be improved for any playback device. In general, the default Film Standard Compression should suffice, although additional options include Film Light Compression, Music Standard Compression, Music Light Compression, and Speech Compression for situations that demand it. For example, audio that is mostly comprised of spoken dialogue and few sound effects may benefit from the use of Speech Compression.

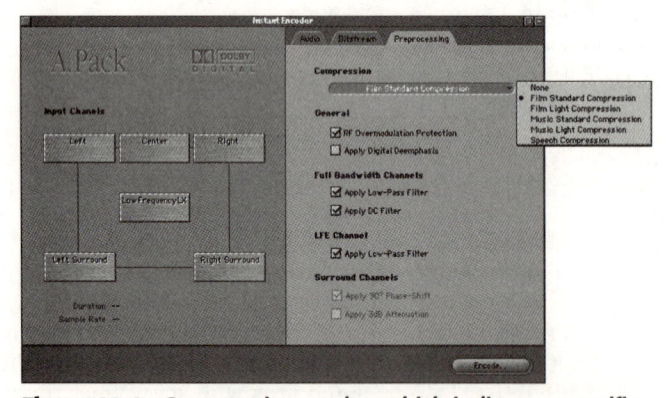

Figure 10-9: Compression modes, which indicate a specific dynamic range as part of the AC-3 standards, are applied in the preprocessing section.

General

Additional parameters may be indicated in this section by checking boxes on and off. Options include RF Overmodulation Protection and Apply Digital Deemphasis. The only option checked as active by default is RF Overmodulation Protection, which indicates whether an RF preemphasis filter should be included to prevent RF overmodulation in a decoder. Apply Digital Deemphasis indicates whether audio should be de-emphasized prior to encoding. Deemphasis is only important if you used another application to emphasize audio before bringing it into the A.Pack application.

Full Bandwidth Channels

This section indicates whether to apply special filters prior to encoding. Both Apply Low-Pass Filter and Apply DC Filter are checked by default. You may choose to turn off a default option if it doesn't apply to your audio files. Apply Low-Pass Filter employs a filter that removes frequencies at or above the given audio bandwidth for your project. Essentially, low frequencies pass unfiltered while high frequencies are removed, ensuring that your audio does not contain any high-pitched whines or other similar interference. You may turn off the low-pass filter if you know that your audio does not contain any frequencies outside of the available bandwidth, because A.Pack automatically determines the available bandwidth when the box is unchecked. The Apply DC Filter option adds a DC high-pass filter to remove DC offsets in material that has not been properly mixed. This is usually not a problem, but if you are unsure that DC offsets have been completely removed from your audio, then you should leave it checked.

Tip If you apply a low-pass filter to audio, and the low pass cut-off is set too low, your high (or treble) signals become indistinct and muddy.

LFE Channel

The Apply Low-Pass Filter option adds a low-pass filter to the Low Frequency Effects channel. This option ensures that unwanted frequencies above 120 Hz are not added to the low frequency (bass) channel. It is turned on by default and can remain checked unless you are certain that the channel does not contain information above 120 Hz.

Surround Channels

Apply 90 degrees is the only option in this section that is checked by default. By turning it on, you are assured that the AC-3 audio you produce can be properly downmixed for Dolby Pro Logic (Dolby Surround) systems. The Apply 3 dB Attenuation option is unchecked by default, but it should be checked if your audio originated on a film soundtrack and is being transferred to consumer video. Surround audio produced for theatres is usually mixed higher by a value of 3 dB and using the Apply 3 dB Attenuation option ensures that an adjustment is made to the audio file, which accounts for the offset in dB values.

The Batch Encoder window

Using the batch encoding options available in the A.Pack application enables you to encode multiple streams of AC-3/Dolby Digital audio at the same time (see Figure 10-10). By doing so, you can save the time it takes to watch over the encoding process, leaving you free time to get a bite to eat or go to bed for a few hours. Imagine falling asleep and awaking to find 10 AC-3 streams encoded and waiting for you! This feature is especially convenient if your project requires several audio streams, such as a DVD with multiple commentary and language tracks.

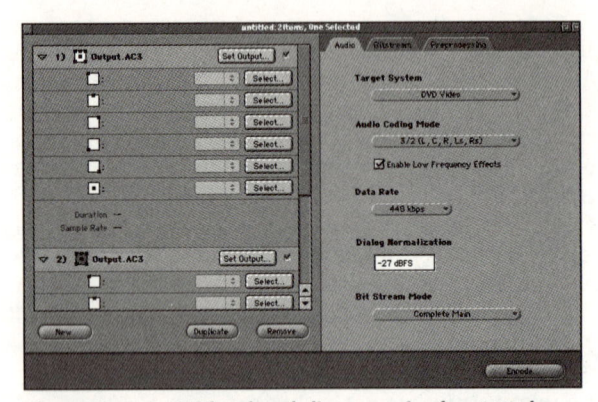

Figure 10-10: With a batch list open in the A.Pack application, several projects can be worked with and encoded simultaneously.

The window is divided into two sections, very similar to the Instant Encoder window. In fact, all settings available on the right of the window are the same as those available with the Instant Encoder. On the left side, you see a scrollable window, the batch list that lists your jobs. Jobs that require encoding may be added to the list by choosing Batch ⇨ New Job from the menu at top or by clicking the New button at the bottom of the window. The batch menu also enables you to remove jobs, duplicate jobs, and flag jobs for encoding (as do the buttons at the bottom). Flagging a job indicates that it should be processed with the batch. You may choose Batch ⇨ Flag For Encoding or Unflag For Encoding from the menu at top or you may simply click the blank space to the right of the Set Output button to place or remove a flag for a job. Encoding is then a simple process of setting the output, by clicking on the corresponding button, followed by pressing the Encode button.

Of course, before encoding you must assign audio files to the proper channels for each job on the list. To do so, expand a job by clicking the triangle and then select an audio file for each channel listed. Additional settings for the job can be set in the tabbed options to the right. After you finish, you can begin the encoding process. In a later section of this chapter, we walk you through the encoding of a project by using the batch list.

Encoding Audio for AC-3/Dolby Digital Using A.Pack

Before working with AC-3/Dolby Digital audio in the A.Pack application, you must mix your audio with six channels in mind. Unfortunately, A.Pack is used strictly for encoding files into AC-3/Dolby Digital, and the task of producing mixed tracks falls outside of its range of capabilities. However, what it does offer is the ability to convert your sound files to AC-3/Dolby Digital and assign audio to six different points in the sound field. In addition, A.Pack gives you the ability to monitor and play back AC-3/Dolby Digital files on your computer system by downmixing the file for stereo — an important feature for determining how the audio would sound on a standard stereo system.

Mixing audio

To produce proper AC-3/Dolby Digital audio, you should have six separate audio files — one for each channel. Generally, the majority of the sound should be placed in the stereo mix (front and center channels) and sound effects should be added to the surround channels. Spoken dialogue is easier to hear and to understand when placed in the center of a mix. Surround channels are best utilized with sound effects, such as explosions and atmospheric audio that accent the main stereo mix. Of course, sounds not included as part of a master mix may be recorded separately and added to the surround channels in a Dolby Digital mix. Working with audio in this way can yield acceptable results at times, but usually nothing as good as mixing in a six-channel environment. Whatever you do, make sure that you are thinking in terms of a surround sound environment and work accordingly.

Preparing audio for import

After you mix your audio in a separate application, you can save it as an AIFF, WAVE, QuickTime, or Sound Designer II file. A.Pack is flexible with the ability to import multiple audio formats as long as all files are using the same sample rate (typically 48 kHz, although 32 and 44.1 kHz are also supported) and the files are the same length. The A.Pack application usually adjusts files that are not long enough by adding silence where necessary, but it is a good idea to cut your audio files to equal lengths in an audio editor first, since the results with A.Pack can be unpredictable.

Using the Instant Encoder to produce AC-3/Dolby Digital streams

By using the Instant Encoder window, audio may be quickly and easily encoded into AC-3/Dolby Digital streams. However, if you want to encode multiple streams at one time, you must use the batch options discussed in the next section.

To encode an AC-3/Dolby Digital file with the A.Pack application, follow these steps:

1. **Click the A.Pack icon found on your hard drive.**
2. **On the left of the Instant Encoder window, click the input channel to which you want to assign an audio file.** You may also drag a file from your desktop and drop the file over the channel you want to assign. You are prompted to choose a file from its location on your computer.
3. **Select the file and click Open.** If the file contains multiple channels, you are asked to specify which channel from your file you want to select for that channel. Select the channel and click OK.
4. **Continue assigning audio files to all of your remaining channels.** Remember, you can set the audio coding mode, in the audio settings to the right, to determine the number of channels you are working with.

> **Tip**
> If your audio files are labeled with the proper suffixes they may be dropped on the Instant Encoder window and automatically assigned to the proper channels. The naming convention is .L for left front, .R for right front, .C for center front, .Ls for left rear, .Rs for right rear, and .LFE or .SUB for the low frequency, subwoofer channel. This naming system is also a useful method for keeping track of audio files and should be used consistently when possible (see Figure 10-11).

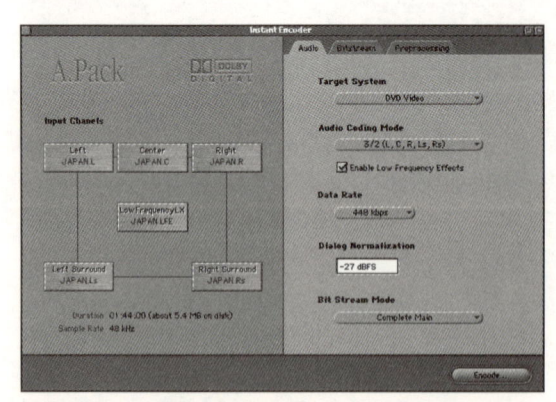

Figure 10-11: Input channels are easily assigned if they are named with the proper suffixes.

5. **Specify special settings for the project by using the settings to the right of the input channels.** After all of the channels have been assigned, and when you are done making changes to the job's settings, you can encode.

6. **Click the Encode button in the bottom-right corner of the Instant Encoder window.**

7. **Specify a name and location for the file and click Save.** The file begins encoding and you can monitor the progress in the box by watching the status bar and making note of the performance speed at the bottom and the time remaining to the right. Encoding can be discontinued at any point by clicking the Stop button (see Figure 10-12).

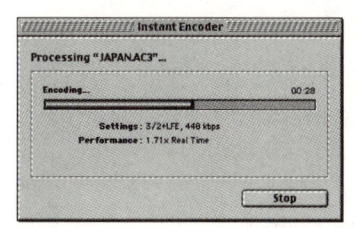

Figure 10-12: The processing window displays the status of the encode with a progress bar, time remaining, Settings, and Performance indicators.

8. **When your file finishes encoding, a Log window opens and displays information about the file you just encoded (see Figure 10-13).** Close the Log window when you are finished viewing it. The Log window can be opened manually at any point by selecting Window ⇨ Log Window for the menu at top.

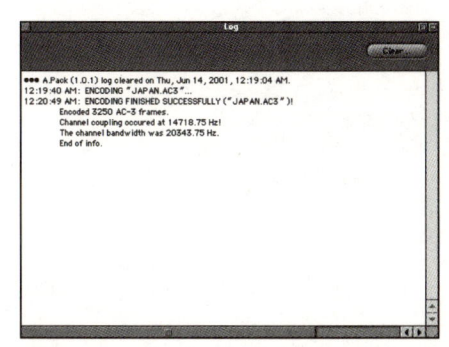

Figure 10-13: The Log window automatically opens following the encoding process and displays information about the file you just encoded.

Encoding AC-3/Dolby Digital tracks in batches

If you have several projects to encode, make your settings in the batch window and let the encoder run.

To set up a batch list for encoding, follow these steps:

1. **Start the A.Pack application by clicking the A.Pack icon found in the DVD Studio Pro folder on your hard drive.**
2. **Choose File ⇨ New Batch List to open the batch encoding window.** The list where you assign your files is on the left side of the window. The small graphics indicate which channel you are working with.
3. **Assign audio by clicking Select next to the proper channel.** Then choose a file from its location on your computer and press Open. You may also assign a file by dragging it over a channel or by dragging a set of files that are conveniently named with suffixes corresponding to the proper channel (see Figure 10-14).

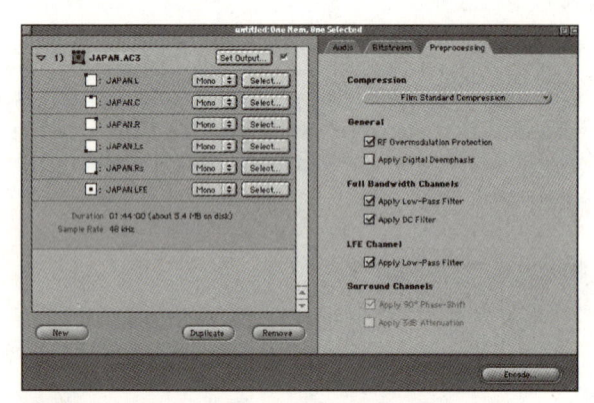

Figure 10-14: Properly named audio files make assigning channels an easy task in the batch list.

4. **Specify special settings for the job by using the settings to the right of the batch list.** The default settings usually suffice, but you have the option of changing any settings that may require fine-tuning.
5. **Specify an output name for the job you want to encode by choosing Set Output and typing the desired name in the Output Name window if it is not already there (see Figure 10-15).**

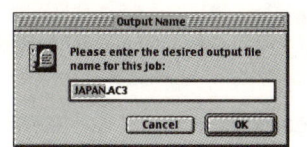

Figure 10-15: An Output Name must be specifed for a job in the batch list.

6. **After you finish working with this particular job, click the triangle on the left to collapse the list of settings (see Figure 10-16).**

7. **Create another job by choosing Batch ⇨ New Job from the menu at the top or by pressing the New button at the bottom of the window.** If you want to duplicate a current job, select the job you want to copy and click Duplicate or choose Batch ⇨ Duplicate Job. Jobs can be removed from the list by selecting the job and clicking Remove or by choosing Batch ⇨ Remove Job.

Tip

Move jobs to a different position in the batch list by clicking and dragging them to a new location.

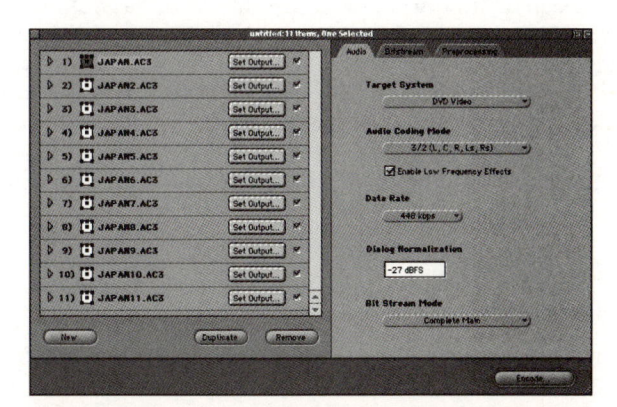

Figure 10-16: Several jobs can be viewed in the batch list by clicking the triangles to the left of the job to expand or collapse each one.

8. **Flag each job that you want to encode by clicking in the space to the right of the Set Output button or by choosing Batch ⇨ Flag for Encoding.** Flagging a job indicates that you want it to be added to the encoding process. Jobs that are not flagged do not process with the batch. You can unflag a job by clicking the mark placed next to the Set Output button again or by choosing Batch ⇨ Unflag for Encoding. After selecting all of your settings, you encode.

9. **Click the Encode button at the bottom of the window to begin the process.** A dialog box opens and asks you to choose an output folder for your batch files.

10. **Select a folder or create a new one and name it.** If any of the files in your batch list contain the same name, you are prompted with a dialog box that asks you if you want to replace or rename any files. Select an option and begin the encoding process (see Figure 10-17).

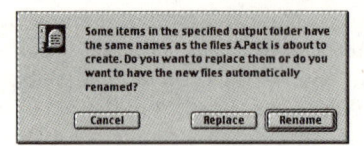

Figure 10-17: If any of the files in your batch list contain the same file names, A.Pack can automatically rename them for you.

Monitoring AC-3/Dolby Digital audio

AC-3/Dolby Digital audio can be monitored with the standard Mac sound manager, without the need for extra hardware. Do so by downsampling the multiple audio tracks into a special stereo mix. Of course, you may also choose to purchase hardware to listen to AC-3/Dolby Digital, which makes the monitoring (and mixing, if you have the applications) more accurate and enjoyable.

To set up A.Pack for the stereo playback of AC-3/Dolby Digital audio files, follow these steps:

1. **Launch the A.Pack application.**
2. **Choose Edit ⇨ Preferences.**
3. **Select Sound Manager as the Output System under AC-3 Monitor.** Your other choice is ASIO — hardware-enabled playback of real 5.1 channels of audio. ASIO requires special hardware and drivers to function but may be worth the price if you need to mix and monitor your own six-channel audio.

Now that we have touched on preparing audio in other applications, it is time to discuss how to use audio in DVD Studio Pro.

Using Audio in DVD Studio Pro

Working with audio in DVD Studio Pro is similar to working with video and other assets in DVD Studio Pro in the way it is used and managed.

Importing audio

As long as audio is in the AIFF, WAV, or AC-3 format, it is acceptable for import into DVD Studio Pro. Importing for audio is done in the same way as importing for other assets, such as video and graphic files.

To import audio into DVD Studio Pro, follow these steps:

1. **With DVD Studio Pro open, choose File ⇨ Import.**
2. **Select your audio file and click the Add button.** If you have more than one audio file that you want to import, select more files by clicking Add or Add All if an entire folder's contents are to be imported (see Figure 10-18).

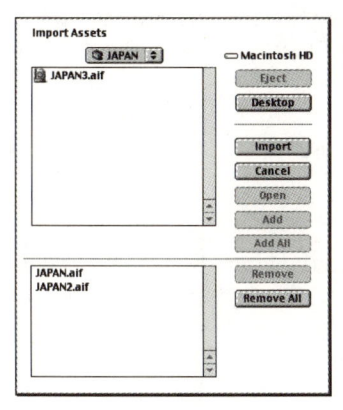

Figure 10-18: The window for importing assets enables you to add individual files or entire contents of folders.

3. **Click the Import button to finally bring the audio files into DVD Studio Pro.** Your audio assets are added to the Assets container and are ready to use as elements in your project (see Figure 10-19).

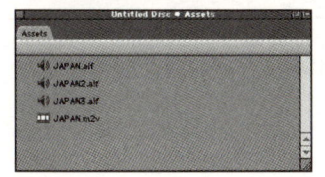

Figure 10-19: The Assets container holds all of your imported files, including audio.

Managing files

Working with audio in DVD Studio Pro also involves keeping track of your assets. The Assets container along with the Property Inspector provide you with information about your files (such as estimated file size, format, frequency, bits per sample and channels) and give you the ability to easily monitor and modify your files as needed. If, for example, you needed to replace an audio file with another file and wanted to avoid confusion without disrupting the order of your project, you can preserve the asset, and switch the file that the asset is linked to. Also, if you simply want to rename an audio asset, you can accomplish that task here.

To rename an audio asset in DVD Studio Pro and replace the file that is associated with an asset, follow these steps:

1. **In the Assets container at the bottom of the screen, click the file you want to modify.** Clicking on an asset to view its properties works the same way for any asset you want to work with, whether audio, video, or a graphic file (see Figure 10-20).

2. **In the Property Inspector, click in the name box and rename the file if desired.**

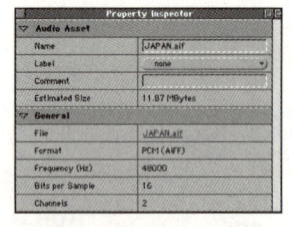

Figure 10-20: The Property Inspector can display detailed information for a particular asset, such as an audio file.

3. **To swap a newer file for an older one, click the blue, underlined file name in the General section of the Property Inspector to open a dialog box.**

4. **Select a new file by clicking it and then click Open to make the change take effect (see Figure 10-21).** The name of the new file assigned to your asset is now listed in the General section of the Property Inspector.

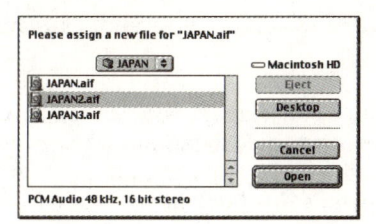

Figure 10-21: Choose a new file to assign to an existing asset.

> **Note** If you attempt to delete an audio asset that is already assigned you receive an error message that states, "Cannot delete Audio Asset" because it is referred to by the following items. This indicates that the asset is already in use and cannot be deleted without potentially disrupting your project (see Figure 10-22).

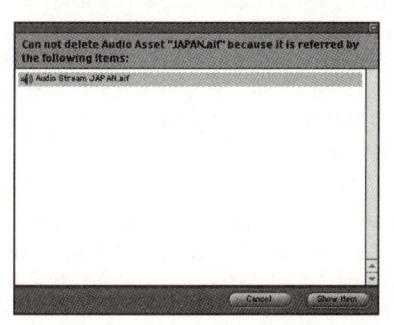

Figure 10-22: An error message appears if you attempt to delete an audio asset that is already assigned.

Adding multiple audio streams to a project

A key feature of DVD is the ability to have multiple audio, video, and subtitle tracks available at any point with the click of a button on the remote control. Fortunately, creating multiple audio streams in DVD Studio Pro is a simple process when your audio files are ready. If you need files to practice with, several video and audio files in the Tutorial section of the DVD-ROM that accompanies this book are included in the Japan folder.

> **Note** The maximum number of audio streams allowed on a DVD is eight.

The following steps describe how to import and add multiple audio streams to a video track:

1. **With the DVD Studio Pro application open, import your assets by choosing File ⇨ Import and selecting the files from their location on your hard drive.** If you are using the Tutorial files provided, they can be found in the Tutorials folder on the DVD-ROM that accompanies this book in the Japan folder. Select at least two audio files and a video file to import.

2. **Click the Add Track button at the bottom of the Graphical View to create a new track tile.**

3. **Drag a video Asset, such as Japan.m2v, over the track tile.**

4. **Drop an audio asset, such as Japan.aif, over the track tile.**

5. **Drop the remaining audio assets over the track tile.** Notice that the number beneath the audio streams icon on the tile increases each time you add a new audio asset. This indicates how many audio streams have been added to the track (see Figure 10-23). Your audio streams are now ready for use with the video track. When users click the audio button on their remote controls, they can cycle through the various audio choices that you created.

Figure 10-23: The track tile contains icons that display information about the number of assets added to the tile, such as 3 audio streams and 1 video marker in this example.

6. **Open the Audio Streams folder by clicking the audio streams icon on the track tile.** Audio assets may be reordered in the folder by dragging them to a new position. The position of audio assets in the folder indicates the order in which they appear when selected by a user pressing the audio button on the remote control. The first item in the list always plays first unless specified otherwise (see Figure 10-24).

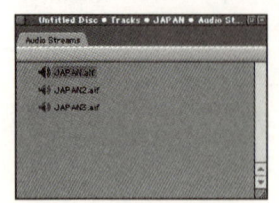

Figure 10-24: The Audio Streams folder displays your audio streams and lets you select them for modification.

7. **Click the audio asset to receive information about it in the Property Inspector.** You can change the name, assign a different asset or specify the language for each audio stream. Select the appropriate language you want to associate the file with from the drop-down menu (see Figure 10-25). You may also choose to leave the language unspecified if your project does not include multiple languages.

Figure 10-25: Assigning a language property to an audio asset is done in the Property Inspector.

8. **After you finish, preview your audio streams by clicking the preview button at the bottom of the Graphical View.** Use the audio button at the bottom of the preview window to select alternate streams. Your project now contains multiple audio for your listening enjoyment!

Summary

✦ PCM and AC-3/Dolby Digital audio are the only formats that work reliably with DVD.

✦ Third-party applications are required for the initial creation and mixing of DVD audio.

✦ A.Pack is a useful application included with DVD Studio Pro that can encode audio of varying formats into AC-3/Dolby Digital streams with up to six channels for DVD.

✦ Using the A.Pack application, AC-3/Dolby Digital audio can be tested and monitored after encoding or prior to encoding.

✦ Up to eight multiple audio streams can be added to project elements in DVD Studio Pro.

This concludes our chapter on working with audio. However, throughout this book you find numerous examples of working with audio in specific circumstances. In the next chapter, we switch gears a bit and begin the topic of understanding video. After the principles of video are covered, we discuss how to work with video, similar to the way we discussed working with audio in this chapter.

✦ ✦ ✦

Understanding DVD Video

CHAPTER 11

In This Chapter

Discussing the nature of digital video

Understanding the differences between various video and film formats

Looking at the file structure of a DVD

Understanding MPEG-2 compression

Encoding with variable bitrates

The emergence of digital video as a viable format has provided the video professional with a high-quality, low-cost medium for creating exciting video content. Of course, as with any new technology, you have new concepts and lessons to learn as well as new problems to be dealt with.

To work with video properly, first you need to have an idea about the many capabilities, limitations, and general terminology related to digital video. In this chapter, we introduce you to concepts that may help you to gain a broader understanding of the medium with which you are working. Depending on your background in video, this information may be completely new or simply a review. In either case, being aware of what the DVD video standard enables you to do is important. The awareness of its capabilities and limitations can help you develop better DVD projects. You also gain a fuller understanding of video production in general.

Enough general technical information about DVD technology is out there to satisfy any hunger. You can find books on the topic, Internet discussion groups and mailing lists, and increasing numbers of seminars — in short, enough information to deprive you of sleep for a long time. What we cover in this chapter could be elaborated on in several volumes, although the most you should need is a basic foundation. No matter what your level of technical interest, just remember that authoring DVDs can be a lot of fun. Working through the technical side frees you for the fun side.

Reviewing Digital Video Basics

Digital video is a relatively new term to most of us. Although digital technology has been around for quite some time, the technology has taken a while to make inroads into the consumer video markets. As computers become faster and software becomes more intuitive, the number of options for working digitally rises dramatically. Consider the first *real* computer you owned (perhaps it was an Apple IIc). The most you could hope to do with it was play basic games (mostly educational titles) and type school reports, all while viewing a *stunning* monochrome display on a system with 128KB of blazing memory. Gradually, the evolution of computers allowed us to create and edit still images, audio, and simple animations. In time, the home video camera and the home computer crossed paths in their development and we received digital video. The acceptance of this technology, and its migration into several areas of daily life and entertainment has led to many improvements in terms of quality and the free transmittal of ideas. However, like any new technology, many questions must be answered, particularly in regards to comparisons with older technologies. In this section, we address some of the basic elements of digital video.

Looking at frame rates

When a new video or film standard is created, decisions must be made regarding the size and motion of the images. Our perceptions of time and space rely greatly on the effect created by a series of pictures in a sequence. Just as different individuals have varying interpretations about what content should be shown so do opinions differ on how to deliver this content. These conflicting opinions led to some problems when trying to agree on exact specifications, particularly amongst different regions of the world.

Fortunately, film is a universal standard when it comes to the motion of images. With little variation, film is viewed at 24 frames per second (fps). Of course, you can find processes that differ from this norm, most of which have since passed into obscurity. For example, the Todd AO format, which ran at 30 fps, gave movies such as *Oklahoma* an added dimension of reality (or surreality), and the familiar 8mm format, spawning countless home movies ran at approximately 12 fps. Video, on the other hand, has always been a bit trickier.

Comparing film with video

Depending on where you live in the world, your television and other home video appliances are subject to standards, such as NTSC, SECAM, or PAL. These video size standards have a wide range of frame rates for displaying video, ranging from 30 (29.97) fps for NTSC to 25 fps for PAL. This one factor makes transferring video material between countries something of a problem. Certainly, we have machines that are capable of converting from one standard to another, but the process is costly and often impractical. See Figure 11-1 for an example of a software solution that tries to simulate the effect of film for video without the inconveniences of other costly processes. Although this way of doing things may not benefit the consumer, film companies do receive the added protection from piracy and security for trade of their product to different regions of the world where films have not yet been released or where rights have not yet been secured. However, the analog film medium is immune from incompatibility, playing on any projector (and screen) worldwide that supports speeds of 24 fps.

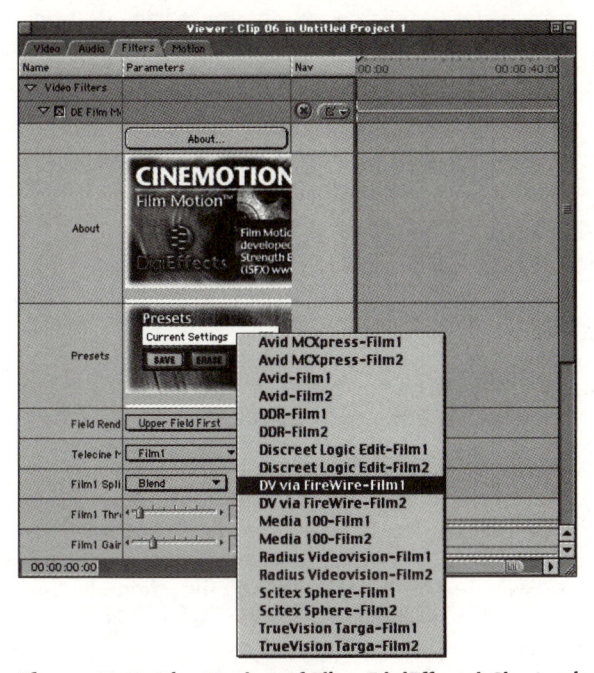

Figure 11-1: The Motion of Film. DigiEffects' CineLook application has a plug-in called CineMotion that adds an effect to simulate the motion of film in a video file.

The difference in frame rates causes a problem when you try to convert a film to video or vice versa. Film runs at 24 fps and does not even come close to the 30 fps required by NTSC video. PAL fairs a little better than its NTSC counterpart, because it runs at 25 fps and nearly matches the rate used by film. Getting film onto video, therefore, requires a special process called *telecine*, which accounts for the difference in frame rates between the two mediums. Many production companies and video post-production houses have suites devoted to this telecine conversion service, which is important to the great number of film and television companies that shoot projects on film before transferring to video. With PAL, the typical telecine process can be averted by slightly adjusting the speed to match the film.

Transferring film to video with telecine

The telecine process of converting film to video is often referred to as *3:2 pulldown* (or 2:3 pulldown), which refers to the technique of distributing single film frames over video fields in blocks of threes and twos (each frame of interlaced video is divided into two separate fields — refer to the section on progressive vs interlaced video for more on this topic). This method is the most reliable way that we have for converting material from a film source to video (see Figure 11-2).

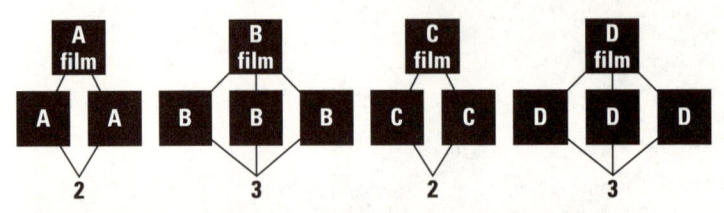

Figure 11-2: 3:2 pulldown is a process of converting film to video by assigning film frames to video fields in alternating blocks of 2 and 3. In this example, four frames of film are spread across five frames (or ten fields) of video.

3:2 pulldown makes use of the way that video processes fields, taking four frames of film and spreading them across five frames, or 10 fields, of video. The first frame is divided between two fields (or one video frame); the second film frame is placed across three fields (1 1/2 frames of video). This succession of frames is repeated until all film frames have been transferred.

Transferring video to film with inverse telecine

Inverse telecine, or transferring video to film, is fraught with many problems that until recently made it a less than desirable option for many filmmakers. However, as digital video technology and post processing applications such as Adobe's After Effects get better, an increasing number of high quality film-to-video options are becoming available (see Figure 11-3). Also, 24p (progressive) video allows for the exact match of frame speed for video to film. The success of features such as Lars Von Trier's *Dancer in the Dark*, and the ongoing support and advocacy for high definition (high definition video (HDTV) is a form of digital television (DTV) that greatly exceeds the resolution and performance of the standard definition television (SDTV) that we are currently used to) video-to-film transfers from people such as George Lucas, are evidence that film to video is a process that should continue to evolve. Many filmmakers now wonder why they should shoot on expensive film stock when they can shoot hundreds of hours of video for a fraction of the cost. Of course, even with all the advances being made with video-to-film transfers, the process is still prohibitively expensive for most low-budget independent filmmakers, not to mention that film still retains the crown when it comes to unmatched picture quality.

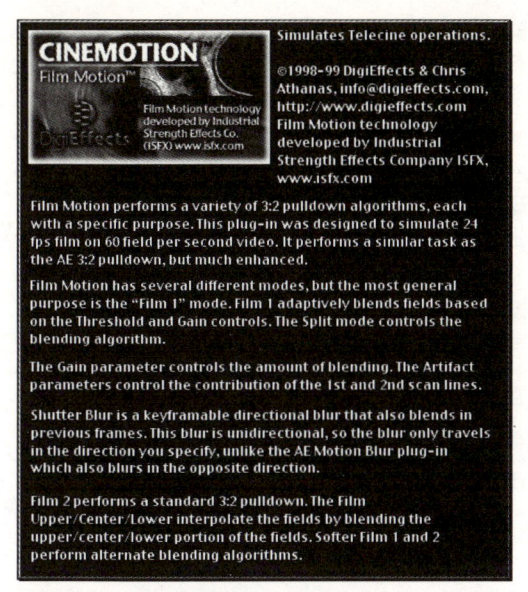

Figure 11-3: DigiEffects' CineMotion is a good choice for achieving smooth film-like motion for video originating from any editing system.

Understanding progressive versus interlaced scanning

The limitations of broadcast technologies have often required interesting workarounds. One of the most important of these workarounds is the way fields in video are dealt with.

For example, when televisions were first built, they were unable to properly handle the speed required for video. As a result, every frame of standard video is composed of two fields displayed in succession. This is known as *interlaced scanning*.

Think of a video picture split up into a series of horizontal lines. A rough metaphor for the ways these lines are drawn is to imagine two painters painting lines on a football field. One painter draws the odd lines, the 5th yard line, the 15th, and so on. The other draws the even lines, the 10th, the 20th, and so on. Each painter paints the lines in the field, but both of them have to do their part for the field to be complete.

When interlaced video is displayed, the first field is drawn, displaying an odd or even number of lines. This is followed by a second field, which displays the remaining alternating lines. The line order in fields is referred to as odd or even or upper and lower to describe which lines show first (see Figure 11-4).

Figure 11-4: Interlaced video frames are composed of two fields that display odd and even lines in sequence to create the illusion of a complete picture.

Thus, for every 30 frames per second of NTSC video, 60 fields are displayed, and for every 25 frames per second of PAL video, 50 fields are displayed. The perceived increase in frame rate caused by these fields creates a sense of video motion, which some think of as being more realistic. Whereas film and its 24 frames per second sometimes seems to have more of a feel of dreams and reverie.

The alternative to the successive fields of interlaced video is called *progressive scanning*. Progressive video is only composed of frames and does not contain fields. Instead, every line of video is displayed at the same time, producing crisp, defined pictures with a perceived frame rate that is a little closer to film. Digital televisions, DVD players, and cameras capable of displaying progressive video are becoming increasingly common, and as soon as you have seen the difference, you may be convinced enough to buy one for yourself. HDTV systems are required to use progressive scan technology, and more DVD players are being made to take advantage of this technology, as well. The good news is that your current DVD discs can almost always benefit from a progressive system. The way that progressive players read the data encoded on most discs is by providing a method for actually deinterlacing the material. This method is particularly effective if the footage on the disc originated from a progressive source, such as film.

Evaluating screen size

Now we need to discuss the differences you may encounter between screen sizes and aspect ratios. Until recently, the proportions for video and television sets have remained pretty standard. However, with the increasing popularity of cinematic mediums and the demand for more and better forms of entertainment, consumers are again shaping the direction of a technology.

Working with different aspect ratios

An aspect ratio is a way to represent the proportions of width and height (see Figure 11-5). For example, a screen that is 16 inches long and 9 inches tall has an aspect ratio of 16:9. Although film comes in a variety of aspect ratios (ranging from full frame 4:3 to 5.5:3 and Cinemascope's 7:3), video primarily comes in two flavors — 4:3 and 16:9. The standard color television set, which we have been using since the first color broadcasts in 1953, uses the 4:3 or 1.33 aspect ratio. Although using this size was somewhat of an arbitrary decision, it was primarily adopted from movies at the time of television's inception, which shared the same 4:3 ratio. As movie producers perceived the television as a potential threat to the future of their industry, they decided to enlarge the scope of the picture in movie theatres, projecting at ever-widening dimensions in an attempt to lure as many patrons to the movies as possible. As a result, today we now have televisions for sale with screens that nearly match the relative dimensions of their theatrical parents.

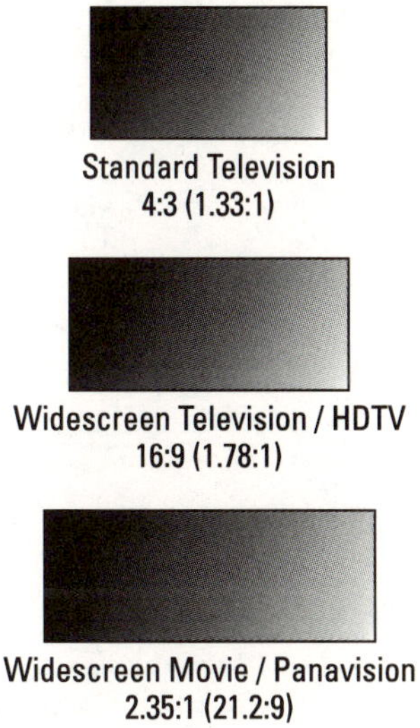

Figure 11-5: Aspect ratios for video and film vary widely, although the two standard sizes for video are 4:3 and 16:9.

Standard Television
4:3 (1.33:1)

Widescreen Television / HDTV
16:9 (1.78:1)

Widescreen Movie / Panavision
2.35:1 (21.2:9)

Deciding between anamorphic and letterbox for widescreen video

At some point you may decide to create a widescreen look for a video or DVD project. Before you decide on how to accomplish this, you should understand exactly what is meant by *widescreen*. Some people may think of widescreen video as a picture that contains black bars on the top and bottom areas of the screen to simulate the look associated with movies. This process of masking the screen is commonly referred to as letterboxing and does not necessarily mean that a video was shot with that aspect ratio in mind.

By applying a black mask to your video, manually or through the use of a special filter, you can crop the vertical dimensions of your picture to produce an effect that was first associated with serious films released on laserdisc. Such an effect is easily accomplished, for example, with Final Cut Pro's built-in widescreen filter (see Figure 11-6). In addition, After Effects plug-ins such as CineMotion by DigiEffects can create more complex and customizable widescreen looks, precisely simulating aspect ratios used by different film formats (see Figure 11-7).

Figure 11-6: A filter supplied with Final Cut Pro easily creates the letterboxed or widescreen look for any video by evenly cropping the vertical dimensions of a picture.

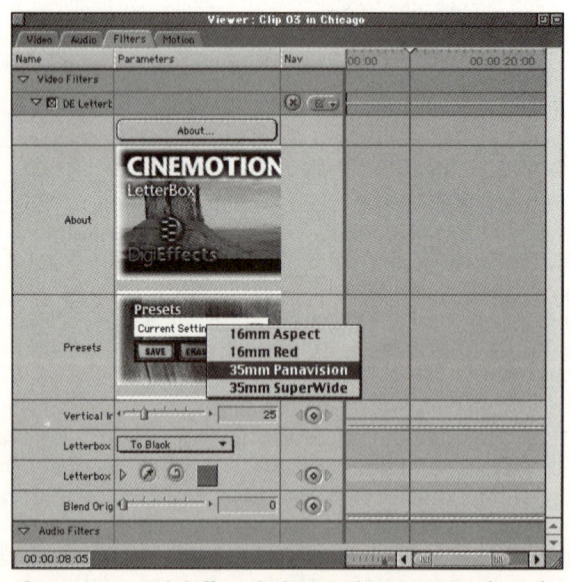

Figure 11-7: DigiEffects' CineMotion can simulate the widescreen look of specific aspect ratios.

Of course, simply cropping the picture does not make it widescreen in essence. It only has the *look* of widesceen. True widescreen results from movies or video that were originally shot at a wide aspect ratio, and then transferred to video without the loss of proportion, thus preserving the original composition intended by the filmmaker. Recently, more video cameras are coming with the option to shoot in 16:9 mode, a unique aspect ratio that does not exactly correspond to common film ratios, but which is a good compromise for next generation television sets and their displays.

HDTV for example, requires a 16:9 display, and the cameras that are used to shoot high definition video contain special 16 × 9 CCDs (charge coupled device) to properly capture that aspect ratio. However, there are some consumer video cameras that offer a 16:9 shooting mode using regular 4:3 CCDs. This is accomplished by squeezing the wider 16:9 picture into the 4:3 space (see Figure 11-8). Later, when the video is played back on 16:9 televisions or monitors, it is unstretched to reveal a picture at the correct proportions. The process of squeezing a wide picture to fit into a smaller space is the principle behind *anamorphic* video. Of course, this process is not the preferred method for working with video, especially when capturing it with a camera, because some detail may be sacrificed. If the video you are shooting is intended for 16:9 playback, try using a camera with a 16:9 ccd to get the best results.

Figure 11-8: True 16:9 anamorphic video looks squeezed when viewed with a standard 4:3 monitor, such as this doorknob — recorded in 16:9 — when viewed with the 4:3 viewer in iMovie.

Determining the Video Safe area

When you view video on a television or NTSC monitor, you are not seeing the complete picture. Beyond the boundaries of the normal viewing area are extra lines of video information referred to as the overscan. This extra space allows for anomalies in the video or display device to ensure a clean picture every time. Material that is always within the normal viewing area is said to be *title safe*, meaning that text is not in any danger of being clipped by the overscanning or the edge of a display.

The title safe area is actually smaller than the *action safe* area — the area of a video image where action occurs that is not in danger of being cut off unless it is viewed on very poorly-calibrated or older television sets that distorts a picture. Approximately five percent (or 36 pixels) of a picture from the edge of the screen is considered to be outside of the action safe area. If you use a computer monitor to preview the video, such as when working in an editing environment like Final Cut Pro, you can see these additional overscan lines. The overscan can be confusing, because what you see is not necessarily what you get in the finished broadcast product. For example, if you have carefully composed a scene to include activity near the edges, or if you have manually added text to a video, the video may appear to be cut off by the edges of your television. Having your video cut off by the television or monitor is not always a bad thing. Sometimes you may want to conceal imperfections such as a boom microphone, or you may want to slightly recompose a scene to center an object. The overscan lines give you a little space to work with the picture.

Video editors such as Final Cut Pro and motion graphic applications such as After Effects come with the option to turn on an overlay that defines the action safe and title safe areas of a video image (see Figure 11-9). This is very helpful for composing shots, such as when you are adding text for credits. To turn on the overlay in Final Cut Pro, choose Title Safe and Overlays from the View menu at top (see Figure 11-10). The innermost rectangle is the title safe area, while the next rectangle out from that is the action safe area. Beyond the action safe area is the portion of the image most probably affected by the overscan.

> **Note** Remember to make sure that all of your video materials conform to the video safe standards described above before importing into DVD Studio Pro or encoding into MPEG-2. DVD Studio Pro is simply an application for assembling assets that have already been prepared and not a video or image editing application where adjustments to the picture can be made.

Chapter 11 ✦ **Understanding DVD Video** 259

Figure 11-9: Title safe and action safe areas can be determined with a simple graphic overlay.

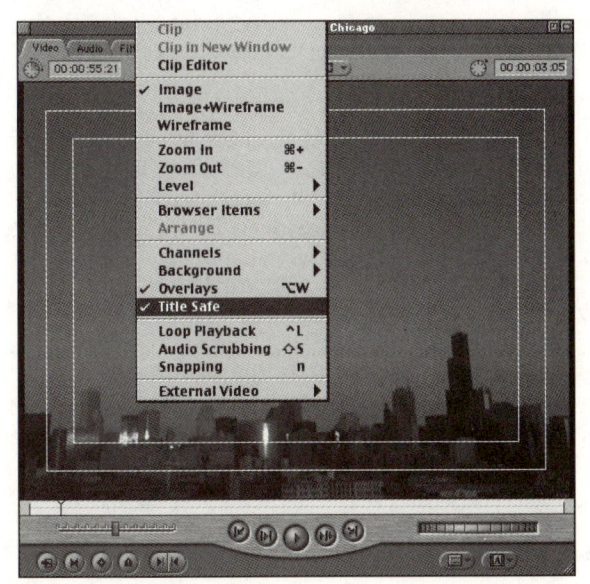

Figure 11-10: Choosing Overlays and Title Safe from the View menu in Final Cut Pro activates the overlay that indicates video safe areas.

PAL versus NTSC

In addition to their differences in frame rates, PAL and NTSC video also exhibit a difference in resolution. Although the aspect and screen size between the two standards remains the same, the number of lines that PAL can display (720 × 576 pixels for DVD) is greater than NTSC (720 × 480 pixels for DVD). In part, the extra bandwidth that PAL's 25 fps has over NTSC's 30 fps allows PAL to focus more attention on increased resolution. The increase in picture quality and PAL's close match to the motion of film makes it the preferred shooting format for video when it is intended for transfer to film. Whether you live in a PAL country such as England or an NTSC country such as Japan, you should consider shooting in PAL if you are producing video as a film print for theatrical release. Doing so makes the conversion job much easier for the production house you choose and gives you an additional increase in picture quality, which can be important for the comparatively low-quality medium of video. You can work in either PAL or NTSC in DVD Studio Pro.

Considering chroma and luminance for digital video

Another quality issue to be aware of when working with video is that video uses a set sampling rate for chroma and luminance values. Digital video for NTSC uses a 4:1:1 sampling rate, meaning that the chroma (color) is sampled once for every four horizontal luminance samples. The *4* in 4:1:1 refers to the luminance signal or Y, and the proceeding 1s refer to the color signal – the first 1 is the actually the color red with the luminance deducted (R-Y), and the second 1 is the color blue with the luminance deducted (B-Y). Together, these values are sometimes referred to as the YUV.

Video sampling rates make use of the knowledge that the human eye places more importance on brightness (or luminance) levels than it does on color. Sample rates vary for different video formats. Certain formats, such as analog Betacam, use 4:2:2 sampling that provides what many professionals consider a better depth of color. Of course, the lower sample values for color in a DV signal allows more space for video on a tape, while almost imperceptibly reducing the quality of color. DVD uses 4:1:0 sampling, which is only a little different than standard digital video.

The DVD video standard uses digital video in ways that differ a little from what we have discussed so far. In the next sections, we talk about what sets DVD video apart from other standard video formats.

Exploring the DVD Video Standard

As a technology and a consumer format, DVD has gained widespread acceptance in a very short period of time. In just a couple years, DVD player purchases have exceeded the demand for the ubiquitous VCR. Of course, no one is throwing out their VCRs just yet, because VCRs are still useful recording devices — although in a few years digital recorders may find their way into the home where DVD discs are used to store episodes of your favorite TV shows (if dedicated boxes based on hard drives or video-on-demand mechanisms don't eclipse them first). So, what is it that makes DVD videos and players so great? Technical specifications for DVDs can help you understand the format even better.

> **Tip** If you are reading this book and haven't ever seen a DVD, go to an electronics retailer and try out some of the systems. Rent a few DVDs and put them in a player or on a computer with a DVD-ROM player. You gain a better appreciation for their common features, and you get a better sense of what people expect in a DVD.

Evaluating image quality

Image quality is the most obvious advancement that DVD has over older video technology. Although DVD is not the most perfect video format available (HDTV and other professional studio formats are superior in many ways), DVD is a terrific compromise on quality versus price of competing technology. For example, the prohibitive cost of Digital Beta decks greatly limit their appeal as home entertainment devices. Even so, the picture quality of DVD can often rival the best studio masters when produced properly. Note that just because DVD is a digital format, the digital nature does not necessarily make it better than an analog medium. Film is an analog medium, but no one would argue that DVD produces a better picture than film.

DVD video is limited to the MPEG format, which includes MPEG-1 and MPEG-2. The most obvious quality difference between the two is the resolution — for NTSC video; MPEG-1 has a resolution of 352 × 240 pixels, and MPEG-2 has a resolution of 720 × 480 pixels. Additionally, MPEG-2 can display true 16:9 video on a widescreen TV while MPEG-1 cannot.

Over two hours of digital video can be placed on a single layer DVD-5 (4.7 GB) disc. Of course, this information is compressed and does exhibit some loss of quality from the studio master. If a disc is poorly produced, such as if the compression process is rushed or if the proper data rates are not used, some visual artifacts, blocks or fuzziness for example, may appear. For the most part, encoding technology has sufficiently progressed so that DVD video is getting better as time goes on.

Evaluating sound quality

Sound quality equivalent to what DVD offers has actually been around for decades. Laserdiscs offered PCM (pulse code modulated), AC-3/Dolby Digital, and even DTS soundtracks before DVDs were even a blip on consumer radar. Videophiles who were already accustomed to high-quality surround sound with their movies expected nothing less than perfection from DVD when it was first released. What they got was the ability for very rich audio, albeit slightly compressed in many cases. In particular, AC-3/Dolby Digital on DVD is an audio format that employs new digital technology to compress large, multichannel files into more manageable sizes. Although a good deal of compression appears to be applied, AC-3/Dolby Digital is actually a very clear and defined format that even most audio perfectionists agree is rather good.

DVD Authoring – Perspective from a Video Professional

Interview with Martin Baumgaertner of Angle Park, a Chicago Mac-based video company specializing in projects for corporate and museum clients (www.anglepark.com).

Have you considered using DVD or DVD-ROM to author video or interactive content? If so, why?

I've only recently considered using DVD for video content, because up until the latest generation of tools the authoring process was convoluted, too slow, and way too expensive to be practical for our clientele. It's also been my opinion that the interactive aspects of the medium have been too limited and difficult to implement to be worthwhile, especially when compared to CD-ROM and internet development tools. But as DVD tools have gotten better and cheaper, my opinion has started to change.

What technical qualities do your clients look for in a completed project?

That's an interesting question. Truthfully, our clients — companies and museums — don't usually consider the technical side of what we do. They're more concerned with the message, so technical considerations are always secondary to a project's communications objectives and creative approach. Technical consulting is just a part of the service that we provide to them.

I think there's a culture of blind technology worship inside the production industry that doesn't necessarily translate into anything meaningful for our customers. Our clients often don't care at all how we're using cutting edge technology; they just want to see the results. Witness the success of "The Blair Witch Project" or the popularity of "South Park" for examples of this in popular culture. These projects certainly have a complex, sophisticated (if maybe a little off-color) message, but they also prove that content is more important than "technical quality." Many people in our industry don't get this. According to everything I've ever learned about production values, "Blair Witch" looked like crap. That didn't seem to hurt it much.

So "technical quality" is a loaded term. That said, for many programs that we produce, technical quality is very important. This may be for purposes of illustration, clarity of meaning, or nothing more than good old aesthetic impact. These can be a critical component for any program, but for us, the medium is not the message. It's just another way to bridge the gap between content and participant.

Is DVD a significant advance in delivering quality images?

DVD's MPEG-2 foundation offers image quality that is far better than VHS. That's a major advance, but that's only part of the "delivery equation." DVD is still a piece of physical media that has to be replicated, packaged, and shipped from point A to point B. Considering these logistical aspects of delivery, it's no better than VHS. In fact, it's harder to work with because it requires authoring and encoding once you've mastered a program.

In addition, there are relatively few replicators that can do DVD when compared to VHS, but that will probably change in short order. There's also the question of compatibility with playback devices that most people have on hand. At this point in time, I can count on just about everyone having a VHS deck, but it's still not guaranteed that every home or office has a DVD player. However, market trends seem to indicate that this will also change in very short order.

What is your opinion of MPEG-2 compression as an alternative to other compression architectures such as Sorensen or RealVideo? Compared to other professional (broadcast) formats?

MPEG-2 is not really an alternative to Sorenson and Real, unless you consider physical distribution of DVD an alternative to Internet streaming. Sorenson and Real are similar technologies to MPEG-2, but developed for low-bandwidth Internet streaming. MPEG-2's bit-rates — usually measured in the megabits — cannot be currently delivered practically over IP. MPEG-2's really designed for a completely different delivery method, such as off of high-end tape machines in the broadcast plant like the Sony IMX lines, or servers, or DVD. And for these applications it's a good match between bit-rate and quality.

I've done LAN experiments with both Real and Sorenson at relatively high bit-rates, such as 2 to 8 megabits/second, and I can report that they don't work very well. Of course, they're not designed for that, either, so it's not really a fair critique. For high bit-rate delivery, such as via servers or DVD, MPEG-2 is a much better choice. But the reverse is true, as well. For applications under one megabit of bandwidth, I'd reach for an internet streaming codec, not MPEG-2.

What do you foresee happening with DVD in the future?

DVD will certainly supplant VHS for movie rental and purchase in the near to medium term. That's definitely going to happen, because it's a good match between medium, content, and consumer behavior. Long term, though, I think that companies like Blockbuster will transition to a combination of streaming media and caching technology to "rent" movies to people in their homes. This won't necessarily be the easiest transition, because they will have to overcome human nature — our attraction to shiny physical objects. People like to hold and own things, and they've been trained for 20 years to go to a store to get a movie. That's not easy behavior to change, but it can be done. Until then, DVD will be the movie rental king.

I don't run a Hollywood studio, though. For my business, DVD offers possibilities now, but streaming media is the end game. For now, though, streaming media has two big challenges. First, ubiquitous broadband connections are still years off. Second, there is a format war between the three major formats. Both of these problems will be addressed in due time. While it's anyone's guess how long that will be, you can't ignore the fact streaming media solves the problem of the cost and logistics of physical media distribution. That will eventually make DVD a niche player.

In the meantime, we try to bring our customers the best possible options to meet their objectives, today. To this end, DVD solves a number of problems, but it doesn't solve them all. The day is soon approaching when we can count on DVD players in every boardroom, living room, laptop, and desktop computer. For us, however, the question will also be: Do those board rooms, living rooms, laptops, and desktop computers have a broadband connection?

DVD also allows for PCM and MPEG audio. PCM audio for DVD, also referred to in this case as *linear PCM* audio, is an uncompressed format similar to audio CDs. Shorter projects or those projects requiring pristine stereo tracks may prefer to work with this format, because it offers 24 bit audio with a sample frequency of 48 or 96 kHz and a maximum of eight simultaneous channels. MPEG audio is not a common choice for DVD audio, because the only video standard that supports this choice is PAL. Essentially, MPEG audio is similar to AC-3/Dolby Digital in sound quality, with the exception that MPEG allows for up to eight audio channels.

Investigating the File Structure of a DVD

A *multiplexed* DVD project is a DVD project that is composed of several files that contain data for a finished project. The organization of these files on a DVD is very important to the proper operation of a disc, which looks for data in a specific order. Essentially, all data is stored in a sequential order, beginning at the top of a standard hierarchy. At the top of the chain is the VIDEO_TS folder that contains all of the multiplexed DVD video files. Within the VIDEO_TS folder, you find Video Title Sets (VTS) and three main types of subfiles, with extensions labeled .VOB, .IFO, and .BUP (see Figure 11-11). Understanding what each of these files does can help you to locate files more easily in a multiplexed project file or disc.

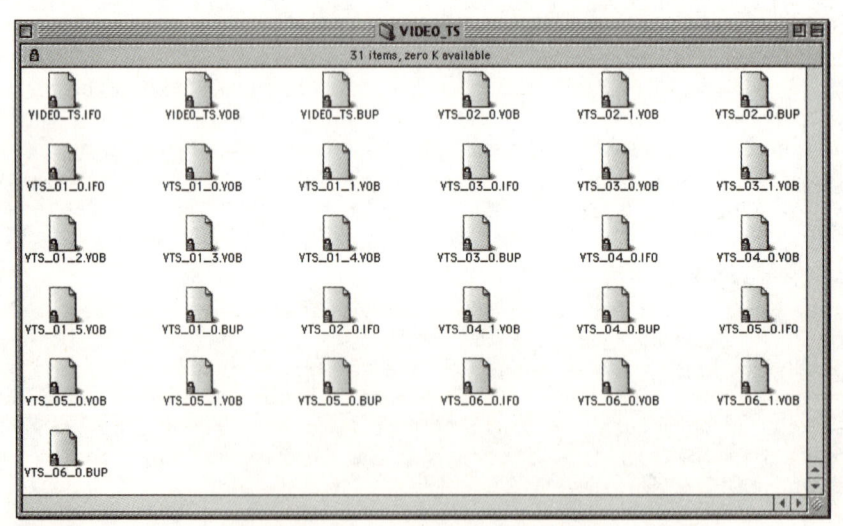

Figure 11-11: DVD projects contain several types of files to organize multiplexed video, including VOB, IFO, and BUP files.

Video Title Set

Within the VIDEO_TS folder, a DVD is further divided into Video Title Sets. The Video Title Set (VTS) contains titles along with information about the titles, such as where they are pointing. *Titles* are the movies or tracks contained on a disc and represent the largest single entities apart from the VIDEO_TS folder. Entire videos or movies are usually contained within a single title, with additional titles used for supplementary materials, such as trailers and documentaries on a studio produced disc. Each section of the title cannot exceed a file size of 1GB (a maximum of approximately .99GB).

Presentation files (VOB)

VOB files, or Video Objects, contain the multiplexed audio, video, and subtitle streams from your project. They are the fundamental elements in a DVD project.

Navigation files (IFO)

IFO, or Information, files contain all of the formatting information for the VOB files in your project. They are responsible for giving a DVD player the information it needs to enable navigation of a disc and to set up display options such as the aspect ratio, subtitles, languages, and menus. The VIDEO_TS.IFO file actually contains all of the data that is required to navigate the overall DVD.

Backup files (BUP)

The BUP, or Backup, files contain duplicate information for the IFO files in case the IFO files fail to function due to a scratch or corruption of a file.

Examining the MPEG-2 Format

Although the DVD specification allows for either MPEG-1 or MPEG-2 video to be used in a project, the quality that we have come to expect from DVDs results from the higher quality MPEG-2 format. An organization that meets under the acronym of MPEG (Motion Picture Experts Group) decided the compression standards that we use for video and audio on DVDs as well as for other delivery mediums such as the Web, where MPEG can be employed. By using existing compression methods as a starting point, MPEG expounded on these methods by introducing a unique frame method that alleviates further redundancy in picture information. See Figure 11-12 for an example of an MPEG encoding solution.

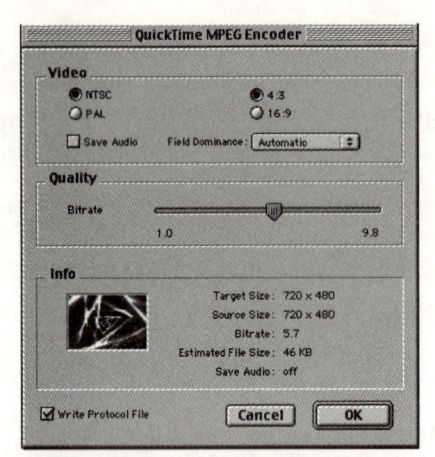

Figure 11-12: The QuickTime MPEG-2 encoder included with your purchase of DVD Studio Pro enables you to produce high quality MPEG-2 video streams for use in a DVD project.

Compressing video with MPEG

MPEG compression uses a unique encoding scheme for working with image sequences. The process begins by employing a technique commonly used with JPEG compression called DCT or *discrete cosign transformation*. What DCT does is break pictures into blocks of information based on the level of detail in the picture. Then the amount of information is reduced by quantizing (standardizing) the resulting values. Manageable file sizes can be produced for single pictures by using this technique; however, this technique is insufficient for long sequences. Fortunately, MPEG developed the process further to include three different types of frames that can be used in a sequence, which reduces the amount of redundancy in the compression of moving images. The three types of pictures (or frames) are intra frames, predicted frames, and bidirectional frames.

Intra frames

Intra frames, or I-frames, are similar to key frames in other video applications. This type of frame contains a complete set of information for a picture and is used as a reference for all other computations. At the beginning of the encoding process, I-frames are compressed by using the discrete cosine transformation we just mentioned. I-frames are usually added to a sequence at regular intervals (typically, at least two per second) or when a major change occurs from one picture to the next, such as when a cut occurs. Complex scenes may require several I-frames, although the more I-frames that are used, the more memory is required to store the resulting video. The need for additional memory is the main hindrance to using only I-frames as a basis for your video. I-frames are the easiest frames to create and can be viewed by themselves without the need for other frames.

Predicted frames

Predicted frames, or P-frames, store only the information that has changed since the last I-frame or P-frame. By first breaking pictures into blocks (macro blocks), P-frames use motion vectors to calculate where these blocks have moved from a previous frame. Then, using DCT calculations, they compress and store this information along with complete sections of blocks that have changed from the last I or P frame. Because P-frames do not contain a complete set of information, they cannot be played without I-frames.

Bidirectional frames

Bidirectional frames, or B-frames, look at frames in front and in back of their position to match blocks of information that have changed. B-frames are similar to P-frames in the way that data is compressed by using vector information. These frames are very efficient at compressing this data, although they require some intense, time-consuming calculations. Some encoders (especially on real-time systems) drop B-frames altogether, because the time required to make the necessary calculations is not possible or practical.

MPEG introduces some new concepts when it comes to encoding video. Another encoding element that you should understand is how choosing one of two types of bit-rates may affect the finished DVD project.

Variable Bit-rate Encoding

Two main types of bit-rate encoding are used for MPEG-2 compression. The first and simplest is constant bit-rate encoding (CBR). This process merely sets a constant bit-rate for the entire project, which it does not deviate from while encoding. The problem with CBR encoding is that it assigns the same amount of importance to every frame or sequence, regardless of complexity, resulting in the inefficient use of storage space and the possible degradation of quality for difficult scenes. Variable bit-rate encoding (VBR) overcomes the problem of encoding long, memory hungry videos by first looking at a scene to determine the complexity and then setting an appropriate bit-rate to encode the scene properly. Complex scenes get more memory (or bit-rate) assigned to them, and simple scenes with little action or variation in motion and picture get a lower bit-rate value, allowing the maximum space on a DVD (for example) to be used as much as possible.

Another form of variable bit-rate encoding is called two-pass variable bit-rate encoding, and this form is the preferred method for producing high-quality MPEG-2 streams. Two-pass variable bit-rate encoding looks at an entire sequence before attempting to encode it. By analyzing the video first (prior to actually encoding), an optimal bit-rate can be set for each section of video, making certain that difficult passages are allotted an appropriately high rate, and simple scenes get less.

DVD video and digital video in general are worth exploring. Hopefully, you now have an idea about a few of the many topics. The following summary reviews the topics that we have discussed in this chapter.

Summary

- Film and video have different frame rates and aspect ratios that can make working with multiple formats problematic.
- Widescreen video can be created either by artificially masking the picture or by working with material that was first shot in the 16:9 ratio.
- When you are composing video and graphics, keep in mind additional lines of video, or overscan, are often cut from the normal viewing area of a television set.
- DVD video is a vast improvement over past consumer formats, but it is essentially a compressed format that is not equivalent to some of the better analog or digital devices available.
- Multiplexed DVD projects are composed of several different file types, containing video and other assets that are arranged in a strict order.
- MPEG compression was developed to encode high-quality video in a relatively small amount of space.
- Three types of frames (I-frames, P-frames, and B-frames) are used in encoding MPEG video. These three frame types provide the ability to remove as much unnecessary information as possible to reduce the overall size of the file.
- Variable bit-rate encoding (VBR) is often superior to constant bit-rate encoding (CBR) because it can fit as much video as possible on a disc, while maintaining higher quality when encoding complex scenes.

In the next chapter, we put some of what we have learned about DVD video into practice. Although you are composing your projects, think about what the different elements you are working with do and how they fit into the scope of digital video. Logically, deconstructing the particulars of the video medium may help you find solutions to problems that you encounter.

✦ ✦ ✦

Working with DVD Video

CHAPTER 12

In This Chapter

Understanding what video formats are compatible with DVD

Working with video in other applications

Using MPEG-2 video encoders

Working with video files in DVD Studio Pro

In the last chapter, we discussed the nature of video as it is used for DVD projects. In this chapter, we discuss the practical side of DVD video — namely, how to create it for use in DVD Studio Pro. It is important to remember that while video may originate from any source, it must first be specially prepared for inclusion on a DVD. Essentially, this involves converting it into MPEG video, the compression standard developed to increase the quality of video while decreasing the memory required to store and move it. Of course, there are other methods for compressing video, but many are not without significant sacrifices in quality. With MPEG we now have a good balance between size and quality.

When working with video for DVD, the most important considerations are which formats are compatible with the DVD standard and which format should you choose for your project. Another consideration is what program to use to prepare the video. While this last consideration is largely based on personal preference and budgetary considerations, we discuss a few programs in this chapter that are representative of the capabilities you would look for in most applications.

In addition to working with video in third party applications, we discuss what to do with the video within the DVD Studio Pro environment. This includes adding video tracks to a project and setting markers to provide instant access to specific scenes or times in the video. Working with video in DVD Studio Pro is actually quite simple once you get the hang of it; the challenging part can be creating quality video to begin with. So, if you haven't already, break out your footage and start creating!

Evaluating Compatible Video Formats

Before thinking about the programs you can use to prepare video, consider which formats are compatible with the DVD standard and decide on a method for creating high quality video. For DVD, your choices are either the MPEG-1 or MPEG-2 format (refer to Chapter 11 for an in-depth discussion of MPEG and other video formats). If video quality is an important concern for you (which I assume it is if you are creating DVDs), then you must use MPEG-2. Almost any DVD that you purchase in a store is composed of MPEG-2 video, with the exception of some bargain non-major, studio titles. MPEG-2 is responsible for the quality that you expect from a DVD versus what you can get from MPEG-1 (MPEG-1 is similar to VHS quality).

If you are trying to squeeze the most video storage space out of a disc, then MPEG-1 may be a consideration, especially for material that is acceptable at lower qualities, such as TV shows or video that originated on CD-ROM or the Internet. VCDs (video CDs), a somewhat popular format for Asian imports or bootleg videos, are examples of discs that use the MPEG-1 video standard. There are players that are made specifically for VCDs, but most DVD players will also play VCDs.

> **Note** If you want an idea about what MPEG-1 quality is like, pay a visit to www.videocds.com to purchase your favorite film on VCD, and check it out on your DVD player. The difference in quality is definitely noticeable, yielding results that approximate VHS, or nearly half the resolution of MPEG-2.

In order to create video for DVD, you must begin with a format that can be encoded into MPEG video. Considering that you are reading a book about DVD Studio Pro, a program that is solely available for use on Macintosh computers, the assumption is that QuickTime is your format of choice for video. QuickTime files can be created on PC platform computers as well and can be transferred to your Mac by any number of large storage mediums such as CD-R or DVD-R, but it is likely that you are doing most, if not all of your video, on a Mac. In this case, you are probably using a video editor such as Final Cut Pro, Premiere, Media 100, or Avid to cut your video before compressing for inclusion on a DVD project. While several very capable programs are available for editing video, we are going to primarily discuss working with Final Cut Pro, Apple's flagship video editor, along with other third-party applications for preparing and compressing your finished video.

Preparing Video in Third-Party Applications

Although DVD Studio Pro provides many tools for creating a finished DVD project, it does not enable you to edit or encode video into other formats (such as encoding RealVideo or QuickTime into MPEG) within the program. These tasks must be accomplished outside of the program before creating a DVD. Fortunately, many great third-party applications are more than capable of handling these tasks. While they are an additional investment (you may already own some of these programs), by purchasing them, they can pay for themselves in other ways, unrelated to the creation of DVDs. For example, Cleaner, one of the programs we focus on in this chapter, is the industry standard application for encoding a large variety of media formats, most significantly video — although it works just as well at converting audio and graphic files.

Final Cut Pro

Apple's Final Cut Pro is the editor of choice for many video professionals working on Macs. It is not uncommon to see Final Cut Pro systems sitting alongside an Avid in many post houses these days. (A *post house* is a company that takes video footage and works on the editing and final adjustments, also known as post production.) Aside from its easy-to-use and feature-packed interface, a large part of Final Cut Pro's appeal is its ability to work with virtually any video format, ranging from DV (Digital Video) to HDTV (High Definition Television).

Final Cut Pro's popularity continues to grow as more real-time options become available. One such option is the RTMac card created by Matrox. This card allows users to work with three layers of video and graphics simultaneously without rendering a sequence for playback. Fades, transitions, and opacities are also available in real-time, greatly reducing the time it takes to put together a video project.

A video-editing program such as Final Cut Pro is important for assembling a DVD, not just to produce the finished video or movie that makes its way onto a disc, but also for creating motion menus through the creation of loops and assembling of layers. In Chapter 6, we discuss how you can use Final Cut Pro for these tasks. Of course, you can use almost any video editing application to create loops, although some of the lower-end programs may not have the built-in effects and compositing options that Apple's Final Cut Pro or even Adobe's Premiere have as standard features. In addition, both Final Cut Pro and Premiere offer the ability to work with many After Effects plug-ins, greatly increasing their abilities for creating specific effects. Working with After Effects plug-ins within a video editing application can reduce the cost, since you wouldn't necessarily need to purchase After Effects, and the time involved with switching between applications when you want to do a simple effect or composite.

> **Note** A 30-day trial of Adobe Premiere 6 is available on the DVD-ROM that accompanies this book.

Cleaner

Terran's Cleaner 5 is an incredible application for working with a variety of media formats. In addition to offering options for encoding MPEG video, it is also useful for creating RealVideo, QuickTime, and AVI files for streaming or downloading over the Internet. For audiophiles, it also includes the ability to encode audio formats such as MP3, a popular format for distribution over the Internet. Users are given the option of working with simple step-by-step processes or choosing advanced features that allow for the modification of every parameter imaginable. Cleaner 5, the latest version of what was once called Media Cleaner, adds the capability of EventStream authoring — a revolutionary process of syncing events to key frames or hotspots in video tracks to activate a variety of functions, including the triggering of pages and links in a Web browser. This feature, borrowed from other Media 100 products, should appeal to multimedia creators looking for greater interactive control over Internet video, much like DVD for the Web.

For our purposes of creating DVDs, Cleaner's video features can prove extremely useful, even aside from its powerful encoding and conversion options. One such feature is the ability to set In and Out Points on video files to create smaller, nondestructive files from a larger original file. A *nondestructive* file is a version of a video clip in which you can make changes without affecting the original file (in other words, nondestructively). This feature prevents you from having to go back into a video editing application such as Final Cut Pro to make multiple versions of a finished video. Instead, you can cut out a scene for inclusion in a motion menu, or make last minute trims to the starting and ending points of a movie. For example, if you have a QuickTime file of a completed video and you want to include portions of it as backgrounds for a few motion menus, don't bother with Final Cut Pro, simply open Cleaner and start making your edits. Of course, you may feel more comfortable working exclusively in a video editor — you may wish to try creating edits in Cleaner and decide for yourself which method you prefer.

To edit a portion of a QuickTime file in Cleaner, follow these steps:

1. **Launch Cleaner.**
2. **Open a QuickTime file to work with by dragging and dropping it onto the Cleaner Batch window or select File ➪ Add Files to Batch and choose the QuickTime file from its current location.** You may add as many files as you like to the Batch window. You may work with each file individually or simultaneously (using the batch options). Because we are performing the task of setting specific In and Out Points, we want to work with only one file at time (see Figure 12-1).
3. **After you have added a QuickTime file to work with, double-click the file name in the Project column.** The Project window opens with your video in it. Note the control and scrollbar beneath the video preview window. They are exactly as you would find in any QuickTime player and they work the same way (see Figure 12-2).

Figure 12-1: The Batch window in Cleaner is where you drop all of the files you want to modify or encode.

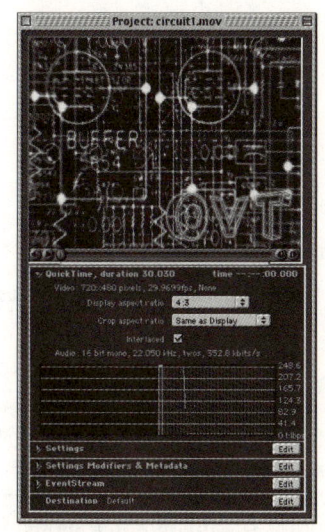

Figure 12-2: The Project window displays your video for previewing and enables you to view your source file and set parameters for the project.
Background image courtesy of OVT Visuals

4. **In the Settings Modifiers and Metadata area, click the Edit button on the right side of the window.** The Settings Modifiers window opens (see Figure 12-3).

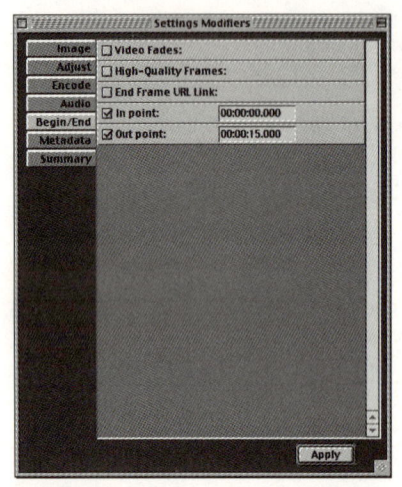

Figure 12-3: The Settings Modifiers window enables you to access specific parameters for your project.

5. **In the Settings Modifiers window, click the Begin/End tab on the left to see the options for setting In and Out Points on your video.**
6. **Select the In Point and Out Point check boxes. Notice that a time code box appears to the right of each option.** In step 7, you will place specific markers on your video file in these time code boxes.

> **Note**
>
> Cleaner uses a time system that displays in milliseconds, instead of the usual frame indicators used by SMPTE (Society of Motion Picture and Television Engineers) time code, which you have been using in ordinary video editors and programs like DVD Studio Pro (the usual frame indicators are divided into frames per second). Cleaner's time system is useful for streaming applications, because multiple events can be added within each frame — a feature that SMPTE does not allow. This time system should not disrupt the editing of your video, because you are navigating and marking points manually in the Cleaner interface. Keep in mind that the frame values you may have do not translate directly.

Chapter 12 ✦ **Working with DVD Video**   275

7. **Return to your Project window with the QuickTime preview by clicking on it.** Navigate through the video file using the scroll bar to find the desired In and Out Points. Make note of the specific time code for each point by looking at the time indicator directly beneath the navigation controls on the right side of the window.

8. **Click Edit to return to the Settings Modifiers window and then type the time code (which you recorded in step 6) for the desired In and Out Points in the time code boxes.**

9. **Click the Apply button on the Settings Modifiers window to add the in and out markers to the video file.**

10. **To verify your In and Out Points, return to the Project window by clicking on it and click the triangle to the left of Settings Modifiers and Metadata to expand the section.** Look carefully at the In Point and Out Point in the Begin/End section. You can preview your choices once more by scrolling through the video if you want. After you are finished, close the Project window (see Figure 12-4).

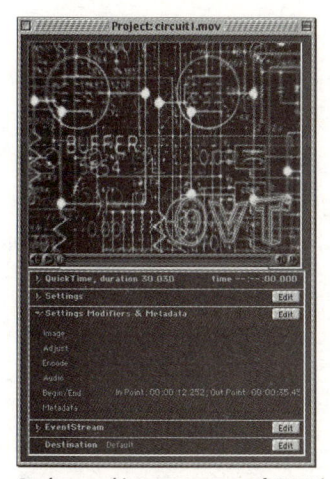

Figure 12-4: Check the status of your settings at the bottom of the Project window — notice the unique time code for the In and Out Points uses milliseconds instead of frames.

Background image courtesy of OVT Visuals

11. **In the Batch window, in the horizontal row for your file, click the box in the Setting column. The Advanced Settings window opens. When you work in Cleaner, you often set parameters for your files in this window (see Figure 12-5).**

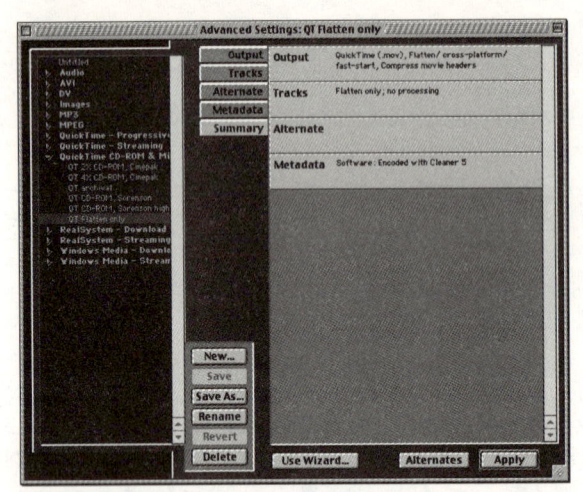

Figure 12-5: The Advanced Settings window is where you can easily access any of the settings for your file prior to encoding.

12. **In the list of formats to the left, click the arrow next to QuickTime CD-ROM & Misc. to expand the number of available format options.**

13. **Double-click QT Flatten to select it.** The Advanced Settings window closes automatically. Flattening a movie prevents it from being processed and renders a copy of your file without additional compression. This is important so that you do not lose any of the quality from the original file. Notice that the box under the Settings column now reads `QT Flatten only` indicating that your setting has been made.

14. **At the bottom of the Batch window, click the Start button to produce a finished file.** A dialog box appears, asking you the name of the file and where you want it saved. Fill in the name and choose a destination; then click Save. Your newly edited file is finished and ready for MPEG-2 encoding.

Toast Titanium

While it isn't strictly a program for the creation of DVD video, Roxio's Toast 5 Titanium (formerly produced by Adaptec) is an excellent tool for creating MPEG-1-based VCDs and for burning finished DVD projects. Along with the ability to burn just about any file or type of disc you might want, Toast's ability to instantly create Video CDs from a QuickTime file is an added feature worth mentioning. A simple drag-and-drop operation is all it takes to begin the process of encoding that turns any QuickTime file into something that can be burned on a CD and viewed on many DVD players. Of course, for creating a DVD project, it also proves useful by enabling you to burn DVD video and data files for DVD-ROM. Toast is an all-around great program that any Mac owner should not be without (see Figure 12-6).

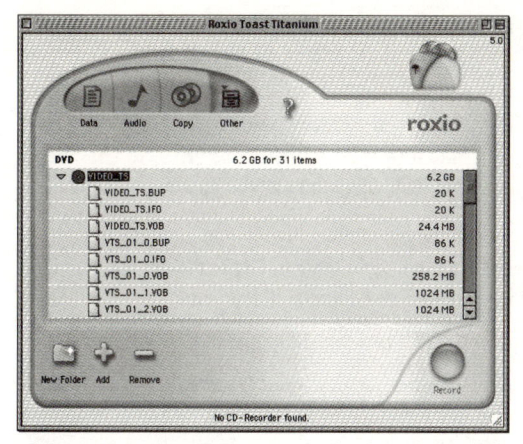

Figure 12-6: Burning DVD and other data with Roxio's Toast Titanium is a simple drag-and-drop process.

Now that we have taken a look at some options for working with video in third-party applications prior to encoding, it is time to actually encode the video for inclusion in DVD Studio Pro.

Encoding Video for DVD

The process of encoding is the crucial link in getting any video on a DVD. Without MPEG or an equivalent compression method, there would be no way of getting two or more hours of video on a single 4.7GB disc. Fortunately, while encoding for the Web can be fraught with many complications, encoding MPEG-2 video for DVD is relatively simple in comparison.

Encoding MPEG-2 with QuickTime Pro

With your copy of DVD Studio Pro, Apple included a high quality variable bit-rate MPEG-2 encoder that works within any application that can export to QuickTime. Even though video editing applications such as Final Cut Pro are capable of exporting QuickTime files to MPEG, encoding is a process that is often most easily accomplished through the included QuickTime Pro. The following steps describe how to open a file and compress it with QuickTime Pro using the new MPEG encoder.

1. Click the QuickTime player icon located in the QuickTime folder on your hard drive to launch the player program.

2. **Open the QuickTime file you want to encode by choosing File ➪ Open Movie and selecting the video file from its current location on your computer.**
3. **Choose File ➪ Export to open the Export dialog box for exporting and encoding of your video file.** In this dialog box, you can choose the file name, destination, and access encoding options (see Figure 12-7).

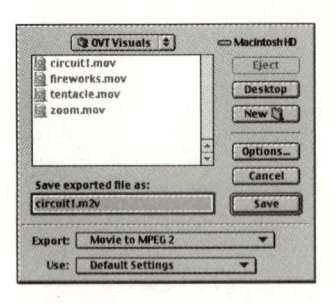

Figure 12-7: You can access Export options through the Export dialog box.

4. **In the Export drop-down menu, choose Movie to MPEG 2.** This option adds the .M2V extension to your file name. You may stick with the default settings provided by the program when you open the window, but check the specific settings for your file first and then make any alterations before completing the export process is better.
5. **Click the Options button in the Export dialog box to launch the QuickTime MPEG Encoder window with detailed options for your MPEG-2 encoding process (see Figure 12-8).**

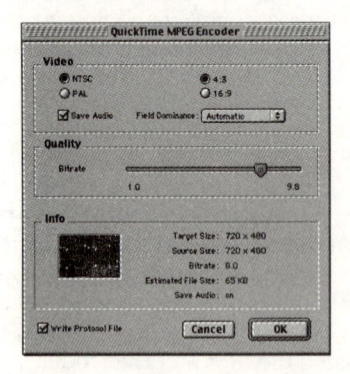

Figure 12-8: There are some standard options for exporting to MPEG-2 in the QuickTiime MPEG Encoder window that can be accessed through the Export dialog box.

6. **Configure appropriate settings for your project in the QuickTime MPEG Encoder window.**
   - Be sure that your Video settings are correct before accepting them. Choices include NTSC and PAL video standards as well as 4:3 and 16:9 aspect ratios. Settings may automatically appear based on the file you

are working with, though it is always important to double-check settings to avoid potential problems down the line.

- The Save Audio checkbox is selected by default. With this option activated, a separate audio file extracted from your video is produced and saved along with the encoded video. If you want both the audio and video for the file you are encoding, this box must be selected.

7. **Choose a maximum bit-rate at which you want to encode your video.** The higher the bit-rate, the better the quality and also the less material you can fit on a disc. Keep your memory limitations in mind and adjust accordingly. Set the bit-rate as high as you can without sacrificing too much space. Every project is different and there is no one right setting. For example, if you're commiting a 5 minute music video to DVD with no extra angles or audio tracks, try setting the bit-rate near the maximum 9.8 Mbits/second or a little under. This provides a maximum data rate and improves the overall quality of the video. For a longer program (say, 1½ hours) with extra angles and/or audio and subtitle tracks, consider setting a lower data rate such as 5.7 Mbits/second — generally considered the default data rate for MPEG-2.

8. **In the Info section, double-check your settings.** After you are satisfied, click OK to open the Export dialog box.

9. **If you haven't already, enter a file name (keep the `.M2V` extension) and click Save.** Your QuickTime video file instantly begins the encoding process to MPEG-2. Your file is ready for importing into DVD Studio Pro!

Encoding MPEG-2 with Cleaner

While the MPEG encoder included with your purchase of DVD Studio Pro is extremely capable and fast, some may prefer to encode their MPEG video in Terran's Cleaner. Cleaner offers many advanced options for MPEG encoding, although some of them are not available until you upgrade to Cleaner MPEG Charger, which is somewhat expensive, but its capabilities provide a significant advance over the base Cleaner 5 program. The standard Cleaner 5 program sets a default bit-rate of 5.7 Mbits/sec, a value that can only be changed with the MPEG Charger upgrade. For some, this may be an acceptable compromise considering the other image parameters allowed for in the program. For example, you have greater control over audio parameters including noise removal, reverb, and other equalization options. Also, you have the ability to set In and Out Points with audio and/or video fades. The following steps describe how to encode MPEG-2 video using the standard Cleaner 5 program.

1. **Launch Cleaner 5.**
2. **Locate a QuickTime file on your computer that you want to encode and drag it onto the open Batch window.**

3. **Double-click the empty box in the Setting column of the Batch window for your project.** This opens an Advanced Settings window where you make modifications to your projects parameters.

4. **In the left side of the Advanced Settings window, click the triangle next to the MPEG option to expand a list of choices (see Figure 12-9).**

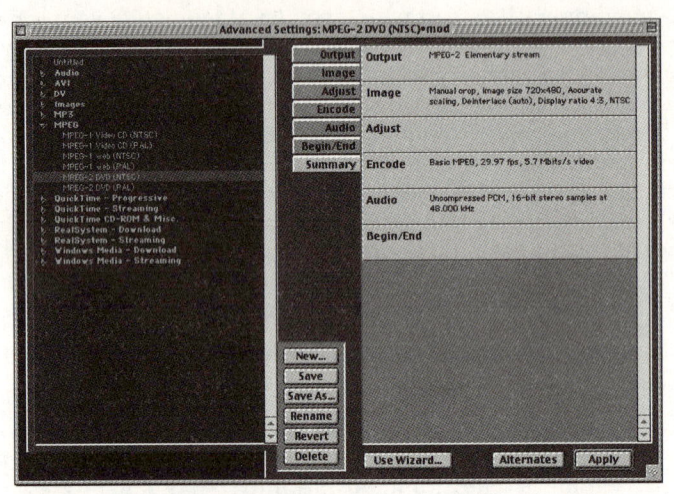

Figure 12-9: With the MPEG option expanded in the Advanced Settings window, you have access to a number of different encoding options, including MPEG-2 for DVD.

5. **Click the MPEG-2 DVD option for the particular video standard (NTSC or PAL) that you are working with.** A new set of parameters appear on the right side of the Advanced Settings window.

6. **Click through the various options (Output, Tracks, Image, Adjust, Encode, Audio, Begin/End, Summary) to see what each one does and make changes where appropriate.** For instance, if you want to sharpen your video image, select Image ⇨ Sharpen and choose values for the radius and amount of the effect. You can even create new In and Out points, as we describe earlier in this chapter, by selecting Begin/End and setting new markers. A number of other possible changes can be made to your video and audio in these menus.

7. **In the Output option, change the File Suffix from** `.MPG` **to** `.M2V`. This is the necessary extension for working in DVD Studio Pro (see Figure 12-10).

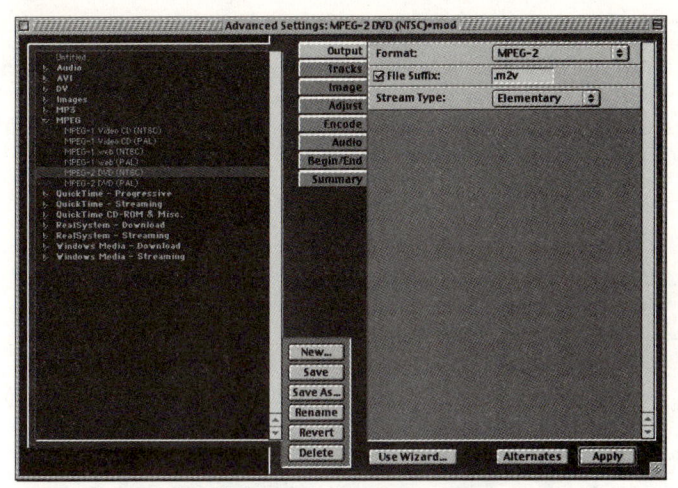

Figure 12-10: The File Suffix should be changed to .M2V in order for the file to work properly in DVD Studio Pro.

8. **After you are finished making changes, click Apply at the bottom of the Advanced Settings window.** The window closes and you are returned to the Batch window (see Figure 12-11). Note that the Setting column now displays MPEG-2 DVD (NTSC or PAL).

Figure 12-11: The Batch window displays your project and indicates if it is ready for encoding.

9. **Click the Start button to begin the encoding process.**

10. **In the dialog box that opens, verify the name of your file, specify a location on your computer where you want it to be saved, and click Save.** The Output window opens and displays the progress of your file as it encodes (see Figure 12-12).

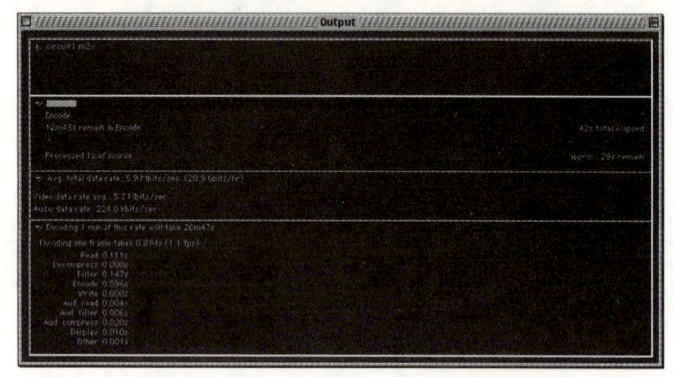

Figure 12-12: The Output window displays information about your file as it encodes, such as data rate and time remaining in the encoding process.

> **Note**
>
> The main drawback to Cleaner is that it has excruciatingly long encoding times, which is why many people prefer the QuickTime encoder included with DVD Studio Pro. By clicking the triangles to the left of an option, the window expands to show more information about that option. Note the indicators that show how much time is remaining in the encode as well as the total elapsed time. This should give you an idea as to whether you should continue sitting at your computer waiting for it to finish encoding or whether you should go do something else for a while. Unfortunately, the current version of Cleaner does not allow you to do anything else on your computer while the file encodes, rendering your computer otherwise useless for several minutes or even hours. (You can "pause" encoding though if you need an e-mail fix or something else.)

Hardware-based MPEG-2 encoding

There is an alternative to the slow encoding times of software encoders. Hardware-based MPEG encoders greatly reduce the amount of time it takes to produce MPEGs from long videos or movies. In many cases, hardware encoders are used to produce MPEG video on the fly, or simultaneously as it plays from a source, such as a live broadcast or directly from an editing deck. The most obvious drawback to using hardware versus software encoders is the price, which is usually much higher. Of course, if you have a business that requires quick turnaround times, which do not allow for several hours of overnight encoding, you may require the hardware to speed things up — especially if you are dealing with live or near live video streams. Digital cable providers use hardware encoders, as do many news agencies that stream events as they happen.

Most home-based project studios that deal in low volume, slow turnaround times need not concern themselves with hardware encoders, unless they need to significantly increase productivity or unless they want some of the other powerful options that these encoders can provide. Some hardware MPEG encoders employ better compression quality, which can lead to a cleaner, improved image with less distortion and artifacts. This is not always the case though, and some real-time hardware encoders actually produce worse results than their software-based counterparts. Not all hardware encoders can produce 2-pass variable bit-rate encoding (discussed in Chapter 11), greatly impeding their ability to produce the highest quality possible. However, when 2-pass variable bit-rate encoding is combined with a hardware encoder, you receive the best of both worlds. If you enjoy working with Cleaner, you might consider the MPEG SuperCharger, an additional option that provides a board to speed up encoding times.

Now that you know how to prepare and encode your video tracks, it is time to work with them in DVD Studio Pro.

Using Video in DVD Studio Pro

Once you understand the interface and how to navigate in DVD Studio Pro, working with video and other assets is actually quite simple. In this section, we detail how to add video tracks to your project and how to use additional DVD Studio Pro features to place markers in your video for easy access to scenes.

Importing video

The simplest task of all to complete in DVD Studio Pro is importing your video. As long as your video file is in the MPEG-2 format with the .M2V extension, it is acceptable for import. Importing for video is done exactly the same as importing for other assets, such as audio and graphic files.

To import video into DVD Studio Pro, follow these steps:

1. **With DVD Studio Pro open, choose File ⇨ Import.**
2. **Select your video file and click the Add button.** If you have more than one video file that you would like to import, select more files by clicking Add one at a time or Add All if an entire folder's contents are to be imported (see Figure 12-13).

Encoding for Visual Quality by Paul Feith

Bio: Paul Feith is a technology consultant to companies facing strategic online issues. Prior to that, he has been responsible for the evolution and implementation of emerging applications and technology within the online division of two of the most highly trafficked Internet brands in the world, Playboy Enterprises and MyPoints.com. He has been highly involved in the streaming media space, and worked closely with RealNetworks, and the ATVEF (advanced television enhancement forum). Through his efforts, Playboy was among the first commercial sites to encode and stream media above 300kbps using 'thinned' mpg.

If you plan on doing your own encoding, there are some general design tips that you should consider when encoding various content. There is no one set of rules to produce visually appealing encoded video. It depends on content of the video (animation, fast motion, talking head) and the upper limit of transmission speed that is available during playback. In fact, it is more of an art form, relying on a number of techniques, and tweaking those techniques to accommodate the media, content, and medium.

The "garbage in, garbage out" principle applies to the media. It is best to start off with high quality media, such as D-1 or Digital Beta. Anything less than Beta SP, DV, DVCAM or M-2 will encode poorly due to video noise in the signal, subsequently getting amplified in the encoding process.

Content remains the only variable left in the pre-encoding process. There are a number of content challenges for encoders, among them — moving water, fast panning/zooming, low light levels, elaborate transitions, and text/line-art. Most video content contains many of these elements, and encoding an entire video using one technique or set of configuration options will produce undesired results. Many of the best encoding service bureaus break up a video clip into several parts, applying a different set of encoding options to each before stitching them back together again.

Video can be captured several ways, and each setting will affect the ultimate quality of the resulting video. Some settings work better than others with various encoding methods. Encoders provide numerous options for encoding at various speeds, levels of image quality and smoothness of motion. At this point it is really a zero-sum game; better resolution means fewer frames per second and vice versa. The connection speed of the user also affects the ultimate viewing quality. A lower speed means less data is transferred per minute of content. Therefore either more image data must be deleted (the resolution drops) or fewer frames per second shown (frames are dropped). Using a higher transmission rate vastly improves both image quality and fps.

Giving the content challenges that exist, here are some encoding techniques that produce the best possible visual quality.

- ✦ When dealing with moving water scenes (such as ocean, waves, waterfalls, pools, etc.), push up the frame rates, lower the resolution, and lower contrast settings.
- ✦ Avoid fast pans and zooms such as a runner against a sideline of spectators. The multitude of high-contrast elements (the spectators) panning at fast speeds will cause the background to become blocky when encoded. If the background quality is just as important as the foreground runner, than lower the frame-rate, increase the data-rate, and lower the contrast settings.

- ✦ Do not encode video noise effects. The noise will look great, but the encoded result will not.

- ✦ Avoid content with low light levels. The video capture device will add video noise. To compensate, you can raise the brightness levels during capture.

- ✦ Avoid complicated transitions if possible. Transitions many times follow the same rules as water. They can be fluid, translucent, and contain a large color palette. Like water scenes, raise the frame rate, and lower the resolution (data rate) and contrast settings.

- ✦ Avoid wire-frame animations and scrolling text. Also, the pixel line size should be a minimum of 2. Test your animations for encoding because you may encounter "stair-stepping" and other problems that will need to be adjusted in the final product. It is best to replace scrolling text and animations with static text or images.

- ✦ To adequately lip sync audio to a "talking head," a range of 7 to 15 fps is recommended.

As for the encoders themselves, there are distinct levels of MPEG-1 and MPEG-2 encoding hardware and software. They range from $75 – $250 software solutions, $200 – $700 single chip encoders, $1,300 – $2,000 single chip encoders, $3,000 – $5,000 dual-chip encoders, and $3,000 to $30,000 plus MPEG-1 & MPEG-2 encoders. The later high quality MPEG-1 and MPEG-2 encoders can include machine control and digital video and audio inputs that render high quality files. This is a must for low bit-rate MPEG-1 files for CD-ROM distribution. These encoders also encode to exacting DVD specifications. Purchase the best encoder you can afford. Remember, image is everything and if your image has artifacts, blockiness, and noise, it will reflect upon the quality of your message. If you cannot afford a high-quality encoder, outsource your encoding to a professional service bureau.

Finally, consider the medium in which the encoded video will be played and decoded. Mediums can be CD and DVD distribution where data rates are not a concern to the bandwidth challenged such as the Internet. Internet distribution is tricky because bandwidth connections vary, as well as media players. It is a good idea to know what connection speed your target audience will be using, and use that as a base minimum to calculate your data and frame rates. An easy way to compensate for these unknown variables is to use an encoding process called "thinned mpeg" whereby you can vary the frame rate and data rate of the encoded video to suit multiple connection speeds. It doesn't end there though.

Media players have their own idiosyncrasies, and for their own reasons. For example, Microsoft's Windows Media Player has a tendency to drop frames before reducing resolution when the playback data rate falls below the encoded data rate of the video. This ensures the quality of the video frames are maintained. Likewise, RealNetwork's RealPlayer does just the opposite, reducing resolution to keep up the frame rates. This makes the video look more fluid, as opposed to a slide show.

With the various media, types of content challenges, encoders, viewing mediums, and players, there is much to consider when encoding your video. Take comfort in the fact that nothing is permanent, and you can try out your encoding techniques over and over until you are pleased with the result. As an overriding guideline, balance quality with file size, and encoding data rates with the medium of distribution. Add to that the techniques for visual quality and you'll have a professional looking encoded video.

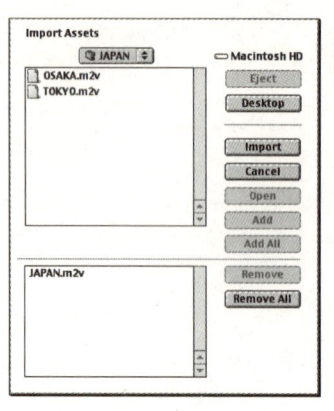

Figure 12-13: The Import Assets window enables you to add individual files or entire contents of folders.

3. **Click the Import button to finally bring the video files into DVD Studio Pro.** Your video assets are added to the Assets container and are ready to use as elements in your project (see Figure 12-14).

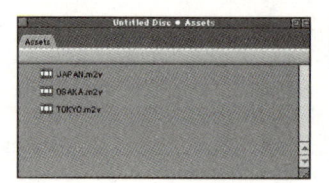

Figure 12-14: The Assets container holds all of your imported files, including video.

Managing video files

Working with video in DVD Studio Pro also involves keeping track of your assets. The Assets container along with the Property Inspector provide you with information about your files (such as file size, aspect ratio, and bit-rate) and give you the ability to easily monitor and modify your files as needed. If, for instance, you needed to replace a video file with another file and wanted to avoid confusion without disrupting the order of your project, you could modify the file using these two windows. Also, if you simply wanted to rename a video asset, using these two windows, you could accomplish that task.

To replace and rename a video asset in DVD Studio Pro, follow these steps:

1. **In the Assets container, located at the bottom of the DVD Studio Pro screen, click the file you want to modify.** This works the same way for any asset you want to work with, whether it's a video, audio, or a graphic file.

2. **In the Property Inspector, click in the Name box and rename the file if desired (see Figure 12-15).**

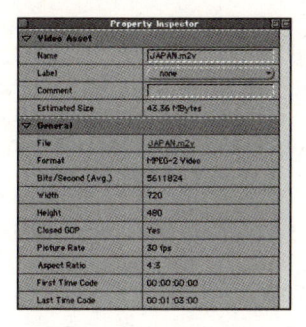

Figure 12-15: The Property Inspector can display detailed information for a particular asset, such as a video file.

3. **To swap a newer file for an older one, click the underlined file name in the General section of the Property Inspector to open a dialog box where you can assign a file.**

4. **Click a new file to select it and then click Open to make the change take effect.** The name of the new file assigned to your asset is now listed in the General section of the Property Inspector (see Figure 12-16).

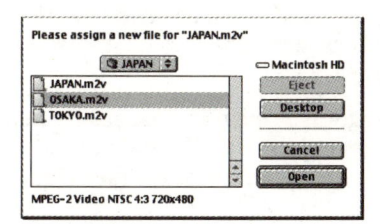

Figure 12-16: Choose a new file to assign to an existing asset.

Note

If you attempt to delete a video asset that is already assigned, you receive an error that states `Can not delete Video Asset because it is referred by the following items`. This indicates that the asset is already in use and cannot be deleted without potentially disrupting your project (see Figure 12-17).

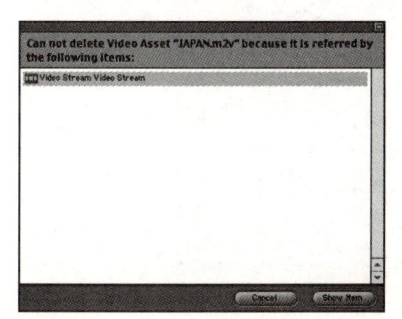

Figure 12-17: An error message appears if you attempt to delete a video asset that is already assigned.

Placing markers in a video track

Markers placed in a video track allow quick navigation between points in a video. Without markers, a viewer would need to fast-forward to get anywhere in a track. Also, markers allow for the creation of *stories*, sequences of markers arranged to play in a certain order.

To create a video track and add markers to it in DVD Studio Pro, follow these steps:

1. **Launch DVD Studio Pro.**

2. **Select File ⇨ Import and choose the files that you want to add by clicking the Add button and then clicking the Import button.** Your video assets should now appear in the Assets container and are available for use in any project element.

3. **Click the Add Track button at the bottom of the Graphical View to place a new track tile in the workspace.** Notice that the color of a track tile is green to differentiate it from other project elements, such as the blue menus and gray slideshows. Also, notice that a track tile has five buttons on it. From left to right these buttons provide access to a folder for Audio Streams, Subtitle Streams, Markers, Stories, and Angles. Because the track is new, these folders are empty with the exception of Markers, which always contains at least one marker to indicate the start of the track.

4. **Add a video asset to the track by dragging it from the Assets container onto the track tile.** A thumbnail image is created to provide a convenient visual reference, indicating which video file has been added to the tile (see Figure 12-18).

Figure 12-18: The newly created track tile contains five buttons and a thumbnail image of the video asset added to it.

5. **Double-click the thumbnail image to launch the Marker Editor (see Figure 12-19).** The Marker Editor is the window that you use to add and move new markers on a video track. In addition to the two buttons, New Marker and Add Button, at the bottom of the window, there is a drop-down menu that currently displays Start of Track. You can use this menu to navigate to new tracks that you create. Also, in the upper right-hand corner is a time code indicator, which tells you precisely where you are in a video track.

Currently, the only marker present is titled Start of Track and is created by default to indicate the beginning of a track.

Figure 12-19: The Marker Editor window is used to add markers (points in a track that indicate where to begin play) to video tracks.

6. **To create a new marker, click the New Marker button at the bottom of the window.**

> **Note:** A maximum of 99 markers can be added to a track.

7. **Drag the new marker to the point you want in the video or click in the time code box in the upper right corner to set a specific time code value.**

8. **In the drop-down list in the lower left corner of the Marker window, title your new marker by clicking in the box and typing a new name where it says Untitled Marker.** You can also name the marker in the Name box in the Property Inspector window. If you wish to change the name of the default marker, Start of Track, you may only do so in the Property Inspector window (see Figure 12-20).

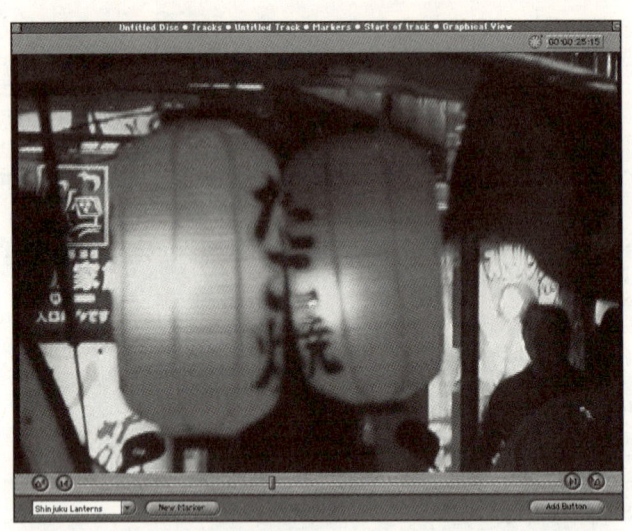

Figure 12-20: A new marker has been added to a video track by clicking on the New Marker button, and the new name is displayed in the lower left drop-down list.

9. **Create a few more markers by continuing to click the New Marker button and typing new names.** After you have created more markers, you can easily jump to them by choosing a name from the drop-down list or by clicking the forward and back buttons at the buttom of the editor (above the buttons).

10. **After you are finished adding markers to the video track, close the Marker Editor.**

Creating stories with markers

In addition to aiding with general navigation of a video track, markers may also be placed in sequences to create *stories* that play back markers in a particular order. You can create different stories for the same track, and you can assign the story to a button in a menu to activate it.

To create a story using your markers, follow these steps:

1. **Create a series of markers in your video track as described in the "Placing Markers in a Video Track" section.** The only difference with creating stories is that the end point of a section must be indicated with a marker. Your stories look for the end point of a clip by stopping at the next marker they encounter. While the markers that indicate the end point of a clip are necessary for a story, they do not appear in your Stories Folder.

Chapter 12 ✦ **Working with DVD Video** 291

2. **After you have added markers with beginning and end points to your video track, close the Marker Editor.**

3. **In the Graphical View, click the Story icon in your track tile.** This opens the Stories Folder where you can create different stories for your project (see Figure 12-21).

4. **With the Stories Folder open, choose Item ⇨ New Story from the menu.** This creates an Untitled Story in the folder.

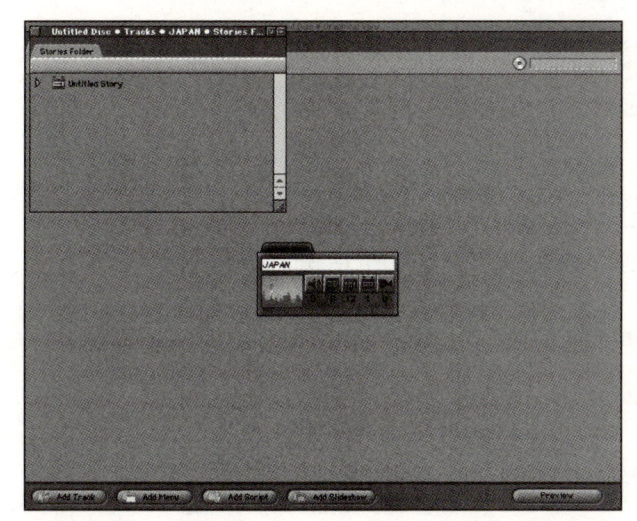

Figure 12-21: The Story icon on the track tile activates the Stories Folder where all of your marker sequences are stored.

5. **Click the Untitled Story once and name it in the Property Inspector.**

6. **In the Project View window, click the Tracks tab and select the track that contains your markers.** You can expand the containers to display their contents by clicking the triangles to the left. Expand the Stories Folder container to reveal your new story. Also, the Project View window may be resized to make it easier to work with and to better see expanded container contents.

7. **Click the Marker icon in the track tile to open the Marker Folder.** All of the markers present in your video track should display.

8. **Drag the markers you want in your story onto the new story you created, located in the Stories Folder in the Project View.** Place the markers in the order you want them to appear, beginning with the first marker at the top of the list. Only use the markers that indicate the beginning of a clip and not the

end of a clip. Markers at the end of clip are present to indicate a stopping point and are not required in the Stories Folder (see Figure 12-22).

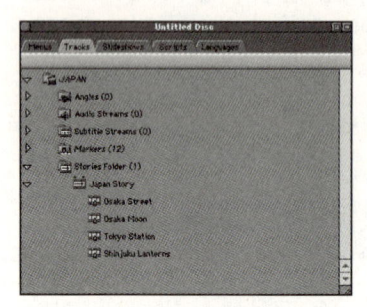

Figure 12-22: The Project View displays the markers that have been added to the Stories Folder.

9. **To link the story you created to a button, double-click a menu tile's thumbnail to launch the Menu Editor.**
10. **Click the menu button that you want to assign the story to.**
11. **In the Action section of the Property Inspector, select the story you want assigned to the button from the Jump When Activated drop-down list.** The button is now set up to advance to the story you created when it is activated (see Figure 12-23).

Figure 12-23: The Property Inspector enables you to link the activated button and the story you created.

12. **Close the Menu Editor window after you are finished assigning stories to buttons.**
13. **Preview the story by selecting the menu tile and choosing Item ⇨ Preview Menu or by clicking the Preview button at the bottom of the Graphical View.** Your story is now complete!

Previewing video

After you add markers or stories to your video, you should view the track by using the standard Preview window. The Preview window has controls at the bottom of the window so that you can navigate the video (these controls mimic typical remote control functions).

To preview your video in DVD Studio Pro, follow these steps:

1. **Double-click the thumbnail image on the track tile to launch the Marker Editor.**
2. **Choose Item ⇨ Preview Marker to open the Preview window.**
3. **Navigate your video by clicking the Next and Previous buttons to test your markers. After you are finished, close the Preview window.**

You have now been exposed to a variety of ways of working with DVD video. The best way to get your feet wet is to come up with an idea for a project, shoot some sample footage, and then spend some time experimenting with ways of preparing video. You may wish to create a super-simple DVD project that enables you to select from a variety of clips, and then try encoding your video at different bit-rates, burn a DVD, and compare the results. Good luck!

Summary

- MPEG is the video format used for DVDs — either MPEG-1 or MPEG-2 video may be used, although MPEG-2 is of much higher quality and is what we generally associate with DVD quality video.
- Many third-party applications, such as Final Cut Pro and Cleaner, may be used in the preparation of your video.
- DVD Studio Pro is shipped with a QuickTime MPEG encoder that works within any application that can export to QuickTime, including QuickTime Pro.
- QuickTime Pro allows you to encode MPEG-2 video quickly and easily using the new encoder.
- Cleaner is another application that can create MPEG files for inclusion in DVD Studio Pro, although, without the additional upgrades, it is rather slow and does not allow you to set your own bit-rate.

- ✦ Hardware-based encoders can greatly speed up the encoding process and are particularly useful for media creators needing to produce files in a short period of time.
- ✦ Video files are imported into DVD Studio Pro and managed similarly to other assets in your project.
- ✦ Markers can be placed in video tracks to make navigating to different scenes easier and also to add stories (sequences of markers that play in a predetermined order) to a project.

While this concludes our chapter on working with DVD video, there are many instances throughout the book where we discuss video in relation to specific project elements. For instance, in the next chapter, we discuss how to create slideshows — primarily with still images, but video may be used as well.

Hold onto your seats, it's time to learn about slideshows!

✦ ✦ ✦

Working with Slideshows

CHAPTER 13

In This Chapter

Working with Still Frame Images

Adjusting Image Sizes in Photoshop

Understanding the Slideshow Editor

Assembling a Slideshow

Using Multiple Languages in a Slideshow

DVD Studio Pro slideshows hold sequences of still images or video clips with or without audio. Remember the days when slide projectors were the best way to share memories of a family trip? Uncle Bob would come over with a box full of slides that he spent all afternoon carefully placing into round trays, while trying not to place any upside down or backward. When the show started, the projector whirred loudly and the bulb emitted a smell of burning dust. With DVD we have a better solution for organizing pictures as well as audio and video content without any of the hassles just mentioned — and in a package occupying a lot less space!

Another way to think of slideshows on a DVD is as a stripped-down version of Microsoft's PowerPoint presentation software. If you are creating DVDs for corporate presentations, you can appreciate slideshows for their ease of use and their capacity for high quality audio and video. Presentations that include extensive use of high quality video clips are usually impractical if not impossible. Of course, slideshows are also good for bringing together extra material in an electronic format similar to scrapbooks and photo albums. Imagine creating slideshows on a DVD and then burning a copy for each family member and friend. Perhaps you have photos and graphics from a wedding or event? Slideshows on a DVD are a nice way to organize extra material into one accessible location.

In this chapter we discuss how to create slideshows for inclusion in a DVD Studio Pro project. In particular, we discuss how to prepare still image materials for a slideshow, touch upon the role of scanners for acquisition, and the role that Photoshop plays by providing tools for resizing and fixing flawed images. Finally, we show you how to assemble a slideshow using the Slideshow Editor and your available assets. It is a simple process as long as you understand the options available to you.

Using Still Frame Images

We begin this chapter by explaining some possible considerations when working with still images. Still images form the core material of most DVD slideshows. Even though video and audio can be included in a slideshow, the majority of slideshows do incorporate some, if not all, still images. They are easy to acquire and are often a plentiful resource of extra material. In DVDs produced by movie studios, slideshows usually consist of behind-the-scenes photographs or galleries of movie posters and promotional graphics. If you think about it, there are virtually endless possibilities for still images in slideshows.

Even though DVD Studio Pro can import any image as long as it meets the required dimensions and is in either PICT or PSD format, it is helpful to understand how to get the best quality out of any image before importing into the program. As far as still images for DVDs are concerned, image quality relates to the resolution, color depth (or bit-depth) and correct proportionality of a picture. You are always rewarded with better results by paying attention to these factors when acquiring and working with images. A common tool for acquiring images is a flatbed or slide scanner, which we discuss in this chapter. Of course, images can originate from anywhere, yet it is often necessary to apply many of the same principles regardless of a pictures origin.

Scanning still images for use in a DVD project

In order to obtain images for your DVD Studio Pro slideshow, such as old photographs, poster art, and other graphics, you may want to consider purchasing a scanner. Flatbed scanners have become increasingly popular and it is now possible to get a decent scanner for around $100. It is also helpful to understand the mechanics of a scanner to make a more informed buying decision.

Flatbed scanners use reflected light on an image, which is read by an array of CCDs (charge-coupled devices) similar to those used in video cameras. These sensors measure light, and color values are assigned for each primary color of red, green, and blue (RGB). This is accomplished by the CCD array making three separate passes over an image to capture the RGB information. Slide scanners employ a nearly identical process that uses light transmitted through a transparency, which is either positive or negative (positive for slides, negative for print film). Flatbed scanners typically scan at lower resolutions than slide scanners, which have the ability to scan at resolutions of up to 4000 lines per inch. Keep the features and limitations of a scanner in mind when looking at various models — remember that video resolutions are not as high as professional quality prints and artwork, yet the color quality is a different issue. Fortunately, due to advances in technology, just about any scanner you choose for use with video projects should yield more than acceptable results.

Determining scanner resolution and bit-depth

Scanning can be a difficult art. Ordinarily, scanning produces somewhat unpredictable results. Trial and error is usually the only method for achieving the absolute best image for your project, although making a few educated settings from the beginning greatly increases your chances of getting the desired scan. Retouching of images for general imperfections and blemishes aside (such as smudges or scratches on a picture), I have found that the best and most manageable image scans are achieved by determining what your final project requires and keeping that in mind from the start. Neglecting to consider the needs for the finished product is where most people go wrong. In particular, I have found that keeping the demands for video images in mind when scanning images for slideshows can also save you a lot of time correcting images later on when working with even a moderate quantity of pictures.

When using a scanner to capture images for your slideshow you need to make decisions about what resolution to scan at and what bit-depth is appropriate. For our purposes of creating video stills we do not need a very large image — only 720 × 540 pixels. By keeping the resolution down while scanning, you can produce manageable file sizes that make working in Photoshop a little easier. Experiment with your scanner and determine the largest size that works with your system. If you have plenty of RAM assigned to Photoshop don't be afraid to scan at high resolutions. Once you have resized the image to fit your DVD project, the file size should be approximately 1 megabyte. When it comes to scanning, pictures can always benefit from higher bit-per-pixel settings. Set the bit depth to 24 bit, 42 bit, or as high as you can as long as resulting file sizes are not too large. If file sizes get out of hand try scanning again at a lower value. Place all finished stills into a separate folder for organizational purposes and delete the original scans to make room on your hard drive.

Note The size of the original scan does not affect the quality of the finished image as long as the image is not artificially enlarged, by increasing the dimensions manually above the original resolution following the scan (for example, scanning at 800 pixels and subsequently increasing the image size to 900 pixels — scanning at 800 pixels and decreasing the image size to 750 pixels is fine). Of course, the quality (most notably the resolution) of the finished image may be affected by the size of the item being scanned — since the smaller the image being scanned the higher the resolution is required to effectively capture it at the desired dimensions.

Detecting and eliminating moiré patterns

Scanned images from printed material such as newspapers, books or magazines, reveal moiré patterns as shown in Figure 13-1 during the scanning process. Images printed on a printing press employ a process that uses rows of tiny dots, called halftone screen patterns. The printed images are entirely composed of these tiny dots that trick your eye into seeing a complete image. When you scan a printed image, you are actually scanning those same dots, causing the scanner to create a crosshatched-looking image with varying lighter and darker patterns. Definitely *not* a good quality image.

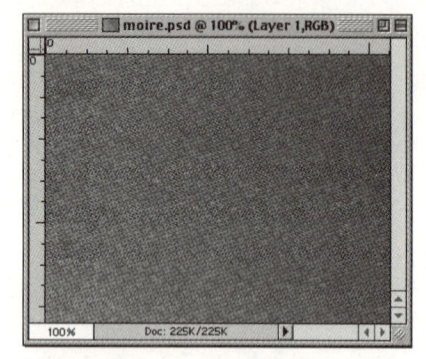

Figure 13-1: An example of a moiré pattern that results from scanning a printed image.

There are a few ways to eliminate moiré patterns. The most common solution is to slightly blur the image, although this can result in a loss of image quality. Use this option carefully. Another possibility is to try tilting the picture at an angle when scanning, perhaps 30 degrees or more, which may deceive your scanner into creating more blended lines. This method is not entirely effective and takes a bit of experimentation.

To remove a moiré pattern by blurring an image, follow these steps:

1. **In Photoshop, open the image you have scanned and want to correct by selecting File ⇨ Open and choosing the image file from its location.**
2. **Choose Filter ⇨ Blur ⇨ Gaussian Blur and blur the image between 1 and 2 pixels, as shown in Figure 13-2.** You want to blur the image enough to make the dot patterns disappear — but not so much that you lose significant image quality.

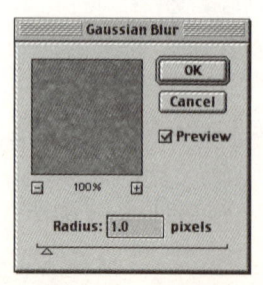

Figure 13-2: The Gaussian Blur filter blurs this image by 1.0 pixels.

3. **Smooth out the dots in the background using the Median filter.** Select Filter ➪ Noise ➪ Median and enter a Radius value as shown in Figure 13-3. Start at 1 pixel and work your way up until you receive your desired result.

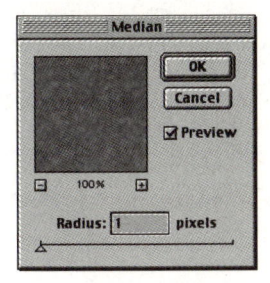

Figure 13-3: The Median filter smoothes out the background.

4. **Click Filter ➪ Sharpen ➪ Unsharp Mask to open the Unsharp Mask dialog box as shown in Figure 13-4.** To make the image sharper, set the Threshold (which affects the sharpness of edges between adjoining pixels — the lower the value the more pixels are sharpened) to 0 levels, adjust the Radius (which affects the thickness of the sharpened edges — low values make crisper edges, while higher values make thicker edges) to a low number of pixels, and set the Amount (which determines the magnitude of the overall sharpness levels — the higher the value, the more noticeable the effect) to a higher value (see Figure 13-4).

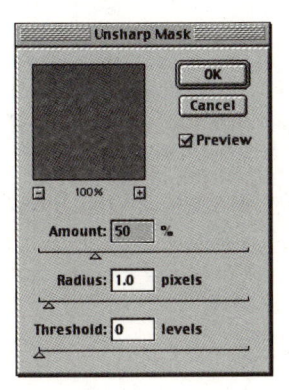

Figure 13-4: Using the Unsharp Mask filter improves your image by increasing the overall sharpness.

If blurring the image does not produce the results you wanted, leave out the Blur filter and simply use the Median and Unsharp Mask filters instead. Your values must change a little to compensate for the lack of blurring that you added in the previous method. By making only a slight increase in the Noise of the Median filter and raising the level of your Radius and Threshold for the Unsharp Mask filter, you can achieve similar results to using a blur, without the probable loss in quality.

1. Choose Filter ➪ Noise ➪ Median and set the pixel Radius to 2 or below.
2. Choose Filter ➪ Sharpen ➪ Unsharp Mask and enter 50% for the Amount, 2 or 3 pixels for the Radius, and set the Threshold to at least 5.

An often-underused feature of Photoshop is the Dust & Scratches filter, which is intended for photographs with small surface defects. You can also use the Dust & Scratches filter to eliminate moiré patterns, although the results vary widely. Give it a try and see if it does the trick. It is a simple and effective solution when it works.

1. Click Filter ➪ Noise ➪ Dust & Scratches.
2. Choose a low value for the Radius, between 1-2 pixels, and keep the Threshold below 40 (see Figure 13-5).

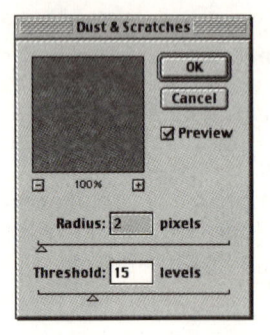

Figure 13-5: Using the Dust & Scratches filter is one way to eliminate moiré patterns.

By applying the techniques we just discussed for material that originated in print, you should notice a considerable difference in your slideshow's image quality. Also, now that you have scanned your images and adjusted them for potential problems, such as moiré patterns, you are ready to prepare them for import into DVD Studio Pro by resizing them to the dimensions required by the program. At this point, the most difficult image manipulation tasks, including the elimination of surface defects and correction of color, should be completed.

Resizing Images in Photoshop

Slideshows in DVD Studio Pro accepts still images in either the PICT format (.PCT) or Photoshop format (.PSD). It is important that all finished graphics match the 720 × 480 pixel image size used for creating DVDs. If your images do not match the correct proportions, you cannot properly use them in DVD Studio Pro. It is like putting on a pair of shoes that are too big or too small — only the right size will do. Your foot is the standard by which you judge the shoe, just as the DVD video specification is the measure by which we judge the size of pictures and other elements. Before working with stills in DVD Studio Pro, prepare your graphics for proper video specs by cropping or resizing them in an image editing application like Photoshop.

To resize an image in Photoshop for a DVD Studio Pro Slideshow, use the following steps:

1. **Select File ⇨ Open and open the photo you want to edit to include in the slideshow.** For organizational purposes, place your original stills in a single folder and create an additional folder for finished images.

2. **To crop the image to a dimension of 720 × 540 pixels (the standard size for creating a DVD image is 720 × 480 pixels, although you want to create all original graphics at 720 × 540 pixels initially to compensate for the non-square pixel ratios used by DVD), select the Rectangular Marquee tool and select Style ⇨ Options ⇨ Fixed Size.** Enter a Width of 720 and a Height of 540.

   If your image is significantly larger than 720 × 540, you may want to resize the image first before cropping. To resize the image, select Image ⇨ Image Size and change the Pixel Dimensions values until you have the size that can be easily cropped to fit the 720-×-540-pixel dimensions as shown in Figure 13-6.

3. **Drag the box around the part of your image that fits best within the required dimensions.**

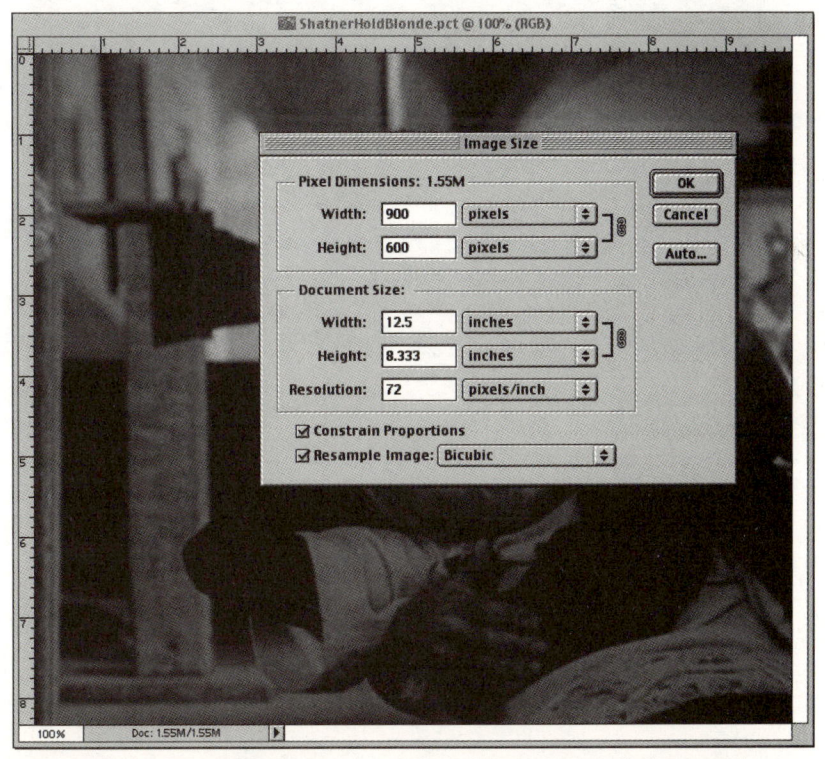

Figure 13-6: Remember to check Constrain Proportions when you resize an image.

4. **After selecting the image area you desire, choose Edit ➪ Copy. Create a new image file by selecting File ➪ New and name the file.** The correct dimensions — 720 × 540 pixels — are already filled in. Photoshop remembers the selection you made when you copied the image and automatically sets those dimensions for your new file.

5. **Select Image ➪ Image Size and set the Pixel Dimensions to 720 × 480.** Make sure the Constrain Proportions box is unchecked at the bottom; otherwise you are unable to alter the dimensions. The graphic looks stretched and wider than normal, as shown in Figure 13-7. This is fixed when the graphics are imported into DVD Studio Pro.

Figure 13-7: The finished graphic looks stretched but it works in DVD Studio Pro.

6. **Flatten the image by selecting Layer ➪ Flatten Image.** Slideshow images are required to have only a single layer, so flattening the image ensures that your layers are not separated when you import them into DVD Studio Pro.

7. **Save your Image by choosing File ➪ Save As and designate a name and location for the file.** Your graphic is now ready to be used in a slideshow project.

Note: When Photoshop files are used in a DVD Studio Pro slideshow, only the background image displays. Since DVD Studio Pro cannot read layered image files correctly, make sure to flatten any images with multiple layers before importing them into the program. This may seem like a peculiar inconsistency, after all, menus can make use of multiple layers. In fact, this inconsistency is the reason that still images must contain only a single layer — because the option to use layers, and to turn them on and off, is not available for slideshow images in the DVD specification, thus rendering Photoshop layers unnecessary.

The next section introduces the interface that is used to add the images you have just created into a slideshow.

Exploring the Slideshow Editor

To assemble a slideshow you use the Slideshow editor in DVD Studio Pro (the only interface you can use for putting together a slideshow), which is accessed by clicking on the thumbnail photo in a slideshow tile. The Slideshow editor (see Figure 13-8) includes two main lists that are used when assembling slideshows. On the left, there is a list of the Slides that are present in the Slideshow (the order of the list determines the order in which the slides display, starting at the top and working down to the bottom), on the right, there is a list for assets that can be used as elements in the Slideshow. Within the Slide list there are columns for displaying information relating to various time and quantity values for that Slide. These columns include Slide, Audio, Time, Duration, and Pause. Also, notice the "Hide Used Assets" box at the bottom. By checking this box, the Assets list on the right no longer displays those Assets that you dragged to the Slide list on the left. Unchecking the box allows all of your Assets — used or unused — to remain displayed. In addition, there are indicators for Total Slides, which indicates how many slides are used in the slideshow, and Total Duration, which indicates the combined duration of all your slides in the slideshow.

The Slideshow editor provides a method for working with numerous assets in a single, easy to use interface. By dragging and dropping assets you can quickly create a slideshow sequence, similar to working with a video editor like Apple's iMovie (instead of working horizontally from left to right, as iMovie does, the Slideshow editor arranges slides horizontally from top to bottom). Working with multiple audio tracks and durations for each slide may seem like a confusing prospect, although this interface makes these tasks relatively intuitive. To add audio streams drop them onto a slide and to set durations simply select a value from the pull-down menu located next to that slide. You should get the hang of the Slideshow editor soon after using it for the first time.

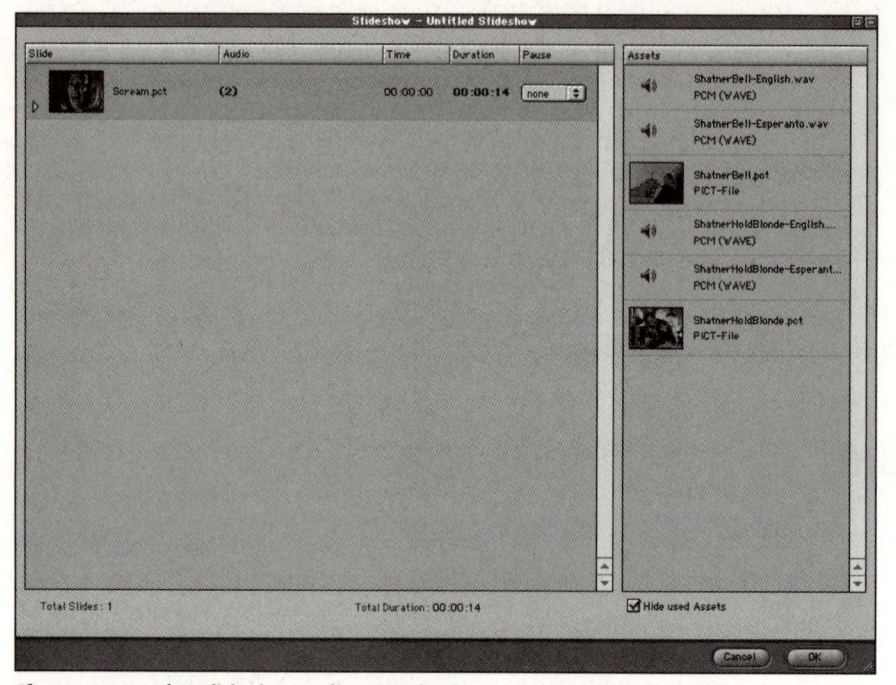

Figure 13-8: The Slideshow editor is where you assemble a slideshow.

Understanding the Slide Area of the Slideshow Editor

The slide area of the Slideshow editor displays a preview of the Slide element in the form of a thumbnail image. By clicking on the small triangle to the left of the thumbnail, a tree expands listing the various Assets assigned to that Slide as shown in Figure 13-9. This expanded menu is particularly useful if you are working with multiple language tracks within a Slide, since it allows you to view the names and values for all the audio streams added to that Slide (which you otherwise would not be able to see). Referring to the expanded list for a Slide provides you with a view for organizing and rearranging elements in logical orders. For instance, if you could not see the order of your audio tracks you may be susceptible to inconsistencies between the various Slides. Viewing your Slide elements ensures that no mistakes have been made while adding assets to your Slides.

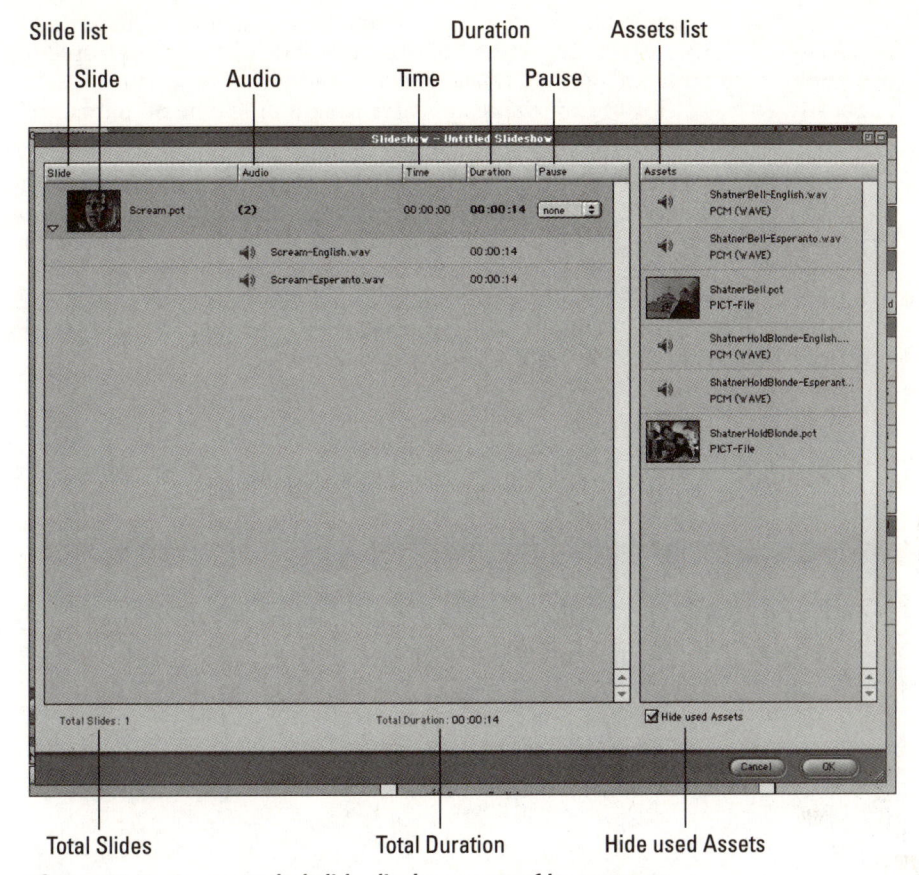

Figure 13-9: An expanded slide displays more of its contents

The following headings are displayed to the right of the slide and indicate values for that particular slide. With the exception of Pause, these values are unchangeable by the user and determined only by the length or number of Assets being used and the value selected for Pause:

- ✦ **Audio.** The value displayed in this column indicates whether an audio Asset has been added to a particular slide and how many audio Assets are present.

- ✦ **Time.** The value displayed in this column indicates the time in the Slideshow when that Slide is first encountered.

Part III ✦ Using Assets in a Project

✦ **Duration.** The value displayed in this column indicates the length of the Assets in the Slide, also known as how long the slide plays uninterrupted. If audio is added to an image in the slide editor or if the slide is video instead of a still graphic, the duration value indicates the running time of that clip.

✦ **Pause.** Determines how long the DVD waits before playing the next Slide. The default value is none. There are additional options for 1, 5, 10, 20, and 60-second pauses by clicking on the arrow and making a selection from the drop-down list as shown in Figure 13-10. Also, there is the ability to set a custom pause length by choosing "other" from the drop-down and setting a value from 0 (no pause) to 255 seconds (infinite pause). Another handy value allows you to set the slide to infinity which advances the slide after clicking the next track button on your DVD player's remote control.

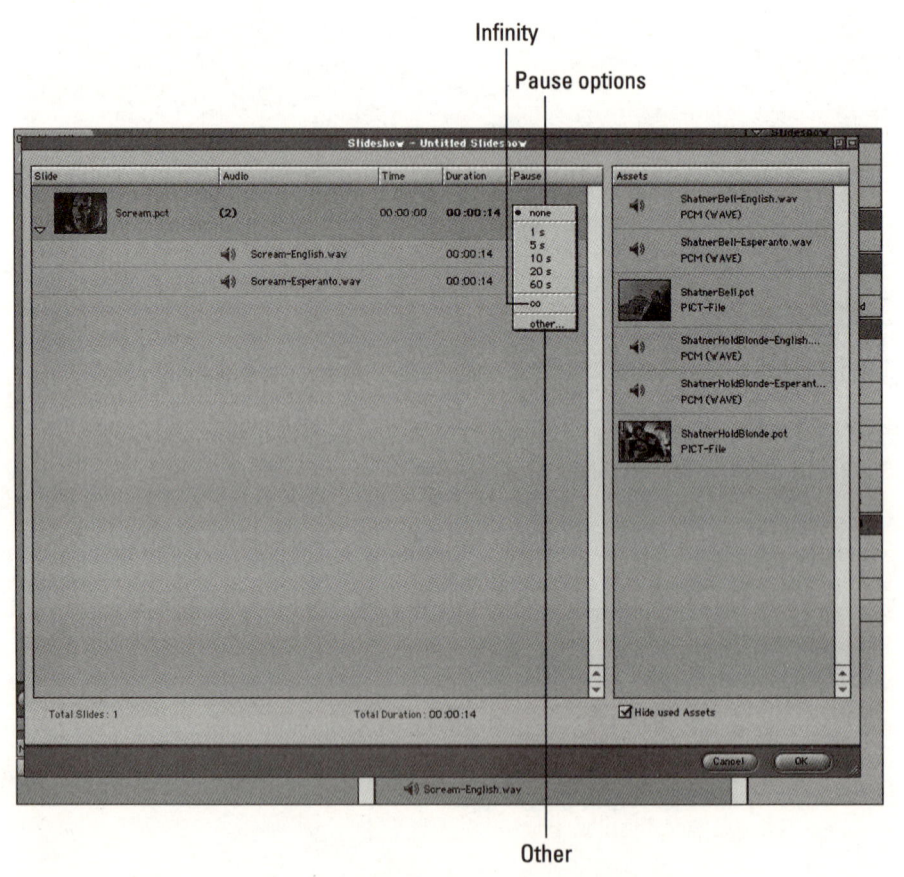

Figure 13-10: Pause options that determine the length between Slides.

Once you have a solid grasp of the elements in the Slideshow editor it is time to proceed to creating a slideshow.

Creating a Slideshow

Now that you have explored the Slideshow editor and have an understanding of how it works and what it does, it is time to create a new slideshow. The creation of a slideshow is quite simple actually, once you realize how to add elements and read the interface. As mentioned earlier, it is essentially a matter of assembling slides in the order that you want them to appear, adding audio elements, and adjusting your durations with the Pause function.

The following steps demonstrate how to create a Slideshow for your DVD project.

> **Note** Slideshows may contain up to 99 assets per slideshow with a maximum of 8 audio streams per asset.

1. **Import (File ⇨ Import) the Assets you want to use.** If you need some images and video to work with for this tutorial, open up the files located in the images and video folders on the DVD-ROM under any one of the Tutorial folders and import them into DVD Studio Pro.

2. **Click the Add Slideshow button at the bottom of the Graphical View.** A new tile is added to the workspace as shown in Figure 13-11. Name the Slideshow in the tile box or in the Property Inspector.

3. **Double-click the tile's thumbnail area and open the Slideshow editor.** This is the part of the interface where you work with Assets to assemble a Slideshow. Notice the Assets list to the right. These are all the Assets in your project that are suitable for use in a Slideshow. For instance, if your PICT files were not created with the correct dimensions they do not display. The Tutorial files included on the DVD-ROM should provide you with adequate resources to test as appropriate Assets.

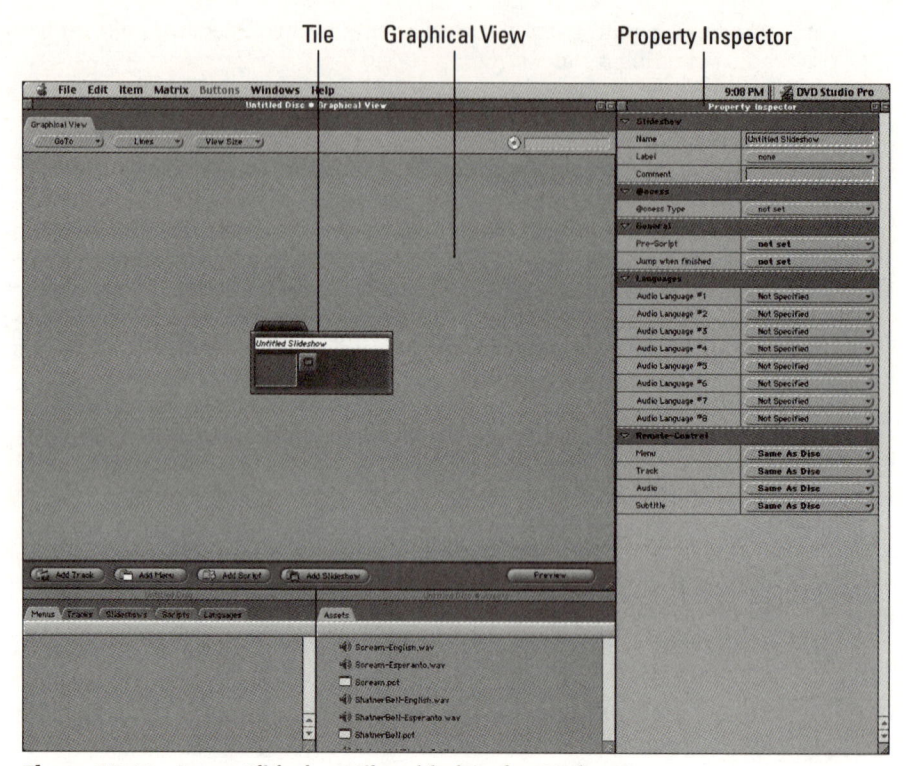

Figure 13-11: A new Slideshow tile added to the Workspace

4. **In the Assets list to the right, locate a PICT file that you would like to display first and drag it into the Slide list.** Again, if you want to use a tutorial file, select one of the tutorial folders on the DVD-ROM, open its images folder and import one or more PICT files. Notice the box at the bottom of the window beneath the Assets list called Hide used Assets (see Figure 13-12). By checking this box the Assets list no longer displays those Assets which you drag to the Slide list. This is an easy way to keep track of which Assets have been used and which are remaining. You can also reuse Assets from the Assets list by leaving the Hide used Assets box unchecked. This is useful when you wish to repeat certain elements. You can check or uncheck this option whenever it is necessary. To the right of the Slide column is a column entitled Pause. If you want your Slide to last for a particular duration click the pause button for that Slide and select a time. Use the infinity option if you want the Slide to advance only when a user clicks the Next Track key on their remote control. Choose other if you want a specific time value which is not listed (see Figure 13-13).

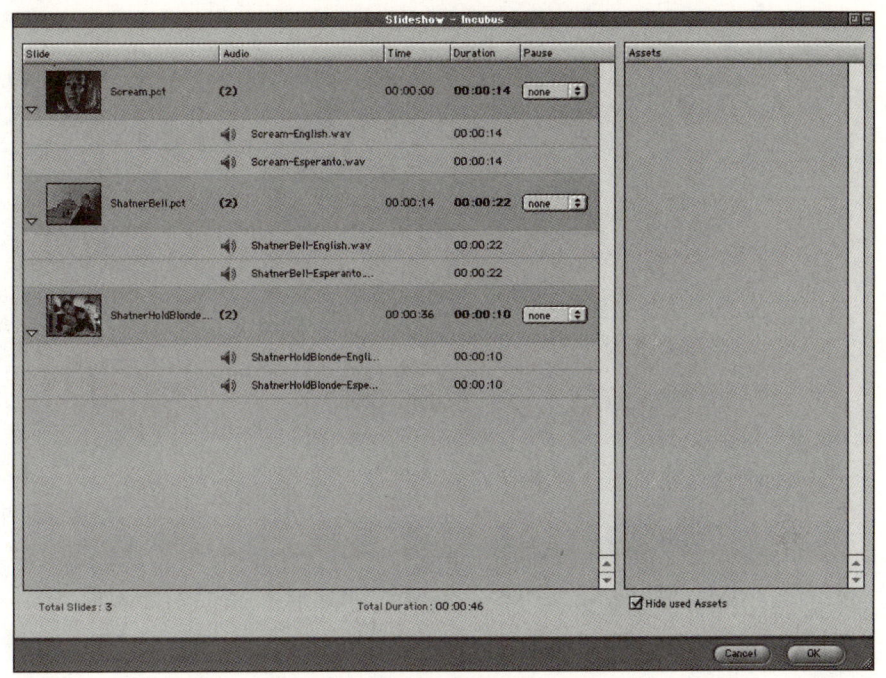

Figure 13-12: In this example, the checked Hide Used Assets box helps you to keep track of which Assets have already been used.

> **Tip** To remove a Slide from the Slide list, select the Slide and choose Edit ⇨ Clear or press delete on your keyboard. Dragging the slide back to the Assets list does not work.

> **Tip** Rearrange your slides by clicking and dragging them to the position where you would like them to appear. Photoshop has a similar way of moving layers in front of each other.

5. **Drag an audio file from the Assets list onto the graphic in the Slide list.** It is important to note that all audio files within a Slideshow must use the same format. For instance, if you decide to use AIFF then all audio streams in that Slideshow must use that format. Also, make sure to match the correct audio files with their slides — otherwise, you are going to have a difficult time later on when you try to rearrange and correct the placement of audio streams. In the worse case scenario, you might not even notice there is anything wrong until you get your replicated disc back from the plant and realize that Girl A is speaking Boy C's lines! Naming Assets consistently from the beginning should help avoid any mix-ups.

Part III ✦ Using Assets in a Project

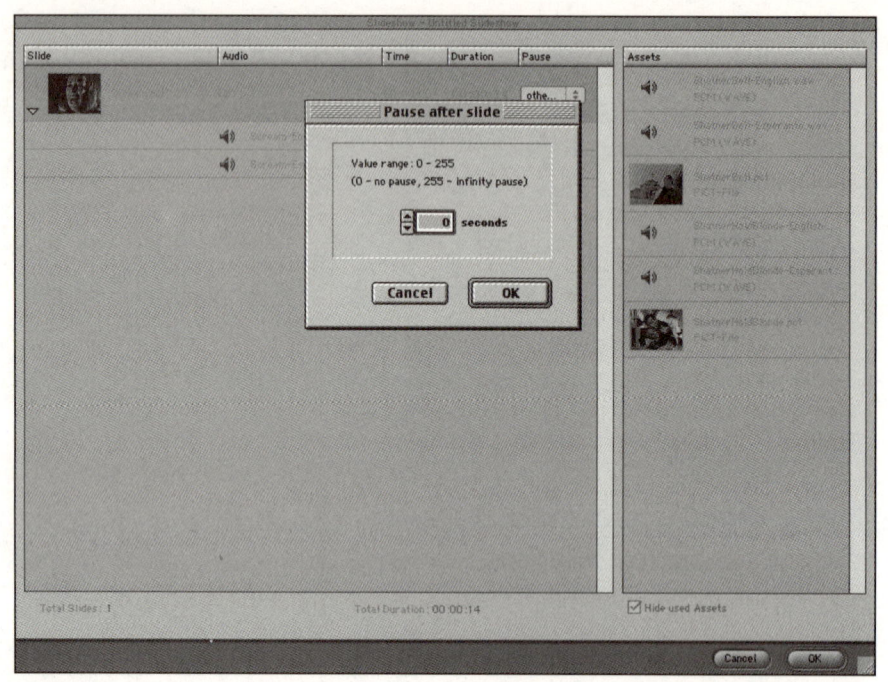

Figure 13-13: By selecting the other option from the Pause pull-down in the Slide editor, you are presented with a dialog box entitled Pause after Slide, which allows you to specify a specific time value from 0 (no pause) to 255 (infinite pause).

6. **If you like, drag a video file from the Assets list onto the Slide list.** Adding a video file is the same as adding a still image. Video does not require audio but you may add it if desired the same as you would for a still image.

Note: When you place an audio file within a Slide or when you add video to the Slide list, the Slide duration automatically changes to match the length of that clip. This can be helpful for Slides with narration that require precise lengths for playback. If an audio clip is longer than the video (or still), the video freezes on the last frame until the audio has finished before playing the next Slide.

7. **Click OK to complete the creation of the Slideshow and close the Slideshow editor.** You are able to go back and edit as much as you like by clicking on the tile's thumbnail as you did in Step 3. (See Figure 13-4)

Figure 13-14: The completed Slideshow tile.

8. **Now it is time to preview the Slideshow. Select the Slideshow tile and click the Preview button in the lower right corner of the Graphical View.** (See Figure 13-15.) Use the Next Track and Previous Track keys on the remote control to move forward and backward between different Slides. Hit the stop button to end the preview and return to the workspace.

Figure 13-15: Previewing a completed Slideshow

Using Languages

If you are creating a multi language title you may wish to include additional languages in your Slideshow. Apple made certain to include every language you can imagine — 136 in all and probably some you have not heard of before. For instance, if you wanted to create a DVD for the movie *Incubus,* you could select Esperanto as a language from the drop-down list.

Cross-Reference

See Figure 1-2 in Chapter 1 for a complete list of available languages.

> **Note:** *Incubus* was a movie made in 1965 starring a young William Shatner speaking his lines in Esperanto — an artificial language constructed from words used by several European languages. The movie recently resurfaced from near extinction and is being distributed on video under the direction of its producer Anthony Taylor (Contempo III). It is a unique movie experience and one that is worth checking out.

To add multiple language tracks to a Slideshow project, follow these steps:

1. **Construct a Slideshow by adding a new Slideshow tile and working with the Slideshow editor as outlined above.** Make certain that you have imported all the audio tracks and other Assets that you are using to create the Slideshow.

2. **In the Property Inspector click Languages ⇨ Audio Language #1.** A dialog box appears that includes a scrollable list of languages arranged alphabetically. Choose the Languages you would like in the order you want them to appear for all Slides in the Slideshow. When a user selects the audio option on their remote control Audio Language #1 is presented first, Audio Language #2 is presented second and so on. (See Figure 13-6.)

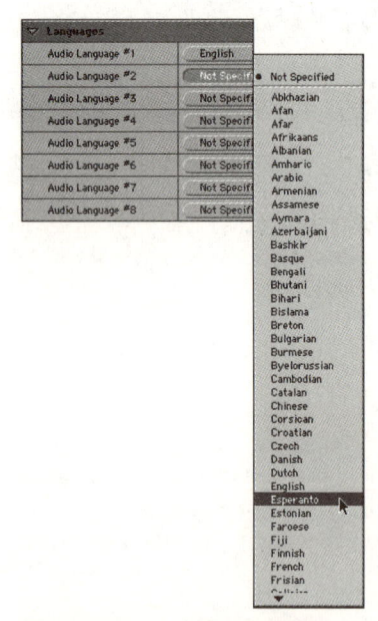

Figure 13-16: Adding Esperanto as Audio Language #2.

3. **Add the audio tracks to the Slide by dragging them from the Assets list onto the Slide list.** The order of your audio tracks is very important and should correspond to the order of the list in the previous step.

> **Note:** When deciding on an order for your alternate language streams, make certain that you use the same order throughout the project. A disc does not allow mismatched audio language streams. For instance, if you have selected English for Audio Language #1 and Esperanto for Audio Language #2 they must appear this way in every Menu, Track, and Slideshow to avoid conflicts.

Summary

You should now understand how to assemble assets and construct a slideshow using DVD Studio Pro. Also, you should have a good idea of the many uses for slideshows in a DVD project. After you have created a few slideshows, you may find even more uses and possibilities that we have not mentioned here. Your imagination (and a few technical considerations) is the only limit to what you can create. If you are looking for ideas, look through your collection of DVDs or try renting some different DVDs to find interesting variations on the type of slideshows that other people are making. Studying professionally created DVDs may also help you with the other topics in this book, particularly when it comes to creating menus. Keep in mind that the benefit of working with Assets in DVD Studio Pro is the way that many of the same Assets are reusable in the various disc elements, including tracks, menus, and slideshows.

- ✦ Slideshows incorporate still graphics as well as audio and video elements.
- ✦ Scanners are useful for acquiring graphics, but also introduce potential problems such as moiré patterns.
- ✦ Slideshow images can be cropped and resized in Photoshop for inclusion in a DVD project.
- ✦ Multiple languages and audio tracks can be incorporated into a Slideshow.

Now that we have concluded our last chapter on using assets in a project, in the next chapter we show you how to assemble and optimize your assets and files when building a DVD project.

✦ ✦ ✦

Bringing It All Together

PART IV

In This Part

Chapter 14
Building the Project

Chapter 15
Preparing for Output

Chapter 16
Outputting a Project

Building a Project

CHAPTER 14

In This Chapter

Setting up sample menus

Setting up sample buttons

Linking buttons to menus and tracks

In this chapter you are going to go through the steps of building a basic DVD project from start to finish, using assets that you created in earlier chapters. The chapter touches on menus, buttons, and adjusting settings for interactivity.

The project you are going to build includes a main menu, a sub-menu, and several additional screens, with links to two separate video segments.

Depending on your level of confidence with the various tasks involved, you may wish to budget up to 3 hours to complete the project and practice. Alternatively, if time presses, you may want to upgrade your bloodstream with a six pack of Jolt cola.

Setting up the project consists of three phases: Preparation, Button Creation, and Linking. The first phase involves opening up a new project file, importing the assets, creating the various Menu and Track tiles and associating the appropriate assets with them. The second phase involves creating the appropriate buttons, and the third phase is where you set up the interactive relationships. Actually, there may be a fourth phase, depending on how things go — either Vexation or Jubilation.

Preparing the DVD Project File

When you are ready to begin, create a new folder, Practice DVD, on your hard drive and copy the PS Records DVD Assets folder into it; the PS Records DVD Assets folder is located in the Tutorial section of the DVD-ROM.

To prepare the DVD project file, follow these steps:

1. **Start DVD Studio Pro.**
2. **Choose File ➪ New to create a new project file.**
3. **Choose File ➪ Import.** The Import Assets dialog box appears.
4. **Locate the PS Records DVD Assets folder and open it.**
5. **Click the Add All button and then click the Import button, as shown in Figure 14-1, to import the assets into the project file.**

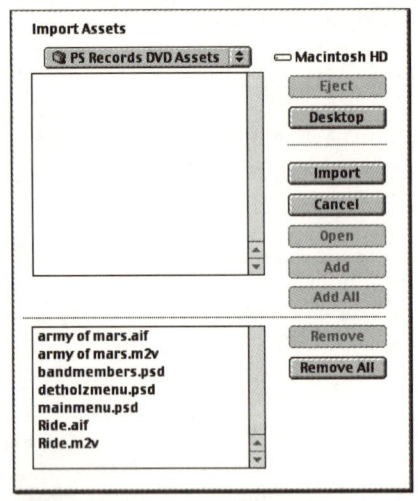

Figure 14-1: The Import Assets dialog box.

6. **Save the project file in the Practice DVD folder.**

Creating the tiles

The project requires two main menus and five additional menus for each of the band members in the Detholz. It also requires two tracks for the two music videos.

To create the main menus, double-click the Add Menu button at the bottom of the Graphical View window. Name the two resulting tiles Main Menu and Detholz Menu, by either clicking in the text area of the tile, or using the Property Inspector with the tile selected. Then add five additional menus and name them Rick, Jim, Andrew, Carl, and Ben, respectively.

To create the tracks, double-click the Add Track button, and name these tiles Army of Mars Video and Ride Video, respectively. After you finish, you can re-arrange the tiles according to taste. The screen should end up looking something like Figure 14-2.

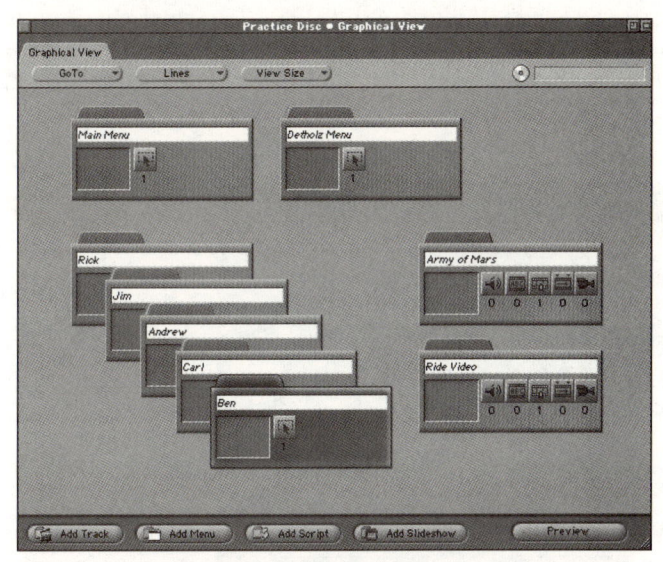

Figure 14-2: This Graphical View shows the newly created tiles. The tile names are in italics, indicating that no assets have been associated with them yet.

The menu and track tiles are ready for you to create buttons and associate the various assets with them.

Associating the assets

You can associate the appropriate assets with the menus and tracks either by dragging and dropping, or by "manually" associating the asset, which in most cases means selecting each tile and then selecting an appropriate asset from the Asset drop-down menu in the Property Inspector. It is a bit more involved to manually associate an audio associate, but fear not, it can be done.

To associate the assets by dragging and dropping, follow these steps:

1. **Drag the mainmenu.psd asset from the Assets Container onto the Main Menu tile in the Graphical View.** (If the Assets Container is not showing in the workspace, choose Windows ⇨ Asset View to activate it.)

2. **Drag detholzmenu.psd onto the Detholz Menu tile and then drag bandmembers.psd onto each of the five band member tiles.**

3. **Drag army of mars.aif and army of mars.m2v onto the Army of Mars track, and drag ride.aif and ride.m2v onto the Ride Video track.** (The AIF files are the audio streams, and the M2V files are the MPEG-2 video streams, all of which were generated using the MPEG-2 encoding function of QuickTime Pro.) After you finish, the Graphical View window should look something like Figure 14-3.

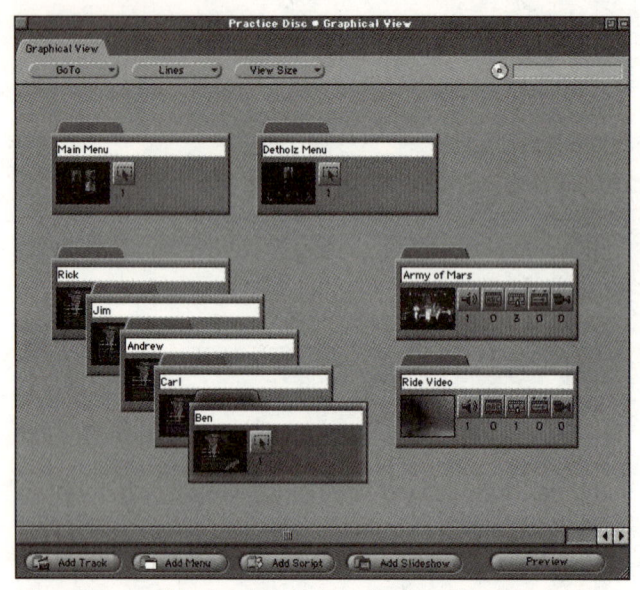

Figure 14-3: This Graphical View displays the thumbnail images in the tiles, reflecting the newly associated assets.

Associating the visual assets

To associate the visual assets using the Property Inspector, follow these steps:

1. **Click the Main Menu tile in the Graphical View window to select it, and locate the Picture section of the Property Inspector.** The Asset drop-down menu will have an initial value of not set.

2. Click the Asset drop-down menu and select mainmenu.psd, as shown in Figure 14-4.

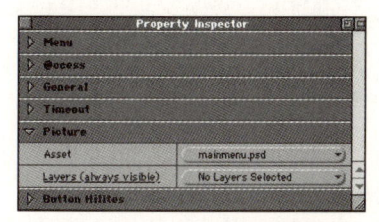

Figure 14-4: The Property Inspector, with the Picture section expanded, shows the mainmenu.psd asset that is selected and associated with the Main Menu tile. When the asset is selected, you are presented with the option to choose which layers in the Photoshop file you want visible; this option comes into play when you are setting up interactivity.

3. **Repeat this process for each of the remaining menu tiles.** The Detholz Menu tile should be associated with the detholzmenu.psd asset, and each of the band member menu tiles should be associated with the bandmembers.psd asset.
4. Click the Army of Mars track tile in the Graphical View window.
5. In the Video section of the Property Inspector, choose army of mars.m2v for the asset.
6. Repeat Steps 1–5 for the Ride Video track tile and select the ride.m2v asset.

Associating audio assets

For those who prefer to manually associate the audio assets, follow these steps:

1. Click the Tracks tab in the Project View window.
2. Expand the Army of Mars track (by clicking the small triangle to the left) and select Audio Streams.
3. Choose Item ➪ New Audio Stream. In the Project View, a new audio stream appears, named Untitled Audio.
4. Select the new audio stream, and in the Property Inspector, name it Army of Mars Audio.
5. In the General section of the Property Inspector, set the asset to army of mars.aif. The name change is reflected in the Project View, as shown in Figure 14-5.

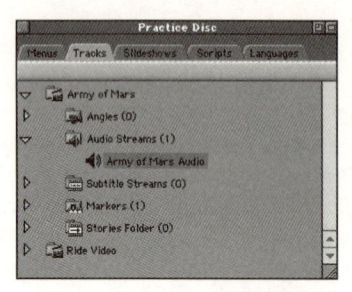

Figure 14-5: The Project View with the Tracks tab selected, showing the expanded view of the Army of Mars track, with the newly associated Army of Mars Audio asset.

6. **Repeat Steps 1 through 5 for the Ride Video track, naming the audio stream Ride Audio and associating ride.aif with the track.**

Whether you used the drag and drop method or associated the assets manually, you are ready to create some buttons.

Creating the Buttons

In the PS Records DVD sample project, there are two different kinds of buttons. The first kind uses the built-in feature of DVD players to create a highlight to indicate when a button is selected. The second kind uses a specially created Photoshop layer to indicate when the button is selected.

Creating the buttons for the Main Menu

Open the Menu Editor by double-clicking the thumbnail area of the Main Menu tile. The first thing you will be doing is setting which layers from the Photoshop file you want visible on the screen.

To make the layers visible, follow these steps:

1. **With the Menu Editor active, locate the Picture section of the Property Inspector.**
2. **In the Layers (always visible) line, use the drop-down menu to select each of the five layers in the Photoshop document, including the background.** Dots appear next to each layer to indicate they are active, as shown in Figure 14-6.

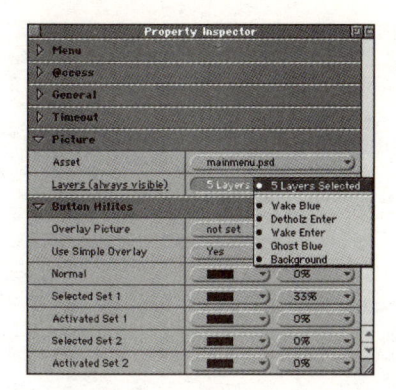

Figure 14-6: The Property Inspector, showing active layers

After the layers are visible you can create the buttons themselves. For the first button, you can use the Untitled Button that appears when you first open the Menu Editor.

To create the buttons, follow these steps:

1. **In the Menu Editor, move the mouse pointer over the rectangular area of the Untitled Button.** The pointer becomes a hand and allows you to click and drag the button. Drag the button down so that the upper left-hand corner is up and to the left of the word ENTER that appears on the left of the screen.

2. **Resize the button by clicking on a corner of the button area and dragging to the desired size, so that the button appears as a rectangle over the word ENTER.** (Then call or e-mail Apple Computer and ask them to put guides and a snap-to-grid feature in version 2.0 of DVD Studio Pro to allow for better alignment of buttons.)

3. **Select the first button you created by clicking on it.**

4. **In the Property Inspector, name the button Detholz Enter.**

5. **Place the pointer up and to the left of the word ENTER underneath the rightmost image, and click and drag a second button.**

6. **Name this button Ride Enter.** The Menu Editor should look something like Figure 14-7.

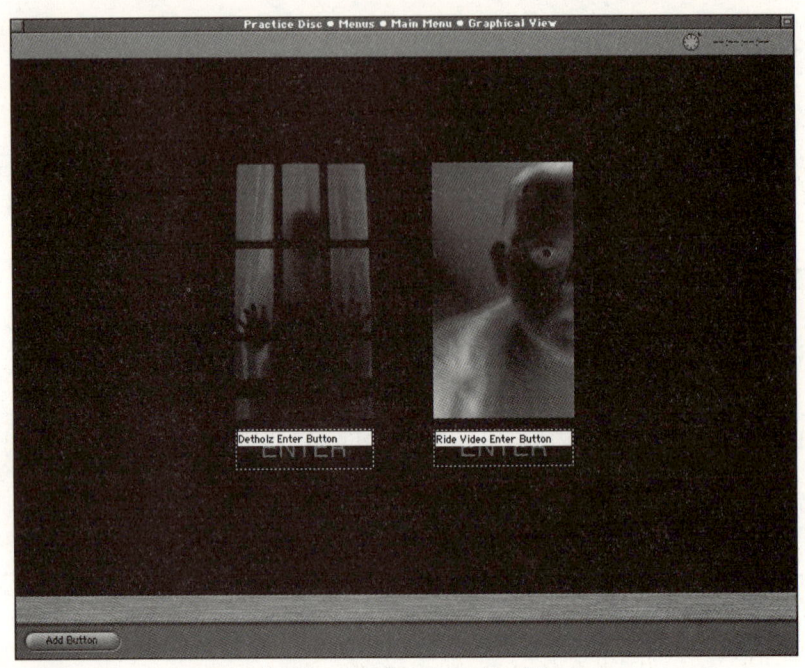

Figure 14-7: The Menu Editor with newly created buttons.

With the main two buttons created, you can now move on to creating the buttons for the Detholz Menu.

Creating buttons for the Detholz Menu

The type of button used in the Detholz Menu is based on a technique which is similar to rollovers on the Web. An image is created that is always visible, and another image is created to only display when you roll the mouse over that button. However, in this case, you are using a remote control to move between the two images. So instead of a "rollover state," you have the "selected state."

In the detholzmenu.psd file, the layers are arranged so that colored versions of the buttons are above the gray versions of the buttons. Within the DVD project, you indicate these colored layers to be displayed for the selected state of the button.

To create the buttons, follow these steps:

   1. Close the Menu Editor if you have it open.

2. **In the Graphical View, double-click the thumbnail area of the Detholz Menu tile to open the Menu Editor.**

3. **Select the Untitled Button and delete it by pressing the Delete key on your keyboard.**

4. **In the Picture area of the Property Inspector, make the background visible, and the gray versions of each button visible, by repeatedly clicking the Layers (always visible) drop-down menu to select each appropriate layer.**
   This includes the Rick, Jim, Andrew, Carl, and Benner layers. The Menu Editor screen should look something like Figure 14-8.

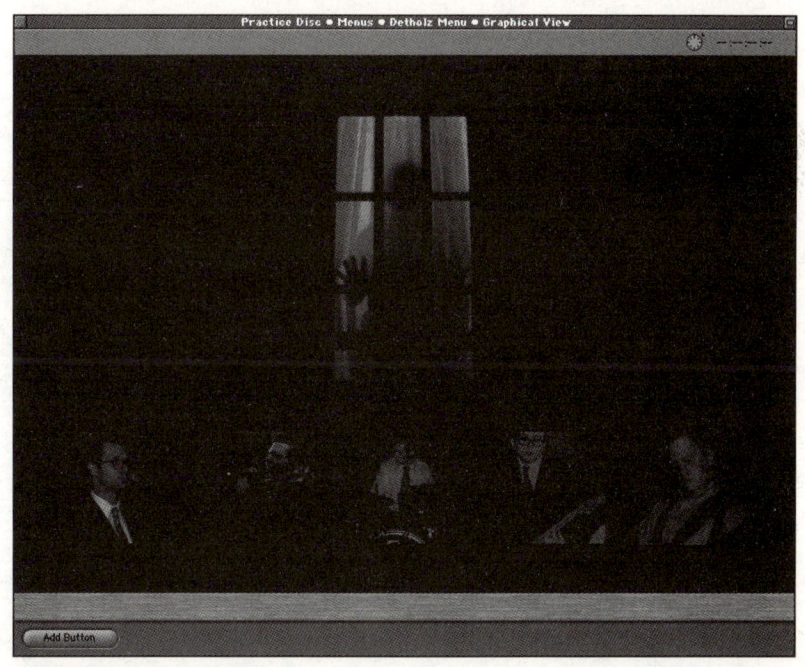

Figure 14-8: The Menu Editor, showing the Photoshop layers that represent the inactive button states.

5. **Choose Item ⇨ New Button (or press ⌘+K) a total of six times to create six new buttons that sit on top of each other.**

6. **One by one, move the buttons down over their corresponding images.**
   Because Photoshop layers are used to represent the selected states instead of hilites, the size of the button outline does not need to be exact. In other words, everything doesn't have to align exactly right.

326 Part IV ✦ Bringing It All Together

> **Note** When you are using Photoshop layers to represent the selected states of buttons, the region defined by the button in the Menu Editor defines the area that will be active when people use the DVD in their computer. When a DVD is played on a computer, people are sometimes given the opportunity to use the mouse to select a button directly on the screen rather than necessarily having to click simulated remote control buttons. At that point, the DVD behaves similar to a Web page; where you roll the mouse over a DVD button to select it.

7. **Name the buttons that appear along the bottom the screen, from left to right: Rick button, Jim button, Andrew button, Carl button, and Ben (or Benner) button.** Name the top button Army of Mars button. The results should be similiar to Figure 14-9.

Figure 14-9: The Menu Editor, showing the newly created and named buttons.

Now that you have created the buttons for the Main Menu and Detholz Menu, you can move on to creating the individual band member menus, which are more like pages than menus.

Creating the band member menus

This is going to get a bit silly. In this section, you will create a Go and Stay button, illustrating the sometimes-humorous question of how to set up navigation for certain kinds of DVD pages. If you feel like getting sidetracked, you could go back into the original Photoshop file, bandmembers.psd, and set things up a bit differently to improve the design. (While you're at it, why don't you see if you can capitalize the rick layer; Rick would surely appreciate getting the same capitalization as other band members.)

Start with the Rick page. Rick Franklin, that is. Also known as Mr. F., Rick has an excellent singing voice, is a good guitar player, and is also a loyal Mac-based graphic designer.

To get Rick going, follow these steps:

1. **Go to the Graphical View and double-click the Rick tile to open up the Menu Editor.**
2. **Make the Stay, GO, rick, and Background layers visible.**
3. **Create two buttons, naming them Go and Stay.** Just do it. You know the drill.

Now you may be sensing that there will be some repetitiousness, some repetitiveness, and some repetitivity in this tutorial because there are four more band members. You're right, and we haven't even gotten into setting all the button states. So you may be asking yourself, is there an alternative?

Why yes Virginia, there is an alternative. We all get to go to bed earlier.

Understanding the beauty of the Duplicate function

In the movie *The Matrix,* at one point our hero Neo is asked to take a leap of faith. Neo is asked to make the jump from the top of one skyscraper, across a vast distance, to another skyscraper. We are in the "Matrix," where everything is just a computer simulation, and yet the leap is quite literal to the senses.

If you would like to take a leap of faith, delete all of the band member tiles that you created, leaving Rick, because you won't need them. No, Rick isn't going for a solo career. The Detholz haven't reached that stage yet. You have the technology to clone Rick, and rename and reconstruct him without needing to set up buttons manually in each new menu page.

Cloning the Rick tile

Okay, so it's not cloning, it's just duplicating. Copying pages, sheep, humans, tiles, what's the difference?

To duplicate a tile, follow these steps:

1. **Select the Rick tile.**
2. **Choose Item ⇨ Duplicate.** A copy of Rick is created. If you open the menu tile, you will see the settings carried over. Woohoo! This means that if you have a series of menus that have similar elements, you don't have to go in and adjust the settings manually for each one. Unless of course, just like one of the authors, you were recently laid off as a result of an Internet workforce reduction and need as much billable time as possible; then you may want to go the manual route. It will be good practice. Yeah.
3. **Delete Rick. With extreme prejudice.**

Wait. We just duplicated! Why delete? Because you don't want to start duplicating menus until you have all the settings just the way you want them. And now you need to set up the interactive relationships between all the buttons you just created.

Linking and Thinking

When you consider a typical DVD, you see that the interactivity is typically not all that complex. You go from one page to another, choose a movie trailer, maybe listen to the director comment on something, and then you watch the movie. Compared to some other forms of interactivity, such as a presentation made using Macromedia Director, or a typical Web site, DVD interactivity is blessedly simple to learn, thanks to the limitations of the hardware.

But in spite of some of the simplicities, you have to keep thinking while you are linking. Because a program like DVD Studio Pro gives you the option to link any button to any menu and to have a variety of button states, visible layers, and so on. So essentially, becoming familiar with the options will help you get to the point where you can be systematic. Being systematic about developing a DVD project will help you to cover all your bases, to make sure all the settings are how you want them.

Let's return to the Main Menu. This is where a storyboard or flowchart could come in handy, to help remind the DVD author of how to systematically set up interactive relationships.

In this case, not only are you setting up the buttons to go from one page or another, you are setting up how the buttons on the remote control work. It's easy to think about getting from one piece of content to another, but the user will appreciate when you've remembered to make the up and down arrows on the remote actually do something, even though the user is supposed to be using the left and right arrows.

Linking the Main Menu

To set up the links on the Main Menu, follow these steps:

1. Click the Main Menu tile in the Graphical View.

2. In the Property Inspector, locate the Button Hilites section and choose a color for the Selected Set 1, preferably something like what is shown in Figure 14-10. What you are doing is setting up the color of the "overlay" for the DVD player to draw to show when a button is selected.

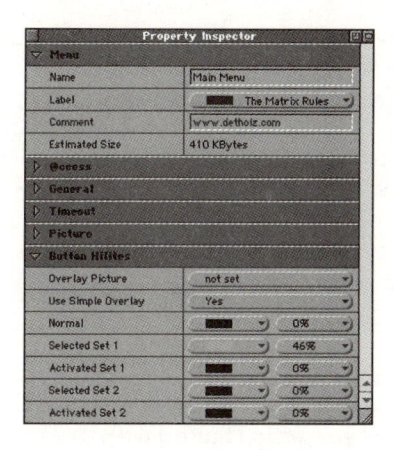

Figure 14-10: The Property Inspector, showing a simple Button Hilites setting, where the Selected Set 1 has a gray color selected with a 46% value.

3. While you're at it, set up the same Button Hilites settings for the Rick tile just as you did for the Main Menu.

4. Go back to the Main Menu tile and double-click the thumbnail area to open up the Menu Editor.

5. Select the Detholz Enter button.

6. In the Property Inspector, set the Jump when activated option to Detholz Menu, and set the Button Links as follows:

    - **Up:** Not Set
    - **Down:** Not Set
    - **Left:** Ride Video Enter Button
    - **Right:** Ride Video Enter Button

The Button Links should look just like Figure 14-11.

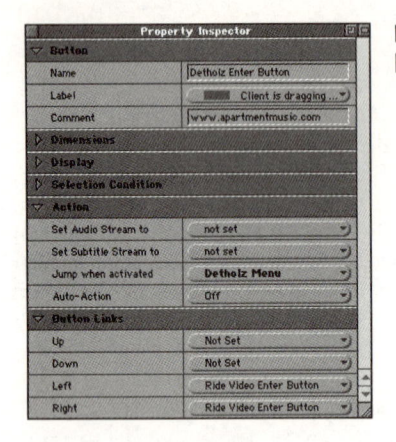

Figure 14-11: The Property Inspector, showing button settings for the Detholz Enter Button.

7. **Select the Ride Video Enter Button.** Because this button leads directly to a video track, set the Jump when activated action to Ride Video.

8. **Set the Button Links as follows:**
   - **Up:** Not Set
   - **Down:** Not Set
   - **Left:** Detholz Enter Button
   - **Right**: Detholz Enter Button

The Button Links should appear the same as in Figure 14-12.

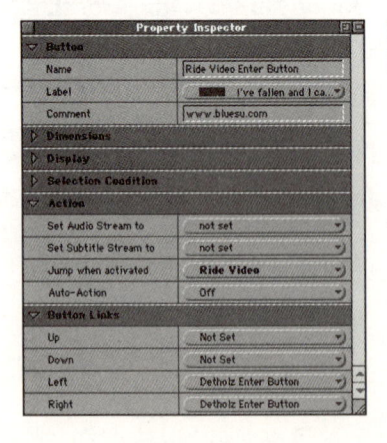

Figure 14-12: The Property Inspector, showing button settings for the Ride Video Enter Button.

If you haven't done so already, you may want to go back to the Graphical View, select the Main Menu tile, and click the Preview button to try things out. You will be able to go back and forth between buttons, as well as go to either the Detholz Menu or the Ride Video track. Woohoo!

Linking the Detholz Menu

Now that you have the Main Menu all linked up using traditional button Hilites, it's time to move on to the Detholz Menu, where the experience of multicolored Photoshop layer-based buttons awaits.

Isn't it nice that you already created the buttons? Now all you have to do is set up the interactivity. And the nice thing is, you don't have to bother with Hilite settings, because you are using the colored Photoshop layers to represent the selected state of the buttons.

To link the Detholz Menu, follow these steps:

1. **Double-click the thumbnail area of the Detholz Menu tile in the Graphical View to open up the Menu Editor.**
2. **In the General area of the Property Inspector, set the Default Button to the Army of Mars button and set the Return Button to Main Menu.** The Army of Mars button will appear active at the top of the screen when a user accesses the menu; when the Return button is clicked on the remote, the user is taken back to the Main Menu.
3. **Select the Rick button in the Menu Editor.**
4. **In the Display section of the Property Inspector, set the Normal State to the Rick layer, and the Selected State to the Rick active layer.**
5. **Set the Jump when activated Action to Rick; this links the button to the Rick Menu.**
6. **Set the Button Links as follows:**
   - **Up:** Army of Mars button
   - **Down:** Army of Mars button
   - **Left:** Ben button
   - **Right:** Jim button

   The Property Inspector should look something akin to Figure 14-13.

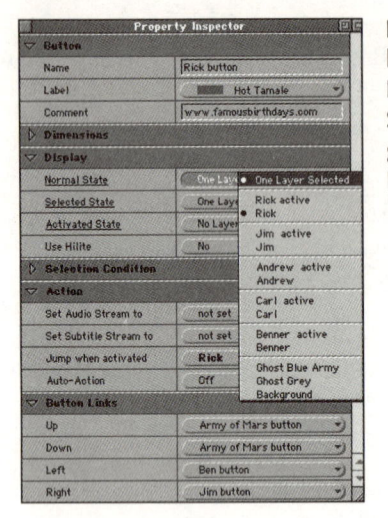

Figure 14-13: The Property Inspector, showing button settings for the Rick button, and the Normal State drop-down menu. In the Normal State, the Rick layer is selected, which displays a gray image. In the Selected State, the colored Photoshop layer called Rick active is selected.

7. **Take a break from the band members for a minute and select the Army of Mars button, and set the Normal State to Ghost Grey, and the Selected State to Ghost Blue Army.**

8. **Set the Jump when activated to Army of Mars option; this will link the menu to the video segment.** Don't forget to set the Button Links! Use your imagination; you are now an experienced DVD button Thinkerlinker.

Sigh. Now you have completed the primary linking for the Detholz Menu. This is where things can get a little wacky, whether it's late at night, or it's so early in the morning that the caffeine hasn't taken effect.

If you took the leap of faith earlier in the tutorial and deleted the various band member tiles, leaving Rick all on his own . . . sorry, Mr. F. What was the author thinking! Anyway, if you deleted those other menu tiles, you won't be able to specify where the other buttons on the Detholz Menu lead to.

So you could re-create the band member menu tiles, Jim, Andrew, Carl and Ben, and when you set the Jump when activated properties for their corresponding buttons in the Detholz Menu, then there would be a place to go. But then you don't get to use that tasty Duplicate feature. Decisions, decisions.

While you are letting the left side of your brain chew on that for a while, take another look at Rick.

Linking the Rick Menu

The Rick menu is pretty simple. All you need to do here in the general menu properties is to set the Default Button to Stay, the Return Button to Detholz Menu, and to also set the color values for the Selected Set 1 Hilites, because the band member pages are using Hilites. (Hint: Use the same settings as you did for the Hilites in the Main Menu.)

For the Stay button, well, there's not much to set, except Up. The rationale behind this silly Go and Stay navigation was simply that there was only going to be one button on the page. Because the Go button looked nice without any adornment, the idea was to have the default button be something other than the Go button — but something a person could see, so they would "get it," as shown in Figure 14-14.

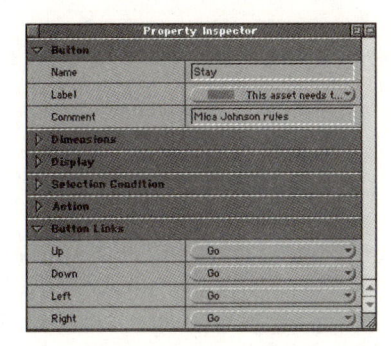

Figure 14-14: The Minimalist button, from the new Philip Glass collection. Go, go, go, go. That's all you need to know.

And for the Go button, it's much the same, except the Jump When Activated Action will lead to _____. Fill in the blank. (Answer: Not the Main Menu. The other one.)

OK, now since the Rick menu is all set up to go, we can proceed to the Clone Wars. As Yoda once said, "Do or do not, there is no try!"

Cloning the Rick Menu and altering its genes

You may want to be sitting down for this, or you might want to go and grab someone and bring him or her over, because it is truly exciting.

Go to the Graphical View, click the Rick menu tile, and choose Edit ⇨ Duplicate to make four copies of Rick, I mean, Mr. F.

Your mission, should you choose to accept it, is to name each tile for one of the four remaining band members, Jim, Andrew, Carl, and Ben. Then, the only thing you have to customize in each menu tile is which layer is visible.

To customize the Jim tile, select the Jim tile. In the Picture section of the Property Inspector, select the Jim layer and then deselect the Rick layer. (See Figure 14-15.)

> **Note** Making layer selections in the Property Inspector's drop-down menus is like a toggle switch; reselecting something de-selects it.

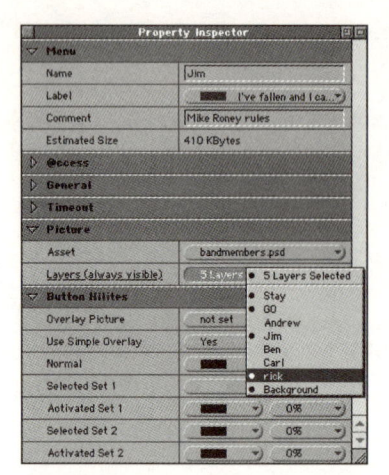

Figure 14-15: The Property Inspector, with Jim layer selected and rick about to be de-selected.

That's right; there was only one step to that. Now you can take care of the three remaining band member menus, Andrew, Carl and Ben. That's the beauty of duplication; it saves time.

Now that the band member menus are taken care of, you can go back into the Detholz Menu and link the remainder of the buttons.

Completing the Detholz Menu

To wrap things up, bring up the Detholz Menu. If you remember, we left off setting up the button for Rick. The other buttons in the Detholz Menu are based on Photoshop layers rather than Hilites.

To set up the other buttons and complete the Detholz Menu, follow these steps:

1. **Select the Jim button.** In the Display section of the Property Inspector, set the Normal state to the Jim layer, and the Selected state to the Jim active layer.
2. **Set the Jump when activated value to Jim.** This will take the user to the Jim menu.
3. **Set the Button Links as follows:**
    - **Up:** Army of Mars button
    - **Down:** Army of Mars button
    - **Left:** Rick button
    - **Right:** Andrew button

The Property Inspector should look something like Figure 14-16.

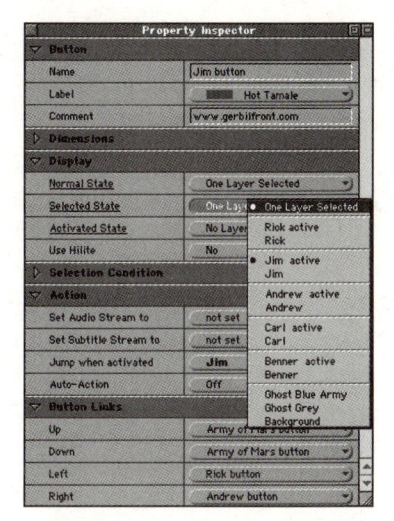

Figure 14-16: The Property Inspector, showing button settings for the Jim button.

Now that you are re-acquainted with the layer-based button states, you can complete the button settings for the Andrew, Carl, and Ben buttons, which lead to their respective Menus. After you do this, you can move on to the Track Settings, which is the final stop on the way to interactive heaven.

Setting the Track properties

Just when you think you are finished building the project, along come those pesky Tracks. If you went back and previewed the project, you may have already discovered something wasn't quite right. You may have tried clicking on the Lines drop-down menu in the Graphical View, and noticed that the lines revealed some ominous or questionable relationships that you overlooked when you went out to buy some more Cheesy Poofs and came back and forgot to set the Jump when activated state of a particular button.

Aha!

At any rate, tracks are people, too. The thing to remember about tracks for a straightforward project is, where do you want someone to go after the track is done playing? And if someone presses a button on the remote control while the track is playing, what happens?

In some cases, leaving the track properties set to the default settings is OK, because they will be the same as the disc properties. But in the case of the PS Records DVD, they need to be customized.

To customize the Ride video track, select the track tile in the Graphical View and set the Jump when finished option to Main Menu, and in the Remote-Control section, set the Return button to Main Menu.

For the Army of Mars track, it's a bit different, because you want to be returned to the Detholz Menu after the video track is done. Set the Jump when finished option to Detholz Menu, and the Return button to Detholz Menu, as shown in Figure 14-17.

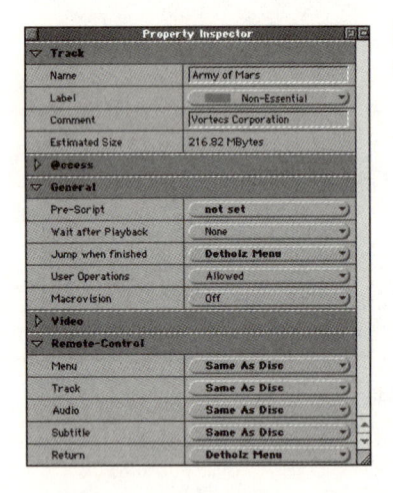

Figure 14-17: The Property Inspector, showing settings for the Army of Mars video track.

Well, you're almost finished. Time for one final check over the Disc properties.

Setting the Disc properties

For many projects, there's not much to do because the defaults often work well. In this case, however, you need to make a few adjustments.

To set the Disc properties, follow these steps:

1. **Click the background in the Graphical View.** Anywhere but on a tile.
2. **In the Property Inspector, set Startup Action to Main Menu.**
3. **In the Remote-Control section, set both Title and Menu functions to Main Menu.** The Property Inspector should look something like Figure 14-18.

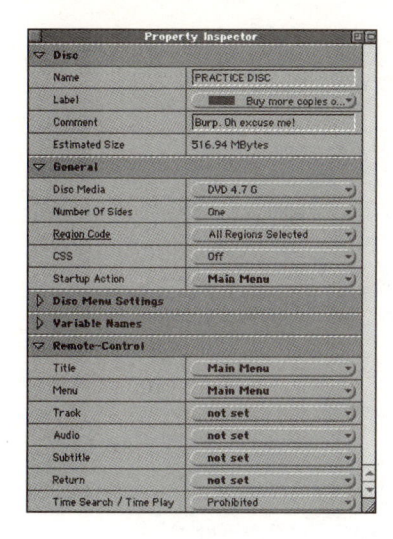

Figure 14-18: The Property Inspector, showing overall Disc settings

So it looks like the PS Records DVD project tutorial is all wrapped up. The approach to building the project was to prepare the tiles, create some buttons, and then do some thinking and linking. As you can see, sometimes things get a little tricky, but developing your own systematic approach can be helpful, as loose ends can tend to accumulate, and where memory may fail, sometimes habit can save your skin. Also, as you develop DVD projects, go ahead and try the Jump Matrix, the Layer Matrix, and the Asset Matrix. They can provide additional insight and help you to see weird things develop and to see gremlins that are trying to destroy your project.

Another thing to consider is building hidden fun into a project. Something like creating an extra Photoshop layer with something interesting on it, and having one of the lesser used buttons on the remote lead to it. (If it's a commercial project and you have an excellent attorney, perhaps you could throw in a small humorous anecdote about your boss, or a picture of your pet python Dennis.)

Since you're reluctant to leave this chapter for some reason, perhaps you'd like to take a nostalgic look back at the interactive relationships and general well-being of the project, by clicking on the Lines menu in the Graphical View and setting it to Always. (See Figure 14-19.)

Ahhh. It all looks so nice. Kind of like walking around in everyday life, trees, people, the sound of music. But then, something doesn't feel quite right. You return to your cubicle and there is a FedEx package waiting for you. Inside is a cell phone, and someone named Morpheus is telling you to sit down at your computer, open up the project file again, and click the Preview button. All of the sudden you realize you need to read the next chapter.

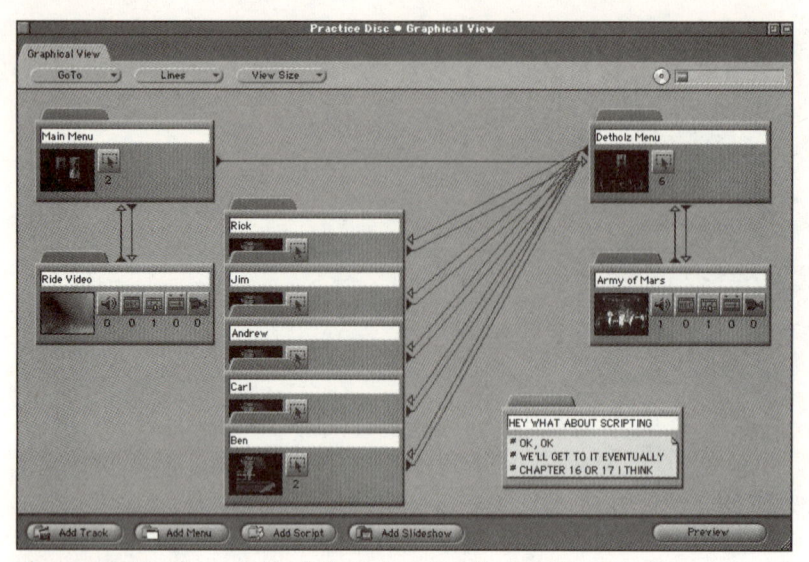

Figure 14-19: The Graphical View, showing the relationships between project elements with the Lines feature set to Always

Summary

✦ You can associate the appropriate assets with the menus and tracks by dragging and dropping, or by manually associating the asset, which in most cases means selecting each tile and then selecting an appropriate asset from the Asset drop-down menu in the Property Inspector.

✦ When you are using Photoshop layers to represent the selected states of buttons, the region defined by the button in the Menu Editor defines the area that will be active when people use the DVD in their computer.

✦ Compared to some other forms of interactivity, such as a presentation made using Macromedia Director, or a typical Web site, DVD interactivity is blessedly simple to learn, thanks to the limitations of the hardware.

✦ Becoming familiar with the options you have in DVD Studio Pro for setting up menus will help you get to the point where you can be systematic. Being systematic about developing a DVD project will help you to cover all your bases, to make sure all the settings are how you want them.

In this chapter, you have developed a full DVD project, including menus, buttons, and video tracks. You have become better acquainted with using various techniques to indicate when a button is selected, using regular hilites as well as Photoshop layers. In the next chapter, you will learn how to prepare a project for output, through a combination of testing and previewing.

✦ ✦ ✦

CHAPTER 15

Preparing for Output

In This Chapter

Testing interactivity

Multiplexing

Disc previewing

In this chapter, you are going to go through the process of preparing a DVD project in DVD Studio Pro for final output. The amount of time you actually have to spend preparing a project is dependent on how thorough you wish to be in previewing and testing the project. The three suggested phases of preparation are Testing Interactivity, Multiplexing and Disc Previewing.

Testing the interactivity of a DVD Studio Project involves systematically checking the settings of your DVD project and utilizing the Property Inspector and other views to see if you can catch any errors. The only kind of preparation that is a prerequisite for actually burning a DVD is the process of *multiplexing*, which generates the appropriate files that a DVD player can read. But as with testing interactivity, before you burn the DVD, it is recommended that you preview the disc after you have multiplexed the files, in case there are any loose ends that you may have missed.

Testing Interactivity

You can take several approaches to testing interactivity, including using the Preview button as you build the project, as well as keeping an eye on the Jump Matrix, and getting used to how things should look in the Graphical View while using the Lines function.

Getting in the habit of testing interactivity with a simple project is worth the investment of time, so that when things get more complex, you have the tools you need to fix — or at least discover — the little bugs that can typically appear in a DVD project, the ones which are foretold in the Loose Ends Statute of Murphy's Law.

Depending on the kind of person you are, you may be able to get away with developing a routine during the creation of a project where you cover all your bases and properly set up all the links out of habit. But there may still come a time when a checklist can be helpful, whether it is employed while you are creating the interactivity or when you are preparing to burn the disc itself.

Reviewing remote-control settings

Most users will be experiencing your DVD through a DVD player hooked up to their television, using a standard hand-held remote control. It may have been a long time since you played make-believe, but in the world of DVD Studio Pro, you can think of it as a professional necessity, where pretending you are the user can lead to astonishing insights and realizations. You might discover that pressing the Menu button on one screen leads to another screen you had entirely forgotten about, where you were taking unlicensed clips of the Teletubbies and experimenting with the multiple angles feature. Best to delete that screen altogether.

Reviewing interactivity using the Preview button

One of the simplest ways to test the interactivity of a project is to use the Preview button function in the Graphical View. This built-in preview function allows you to click on any individual menu or track tile and preview functionality as if you were jumping to that particular section of the DVD. (See Figure 15-1.)

Figure 15-1: The simulated remote control buttons at the bottom of the Preview window

When the project is done, the best approach is to click in the background of the Graphical View window with no particular tile selected, and then click on the Preview button at the bottom of the Graphical View window. For the purposes of the Preview button, this simulates the DVD being inserted into the player. If you don't get a preview of some kind, you may not have the Startup Action set in the Disc section of the Property Inspector (when the Property Inspector is a gray color, representing the overall Disc Properties).

At this point, when you use the Preview function, you might want to click on each remote control button, both to test what they do and to remind yourself of their existence. You may wish to go to each screen in the project and systematically click on every one of the simulated remote-control buttons, and even go and activate each audio and/or video segment in your project and click on the buttons there, too.

Keep in mind that you can set interactivity in every one of the following places in your project:

✦ The main Disc Properties area of the Property Inspector

✦ Every individual button that you have created for a menu by using the Menu Editor

✦ Every individual video or audio track

Checking Interactive settings in the Property Inspector

Another approach for reviewing interactivity is to systematically click on each menu and track tile in the Graphical View and review the interactive settings in the Property Inspector. Remember that you can set the Lines drop-down menu in the Graphical View to give further visual indication of how a user can travel between project elements, given the current settings.

Reviewing Disc settings

You can think of the Graphical View window as a "mega-tile" for the overall disc. Clicking on the background of the Graphical View window brings up the general Disc Properties, where you can make some overall interactive settings. (See Figure 15-2.)

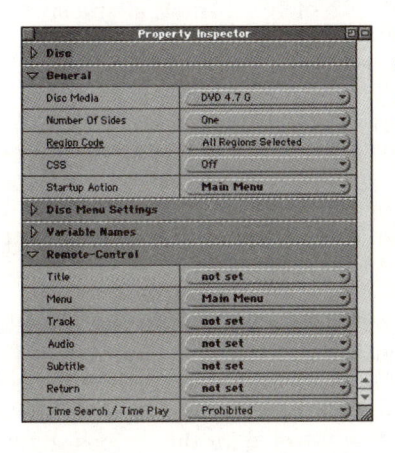

Figure 15-2: Overall disc properties displayed in the Property Inspector. The settings that directly relate to interactivity are Startup Action, Title, Menu, Track, Audio, Subtitle, Return, and Time Search/Time Play.

You'll want to make sure to have something set for the Startup Action. Beyond the Startup Action, the settings you will want to adjust will depend on the project, but you will probably want to have at least the Menu button do something.

Reviewing menu settings

To review interactive settings for an individual menu, select the menu tile in the Graphical View, and its settings come up in the Property Inspector. (See Figure 15-3.)

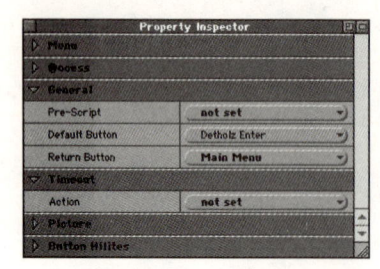

Figure 15-3: Overall menu properties are displayed in the Property Inspector. The settings that directly relate to interactivity are Pre-Script, Default Button, Return Button, and the Timeout Action (use this setting if you want the user to be taken somewhere automatically after a certain number of seconds).

- **Pre-Script:** The Pre-Script option allows you to select a script that will run before anything else happens in the menu.

- **Default Button:** The Default Button is the button that will initially be selected when you get to a particular screen.

- **Return Button:** With the Return Button, you want to imagine where a person may have come from to get to that particular screen. If it could have been from only one place, you will probably want to set it to that previous menu. In other words, if you are adjusting the Return Button setting for Screen 3 of a DVD project and the only place you can get to Screen 3 is from Screen 2, then the Return Button on Screen 3 should send you back to Screen 2. For a screen where there could be more than one point of origin, you may want to set the Return Button to the first menu in the DVD project.

- **Timeout Action:** The Timeout Action allows you to automatically take the user to a particular project segment such as a menu or track, after a specified delay. It is discussed in Chapter 6.

Reviewing button settings

After you have reviewed the settings for the overall menu, it is time to go into each button in the menu. To review interactive settings for an individual button, locate its corresponding menu tile in the Graphical View and double-click on the thumbnail area of the tile to bring up the Menu Editor. Then select each individual button to review its settings in the Property Inspector. (See Figure 15-4.)

The Display settings are important to review so that your buttons display properly on the screen. If you are using Hilites, the Normal, Selected and Activated States may not apply, as long as you remember to set the Use Hilite setting to Yes. Conversely, if you are using Photoshop layers for the various states of a button, Use Hilite should be set to No, and the appropriate layers should be selected for the various states.

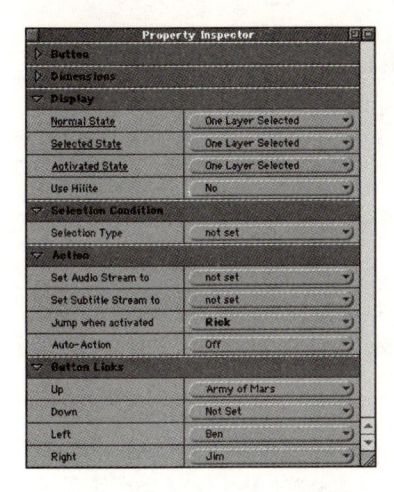

Figure 15-4: Overall properties are displayed in the Property Inspector. The settings that directly relate to interactivity are all of the settings in the Display, Selection Condition, Action, and Button Links areas.

The Selection Condition setting only comes into play if you want that particular button to be in a selected state after the track it leads to is viewed, instead of having the default button for the menu selected. For example, you might have a screen where the default button is set to a button which leads to Special Features. On the same screen, there is a button named Play Video which leads to a video track. So normally when a person accesses the screen, the Special Features button is selected, because it is the default button. But if you set the Selection Condition, you could have the Play Video button show up as selected when a person is returning from the video.

You are almost *always* going to want to have something set for the Jump when activated action. That is, unless you are doing an avant-garde DVD menu as an artistic statement where certain buttons lead nowhere, or if certain buttons on a DVD screen exist only to display information through Photoshop layers on other parts of the page. One of the freedoms afforded by the way DVD Studio Pro works with Photoshop layers is that you can have the selected state of a button display a special Photoshop layer in some other spot on the page so that each button on a page can cause a particular picture to display, for example. OK but let's not get distracted.

The Button Links settings are pretty straightforward; just make sure they make sense relative to the menu in question, and that they are consistent from button to button. For example, if you have a row of buttons at the bottom of a menu, you will probably want either all of them or none of them to have the Down arrow on the remote control activated. Let's say you have four buttons at the bottom of the screen. With one of them, the user can press the down arrow on their remote and the user jumps to a button at the top of the screen. But with the other three buttons, the down arrow on the remote control is not set. Just make sure you try all of the button arrows for every button on your screen, to make sure the available actions make sense to the user.

Reviewing track settings

After reviewing disc, menu, and button settings, you need to deal with the track settings. The most common final touch you might want to make here is to adjust the amount of waiting time after a track ends, and to determine where the Menu and Return keys lead to. (See Figure 15-5.)

Reviewing track interactivity settings toward the end of a project is good, because some of the settings can only be selected after the intended "destination" menus have been created. In other words, the menu you want to send a person to will only appear in the Jump when finished drop-down menu after you have created that menu, so if you create your tracks first in a project, you need to come back and adjust the settings later.

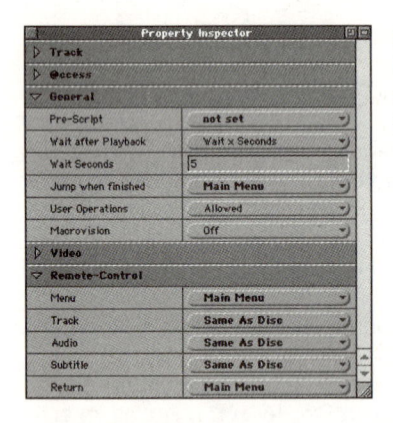

Figure 15-5: Overall track properties displayed in the Property Inspector. The most typical settings to adjust for interactivity are Jump when finished, Menu, and Return buttons.

Developing a checklist

Depending on the type of people you are dealing with, you may want to develop a simple form for a client, colleague, or yourself to fill out as the project progresses, to indicate exactly what each button does. This form could relate directly to the Property Inspector as well.

Figure 15-6 shows a suggested table that was created in Microsoft Word. It consists of a series of tables that you can duplicate and customize. You can either fill it out on-screen or print it out and complete it by hand. Ultimately such a form can serve as a good checklist to confirm that interactive settings have been made properly, or at least according to the original plan.

The form/checklist is available on the DVD-ROM in the CH15 folder in the Tutorial section, as checklist.doc. Bon appétit!

Disc Name	PS Records DVD
Startup Action	Main Menu
Remote	
Title	–
Menu	Main Menu
Track	–
Audio	–
Subtitle	–
Return	–
Time Search	–

Menu Name	Detholz Main Menu
Picture	detholzmenu.psd
Remote	
Default Button	Rick Button
Return Button	Main Menu

Button Name	Rick
Normal State	Rick (grey layer)
Selected State	Rick Active (color layer)
Remote	
Up	Army of Mars track
Down	–
Left	Ben button
Right	Jim button

Figure 15-6: A suggested form/checklist, modeled after various sections of the Property Inspector, in which the elements can be customized and duplicated according to taste. You can construct a guide for both creating and checking the interactivity of a DVD project.

Multiplexing

Multiplexing is the final stop on the way to DVD burning. It is the process of combining the various assets and settings in a DVD Studio Project file and creating the final DVD files, characterized by the VIDEO_TS folder. When you multiplex the project, the end result is a VIDEO_TS folder.

Note: When you multiplex a DVD project, DVD Studio Pro creates both a VIDEO_TS and AUDIO_TS folder. The AUDIO_TS folder will be empty, and is not strictly necessary to have on a DVD disc, but some people recommend leaving it when you burn your disc, in case some DVD players are confused by its absence.

When you insert a DVD into your computer, you may have noticed the VIDEO_TS folder. The computer inside the DVD player looks inside this folder for all the appropriate ingredients, in the proper format, and the process of multiplexing is what generates these files.

Apple recommends having fast hard drive mechanisms for multiplexing, and while it is not strictly necessary, it can be helpful, especially the more complex your project is, when you start getting into multiple angles, streams, and so on.

There are three kinds of multiplexing in DVD Studio Pro:

- **Preview multiplexing:** Preview multiplexing is the built-in process that allows DVD Studio Pro to give you a real-time preview of the DVD, including the video and audio tracks, on the fly.

- **Build multiplexing:** Build multiplexing is the process of "building the disc" from the DVD Studio Project file and assets, and generating the VIDEO_TS folder, without actually burning a DVD-R.

- **Build and format multiplexing:** Build and format multiplexing is essentially the same as build multiplexing, except you end up burning the DVD-R. Burning a DVD-R disc is discussed in Chapter 16.

Note To avoid errors when burning a disc, Apple recommends (in the Read Me file that comes with DVD Studio Pro) quitting all other open applications (that means AOL instant messenger, too — sorry), setting the Energy Saver control panel so the hard disk will never sleep, and turning off AppleTalk, File Sharing, Web Sharing, hard disk indexing, and network time server access. Read the Read Me. Read me? Doing all these things would probably be a good idea even if you're only multiplexing without burning a DVD disc.

In general, the reason why you would want to build a disc without burning is if you were just previewing the disc or if you were sending the VIDEO_TS folder off on DLT tape or another format to have a DVD manufactured. DLT tape has been the standard way to send projects off for replication at a plant, but the available range of media that manufacturers will accept is sure to expand.

You can actually set up the Apple DVD Player program so that in addition to playing standard DVD discs, you can open a VIDEO_TS folder, thereby previewing the final project. This method is discussed later in this chapter.

Multiplexing with Build Disc

Multiplexing is fairly simple. You have all of your files together, the project file is built in DVD Studio Pro, and you've spent at least a modest amount of time testing the project, but you're eager to more thoroughly test the DVD.

Basically, when you have a project ready, all you do is choose File ⇨ Build Disc.

Chapter 15 ✦ **Preparing for Output** 347

To practice multiplexing with the Build Disc menu option, follow these steps:

1. **Locate the PS Records DVD project file in the Tutorial section of the DVD-ROM.** If you've been tinkering already with any of the files, you may wish to copy both the PS Records DVD project file and the PS Records DVD Assets folder over to your hard drive to make a fresh copy.

2. **Open the PS Records DVD project file and choose File ➪ Build Disc.** The Build Disc dialog appears.

3. **Select the folder in which you want to save the multiplexed files.** A Progress window opens up to give an account of how things are going. (See Figure 15-7) This window is discussed further in Chapter 11.

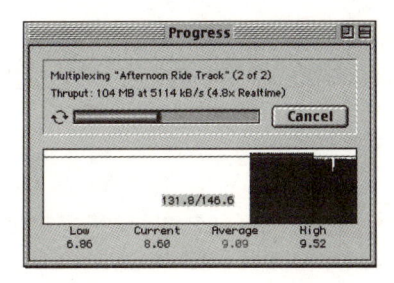

Figure 15-7: The multiplexing progress window, giving a play-by-play account of bit-rates and throughput.

4. **Locate the folder in which you saved the multiplexed files, and find the VIDEO_TS folder.**

5. **Open the folder.** Voilá! All of the files that tell the DVD player what to do are inside. (See Figure 15-8.) Don't mind the AUDIO_TS folder that was also created. It isn't used, but is created automatically as part of the process. Some recommend burning this empty folder onto the DVD along with the VIDEO_TS folder in case some DVD players look for it.

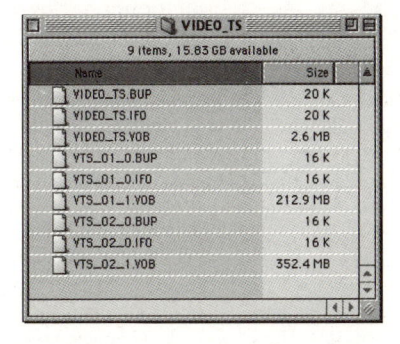

Figure 15-8: The mysterious files are created in the VIDEO_TS folder as a result of multiplexing, which only DVD players and DVD authors dare to understand.

When the VIDEO_TS folder is burned onto a DVD-R media, or manufactured on a DVD disc, the DVD player's computer interprets them into the interactive experience for the user. For more information on these items, check out the Investigating the File Structure of a DVD section in Chapter 11.

Congratulations, you have just built a DVD project! Now, what good is a stupid VIDEO_TS folder sitting on your hard drive? Read on!

Disc Previewing

To preview a DVD disc that has been built and thereby multiplexed but not burned to DVD-R, you need to have Apple's DVD Player program on your computer (Version 2.3 or above, according to the documentation).

First, you need to enable Apple's DVD Player program to be able to open a VIDEO_TS folder directly. Next, you can play the DVD through the Open VIDEO_TS option in the File menu.

To try it out, follow these steps:

1. **Open the program and choose Edit ⇨ Preferences (see Figure 15-9).**
2. **Select the Advanced Controls tab and click the Add Open VIDEO_TS Menu Item to File Menu checkbox.**
3. **Click OK.**

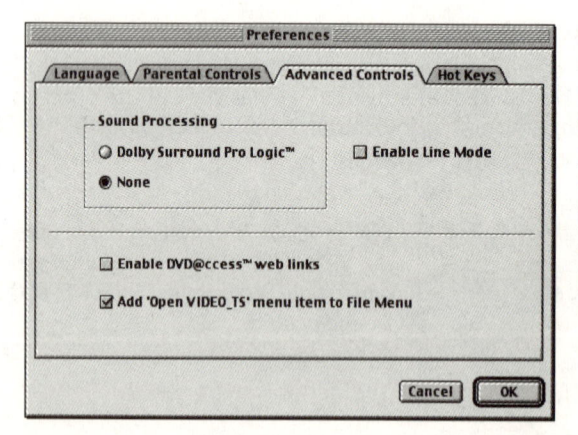

Figure 15-9: The Advanced Controls tab in the Preferences area of Apple's DVD Player program, Version 2.5. Selecting the Add 'Open VIDEO_TS' Menu Item to File Menu option allows you to open a multiplexed DVD project with the File menu in the Apple DVD Player program.

4. **In the Apple DVD Player program, choose File ⇨ Open VIDEO_TS and locate the VIDEO_TS folder that you recently created.** It is also in the multiplexed DVD folder in the CH15 folder in the Tutorial section of the DVD-ROM.
5. **Select the VIDEO_TS folder and click the Choose button.**
6. **Try the DVD out through the Apple DVD Player program.** (See Figure 15-10.) The mouse pointer itself is substituted for arrow keys on a remote control;

you simply move the mouse over a button on the DVD menu, wait for it to be selected, and then click to activate. (If you don't see the Controller or Viewer window, choose Window ⇨ Show Controller or Window ⇨ Show Viewer.)

The floating "controller" can be moved around the screen. To access the rest of the typical remote control buttons, click on the downward-facing triangle just below the Pause button on the controller. If the Menu button doesn't work, you may just need to press Play. To end your DVD joyride, click Eject.

Figure 15-10: A DVD Menu as seen through the Apple DVD Player program.

Hopefully everything works properly. Don't be disappointed if the performance is not perfect with the video playback, or if the menus do not seem to function the same way they do in the Preview function within DVD Studio Pro. The final test is to actually burn a DVD-R disc.

Summary

In this chapter you learned to prepare a disc for output, through the process of multiplexing. You also learned how to test and preview a project. While testing and previewing is not strictly necessary, it can help you avoid headaches down the road.

✦ To be a true DVD author, you must always test your interactivity before burning a disc. If you do not test your interactivity, your DVD authoring license may be revoked. Just kidding.

✦ The easiest way to test interactivity is to use the Preview button in the Graphical View.

- ✦ The most thorough way to test interactivity is to systematically review the settings for the general disc, the menus, every individual button, and each track, using the Property Inspector.

- ✦ For the overall disc properties, make sure to set a Startup Action, and in the Remote-Control section, set something for the Menu button.

- ✦ For a menu's general properties, you will probably want to set the Default Button and Return Button values at the very least. The Default Button is the button that will be selected when a user first visits the screen. The Return Button is the menu they will be taken to if they press the Return Button.

- ✦ When you are not using hilites, the Use Hilite value in the Property Inspector for an individual button is set to No, and Normal and Selected States are typically set to appropriately created Photoshop layers.

- ✦ Hilites are generated by a DVD player to indicate that a button is selected. A typical situation where you are not using hilites is if you create a Photoshop layer which represents the Selected State of your button.

- ✦ When you are using hilites, you don't need to necessarily specify a layer for the Normal, Selected, or Activated state of your button.

- ✦ For a Button's properties, make sure you review what the Direction buttons are doing.

- ✦ In a track's general properties, make sure to set something for the Jump When Finished setting. This is where a user will be taken when the track is finished playing. You will also probably want to set something for the Menu and Return buttons as well, at the very least, and then test them, to make sure they make sense.

- ✦ When developing or testing the interactivity of a DVD project, a checklist can be your best friend.

- ✦ There is a helpful form/checklist document entitled checklist.doc in the CH15 folder in the Tutorial section of the DVD-ROM.

- ✦ In order to prepare a DVD Studio Pro project, the only thing you really *have* to do is multiplex your files prior to burning the disc.

- ✦ Multiplexing is the process that DVD Studio Pro goes through to combine the various elements of a DVD project and generate the appropriate files in a format the DVD players will understand.

- ✦ You can use Apple's DVD Player program to open a VIDEO_TS folder directly, and thereby preview a project that has been multiplexed using the Build Disc option in DVD Studio Pro. To enable the Apple DVD Player Program, choose Edit ➪ Preferences, select the Advanced Controls tab, and click the Add 'Open VIDEO_TS' Menu Item to File Menu check box. Then click OK.

In the next chapter, I take you through the process of burning a DVD-R disc, and discuss the various kinds of burners that are available. I also include a discussion of the various methods of getting a DVD manufactured on a mass scale.

✦ ✦ ✦

Outputting a Project

CHAPTER 16

♦ ♦ ♦ ♦

In This Chapter

Understanding multiplexing

Burning a disc

Understanding DVD-R media

♦ ♦ ♦ ♦

In this chapter, you will learn about the various ways you can output a DVD Studio Pro project, including the most common option of burning a DVD disc directly from DVD Studio Pro.

If you are using DVD Studio Pro on a G4 with a built-in SuperDrive, the burning process is fairly straightforward using the Build Disc and Format command. If you have an external DVD-R burner, you will still probably be able to burn a disc right out of DVD Studio Pro, but you might need to use a program like Roxio's Toast Titanium. (www.roxio.com)

In order to burn a hybrid disc that has both DVD and DVD-ROM content (where you might want to include files such as HTML documents or an interactive project made in a program like Macromedia Director), you will definitely need to use a program like Toast. The current version of DVD Studio Pro does not support directly burning a hybrid DVD disc.

Finally, those who wish to get a DVD project manufactured will probably need to output directly to *DLT* (Digital Linear Tape), a tape format which is used at manufacturing plants to make finished DVD projects.

Regardless of your final output media, multiplexing is one of the required steps.

Understanding Multiplexing

Burning a DVD disc is much like burning a CD-R disc, except that part of the process involves encoding all of the files into a special DVD format that is based around a VIDEO_TS folder. This encoding process is also known as multiplexing. When you multiplex a project and generate a VIDEO_TS folder, you can't open the folder in DVD Studio Pro. The only way to open the folder is by burning the VIDEO_TS folder to a disc or by choosing File ➪ Open VIDEO_TS in the Apple DVD Player.

Note   The Open VIDEO_TS function does not automatically appear in the File menu of the Apple DVD Player program. A preference needs to be changed in the program; setting this function up is discussed in Chapter 15.

When you are working on a project in DVD Studio Pro and you use the Preview function in the Graphical View window, DVD Studio Pro is actually multiplexing your project on the fly — taking the video, audio and other project elements and temporarily combining them to allow you to see how the multiplexed project will run.

Using the Build Disc command

Choosing the Build Disc command from the File menu in DVD Studio Pro does not burn a disc, it simply allows you to multiplex the project. It generates a VIDEO_TS folder, which can be previewed with a software-based DVD player, such as Apple's DVD Player.

A multiplexed project is a finished product; the only thing that remains once a project has been multiplexed is to burn it to a disc. The Build Disc command allows you to multiplex, and then either store the resulting VIDEO_TS folder for later, or burn it to a disc using your own program such as Toast Titanium. For example, if you are using DVD Studio Pro 1.0 (which doesn't support burning directly to an external DVD-R burner) and have an external DVD-R burner, you will need to first use the Build Disc command to generate the VIDEO_TS folder, and then you can burn the VIDEO_TS folder to disc with a program like Toast Titanium. (DVD Studio Pro 1.1 adds support for external drives so that you can use the Build Disc and Format command to multiplex/burn the disc all in one step.)

Using the Build Disc and Format command

Choosing File ➪ Build Disc and Format allows you to multiplex the project and burn the disc all in one process. Depending on what kind of devices you have connected, it can allow you to save the project to DVD-R, DVD-RAM, or DLT.

When you choose File ➪ Build Disc and Format, the Format Disc window appears (see Figure 16-1). If you do not have any devices hooked up, the lower area of the window is blank.

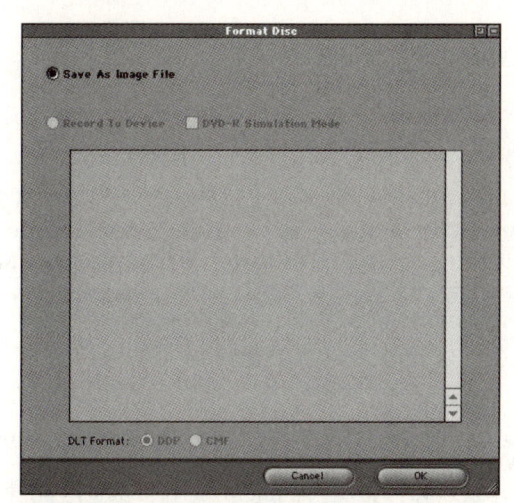

Figure 16-1: The Format Disc window, with no external or internal devices available. The only option in this scenario is the Save As Image File, which generates an `.IMG` file, an archive file that could be conveyed to another party.

When an external or internal device is available, the Format Disc window allows you to select a particular device from a list of those available. If the device is a DVD-R drive, you can select the DVD-R Simulation Mode checkbox in the upper portion of the Format Disc window; this function checks for potential errors before burning the disc.

Note
DVD Studio Pro 1.0 users may not be able to "see" external DVD-R burners in the Format Disc window. DVD Studio Pro 1.1 adds support for external DVD-R burners. If you have an external device that you want to write to and it doesn't show up in the Format Disc window, make sure you have it installed correctly with the latest available drivers. (*Drivers* are files that come with the drive; the latest drivers are usually available for download from the manufacturer's site.) If nothing seems to work, contact the manufacturer of the drive directly and ask for an updated driver; you might also want to scan the discussion groups on Apple's site to see if anyone else is encountering the same difficulty. (`http://discussions.info.apple.com`)

Burning a DVD-R in DVD Studio Pro

Burning a DVD-R in DVD Studio Pro is a snap. Once you have your project ready (and tested, hopefully), grab a blank DVD-R disc and get ready to rumble.

To burn a DVD-R disc, follow these steps:

1. **Press the Open button on the DVD-R drive and insert the blank disc.**
2. **In DVD Studio Pro, choose File ⇨ Save to save the project.** It's always a good idea!
3. **Choose File ⇨ Build and Format Disc.** A dialog box appears (see Figure 16-2), allowing you to select a location for the VIDEO_TS folder. When you use the Build and Format Disc command, a VIDEO_TS folder is written to your hard drive as a preliminary step to burning the disc. (You can then go back and delete the VIDEO_TS folder if you like after the process of burning your disc is done.)

Note: DVD Studio Pro doesn't seem to like the idea of letting you create a new folder on the desktop within the Build Disc or Build Disc and Format command, so if you want to create a new folder to place the VIDEO_TS folder in, navigate to somewhere within a hard drive rather than on the desktop.

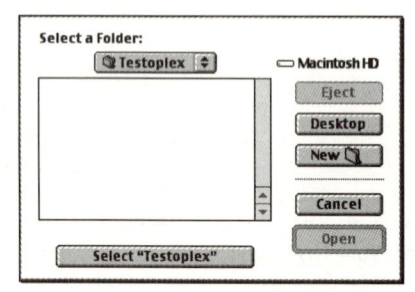

Figure 16-2: The Select a Folder dialog box for the Build Disc and Format command. The location you choose is where the VIDEO_TS folder will be written. If you have a few different hard drives to choose from, choose whichever is the fastest, because the performance of multiplexing is enhanced when writing to a faster hard drive, such as an Ultra Wide SCSI drive. The built-in hard drive on a Mac can be used without any adverse effects, as long as you have the space!

4. **Choose your location and click on the Select button to open the Format Disc window. (See Figure 16-3.)**

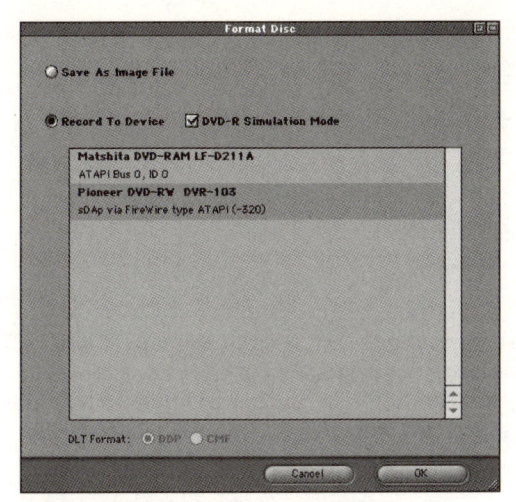

Figure 16-3: The Format Disc window, showing an internal DVD-RAM drive and an external FireWire DVD-R drive. The Simulate DVD-R check box becomes active when you select a DVD-R drive.

5. **In the Format Disc window, select the appropriate drive and click OK.**
   The project starts multiplexing (Figure 16-4 and 16-5).

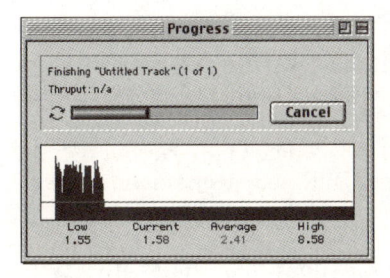

Figure 16-4: The Progress window, showing a play-by-play account of bit-rates, stock prices, the relative ornithological velocities of various African Swallows, and other useful information. The Untitled Track is the project element currently multiplexing; the progress window goes through individual tracks, files, and so on.

6. **While the project is multiplexing, take a break.** Get a drink of water. Visit www.puzz.com. Better yet, go outside and jump up and down for a bit. If it's a really big project, go rent the movie *Antitrust,* and be amused by the sight of a pseudo Bill Gates (Tim Robbins) using a Powerbook Titanium.

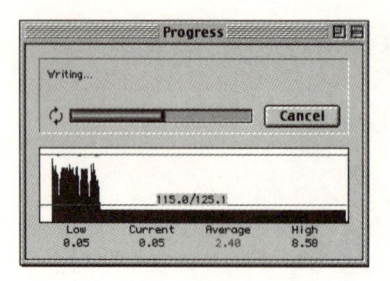

Figure 16-5: The Progress window, showing a disc being burned. Burn all your DVDs! DVDs are evil! (If you play a DVD backwards, it gives an animated annual report for Microsoft.)

7. **After the project is done, take the DVD-R disc to a retailer such as Best Buy and annoy the salespeople by trying your disc in every possible DVD player and computer with DVD-ROM, including the large home theater system with the plasma screen.** OK. So who said burning a DVD-R disc couldn't be fun!

Burning a DVD-R in Toast Titanium

Ah, Toast, Toast, Toast (Figure 16-6). You've come so far, and now we're burning DVDs.

Toast is cool. If you get it, make sure you download the latest update (such as 5.0.1) before you try to use it with an external DVD-R drive. Before you purchase it to use with a built-in SuperDrive on a Mac, check and see if Apple supports the use of Toast with an internal SuperDrive.

Toast will allow you to burn to those beautiful FireWire external DVD-R drives that are coming on the market from companies like all4dvd, LaCie, APS, and so on. Maybe by the time this book comes out, DVD-R drives won't be backordered anymore! The Cyclone from www.all4dvd.com is a nice unit — about a grand when first on the market — and uses the Pioneer DVR-A03 mechanism, same thing as you would find in the first G4's with the SuperDrive; it does nicely.

The other nice thing that Toast allows you to do is to burn data to a DVD-R, in essence allowing you to back up mass amounts of data, such as all the sedimentary digital build-up of the last year that you keep putting off. And making a hybrid DVD/DVD-ROM is as simple as dragging and dropping a VIDEO_TS folder, dragging your files and folders, and burning away.

Q&A with Jim Taylor, author of *DVD Demystified*

BIO: Jim Taylor is Chief of DVD Technology at Sonic Solutions, the leading developer of DVD authoring systems. He is the author of *DVD Demystified*, the best-selling book about DVD technology, published by McGraw-Hill. Called a "minor tech legend" by E! Online, Jim created the official Internet DVD FAQ, writes articles and columns about DVD, serves as President and Technical Director of the DVD Association, and sits on advisory boards of various companies in the DVD industry. Jim received the 2000 DVD Pro Discus Award for Outstanding Contribution to the Industry, and was also named one of the 21 most influential DVD executives by *DVD Report*. He has worked with interactive media for over 20 years, developing educational software, laserdiscs, CD-ROMs, Web sites, and DVDs, along with teaching workshops, seminars, and university courses. Before joining Sonic in 2001, Jim was DVD Evangelist at Microsoft for two and half years, and was formerly VP of Information Technology at Videodiscovery, an educational multimedia publishing company.

How did you originally get involved in DVD authoring?

I wrote *DVD Demystified* and decided that it ought to have a disc to go with it. So I chose what should go on it, then I got nice people like Ralph LaBarge (then at NB Digital) and Randy Berg (then at Rainmaker) to help me produce it.

How much did the original authoring systems go for?

I vaguely remember that the Scenarist SGI system we first authored on cost $75,000 just for the software. Hardware and other goodies added another $30,000 or so.

What are the current price tags on the high-end systems?

Much lower!

How does a Scenarist or DVD Creator system compare to DVD Studio Pro?

Scenarist and DVD Creator are high-end systems that give an author the power to precisely control all details of a project, and to use specialized features of DVD such as text descriptions of disc content, jacket pictures, CSS encryption, and karaoke.

(Author's Note: If someone knows of a good comparison between DVD Studio Pro and Sonic's high-end systems, check `www.dvdspa.com` to see if it's up yet, and if not, please e-mail one in or a link to one.)

Do you think InterActual and Apple will get together so that DVD Studio Pro users will be able to author PC Friendly content?

Who knows?

Would there be any way for a DVD Studio Pro-based DVD author to work with a Sonic or Spruce PC-friendly-capable DVD author to generate PC Friendly content, i.e., to give the PC-friendly author the correct encoded video streams to make reference to?

Continued

Continued

A disc authored with DVD Studio Pro can have InterActual Player features added to it using InterActual's developer tools. It doesn't require that there be an integration into the authoring system. (Note that PC Friendly has been replaced by InterActual Player 2.0.)

How do you think things will shake out between Apple, Spruce, and Sonic?

After the recent mergers in the DVD authoring industry, Sonic and Apple are the remaining companies with mature, robust commercial authoring systems. Sonic's products, including OEM products, are based on its third-generation DVD-Video authoring engine. Sonic and Daikin spent years improving the core engines to improve efficiency and compatibility of playback on the wide variety of DVD players in the marketplace. It's difficult or impossible for newcomers to the market to match the compatibility and richness of features provided by Sonic software. Sonic has already shipped over a million copies of its entry level authoring applications: DVDit and MyDVD.

Do you think most of the DVD authoring going on is being done on Mac or PC systems?

Hard to say. Sonic has sold over a million copies of DVDit and MyDVD, both of which run only on Windows. Most of these are bundled with computers or video hardware, so not all of them are being used.

From your perspective, are either Mac or Windows NT better suited for DVD authoring in terms of their strengths, or does it depend primarily on how they are configured?

There are few differences these days between operating systems. Macs still have a bit of an edge in ease of moving assets from one system to another, but Windows systems are cheaper and more available. Windows 2000 and Windows XP are very stable, which is important in a production environment. Mac OS X will also provide this advantage once DVD Studio Pro is updated to run on OS X.

If you're putting together a Mac-based DVD authoring system and have a particular budget for gear, what are the most important things you would recommend concentrating on? RAM? Hard drive space?

No question: hard drive space. Get as much as you possibly can, because you'll use all of it.

With the emergence of DVD-R General Media, is there really any reason for someone to get a DLT drive or DVD-Authoring drive?

DLT is more reliable and accepted by more replicators. It's also good for backing up large projects. Anyone doing a lot of DVD authoring should have DLT. Adding CSS encryption to a title requires DLT or DVD-R(A). But for casual authoring, DVD-R(G) works just fine, since many replicators will now accept DVD-R(G) discs as replication masters.

If you're on a budget, what DLT drive would you recommend getting?

A used one. (Author's Note: Try looking for old Compaq DLT drives on eBay or elsewhere, some of which can be had for only several hundred dollars.)

How do you see consumer adoption of DVD-R format developing in comparison to how CD-R format was developed?

About the same, but faster since we are used to CD-R.

What do you think Hollywood's attitude is about the DVD-R General format? How nervous do you think they are?

Hollywood is very worried about recordable discs, especially once they become extremely cheap. However, one of the best defenses is to put movies on dual-layer discs, which won't fit on recordable discs.

What makes one kind of DVD-R General Media better than another?

Differences in dye formulation, accuracy of spiral and wobble molding, quality of production, etc.

What kind of compatibility does DVD-R media have in DVD-ROM drives?

Almost all new DVD-ROM drives read DVD-R media, although some drives still get errors reading the data.

What would you recommend as being the best General DVD-R media, in terms of compatibility issues? Is Mitsui's "gold-dye" media really more compatible?

Reports on media quality vary, and manufacturers change their formulations, so there's a kind of "leapfrog" game for which media is the best at any point in time.

If you turn in a DVD-R General disc, can a replicator add CSS copy protection and region coding?

In general, no. However, if the replicator is willing to go through the work of re-imaging the disc to add CSS encryption, then it can be done.

In a Pioneer white paper, it says the following: "Either type of DVD-R media can be used for DVD video authoring, which is the process of preparing video content for use in DVD video players. It should be noted, however, that CSS encryption cannot be used with either type of DVD-R media." But various sources will give conflicting answers to this question, as to whether DVD-R (A) discs can carry the data. So can CSS be put on DVD-R (A) media?

Creating a disc with CSS encryption requires that sector encryption flags be provided to the replicator in a DDP or CMF file. These files are stored on a DLT or in the control area of a DVD-R(A) disc. Since DVD-R(G) has a non-writable control area, it can't be used to store CSS replication information.

CSS can't be directly used on DVD-R(A) or DVD-R(G) discs, since the keys can't be written to the control area. In the case of DVD-R(A), the media is actually capable of holding CSS keys but the drives block it from being written.

What are a few of your favorite things? I mean, DVDs?

Continued

Continued

I'm always impressed by amazingly creative visuals, often short ad spots, that create a deep visual or emotional impact on the viewer. One of the things I love about DVD is its ability to preserve the carefully crafted details of a piece of video — things like curling smoke, textures, shadows, dust particles drifting in sunlight, and sweat beading on someone's forehead — the details that make all the difference in how viewers are affected when they see the video.

What do you think some potential business opportunities would be to consider for independent DVD authors?

Unlimited. See Chapters 1, 10, and 12 of *DVD Demystified*. ;-)

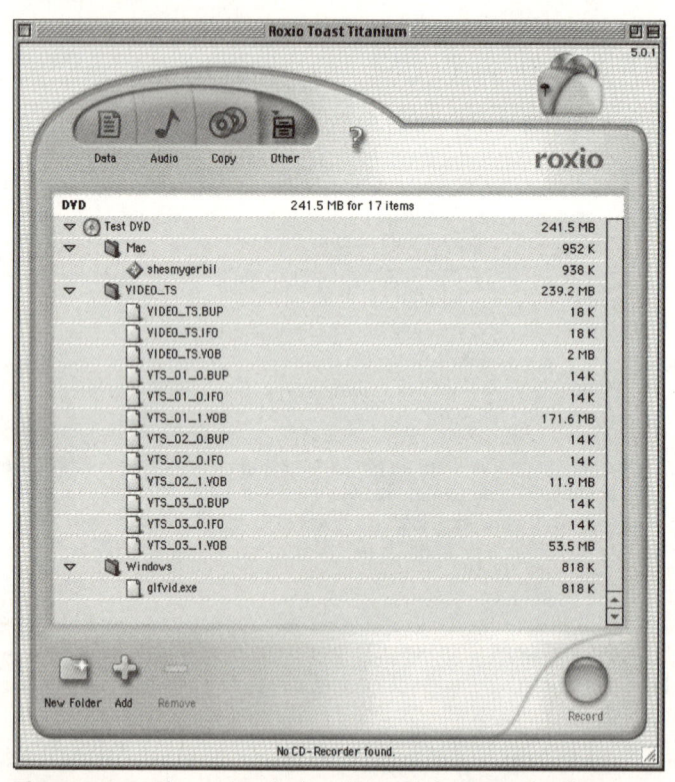

Figure 16-6: The main window for Roxio's Toast Titanium, showing a sample DVD, with the VIDEO_TS folder and its BUPs, IFOs, and VOBs, as well as two additional folders, representing Windows and Mac versions of a standalone Flash file that will be accessible as DVD-ROM content when the disc is inserted into a computer with a DVD-ROM drive.

Outputting to DLT

Don't feel bad if you have no idea what DLT is. *DLT* is a tape backup drive that can store an insane amount of data on one cartridge. Typical formats for DLT include DLT 4000 (40GB) or DLT 8000 (80GB), and the amount you can store on an individual cartridge depends on whether you use the Compression feature or not.

For people who work with video, DLT could be something to seriously consider as a means of backing data up or if you have video or DVD-related projects or single video files that are so big they have trouble fitting on a hard drive, much less a 4.7GB DVD-R disc.

Other than as a backup solution, DLT is the standard for DVD manufacturing, also known as *replication*. It also opens up the world of dual-layer or dual-sided DVD projects (DVD-9, DVD-18 formats), which require more space than a 4.7GB DVD-R disc could ever afford.

Manufacturing a DVD

In the age of DVD video, the way manufacturing a DVD has evolved is that the people who have the ability to churn out DVD discs in a factory often have systems that run directly off of a DLT tape. Some may accept DVD-R media or you might even be able to drop a FireWire hard drive in the mail, but DLT is still the standard.

DLT drives can get pricey, but you can find them for as little as $1,600 from companies such as APS Technologies (www.apstech.com). A single DLT tape can go from anywhere between $50–100 depending on the brand and capacity. But when it's all hooked up, the throughput can be impressive for tape, saving something like 3MB per second.

Different types of DLT capacities exist, but the most common for turning in a tape to a manufacturer is known as DLT 4000, also referred to as Type 3 media. If you decide to go for a DLT drive, make sure you get one that is compatible with the replicator(s) you intend to work with. (Some corporate environments have a DLT drive for network backup, and it might be possible for the enterprising DVD author in such a place to borrow one and hook it up to his Mac with the right drivers and an Adaptec SCSI card, or even do it over the network somehow.)

Note The DVD-R format does not necessarily support disc-level features, such as CSS, Macrovision or Region Coding. If you are getting a DVD manufactured and are sending the data on DVD-R or getting the DVD-R converted to DLT, make sure it is possible for the manufacturer to accomplish the disc-level features you want.

Several helpful sites provide information on DVD manufacturing; for a current list of links, visit www.dvdspa.com.

Outputting to DLT in DVD Studio Pro

The process of outputting to DLT tape in DVD Studio Pro assumes that you have a DLT drive properly installed and connected.

It is basically the same as outputting to other formats, except you select the DLT drive in the Format Disc window (Figure 16-7). When you do this, the DDP and CMF radio buttons become active in the bottom of the window, and you have to choose either DDP or CMF format.

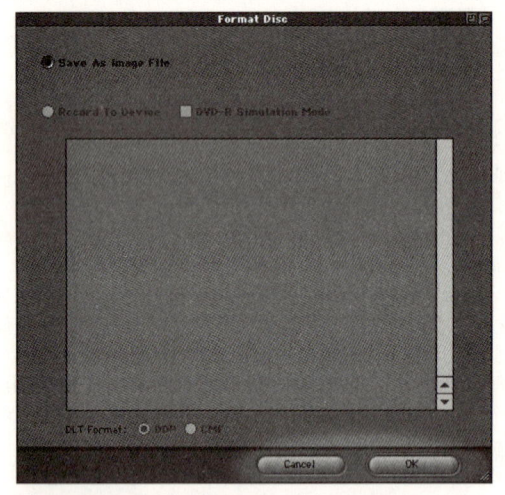

Figure 16-7: The bottom of the Format Disc window

The DDP (Disc Description Protocol) format for storing data on a DLT is required on certain DVD replication systems. It is the older of the two formats, but is still used by many replicators.

The CMF (Cutting Master Format) format is newer. Before CMF, you could prototype a project on DVD-R, but you still had to hand in a DLT to the replicator. With CMF-capable systems, a project could be turned in on a DVD-R Authoring disc (not DVD-R General — see the section below on DVD-R media). The CMF standard allows a DVD-R to theoretically serve as Press Cutting Master Disc.

So why turn in a DLT with a CMF system? Ask the replicator, but if they tell you to, likely their reasoning will be that DLT is more reliable. Beyond that, certain features such as CSS, Macrovision, and Region Coding may not be possible with a DVD-R master.

Both DDP and CMF formats typically represent a DVD by having an overall disc image with the general DVD information augmented by formatting data that tells the replication machine what to do.

To get even more technical, it's like Grand Moff Tarkin telling the Deathstar flunkies how to get busy with their lasers. Only in this case, the DDP or CMF file is telling the LBRDF (Laser Beam Recorder Data Formatter) how to process the data and get busy with its laser beams to burn a DVD instead of a rebel ship. Of course the RBL (reader boredom level) has reached the EGO (eyes glazing over) stage, so now you can proceed to the NAS (next available section).

Chances are you will be using the DDP format, but check with your replicator to make certain either way.

> **Note** DVD Studio Pro 1.0 had a bug when outputting in the DDP format to DLT. In a technote on the Apple site, version 1.0.1 was mentioned as fixing the bug, but no information was available as to how to get 1.0.1. But it may be because version 1.1 was about to come out, and that version takes care of the problem, so the best thing to do — if you haven't already — is to just download the free 1.1 upgrade (or whichever is the latest version) from www.apple.com/dvdstudiopro/update. Do it! For more information on the DDP bug, check out article 31343 in Apple's Tech Info library, which may be accessible from the following link: http://til.info.apple.com/techinfo.nsf/artnum/n31343/.

DVD-R Media

There are two general types of DVD-R media, Authoring and General. Back in the days before the SuperDrive DVD-R burner came about, about the only thing one could burn DVD-R discs on were burners from Pioneer's professional series, such as the DVR-S201 drive, which may still run about 4 grand or so. DVD-R Authoring media, also known as DVD-R (A), initially had a capacity of 3.95GB, and later a variation came out with 4.7GB.

DVD-R General and DVD-R Authoring media are very similar, but DVD-R Authoring media is only compatible in DVD-R Authoring burners such as Pioneer's DVR-S201. Practically speaking, DVD authors simply need to make sure they are ordering or purchasing DVD-R General media when they walk into a store or order online, unless you happen to have one of the high-end burners.

> **Note** Some manufacturers may need to convert a DVD-R General disc to DVD-R Authoring in order to make a glass master for manufacturing, and they may charge you for it.

Burning a DVD Master: Issues to Consider

An article by Lee Purcell, author of *CD-R/DVD Disc Recording Demystified*

BIO: From a country outpost on the edge of old-growth Vermont forest, Lee Purcell authors computer books and telecommutes while working on writing projects for Silicon Valley and Boston metro area clients. His latest book is *Flash Character Animation: Applied Studio Techniques*, published by Sams Publishing in August 2001.

Technology news sources increasingly point to DVD authoring as the next killer app — the program that will revive sluggish computer sales and spawn a renaissance of desktop video production, capturing the hearts of video professionals and home moviemakers alike. Apple Computer Inc. figures strongly in this burgeoning movement. The processing subsystem in the Motorola PowerPC G4 microprocessor, known as the Velocity Engine, makes it possible to substantially reduce the time required to encode the MPEG-2 video used for DVDs without requiring expensive add-on hardware. Instead of taking ten to twenty hours to encode one hour of MPEG-2, the 766MHz Power Mac G4 cuts the encoding time to about two hours. Equip yourself with a SuperDrive-equipped Power Mac G4, Final Cut Pro (for video editing), and DVD Studio Pro (for DVD authoring), and suddenly you have the tools to start your own desktop movie studio, burning DVD-R masters to hand off to your local replicator. In a couple of weeks, you can be competing alongside Columbia and Universal to gain shelf space at DVD outlets around the world. Right? Well, not exactly. Here is the chink in the armor.

The new inexpensive breed of DVD-R recorders, being used by companies such as Apple and Compaq, work only with one type of recordable media, known as DVD-R General Purpose media with a capacity of 4.7GB. The general type of recordable media has a recording sensitivity keyed to a laser with a 650-nanometer wavelength and can't be used in professional DVD-R recorders that use a laser with a 635-nanometer wavelength. These high-end recorders, including Pioneer's DVR-S201, use another type of recordable media, known as DVD-R Authoring media. Two capacities, 3.95GB and 4.7GB, are available.

So, why can't you bring your replicator a DVD-Video disc master created on DVD-R General Purpose media? Aside from the differences in the sensitivity of the dye layer, General Purpose media lack a provision for recording the Disc Description Protocol (DDP), a header file that provides essential information about the DVD content so that the replicator can create a DVD glass master — the first stage in the replication process. DVD-R Authoring media supports recording of the DDP in the lead-in area of the disc. This type of media also supports a new feature called Cutting Master Format (CMF) that provides similar information to DDP, although support among replicators for this data feature is still limited. Either of these data regions will allow a DVD-R one-off to be used as a master instead of Digital Linear Tape (DLT), the prevailing standard for master submissions. Can you use Authoring-grade media in a SuperDrive or other inexpensive drive to create a master? Unfortunately not — the media type has to correspond with the recorder type. Pioneer produces two inexpensive DVD-R recorders, the DVR-A03 and the DVR-103, which are commonly bundled in computers — both these drives use General Purpose media only.

For playback, General Purpose media discs work as well in any playback device that can accept DVD-R media — whether a computer-installed DVD-ROM drive or DVD set-top player. This includes most recent generation drives and players. You can use your SuperDrive to create movies to entertain your friends and family, or to author business presentations that are playable on a standard DVD player. You just can't take the DVD-R disc to a replicator for use as a master. Industry players are scrambling to develop services that will allow General Purpose media to be submitted for high-volume replication, but so far these conversion services are not widely available. A more expensive route is to submit the video files on a DVD-R General Purpose disc and then pay a developer or replication service to copy the files, construct the DVD framework, and then transfer the content to DLT or DVD-R authoring media.

Copy protection is another wrinkle to consider when recording masters to DVD. Copy protection schemes, such as Macrovision and Content Scrambling System (CSS), prevent illegal replication of DVD-Video content. While some software, such as Apple's DVD Studio Pro, can encrypt the disc master image using CSS, the DVD-R Authoring media cannot accept CSS-encrypted data. If your DVD title contains video that must be protected through encryption, your best option is to submit the master data on a DLT cartridge. DVD-R just won't handle the encrypted content. This represents a potential problem to video producers who may be producing training content or other video material that could be pirated. No one wants to give away their video content, so copy protection requirements may force you to adopt DLT as your submission media.

The least expensive solution to this problem may be to acquire a DLT drive and submit your DVD master data on tape cartridges for replication. These drives double as high-volume backup storage systems, typically using LVD SCSI or Wide Ultra SCSI interfaces, with transfer rates ranging from 3MB/sec upwards to 11MB/sec. However, the standard DLT drive that replicators accept tapes from, the Quantum DLT 4000, uses a SCSI-2 interface, which is less expensive and more pervasive. Tape cartridges come in several different capacities; a DLT IV cartridge holds up to 40GB of data native and 80GB compressed. Prices for base model DLT backup systems start at around $1500 at outlets such as APS Tech (www.apstech.com) and Computer Discount Warehouse (www.cdw.com). For a less expensive approach, you can often find used units for around $300 to $400 through online auctions, such as eBay (www.ebay.com). Make sure that you budget the cost of the high-speed SCSI host adapter, if your system does not have a SCSI board installed. While DVD-R authoring drives have dropped in recent months, you can still expect to pay about three times more than what a DLT backup system will cost for one. For example, the street price for a DVR-S201 is about $4,000 at press time.

While the new killer app on the street shows promise, it's not always a straight line path from setting up a desktop digital video studio to producing a master for replication. But, if you know your way around the obstacles, you'll find that DVD replication prices have become particularly attractive in recent weeks. Many companies are scurrying to convert existing training content and marketing materials from conventional video to DVD-Video, offering many potential projects for developers with skills in this area.

Continued

> *Continued*
>
> *This article was reprinted with permission from Discmakers. Discmakers is a full-service disc manufacturer, offering a variety of products and services.* (`www.discmakers.com`).
>
> *Author's Note (Todd): DVD-R General does have its limits, but I agree that you will see increasing numbers of DVD replicators accepting DVD-R General media. Depending on the manufacturer, some of the conversion costs from DVD-R General may simply be absorbed in their overall pricing. At time of writing, I just sent off a DVD-R General disc to Metatec for a short manufacturing run, albeit without needing copy protection. For me, in some cases copy protection is optional — I have some promotional projects that I would like seen by as many people as possible, and I would be glad if people copied them and passed them along. The SuperDrive/A03/103 can write to DVD-RW media.*

DVD-R Media in DVD Players

If you haven't gathered this already, it is important to realize that DVD-R media is not 100 percent compatible in all DVD players or DVD-ROM drives. In other words, if you make a project in DVD Studio Pro, the only way to guarantee 100 percent compatibility is to have a project manufactured, and until prices come down, manufacturers will probably have minimum runs of at least 1000 discs.

Apple maintains a compatibility list at `www.apple.com/dvd/compatibility/`, which has compatible/not-compatible listings for consumer DVD players, but it is not necessarily exhaustive. (Additional resources and links in this area are available on the `www.dvdspa.com` site.)

DVD-R media is not 100 percent the same from manufacturer to manufacturer, due to varying manufacturing processes, but with time, the situation will improve. Allegedly, DVD-R media from certain manufacturers has better compatibility than others, but there is not a clear best brand.

Trusting no one

If you have a project that you want to release on DVD-R media, the best thing to do is to think like Fox Mulder and not trust anyone, but to test the media on the specific players or DVD-ROM drives your audience is known to have. With DVD-ROM drives, the internal mechanism could conceivably vary from computer to computer, even if it is the same model, because manufacturers will sometimes use different DVD-ROM mechanisms in different production runs of a particular computer.

You may wish to consult the various compatibility charts, do some research on the media itself, make a project, and then "hope" that it will work, but if it absolutely has to work, then you need to test the scenario ahead of time, or at least ask what kind of player a person has access to.

Sharing DVD-R-based projects

If you are sharing a DVD prototype with a limited audience, you may wish to print out one of the online compatibility lists and include it with the disc, mentioning that DVD-R media is compatible with most players. It is likely that for a while, a DVD-R could be such a novelty that a person will find a player to try it on one way or another.

The G4's internal SuperDrive uses Pioneer's A03 drive at present, and many of the external FireWire-based DVD burners use Pioneer mechanisms, although this will change over time, as Panasonic and others start entering the market. But if you are burning discs on a burner that is based on the Pioneer mechanism, it is fairly certain that the discs will be 100 percent compatible with Pioneer players.

Another way to convey a DVD-R disc is to say that not all players are compatible with DVD-R media, which puts a more positive light on the situation. According to some, DVD-R discs work in up to 90 percent of the players out there.

Outputting a Project to CD-ROM

Yes. You can actually put a VIDEO_TS folder on a CD-R. Try it, it's fun. You just use the Build Disc command to multiplex the project, and the only rule is it has to fit on the CD-R.

Sorry, the CD-R won't play in a DVD player. But the native file format for CD-R is not much different than the file format a DVD player requires for a DVD disc. The DVD player uses a file format called UDF bridge, which is a cross between UDF and ISO 9660. (To delve into the particulars of these technical tasties, read Jim Taylor's *DVD Demystified* published by McGraw-Hill, visit www.dvddemystified.com, or read Lee Purcell's *CD-R/DVD Recording Demystified* also published by McGraw-Hill. There are also reviews of these and other available DVD book titles on the market at www.dvdspa.com.)

Cross-Reference: More information is available in Chapter 22.

Perspective on DVD-R by Andy Marken

BIO: Andy Marken is President of Marken Communications, Inc. (www.markencom.com), based in Santa Clara, CA. In his more than 25 years in communications, Andy has been involved with a broad range of corporate and marketing activities. His experience includes strategic and market planning and execution for more than 10 years with communications and Internet firms including AT&T and CERFnet as well as more than 15 years in storage and storage management with firms including Philips and Panasonic. Prior to forming Marken Communications in mid-1977, Andy was vice president of Bozell & Jacobs and its predecessor agencies (Hal Lawrence Inc. and Lawrence & Lierle). During his 12 years with these agencies, he developed and coordinated a wide variety of marketing and promotional efforts for firms including Atari, Amdahl, and Control Data. A graduate of Iowa State University, Andy received his Bachelor's Degree with majors in Radio & Television and Journalism. Widely published in the industry and trade press, he is an accredited member of the Public Relations Society of America (PRSA). Marken Communications is a full-service agency that concentrates on business-to business market planning, positioning, development, and communications.

What is the demand like for DVD-R burner mechanisms right now?

Demand is quite different from what will be shipped. This year more than 5 million recordable DVD drives will be shipped. By the end of next year IDC estimates that 15 million recordable DVD drives will be shipped. The attach rate (use of media) is presently 6 R discs for every rewritable disc. By the end of next year it is estimated that this will rise to 10-1.

What is the Panasonic mechanism? How does it differ from Pioneer's?

Both units deliver the ability to write to R media which can be played anywhere. The Pioneer unit incorporates DVD Forum standard -RW technology which is a streaming tape type of implementation which allows the user to develop a complete file and stream it to the media. He or she must then also access it sequentially (as with tape). There is no error correction or defect management designed into the technology which means any error during transfer will result in lost data. Any writing to a defective error of the disc will also result in lost data. The -RW media can be overwritten about 1,000 times. The unit also includes CD-RW capability (8x). Which means that with a single drive users can write both DVD and CD media. Pioneer's drive cost at the present time (street price) is about $750. The Panasonic drive does not include CD-RW technology because of the rapid advance of read/write speeds with CD-RW drives. CD-RW standard is already at 12x, most drives are 16x and 24x recently began shipping. The drive writes to 4.7GB single sided and 9.4GB double sided RAM media which can be overwritten 100,000 times. The RAM technology is random-access just as with a system hard drive which means users simply drag-and-drop data rather than open a session, write and close a session. RAM technology also incorporates comprehensive error correction and defect management technologies to protect data. Sectors can be edited and overwritten in 2KB sectors as opposed to 32KB sectors for -RW. Street price of the Panasonic drive today is about $530. Both RAM and -RW are DVD Forum standards. DVD-RW media is becoming increasingly popular.

What kind of drivers would someone need to get an internal Panasonic DVD-R mechanism to run on a typical G4 or G3, and what software could they use? Toast?

Drivers for any internal drive need to support the computer's operating system. Exact implementation of third-party software support for DVD drives is up to the individual companies. All leading ISVs (independent software vendors) have or will shortly implement support for DVD-R, DVD-RW and DVD-RAM. Toast support is already provided.

How do you see things developing with DVD-authoring from your perspective?

There are three levels of DVD authoring. First there is the very high-end area which has two segments. One is Hollywood using Premiere/Avid technologies with multi-layer technologies with the major focus on streaming content and subtitle areas. Then there are the pure content developers — games, business video solutions which are multi-level, multi-track options. Next there is the professional/prosumer area which uses mid-level video options such as Pinnacle Systems Studio 7 and Pro-One. These individuals include professional videographers, video production labs and high-end consumer videophiles. Finally there is the general consumer video arena which is people who have video camcorders (analog or digital) and use products like Pinnacle Systems Studio 7 or Express.

Will the market roughly follow the growth of people doing their own digital video editing?

While there is some parallel, the market for recordable DVD drives is much larger than simply the camcorder/video authoring solution market. It will more closely track PC sales.

Who are the players in making external DVD-R drives? Is a FireWire any different than an IDE?

External drives are offered by QPS, LaCie, Formac, Memorex, and perhaps a dozen other manufacturers who take the basic drive and incorporate it into their mechanisms. IDE is strictly an internal drive solution. FireWire is only external.

Is Toast the best software to use with an external DVD-R burner?

If the user has a Mac system then Toast is perhaps the leading solution. If it is a PC environment, Roxio and Veritas are the leaders.

How do you think things will shake out between Apple, Spruce, and Sonic?

Apple has acquired Spruce Technologies and is in the process of absorbing it into their organization. Spruce's technology will be added to Apple's rich Mac-only environment and will disappear from the scene. Taking market share from Sonic are firms such as MedioStream, Gear, and Pinnacle Systems all of which have a much more feature-rich and elegant (from the user's perspective) solution. Roxio, Veritas, and others are expanding their product lines also to challenge Sonic's offerings.

Continued

> *Continued*
>
> **What kind of markets do you see for independent DVD authors?**
>
> We assume when you refer to independent DVD authors you are referring to individuals who actually author their own content. Camcorders, low-cost computing power and professional quality/economic authoring solutions can now enable almost anyone to produce a theater quality video. This will mean we will see a proliferation of video content that is produced by "low-budget" operations that will be marketed through multiple channels — internet, vertical niche cable operators, special arts theaters, etc. Independent authors will also find markets for themselves in the business video arena — training, HR and corporate communications as well as healthcare and educational videography, documentaries and training including video-on-demand. There are also the wedding and special events videographers who are upscaling their activities.

Summary

The goal of DVD authoring is to end up with a final product. In this chapter, you have learned about some of the ways you can output a project, including multiplexing, which takes all of your files and encodes them in a way that a DVD player can understand. You have also taken a closer look at DVD-R media, the revolutionary format that makes independent DVD authoring and prototyping possible.

- You can burn a DVD-R disc directly from DVD Studio Pro using the Build Disc and Format command.

- Multiplexing is the process of encoding a DVD project, and it produces a VIDEO_TS folder.

- The free DVD Studio Pro version 1.1 upgrade adds support for some external DVD-R burners so that you can burn to them directly from DVD Studio Pro.

- Toast Titanium from Roxio (formerly Adaptec), often bundled with external DVD-R drives, allows you to burn hybrid DVD/DVD-ROM content to DVD-R discs with a simple drag-and-drop process.

- Jim Taylor is a nice guy. He is a good author too. And he has an extensive Web site with DVD-related information (The DVD FAQ) that can provide hours of fun for the whole DVD authoring family (www.dvddemystified.com). It is a valuable online DVD reference. (The DVD FAQ are also available in HTML form on the DVD-ROM that comes with this book.)

- DDP is still a common format for turning in a DVD project on DLT tape for manufacturing. Version 1.0 of DVD Studio Pro had a bug writing to DDP format; version 1.1 fixes the problem.

- CMF is another format for turning in a DVD project on DLT tape. Check with the manufacturer to see which one you need to use if you are burning a DLT.

- If you are turning in a project on DVD-R General media to a manufacturer, don't expect to necessarily be able to use CSS, Region Coding, or Macrovision.

✦ ✦ ✦

Advanced Interactivity

PART

V

In This Part

Chapter 17
Understanding Scripting

Chapter 18
Adding Multiple Angles to a Project

Chapter 19
Working with Multiple Languages

Understanding Scripting

CHAPTER 17

In This Chapter

Understanding the concept of scripting

Understanding the Script Editor

Scripting limitations

In this chapter, you will learn about the scripting capability in DVD Studio Pro and various ways you can use scripting to add additional interactivity to your projects.

If you already own DVD Studio Pro, you may have glanced at the Scripting chapter in the manual. Don't be discouraged if you found all the references to programming confusing. You are not alone; most DVD authors are not programmers. This chapter introduces some basic concepts and walks you through a few examples that will help give you some ideas.

Alternatively, you may feel confident about your programming skills, tried writing some scripts, and inadvertently found some of the bugs that exist in version 1.0 of DVD Studio Pro (the Return function, for example).

Now it's time to pack your bags and roll up your sleeves. Prior to departure, you may wish to visit the online store at www.joltcola.com and order a case or two. For the first exercise in the chapter, you will also need some M&M's, two paper or styrofoam cups, and a felt-tip marker that you can use to write on the cups. If you have some dice, that would be nice; a pen and paper will suffice for dice. Seriously. Go get this stuff and then come back. If you are at work, you can blame it on the authors. Only the Jolt cola is optional.

Understanding the Concept of Scripting

DVD players are like elephants. They are powerful, have a lot of storage capacity, and love to eat peanuts.

You may have seen an elephant appear in a movie or in an episode of "Animal Kingdom." Usually in a movie or television show there is some kind of script, and most of the time, the action follows the script. Sometimes when animals are being filmed, they do what they feel like doing, not what the script says. Especially if a bug of some kind lands on their nose. But if you're nice, they will often be reasonable.

If you want to write a script for a DVD player, there are certain rules but it's doable, even if you're not a programmer. Think of a script as a "movie script" for a DVD player, where the script is telling the DVD player what to do.

In the same way scripts in Hollywood are written in a certain language, DVD Studio Pro gives you a language to write a script in that the DVD player will understand. DVD Studio Pro's scripting language has some similarities to other DVD authoring programs, but it is essentially only for DVD Studio Pro.

A DVD player's brain is not so big, but it can do some fun tricks. It can pick random numbers, it can go to different parts of your DVD such as a certain menu or track, and it can set which audio or subtitle streams are going to play next. A DVD player's ability to remember things is very limited, but variables are one way it can remember.

Comprehending variables

OK, it's time to get out all the stuff and break open the M&M's. You'll only need a few.

When you tell a DVD player what to do with a script, it likes to work with numbers. If you want it to randomly pick which track to play next, you tell it to pick a number. After it picks the number, it remembers the number by putting it in a variable.

A *variable* is a container for storing information. If you put information in a variable, it will stay there so that you can come back and look at it later. You can add new information or change existing information.

So take one of the cups and write an A on it with the marker. Now choose a number, either 1 or 2. If you chose 1, put one M&M in the cup. If you chose 2, put two M&M's in the cup. (The color doesn't matter, as long as you take out all the brown M&M's.)

What you have just done is selected a random number and stored the information in a variable named A. If DVD Studio Pro scripts worked with humans, the script would look like Figure 17-1. The script is instructing the reader to pick a random number, and then to store the information by putting that number of M&M's in the cup marked A. The cup represents the variable A. It is a "variable" because the information can change or "vary" from time to time.

Figure 17-1: A DVD Studio Pro script tile.

So what the heck is the ?= stuff in the script? It's called an assignment. Kind of like a homework assignment. The ?= is your homework, and it means to pick a random number. Assignments are one of the several programming ingredients you can use to make a script.

Now that you have selected a random number, what are you going to do with it?

Playing with menus

Let's pretend for a moment that you're in a restaurant, one that serves three different kinds of cuisine; Cajun, Catalan, and Klingon. The waiter sees the Federation markings on your uniform and gives you the Klingon menu first. You pass on the Klingon delicacies, and the waiter says they only have two other menus and asks which you would like to see.

Choices, choices. But lucky for you, you are prepared. You brought your paper cup random number generator with you, and since you forgot the number you picked before (having been distracted by a random laser blast), you use the random number generator to pick another number. The waiter is waiting patiently. To help you make a decision, you say to yourself, if I pick 1, I choose the Cajun menu, and if I pick 2, I choose the Catalan menu. Your random number generator yields the value of 1.

And you say, "I will take the Cajun menu please."

Now, instead of engaging in risky interstellar travel, let's say you just want to write a script involving variables and menus in Studio Pro. Something very simple, just to get the hang of it.

No problem! A folder called Simple Random is in the Tutorial section of the DVD-ROM in the CH17 folder, which contains a file called randomscreens.psd. To work through the Tutorial, you need to know how to make buttons and how to adjust project element properties in the Property Inspector. If you don't know how to do this yet, you can just open up the project file in the Finished Files folder within the CH17 folder.

To make a simple DVD project with a script, follow these steps:

1. **Open a new project in DVDSP and create three menu tiles, naming them Klingon Menu, Cajun Menu, and Catalan Menu, respectively.**
2. **Set the Startup Action of the project to Klingon Menu.**
3. **Import the randomscreens.psd file from the CH17/Simple Random folder, and select each menu tile.** Set the Picture - Asset setting to randomscreens.psd, and set the Picture - Layers (always visible) value to include background and a single appropriate "foreground" layer (Klingon, Cajun and Catalan) for each different tile. Each tile should have two layers always visible, a background and foreground layer. Tile 1 has background and Klingon, Tile 2 has background and Cajun, and so on.

4. **Create a script tile by clicking the Add Script button at the bottom of the Graphical View window.** A script tile can be renamed the same way as any other tile.

5. **Name the script tile** Choose Menu.

6. **Create a button named Mike Roney over the Klingon Menu text in the Klingon Menu tile, and with the button selected in the Menu Editor, set the Jump when activated value to Choose Menu.** This sets things up so that when a person activates the Klingon Menu button, the script will run.

7. **Set the Button Hilites values for the Klingon Menu tile. Setting Selected Set 1 to a gray color, 33%, works fine.** Also, set the Default Button value to Mike Roney, so that the button will be selected when the menu is previewed.

8. **Because you won't need return buttons in the Cajun and Catalan menus, set the General - Return Button value in the Property Inspector to Klingon Menu.** Alternatively, set the Timeout Action to Klingon Menu with a duration of one second to return to that menu without having to click on the Remote. Depending on how you have the Lines drop-down function set, the Graphical View should look something like Figure 17-2.

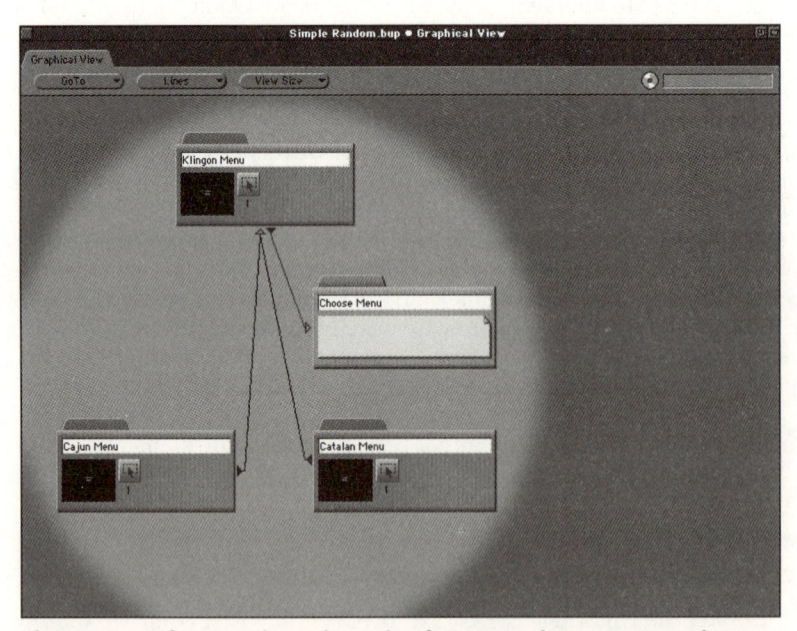

Figure 17-2: The stage is set for a simple DVD project to use a script. The button in the Klingon Menu triggers the script, and once the script chooses either menu and takes the user there, the user can get back to the Klingon Menu, either with a Timeout Action or by clicking the Return key on the Remote.

9. **Open the Script Editor by double-clicking on the rectangular area for the script tile.** When there is a script in the tile, it shows a preview in this area. The Script Editor is essentially a simple text editor.

10. **With the Script Editor open, click in the large white area as if you were in a word processing program writing a chapter for a book about DVD authoring only days before the deadline.**

11. **Refer to Figure 17-1 and look at the line of script that told you to generate a random number.** Type that line in the Script Editor, press the Return key, and create three more lines of script to match those that appear in Figure 17-3. Don't worry about indenting the text; when you click the OK button in the Script Editor, the editor automatically indents the text. If you feel bold, try using the drop-down menus in the Script Editor to type some of the text for you. That's all they do.

In Figure 17-3, the first line in the script selects a random number between 1 and 2 and stores the information in the variable A. The subsequent lines check the contents of the variable "A" and, depending on the contents, cause the user to go to a particular menu. The A == 1 and A == 2 phrases must have two equal signs.

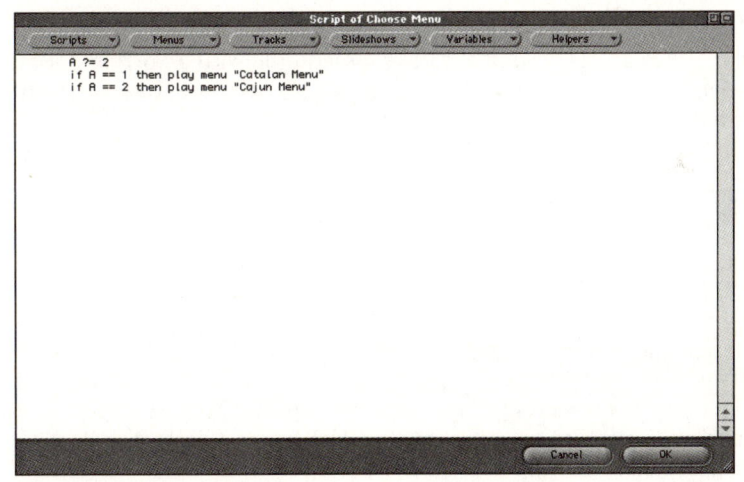

Figure 17-3: The upper-left corner of the Script Editor, showing the Choose Menu script.

If you preview the project and follow the instructions correctly, you first encounter a screen with a button. When the button is activated, it runs the script, and either the Catalan or Cajun menu appears.

Congratulations! You wrote a script!

Assigning scripts

When you work with scripts in DVD Studio Pro, you can assign them as the "action" for a button, track, menu, marker, story, slideshow, or remote control key.

Once you have created a script, it will show up as an option in the Jump when activated, Jump when finished, or the Pre-Script area of the Property Inspector with a project element selected. You could have the activation of a button run a script, or you could create a story and assign the script to the Jump when finished or Pre-Script value.

Assigning scripts to markers

To assign a script to a marker, first you need to create the marker in the track tile using the Marker Editor. Then you need to create a story for the track and assign the script to the marker within the story. For example, you could insert a marker ten minutes into a video segment that triggers a script. Perhaps such a script would pick a random number, and depending on that number, would jump ahead to a different point in the track, or another track entirely.

Understanding pre-scripts

Pre-scripts, such as the pre-script in Figure 17-4, run before a menu, track, or slideshow. For example, you might have a menu tile with a corresponding pre-script that determines which button will be selected in the menu, based on which track the user played last:

```
A = getLastItem()
if A == track "Track One B" then play button "Choose Track One Button" of menu "Which Button example"
if A == track "Track Two" then play button "Choose Track Two Button" of menu "Which Button example"
```

Figure 17-4: Which Button script.

The above script is incorporated in a project file Which Button in the CH17 folder on the DVD-ROM. The purpose is basically to have a particular button selected on the menu, based on where a person is coming from. This can be a nice user experience, as an improvement over having to select a single default button each time a menu displays.

Understanding IF-THEN statements

In the second and third lines of the Choose Menu script (see Figure 17-3), there are IF-THEN statements, which are basically a computer's way of deciding what to do. In the second line of the script, the DVD computer brain will check the contents of the variable A, and if A==1, it goes and plays the appropriate menu.

WRONG: If A = 1 then play menu "Catalan Menu"

RIGHT: If A == 1 then play menu "Catalan Menu"

A computer is very literal. If you tell it A=1, it thinks A=1. Think of it word by word. The computer encounters the word `if` followed by A=1; therefore, the computer thinks you are telling it that A=1, rather than asking `if` A=1. So you just need to differentiate telling from asking, and the way you "ask" is to use two equal signs. In general, with programming or scripting, a computer doesn't assume anything; you have to tell it.

Understanding the Script Editor

The Script Editor is actually pretty simple. But the drop-down "typing" menus are nice and can be helpful, especially if you are writing a script where several lines are similar or where you can't quite remember the name of a particular project element that you want to attach a script to.

Understanding the general drop-down menus

Each of the drop-down menus at the top of the Script Editor window type text in the Script Editor window for you. You can then delete parts of the text and replace it. Playing with these drop-down menus can also help you get acquainted with some of the things you can do with scripts.

Scripts

This Script Editor drop-down menu displays a list of any current scripts that have been created, and if you select a script, its name is typed in the Script Editor window.

Menus

This drop-down menu will display a list of any current menus in the project, along with the names of any buttons in that menu. It is especially helpful if you want to jump to a particular button on a particular menu but can't remember the name of the button. Even if you have a menu but aren't using any buttons, there might be an untitled button in the menu, and this drop-down menu will allow you to select it. (See Figure 17-5.) The Play function in a script allows you to take a person to a particular button on a menu, where that button will show up in a selected state, ready for a person to activate it. But all this drop-down menu is doing is making it easy to type in the right thing.

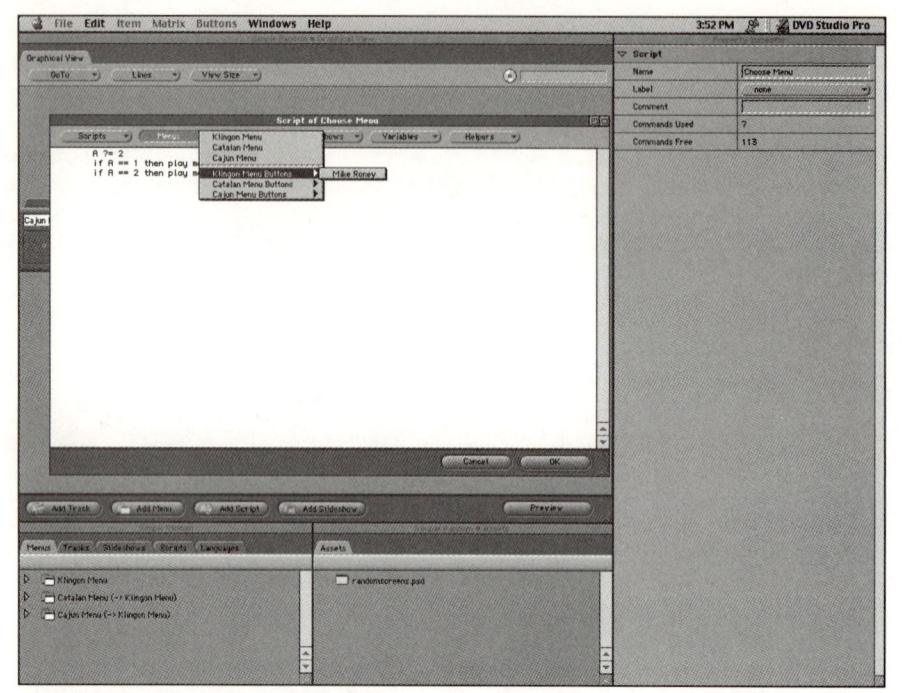

Figure 17-5: The Script Editor, with the Menus drop-down menu activated, allowing instant access to menus and buttons.

Tracks and Slideshows

These drop-down menus are similar to the previous menus, allowing easy access to created project elements. If you don't have a track or slideshow in the project, nothing will happen when you click on these drop-down menus.

Variables

This drop-down menu seems almost pointless, selecting a variable effectively types an individual letter for you. But if it helps you to keep things clear, an individual variable can be renamed in the Property Inspector when the Disc Properties are showing (if you click on the background in the Graphical View window). A variable is also sometimes referred to as a Global Variable.

> **Note**
>
> In DVD Studio Pro, you can have up to eight different variables, also known as Global Variables. You can change the variable name in the Variable Names section of the Property Inspector. Even if you have already created a script, the variable name will be updated in the script. This holds true for other project element names as well. If you mention a particular menu in a script and later change that menu's name, the name is automatically updated in the script. Thanks Apple!

Helpers

The last drop-down menu on the far right of the top of the Script Editor window is the Helpers menu.

Understanding the Helpers drop-down menu

The Helpers menu gives instant access to any of the special actions you can include in a script, and is broken down into several categories. Just like the other drop-down menus, all it really does is type text for you. (The manual that comes with DVD Studio Pro has a handy reference to all of these actions.) Programmers will find those pages to be a convenient, concise reference to the scripting capability of DVD Studio Pro. The eyes of non-programmers may start to glaze over. Join the club. It's like the Wicked Witch of the West cast a spell of intimidation, and words like *operators* and *parameters* and *functions* are like little monkeys flying all over the page, hindering your passage to the Emerald City.

But if the reading of that section in the manual was a traumatic experience, or if you've never read it, maybe it's best to leave it for another time. We're off to see the Wizard!

> **Note** In the CH17 folder in the Tutorial section of the DVD-ROM, a Simple Script Samples folder contains a DVD Studio Project file named Samples, which has a few script tiles that correspond to some of the concepts mentioned in this chapter.

The Helpers menu is broken down into categories, corresponding to the various things you can do with scripting; playing, commenting, commanding, assigning, comparing, and functioning, as shown in Figure 17-6.

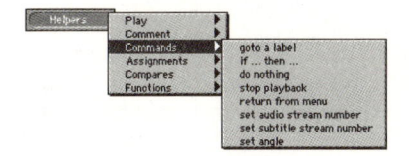

Figure 17-6: The nuts and bolts of scripting using the Helpers menu. Hours of fun for the whole family!

As you become familiar with the Helpers drop-down menu, you will see that it is a convenience, not a requirement. Also, what you see in the menu isn't necessarily what appears in a line of script. Part of the purpose of the Helpers is to represent a programming concept in a more familiar way, and then type the appropriate code for you. For example, in the Functions sub menu of the Helpers menu, there is an option A greater than or equal B. When selected, the following code is typed:

```
A >= B
```

So the menu choice becomes your "helper," making scripting easier to do, and presenting you with a translation of what the code is doing. You could write a script without ever accessing any of the drop-down menus, including the Helpers, but they can make life easier.

Playing DVD project elements in a script

Inserting a Play statement in a script will cause the DVD player to go to that project element and activate it. (See Figure 17-7.)

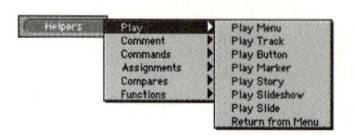

Figure 17-7: Options for playing project elements in a script. The Return from Menu function may not work in version 1.0 of DVD Studio Pro.

When you choose Helpers ⇨ Play ⇨ Play Menu, it types out the following line of script for you:

```
Play Menu "My Menu"
```

In the above text "My Menu" is a placeholder. If you actually named a menu tile My Menu, this line of script would work; otherwise, the Script Editor won't allow you to leave without putting the name of an actual menu tile from your project in between the quotes.

The Return from Menu option in this drop-down menu is the equivalent of the Return from Menu command in the Commands section of the Helpers.

```
return
```

(Doh! They're not 100 percent the same, one of them types return with a capital R, the other is lowercase. But it doesn't matter; they're the same thing.)

Choosing a scripting technique

You may be tempted, as you're building a script, to refer to project elements that you haven't created yet. But when you click OK in the Script Editor, the editor wants to refer to project elements that are not there and causes an error. The best technique then is to create all the menus and tiles ahead of time so that when you refer to them in the script, you can click OK with confidence.

However, if you find yourself in that kind of situation or some other situation where you can't seem to figure out why the Script Editor is being silly after you click OK, simply select all the text in the script, choose Edit ⇨ Cut so there's nothing for the Script Editor to get in a tizzy about, and click Cancel. This gets you out of the Script Editor to take a break or do some troubleshooting. You might wish to access the Stickies program and choose Edit ⇨ Paste to leave your script in a note to make sure you don't lose it. Remember, if you get in this situation in the Script Editor and you cut the text, you need to put it somewhere, or lose it forever! Remember the Disney movie *Black Hole*? Yeah, that's where it will go if you don't save it. The Mac's built-in SimpleText program may also be an option for preserving the script.

If you are writing a line of script and have already created a menu tile, it can sometimes be helpful to use a select-and-replace technique. Say you are looking in the Script Editor at the following line:

```
Play Menu "My Menu"
```

If you have a menu called Main Menu that you want to play in the script, you could type the name in, or you could select the following text from that line: Menu "My Menu".

After you have this portion of the line selected in the Script Editor, you can then select the Menu's drop-down menu in the Script Editor and replace the text "My Menu" with "Main Menu." (See Figure 17-8.)

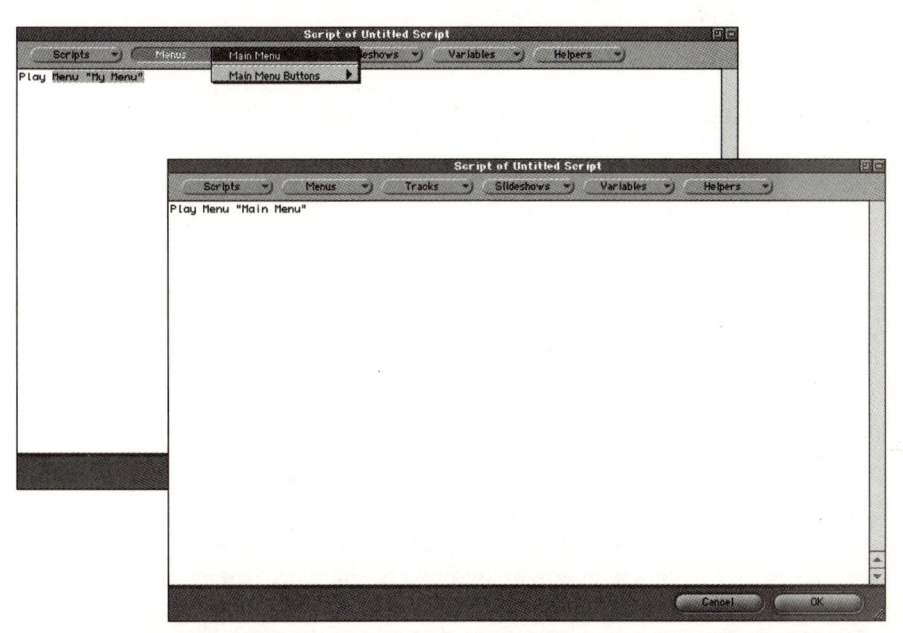

Figure 17-8: Before and After. A technique for selecting part of a script line (upper diagram) and accessing the Menu's drop-down menu to replace the text with the proper menu name.

For short scripts, it may be more convenient to just type everything in. But when it's late at night and you're working on a number of lines of script, it can be helpful to use the select-and-replace technique.

Adding comments in a script

A *comment* is a way of leaving a note to yourself or someone else in the script. It is customary with many programmers and Web developers to leave comments in code that explains how a script works.

When you select the Comment option, it types the following line of text into the Script Editor:

```
# This is a Comment
```

The idea is that you replace `This is a Comment` with whatever you wish. If you type text on the next line, the Script Editor will reject it. In other words, any comment text on an individual line in the Script Editor needs the # sign before it. The # sign is what makes a line a comment. If you put a # sign at the beginning of any line of a script, it "disables" that line of the script and it becomes a comment. (See Figure 17-9.)

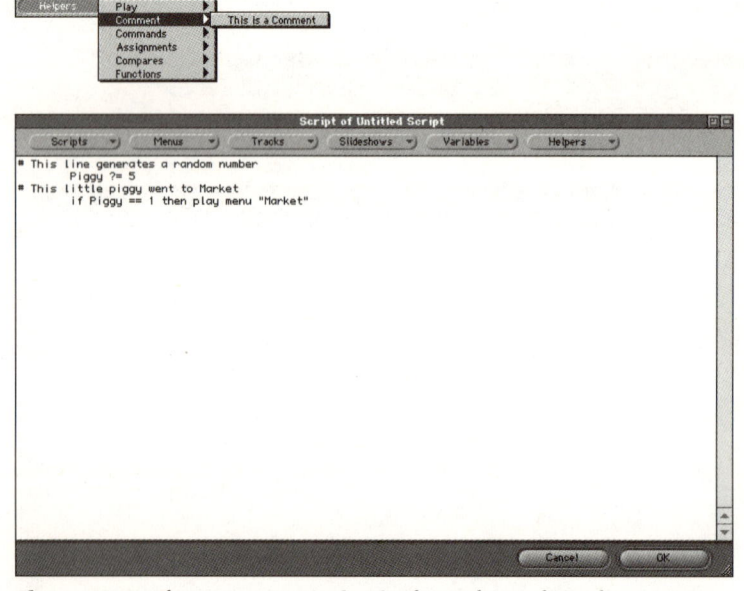

Figure 17-9: The Comment option in the Helpers drop-down menu and an example of comment lines

The Comment option could be used as a way of temporarily bypassing the error-checking feature of the Script Editor, if you're working on a script and you end up with a *syntax error*.

Note When you are working on a script and you type your code incorrectly, you will get an error message when you click OK in the Script Editor. The error message will state that there is a Syntax Error, which means there is something wrong with the script. Sometimes the Script Editor will tell you what the problem is, such as if you are referencing a project element that doesn't exist. Other times it will just say Syntax Error, displaying the line of script where there is a problem, without telling you what is wrong. It could be that you typed an extra letter or punctuation symbol that the Script Editor doesn't understand, or that you typed a comment without putting a # sign at the beginning of the line. The Script Editor can be feisty if you try to argue with it.

To bypass the error checking feature of the Script Editor so that you can exit the Script Editor to take care of needed business elsewhere in the DVD project, you can insert # signs at the beginning of every line in the script. But beware, this entirely disables the script; it only allows you to temporarily get out of the Script Editor.

Adding commands to a script

A command is sort of like when you "play" a project element; you are commanding something to happen, but you are not necessarily making reference to a particular project element; it is more general. One of the most common commands is the IF-THEN statement. (See Figure 17-10.)

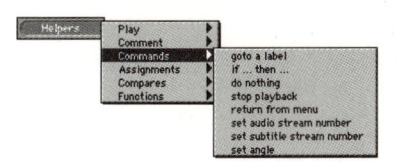

Figure 17-10: The Commands option in the Helpers drop-down menu in the Script Editor

goto label command

Labels are like Internet bookmarks within a script. When you are browsing the Web, you can use bookmarks (in Netscape), also known as favorites (in Internet Explorer), to mark a specific place you want to go back to on the Internet.

In a script, you can "label" a line or section of script so that you can go to that specific section of the script if you want to. If you don't use labels, the script is processed line by line, one after the other. Using labels allows you to have "compartments" within the same script, sub-sections that are only accessed under certain conditions.

Here is a very simple example of the use of labels:

```
A ?= 2
   if A == 1 then gotoLabel labelone
   if A == 2 then gotoLabel labeltwo
labelone:
   C = 3
   play menu "Main Menu"
labeltwo:
   C = 4
   play menu "Market"
```

In the example, a random number is generated. If the chosen random number is 1, then the script goes to the label one "compartment" in the script. If the chosen random number is 2, it goes to a different compartment. Essentially labels allow you to keep different sections and actions of a script completely separate.

IF-THEN command
`IF-THEN` statements are discussed in the beginning of this chapter and are easy to understand. With an `IF-THEN` statement, the DVD player is saying, "if this happens, do this" and "if something else happens, do this instead."

Do Nothing command (nop)
The Do Nothing command types the letters `nop` in a script.

Remember, in history class, learning about the Do Nothing political party? The do nothing command is kind of silly, but if you're working on a script, and need to exit the Script Editor, you can put the Do Nothing command as a placeholder, or you could just type `nop`.

If you were constructing an `IF-THEN` statement and had the `if` part figured out, but not the `then` part, you might end up with a line like this if you needed to exit:

```
If a == 1 then nop
```

Stop playback command (stop)
The `Stop` playback command stops the playback of a DVD project.

Example:

```
If a == 1 then stop
```

Return from menu command (return)
The `Return` from menu command returns the user to the track that was previously playing.

Example:

```
If a == 1 then return
```

> **Note:** The `Return` from menu command may not work properly in version 1.0 of DVD Studio Pro.

setAudioStream number command (setAudioStream)
`setAudioStream` allows you to select which audio stream will be active in a track where there are multiple audio streams.

Example:

```
setAudioStream 1
play track "Track One"
```

setSubtitleStream number command (setSubtitleStream)
`setSubtitleStream` allows you to select which subtitle stream will be active in a track where there are multiple subtitle streams.

Example:

```
setSubtitleStream 1
play track "Track One"
```

setAngle command (setAngle)

`setAngle` allows you to select which multiple angle will be active in a track where there are multiple camera angles.

Example:

```
setAngle 2
play track "Track One"
```

Assigning a value to a variable in a script

An assignment is when you take a value and assign it to a variable. A single equal sign is used. What appears on the left of the single equal sign "takes on" the value of whatever is on the right side of the equal sign. Figure 17-11 shows the four assignments. The easiest one to play with is the random number generator. Remember that as with any other use of variables, you can rename variables in the Property Inspector when the overall disc properties are selected. Hence `A = B` could become `Apocalypse = B` or however else you wanted to name the variables.

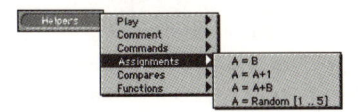

Figure 17-11: The Four Assignments of the Apocalypse.

A = B assignment (A = B)

`A = B` is an example of one variable taking the value of the other. You might have a certain value in `B` from another script, and you want to take that value and work with it in `A`, but then leave `B` alone.

Example:

```
A = B
A = A+1
```

A = A+1 assignment (A += 1)

`A = A+1` is an incremental counter. Each time the line of script runs, the value of the variable increases by one. The way to think of it is A=A+1, but the way it needs to be written in code is `A +=1`. It is an example of addition.

Example:

```
A += 1
if A == 10 then play track "Track One"
```

A = A+B assignment (A += B)

A = A+B takes the value of two variables, adds them together (A+B), and places the resulting value into a variable (A). The way to think of it is A=A+B, but the way it needs to be written in code is A += B.

Example:

```
A += B
if A == 10 then play track "Track One"
```

Note: In the Script Editor in version 1.0 of DVD Studio Pro, choosing Helpers⇨Assignments⇨A=A+B will result in the computer typing B += B, rather than A += B.

A = Random [1 .. 5] assignment (A ?= 5)

This is the random number generator. Choosing this assignment results in A ?= 5, where the 5 can be replaced with anything you want.

Example:

```
A ?= 2
if A == 1 then play track "Track One"
if A == 2 then play track "Track Two"
```

Note: Just because something you want to do doesn't show up in the Helpers area of the Script Editor doesn't mean you can't do it. As you become more familiar with programming concepts, you may wish to try variations on what appears in the Helpers; this is where reviewing the reference section in the Script chapter of the DVDSP manual will be helpful. For example, The A = A+1 (A += 1) assignment is an example of addition, which is mentioned in the DVDSP manual. You can also do subtraction, even though it isn't a Helper. If you wanted to count in reverse, You would simply have (A-=1). Remember, Helpers are only there to help you type things out; they don't necessarily represent the limits of what you can do with scripting.

Woohoo! Done with all the assignments!

Comparing Variables in a script

Comparisons in a script allow you to compare the value of a variable to another variable or a number. (See Figure 17-12.)

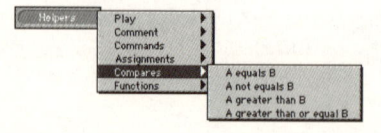

Figure 17-12: The Compares option in the Helpers drop-down menu in the Script Editor

Comparisons are typically used as part of an `IF-THEN` statement.

A equals B comparison (A == B)
Example:

```
if A==B then A += 1
```

A not equals B comparison (A != B)
Example:

```
if A!=B then A += 1
```

A greater than B comparison (A > B)
Example:

```
if A>B then A += 1
```

A greater than or equal B comparison (A >= B)
Example:

```
if A>=B then A += 1
```

Here's a list of the comparisons, using numbers instead of variables:

```
if A==1 then A += 1
if A!=10 then stop
if A>10 then play track "Track One"
if A>=2 then nop
```

Using functions in a script

Functions are the specialized parts of script that relate to the DVD player's ability to remember certain things, like which track was playing last, which project element is currently active. Functions allow you to ask the DVD player a question and then instruct it to do something based on what it told you. (See Figure 17-13.)

Note: Not all the functions in version 1.0 of DVD Studio Pro work. You may think they work; and they may work in Preview mode. But some people have experienced problems when using such functions in final projects burned to DVD-R disc. Other people had problems with scripts in Preview mode or when previewing a multiplexed project by opening the VIDEO_TS folder on their hard drive in an Apple DVD player, only to find that the manufactured DVD did fine. The jury is not out yet. In short, the best thing to do if you have version 1.0 of the program is to upgrade to DVD Studio Pro 1.1 (www.apple.com/dvdstudiopro/update), or whatever the latest upgrade is.

It's pretty cool how much you can do.

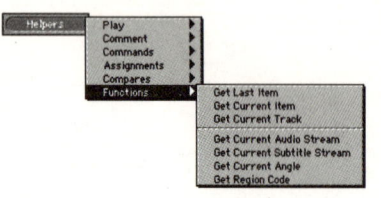

Figure 17-13: Conjunction junction, what's your function. The functions allow you to "Get" information.

The way you use a function is to "Get" the information and put it in one of your global variables. (If you haven't renamed any of them, they are A, B, C, D, E, F, G, and H. Remember, you have eight variables to work with, that's it.)

getLastItem() function
Example:

    A = getLastItem()A

getCurrentItem() function
Example:

    A = getCurrentItem()

getCurrentTrack() function
Example:

    A = getCurrentTrack()

getAudioStream() function
Example:

    A = getAudioStream()

getSubtitleStream() function
Example:

    A = getSubtitleStream()

getAngle() function
Example:

    A = getAngle()

getRegionCode() function
Example:

```
A = getRegionCode()
```

Befriending the Log window

When working with scripts in DVD Studio Pro, the Log window can be your friend — a really good friend. You may have noticed when you multiplex a project (using Build Disc), whether you are saving it locally or burning it to a disc that the Log window comes up.

It can be especially helpful when working on scripts. If you preview a project that involves scripts and then check the log, it will give you an instant replay, a blow-by-blow account of what just happened, including what values a function produced, what random number was chosen, and so on.

To see the effect, you can go back to the Simple Random project file mentioned earlier in the chapter (in the CH17 folder in the Tutorial section of the DVD-ROM), preview the project, and then choose Window ⇨ Log, and you see something like Figure 17-14. The first three times the button is activated, the random number ends up being 2, and the Cajun menu comes up. The fourth time around the random number is a 1, and the Catalan menu is selected.

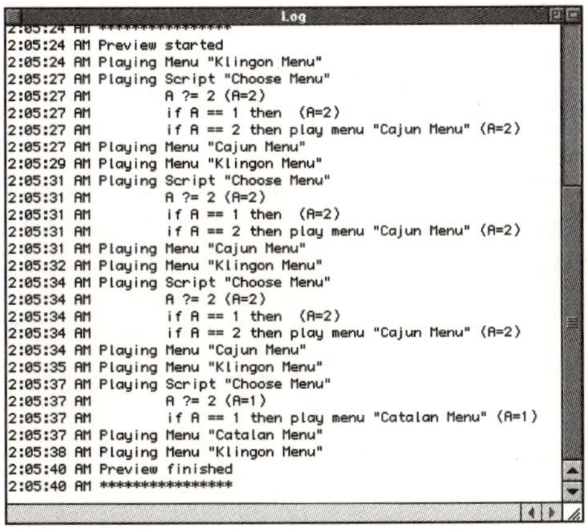

Figure 17-14: The Log window, accessed after previewing the Simple Random project.

One logging technique to consider is to resize the Property Inspector so the Log window always appears. It's an interesting way to see what's going on in the project, and prevents you from always having to select the window and having it disappear again. (See Figure 17-15.)

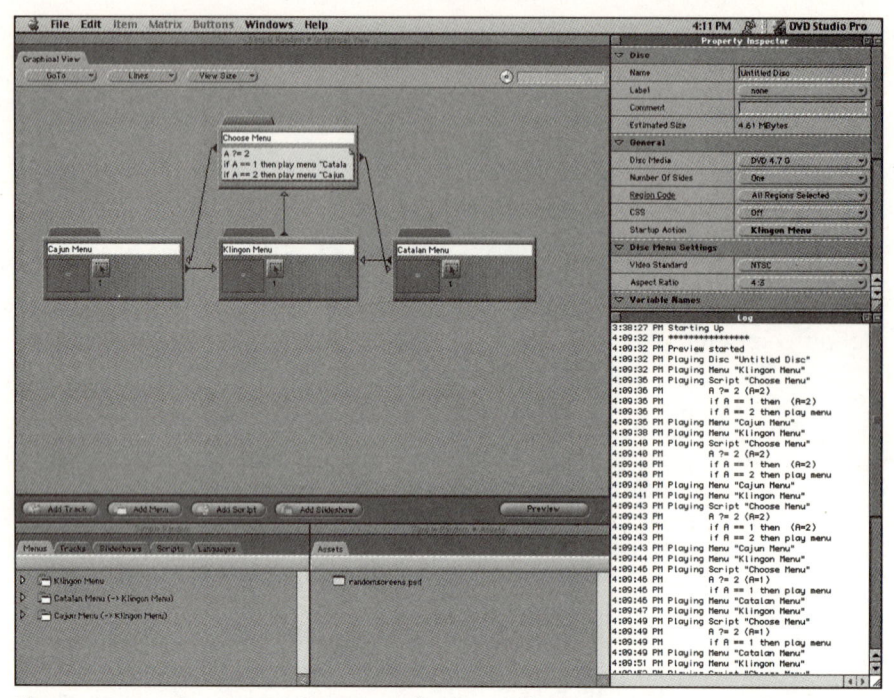

Figure 17-15: The DVD Studio Pro workspace, with the Property Inspector resized so the Log window always appears in the lower right-hand corner.

Scripting Capability

The scripting capability of DVD Studio Pro is based on the way it was programmed, and the limitations of DVD players in general. It's a super-powerful program for a really good price, considering what you'd have to spend with other programs to get the same features. For example, at the time of writing I don't believe you can find another program that gives you CSS and Region Coding for a grand. Go Apple!

One thing you may have noticed is that you only have eight variables to work with. But those eight variables can conceivably be re-used, over-used, abused, and amused in so many different ways within a project that only having eight is not so bad. If you run out of variables, you may be able to "compartmentalize" variables, using them in more than one place for different purposes. This can be tricky, but if you keep an eye on the Log window, you will know for sure whether or not it will work.

Another limitation is the number of commands that can be used in a script — 128 to be exact. Whenever you click on a script tile, the Property Inspector will tell you how many commands you've used, and how many you have available. It displays the name of the individual script tile you have selected, but gives an overall project count for the commands. The Quiz project in the CH17 folder on the DVD-ROM is an example of a reasonably hare-brained and semi-complex scripting adventure that attempts to do a lot of silly things, but only uses 64 commands after a while. (See Figure 17-16.)

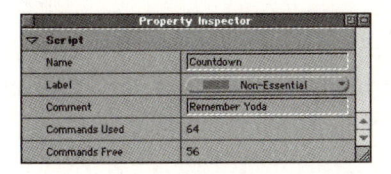

Figure 17-16: The Property Inspector, with the Countdown script tile selected in the GLF Quiz DVDSP project file.

If you end up butting up against the 128 command limit, seeing if there is a way to consolidate some of your scripting. Someone who is familiar with programming may be able to take a look at the script and identify a way of doing the same thing with fewer commands.

> **Note** If you are using Version 1.0 of DVD Studio Pro and wish to get into scripting, you should at least get the free upgrade to Version 1.1 from Apple, or any other free upgrades that are released in the future. (Hopefully the future will bring some interesting new purchasable versions of the program as well.) Aside from introducing several fixes to potential problems with scripting, Version 1.1 includes an expanded manual and several other enhancements. You can download the free upgrade from: www.apple.com/dvdstudiopro/update.

Summary

This chapter wraps up the section on advanced interactivity. So far you've been exposed to scripting, multiple angles, and multiple languages. All of these features are worth some experimentation since they have the potential to bring a DVD project to an entirely new level. Becoming familiar with advanced interactivity allows the DVD author to make a unique contribution to collaborative projects from the beginning. Awareness of what you can do with a DVD can open up a new world of possibilities with the way video is shot, the way audio is recorded, and the way graphics are prepared.

✦ A variable is a container for storing information. If you put information in a variable, it will stay there so you can come back and look at it later. You can add new information, or change existing information.

✦ A DVD Studio Pro project can utilize up to eight variables, and each of them can be renamed. The default names are A, B, C, D, E, F, G, and H.

✦ A DVD Studio Pro project can use a total of 128 scripting commands, whether they are all in the same script tile, or scattered in different script tiles. You can keep track of how many commands are used by consulting the Property Inspector with a script tile selected.

- ✦ A script can be assigned to a button, track, menu, marker, story, slideshow, or remote control key.
- ✦ The Script Editor has a series of drop-down menus that type information in a script for you. Accessing these drop-down menus is the equivalent of typing on the keyboard; it is a timesaving feature, which makes it easy to reference project elements in a script.
- ✦ The Log window can be a useful way to troubleshoot a script, or to find out what values variables have at different times during the script.

Chapter 18 discusses methods for utilizing multiple video tracks in a DVD project and explains how this allows the DVD user to experience the same scene from different camera angles. There is a discussion of alternative usage of the multiple angle feature in commercial DVDs. The DVD user can watch a movie and select between multiple video tracks, such as the movie or an interview with the director discussing a particular scene.

✦ ✦ ✦

Adding Multiple Angles to a Project

In This Chapter

Understanding the uses of multiple angles

Organizing project elements

Creating angles in a video editing application

Working with multiple angles in DVD Studio Pro

Budgeting for DVD bit rates

Using the DVD Studio Pro Preview function

Of all the new options provided by DVD technology, none has more interesting implications than multiple angles. With this capability, DVD has the potential to include a greater degree of interactivity and a less linear approach to story telling. Imagine, watching a movie from different characters' points of view. Perhaps you are watching an event take place and decide that you want to see it from another location in the room. With multiple angles you could do that and much more. Whether you watch a long shot or skip to a close up, the action and story are no longer limited by a singular perspective — freeing the creator to stretch the limits of audience participation.

DVD Studio Pro lets you take advantage of multiple angles (a maximum of nine video streams), something which most mid- to low-price DVD authoring solutions have not offered in the past. Now the problem becomes deciding what to do with all of those extra tracks and how to create them effectively. Also, it is important to consider managing these assets carefully to produce a disc that not only looks good but works well within the limits of DVD specifications.

In this chapter, we discuss how to create multiple angles and provide you with some methods for easily creating alternate video tracks in a video editing application.

The Importance of Multiple Angles

When the DVD standards were being developed, it was realized that folks were demanding more from their home entertainment experiences. By this point in time, people were used to the interactivity that was provided in video game systems,

CD-ROM and DVD-ROM computer programs, as well as the Internet. These factors could not be ignored if the DVD-Video format wanted to draw the interest and dollars of technology-savvy consumers. After all, previous attempts at introducing new video formats had only found a limited niche among cinephiles who demanded the best available picture and audio. DVD had to offer more than improved picture and audio, as compared to other formats like laser disc. The answer was in the development of interactivity.

The term *multi angle viewing* brings to mind the most obvious use of this technology, which is to present additional views of an action or scene occurring on screen. This could be a car chase sequence filmed with several cameras, or a dramatic interlude from the viewpoint of different characters. In either case, the idea is to put the viewer in control of what he wants to see in a scene.

Of course, angles are not limited to simply displaying alternate views or camera locations within a scene. Multiple angles can also be used to add material that augments the experience provided by the main video track — including behind the scenes set construction, dramatic coaching, or interviews, which run concurrently. These are only a few ways to make better use of the technology.

Preparing Tracks

Before you can begin adding several angles to a DVD project, you must first create video tracks that properly sync with each other so that switching from one to another is seamless and logical. Track construction usually occurs in a video editing application, such as Final Cut Pro, that allows you to work with multiple tracks at one time. By working in this way, you can more easily preview and line up tracks that are intended to be (in many cases) nearly frame accurate in their positioning.

Managing multiple video files in DVD Studio Pro

Careful management of your video tracks is important if you are to maintain order in a complex project. You can easily lose track of multi-angle assets if they are not properly dealt with. You may want to consider keeping a written log (which is also a good idea for your overall DVD project) that details what angles go with a particular track, the order the angles should be placed in, and the time codes where material appears.

DVD Studio Pro also provides a few methods for keeping tabs on your multi-angle assets and their placement in a project. The Matrix views are helpful in this regard, giving you a visual approach for checking and managing your files. Also, the Project View window is useful for easily viewing, organizing, and modifying your assets.

Using the Matrix views

Checking your multi-angle assets with the Matrix views may help you determine potential mistakes when working with several video assets. To use the Matrix views, follow these steps:

1. **Open the project in DVD Studio Pro that you want to check.** This usually includes a complete project with menus, tracks, slideshows, and scripts already created and linked together.

2. **With your project open, you may view the first visual matrix named the Asset Matrix by choosing Matrix ⇨ Assets of Disc (Figure 18-1) from the menu at the top.** This view provides you with a look at the assets available for inclusion in your video track or other disc element. Notice that the primary video stream is the asset marked as being assigned to your track. The additional video tracks available are listed along the top of the Asset Matrix window. If you wish to switch the primary video stream for another, simply place the dot in the matrix that corresponds to the intersection of the elements you want to link. Your project is instantly updated with the new connection.

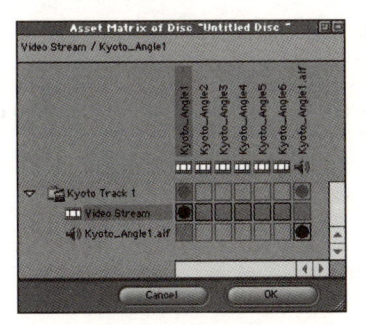

Figure 18-1: The Asset Matrix allows you to view the relations between disc elements in a visual way.

3. **When you are finished using the Asset Matrix, close the window by clicking OK to accept any changes you have made, or click Cancel if you do not want the changes you may have made to take effect.**

4. **Next, take a look at the Jump Matrix by choosing Matrix ⇨ Jumps of Disc (Figure 18-2) from the menu at the top.** The Jump Matrix is a visual representation of the possible actions you may assign to disc elements. While it is not especially important to working with multiple angles, it is important to the interactions with video tracks in general. You may alter links between actions and other disc elements by changing the position of the dot that occurs at the intersection, in the same manner as described for the Asset Matrix view.

5. **After you have completed viewing the Jump Matrix, click OK or Cancel to close the window and return to the workspace.**

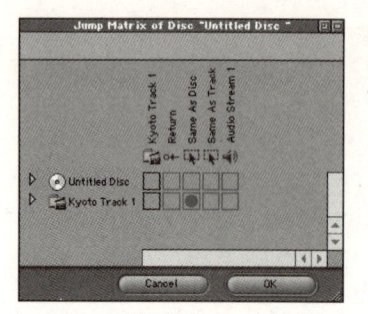

Figure 18-2: The Jump Matrix is another visual method for viewing and managing project elements.

Using the Project View

Probably the most useful method for managing your containers and disc elements is the Project View. From this window, it is easy to see all of the angles you have added to a particular track without the need for opening multiple windows or disturbing other elements in the workspace. Accessing disc elements in order to modify properties is as easy as selecting a tab and accessing a file alias through a folder hierarchy similar to viewing files as a list on your Macintosh hard drive. To use the Project View, follow these steps:

1. **With your disc project open, select a tab for a container from the Project View window in the lower left-hand corner of the DVD Studio Pro interface, as shown in Figure 18-3.** Because we are discussing how to work with multiple video assets for angles in a project, choose the Tracks tab. Notice that the tracks you have created are listed as a folder in this container.

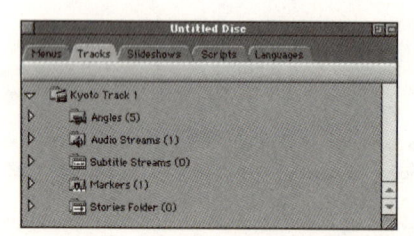

Figure 18-3: Select a container from the tabs in the Project View in order to access its contents.

Tip: You may create a separate window for the container you want to work with by dragging the tab from the Project View window and dropping it anywhere on the desktop. You cannot return the container to the Project View without first closing the Project View window and reopening it by choosing Windows ⇨ Project View from the menu at the top. After the Project View is reopened, the container is returned to its original location. You may also open a separate window for a specific subcontainer (such as a particular track name) by double-clicking on the folder icon to the left of it.

2. **Click the arrow to the left of the Track folder.** The various containers, which contain the assigned assets (labeled in parenthesis to the right), should be

displayed. Expand the Angles folder to see the angles you have created and added to the track so far. (See Figure 18-4.)

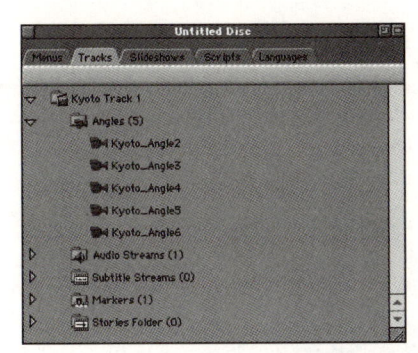

Figure 18-4: Items and subcontainers in the Project View are similar to an alias for a folder on a hard drive.

3. **Click on an item to view its properties in the Property Inspector.** For example, choose an angle that you have created by clicking on it in the Project View window. By doing this, you may modify any available settings, including the name and the asset assigned to it. If you wanted to switch one angle for another, this is a quick and painless way to accomplish that task.

4. **You may also create new items in the Project View.** This is done by selecting an item or subcontainer and choosing the item you want to create from the Item menu at the top.

Preparing angles in Final Cut Pro

While there are many video-editing applications that can successfully work with multiple tracks of video (such as Adobe Premiere), we are going to discuss Final Cut Pro, because it provides a logical link with Apple's DVD Studio Pro. Also, hardware solutions, such as the RTMac card from Matrox, which Final Cut Pro is capable of using, make working with multiple tracks a real-time experience, greatly reducing the amount of time and stress involved in the creative process. Whether you are new to video editing or a seasoned professional, Final Cut Pro is an editing tool that is easy to learn yet powerful enough to meet the needs of any production, whether DV, HDTV, or film. Check out Zed Saeed's *Macworld Final Cut Pro 2 Bible* also published by Hungry Minds, Inc. for detailed information on Final Cut Pro.

Working with multiple video tracks is one of the most basic and necessary features of any professional video editing application, whether Final Cut Pro, Avid, Media 100, or Premiere. In the following steps, we detail a process that could just as easily be applied to another editing application. The principles of working with layers of video when editing are generally the same no matter what price range or platform you are working under.

The following steps show you how to create multiple tracks in Final Cut Pro, which can be exported for use as angles in a DVD Studio Pro project:

Note: The maximum number of simultaneous video streams allowed on a DVD is nine. DVD Studio Pro allows you to use all eight alternate video streams as multiple angles.

1. **Launch the Final Cut Pro application.**

2. **Import the video tracks you plan on working with into the Browser by choosing File ⇨ Import ⇨ File and selecting the video tracks you want to use as multiple angles. (See Figure 18-5.)**

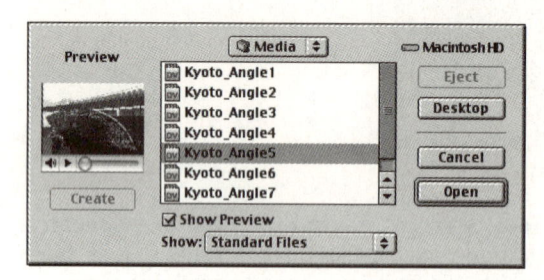

Figure 18-5: Import the video files that you want to use as multiple angles into Final Cut Pro.

3. **Open the clip you plan on using as your primary video stream by double-clicking it in the Browser window.** Preview your other tracks as well and make certain that they are labeled correctly according to the order in which you want them to appear. (See Figure 18-6.) By doing this, you may avoid confusion later when importing or exporting tracks for use as angles in your DVD project.

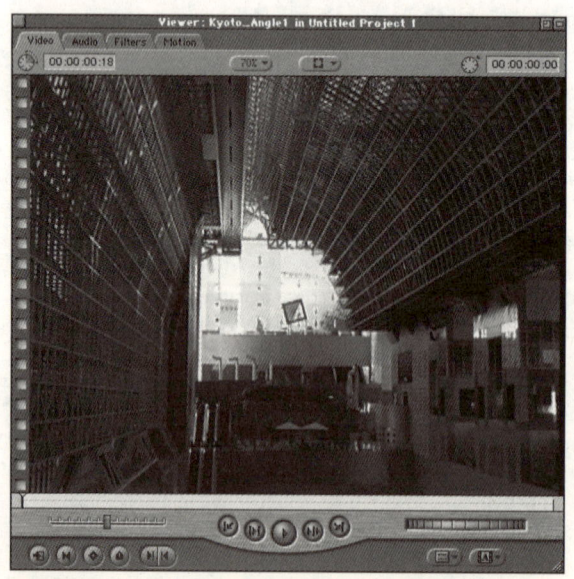

Figure 18-6: Preview and label your video files correctly before adding tracks to the Timeline.

4. **Insert your main video track into the current Timeline by clicking and dragging it from the Viewer window onto the Canvas window, where the option to insert is presented. (See Figure 18-7.)**

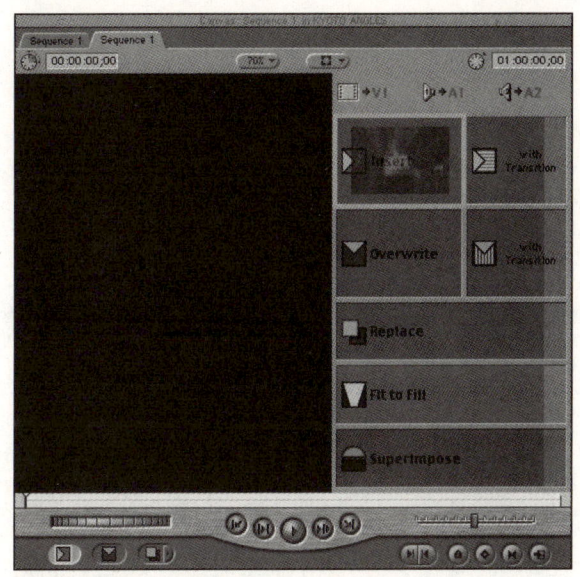

Figure 18-7: Inserting video clips into a track is a simple drag-and-drop process.

5. **Add additional tracks to the Timeline (up to 99 video tracks or layers may be inserted) so that you can add the remaining angles for your project and work at syncing them with the main video track.** With the Timeline window selected, choose Sequence ⇨ Insert Tracks. A dialog box opens that prompts you to select how many tracks you want to add. (See Figure 18-8.) Type the number of extra tracks that you want to add into the Insert Video Tracks box. Choose After Last Track to insert the new tracks above video track number one.

You may create the tracks in any order you want, but this is generally the default method for working with several tracks, leaving the first layer on the bottom, similar to adding layers in Photoshop. Consider whether you want to work with multiple audio tracks for these angles at the same time or whether you do not require additional audio specific to each track. If you are planning to have different audio for each of these new tracks, it is a good idea to work with them at the same time as you are creating the new video tracks (or angles), ensuring proper syncing and generally making your work a lot easier and efficient.

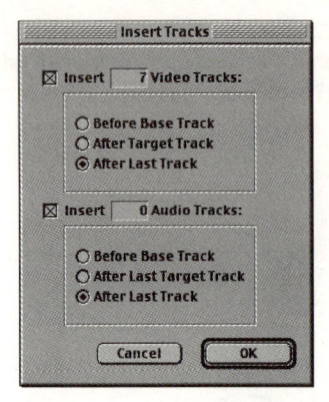

Figure 18-8: Type the number of additional tracks you want to add to your project and choose the order you want the tracks added to the Timeline.

6. **Expand your Timeline window horizontally to view all of your tracks at once, or use the scroll bar on the right side to see certain tracks.** (See Figure 18-9.)

Figure 18-9: The Timeline window may be expanded or the scroll bar may be used to view multiple tracks of video and audio.

7. **Now it's time to add your clips to the appropriate video layers.** Before adding video to track number two, be sure to lock the audio tracks from the previous layer that you do not want to overwrite (see Figure 18-10). Also, be sure to place the playhead (the marker that signifies your position when you move through a clip or sequence) at the beginning in the Timeline window, otherwise material may be added to the wrong part of your sequence.

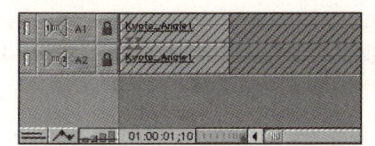

Figure 18-10: Lock the audio tracks that you don't want to accidentally overwrite when you add material to new tracks.

8. *(Optional)* **You may skip this step, but it is often a good idea to preview or make basic edits to the clip before adding it to the sequence.** In the Browser window, double-click on the next video clip (angle) you want to work with. It should be displaying in the Viewer window.

9. **To add this video clip (angle number two) to a new track, click on V2 in the Timeline window.** This selects that track so that you can add new material.

10. **With the new track selected, drag the image from the Viewer window onto the Canvas window and select Overwrite when prompted to make a choice (see Figure 18-11).** This adds the clip to video track 2, above the first video track. You may also accomplish this by dragging the clip directly from the Browser window.

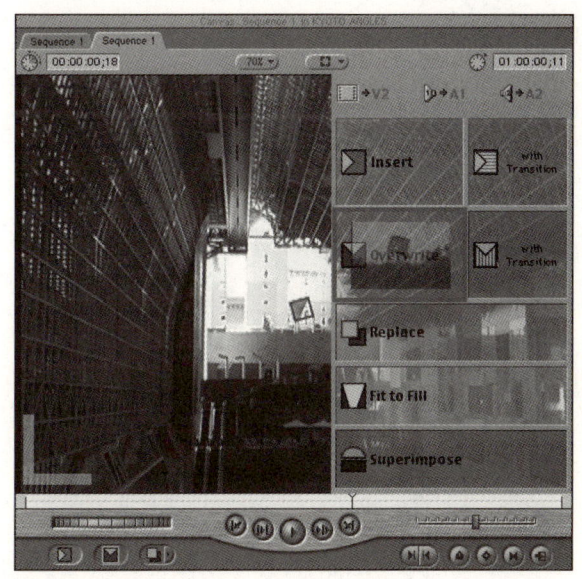

Figure 18-11: Place a new video clip into another track by using the Overwrite command, while making certain not to affect the video or audio on other layers.

11. **Repeat Steps 7 through 10 above for each additional video track you want to add.** Each of your angles should be placed on a separate track.

12. **After you place your clips on separate tracks, it is time to make certain that each track is exactly the same length, as shown in Figure 18-12.** In many cases, your video tracks differ (at least slightly) in length. You can add filler to portions of the track that require it (usually at the beginning or end) or you may leave gaps in the Timeline where there is no material. Use all of the editing tools at your disposal in Final Cut Pro to make the tracks equal in length as well as synced to the one below it. If you are attempting to sync video tracks, choose the first track as your reference, since this should provide a consistent base for aligning each angle without the deviation that can occur through successive generations of comparisons.

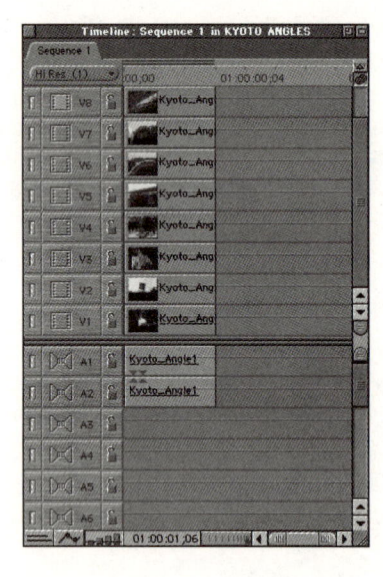

Figure 18-12: Your tracks should all be the same length and synchronized to each other.

After you have completed the editing process of aligning the material in your tracks, you need to export individual tracks, which later are assigned to their own angles in DVD Studio Pro (discussed later in this chapter in the section "Creating Multiple Angles in DVD Studio Pro"). If you have several tracks to export, it is easiest to create a batch export list and encode all of your files to MPEG-2 in a few easy steps.

1. **Creating a new bin for your files.** Choose File ⇨ New ⇨ Bin.
2. **After your new Bin window appears, select all of your clips from the Timeline by Shift+clicking each one, or by choosing Edit ⇨ Select All.**
3. **Drag these files and drop them onto the new Bin window.** All of your tracks should appear in the Bin window, as shown in Figure 18-13. Again, make sure that they are all selected by choosing Edit ⇨ Select All.

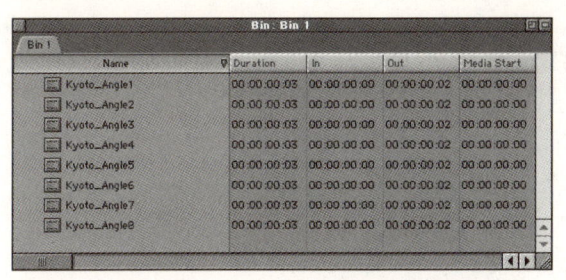

Figure 18-13: After you have selected all your tracks and dragged them to a new Bin window, you are ready to begin the process of batch exporting.

4. **When all the files you need to export have been selected in the bin, choose File ➪ Batch Export to open the Export Queue window (see Figure 18-14).**

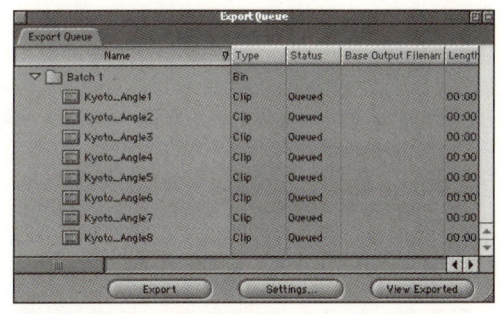

Figure 18-14: When you select the Batch Export option from the File menu, the Export Queue window opens, which lists the files that are ready for batch export.

5. **Select all of your files in the Export Queue window and choose the Export options for all these files by clicking the Settings button at the bottom of the window.** A new window opens that enables you to make export settings for your files. Begin by setting a destination (location on your hard drive) for the files, as shown in Figure 18-15.

6. **Choose MPEG-2 as the format and, with the options button, make your bit-rate settings.** Click OK when you are ready to return to the Export Queue window.

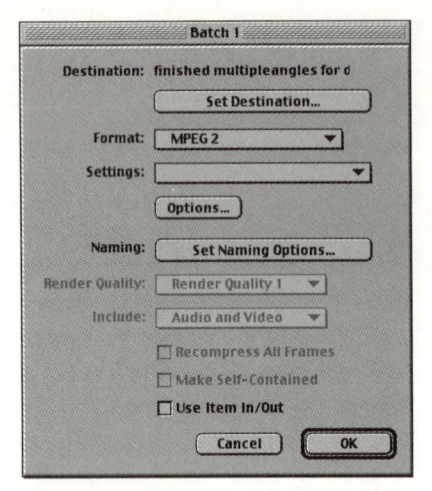

Figure 18-15: Adjust the export settings for your batch by clicking the Settings button, which opens a Settings window.

7. After you have made all of your settings, click the Export button at the bottom of the Export Queue window.

After you have prepared your angles for inclusion in your DVD project, you need to open DVD Studio Pro and create a video track with these alternate streams.

Creating Multiple Angles in DVD Studio Pro

Working with multiple angles in DVD Studio Pro is easier than it may seem. You may think that adding angles to a project would be similar to working with tracks in a video editor like Final Cut Pro, but it is actually easier than that. If you are familiar with adding multiple audio streams to a track, then you already know what is involved with adding additional angles. Essentially it is a matter of dragging and dropping your prepared angles onto the appropriate track tile.

Note: All video streams used as multiple angles in a track must be identical in length and encoded with the same settings. Be sure to check your angles before adding to a track in DVD Studio Pro.

To add multiple angles to a track in your DVD Studio Pro project, follow these steps:

1. **Open DVD Studio Pro and import the video streams, audio streams, and other assets that you want to use in your project (see Figure 18-16).** Select File ➪ Import, then choose the assets you want to import by clicking Add or Add All. After you are finished selecting the assets you want to import, click the Import button to bring them into the Assets container in DVD Studio Pro.

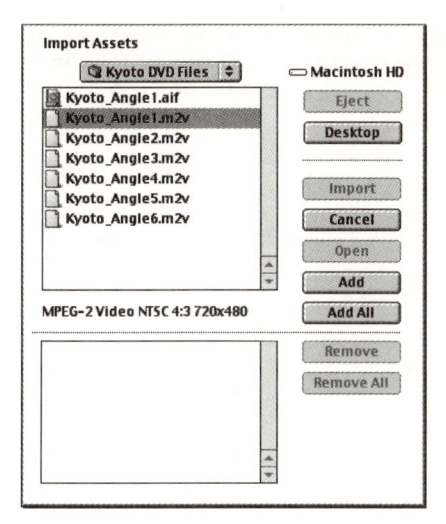

Figure 18-16: Import the video files and other assets you want to use in your DVD Studio Pro project.

2. **In the Graphical View, click Add Track to place a new track tile in the workspace (see Figure 18-17).** This is the track you are going to use to add multiple angles.

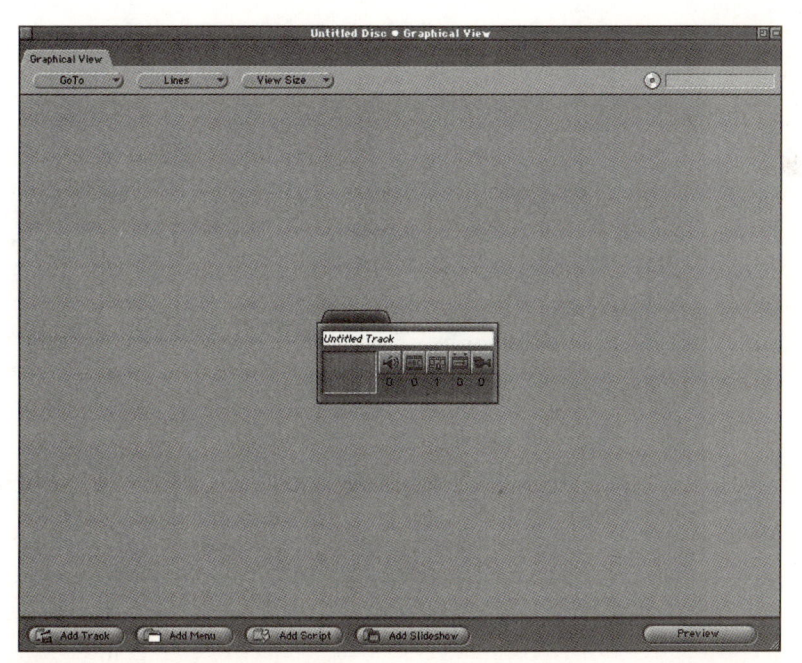

Figure 18-17: Begin your multi-angle project by adding a new track tile to the Graphical View workspace.

3. **Name the track by highlighting the text box at the top of the tile which says "Untitled Track" and typing a new name (see Figure 18-18).** You may also name the track by filling in the Name box located in the track section of the Property Inspector.

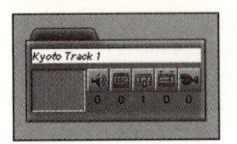

Figure 18-18: Name your track tile by typing in the text box located on the tile or in the Property Inspector.

4. **Next, add the video stream that should act as the main video track (see Figure 18-19).** The video steam that you add is the video that plays first by default. It is the primary video that is supplemented by the alternate angles. DVD Studio Pro differentiates between your main video stream and the alternate video streams (or angles) that you add to the track. Keep this in mind when adding video to your track, so that all video assets are added in the order that you want them to appear. Drag your main video asset from the Assets container onto the track tile.

Figure 18-19: Add your primary video stream to the track tile by dragging it from the Assets container.

5. **At this point, you may add any other elements that go along with your main video stream, such as an audio track or subtitle stream.** You may add more elements later on, although it is often easier for organizational purposes to add the main elements of a project first, before supplementing them with additional material.

6. **Open the Angles container by clicking the angle icon located on the track tile.** Every track tile contains several icons that access the containers for subtitles, audio streams, angles, and scripts (see Figure 18-20). These containers are where elements are organized and added to a track.

7. **With the Angles container open, add an angle to the track by choosing Item ⇨ New Angle from the menu at the top (see Figure 18-21).** This is the video stream that appears second in your project. This is the first angle a viewer sees when he presses the Angle key on the remote control.

Tip

You can also drag angles from the Assets container onto the open Angles container as long as all of your tracks are of equal length.

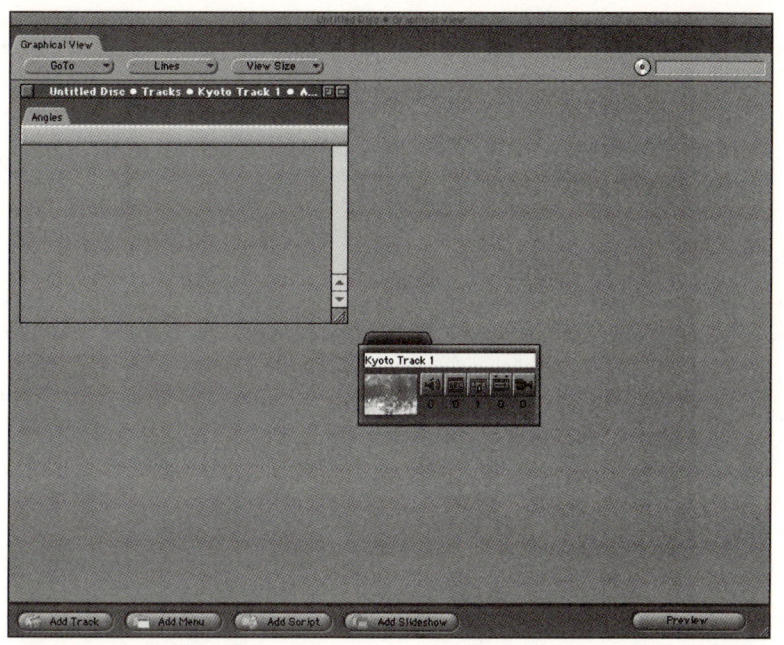

Figure 18-20: The Angles container can be accessed by clicking the angle icon located on the track tile.

Figure 18-21: An untitled angle is created in the Angles container by choosing New Angle from the Item menu.

8. Name the angle by first selecting it and then typing a name in the box located in the Property Inspector, as shown in Figure 18-22.

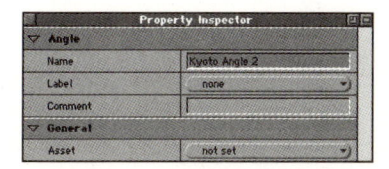

Figure 18-22: New angles are named in the Property Inspector.

9. **With the angle still selected, assign a video stream to this angle by selecting an asset from the drop-down menu in the General section of the Property Inspector (see Figure 18-23).** While the Asset menu lists all the available video files, only choose streams that are identical in length and encoding parameters to the primary video stream. This is important to the proper operation of discs with multi-angles. If angles do not match, the multi-angle DVD Studio Pro project may not function correctly.

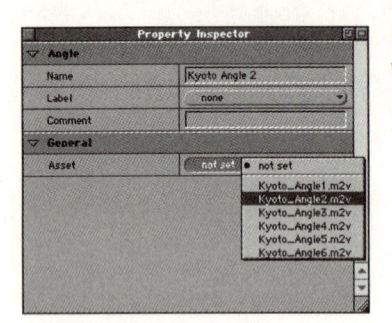

Figure 18-23: Select a video asset that you want to assign to an angle from the Asset menu in the Property Inspector.

10. **To create additional angles, repeat Steps 7 through 9 and label each angle differently (Angle 2, Angle 3, and so on), assigning the appropriate video asset to each one (see Figure 18-24).** The track you created first is the angle that displays first when a user presses the Angle button on a DVD remote control. Place any additional angles in the order that you want them to appear when viewed on a disc, listing the first angle at the top, the second angle beneath, and so on. Carefully consider the user experience, creating an order that makes sense and provides a logical path from one to the other.

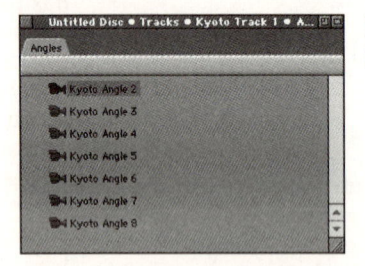

Figure 18-24: Add up to eight video streams to the Angles container for each track.

Tip

If you are using multiple angles on more than one track in a project, make sure that you place your angles in a consistent order. If you have an angle that contains interviews or direct commentary and another that includes behind-the-scenes footage, make sure that these angles appear in the same order on each track in your project. A viewer should be able to assume that the order of elements remain consistent throughout a disc so that Angle 1 is always director commentary while Angle 2 is always behind-the-scenes footage.

After you have created your angles in DVD Studio Pro, it is important to consider how they are going to affect the overall project, particularly as they relate to a disc's maximum bit rate.

Adjusting Bit Rate

As with any DVD project, it is vitally important to budget for the bit-rate you want to use or are capable of using. If you do not plan ahead, you may find in the end that you need to re-encode material or completely rethink your use of additional project elements, such as subtitles, audio, and video streams. While DVD Studio Pro is capable of using many different elements that occur simultaneously, you must be aware that every one of these elements increases the amount of memory required to read the disc, bringing you closer and closer with each element to the maximum bit-rate allowed by the DVD standards. If you intend to use several angles for your project, be sure to encode all of your video assets at a slightly lower bit-rate (remember to use the same settings for each video stream).

You may have to re-encode some of your assets if you intend to use many simultaneous project elements. For example, if you were using eight audio tracks, eight video tracks, several subtitle tracks, and other elements at the same time, you are limited in the quality (maximum bit-rate) that is allowed for each element. In fact, you can have serious problems when it comes time to multiplex the project if you do not budget enough space for adequate throughput.

Previewing a Multi-Angle Project

When it comes time to test your multi-angle project, DVD Studio Pro has the capability to preview it as though you were watching the DVD on a television and operating a remote control. Previewing is an important step in the process of creating a DVD, particularly when creating a project with multiple angles since many potential problems can occur. By thoroughly playing the project, switching between angles and skipping tracks, you may encounter potential problems that can be fixed before sending the project off to a replicator, thus beginning the expensive process of producing a manufactured disc.

Previewing a completed multi-angle project in DVD Studio Pro is similar to previewing other disc elements, such as alternate audio tracks and menus. To preview a project, follow these steps:

1. **With your project open in the workspace, click on the track tile you want to preview in order to select it.**

2. **With the track selected, click the Preview button located at the bottom of the Graphical View, as shown in Figure 18-25.** You may also choose Item ⇨ Preview Track from the menu at the top to select the track you want to preview. As soon as you select the Preview function, your video track begins to play.

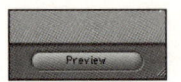

Figure 18-25: Activate the Preview function in DVD Studio Pro to test your multi-angle project.

3. **At this point, you may switch between the various angles that you have created by clicking the Angle button (Figure 18-26) at the bottom of the Preview window.** Each time you click the Angle button, another angle is displayed, cycling through the angles until they return to the first video stream. Notice the display (Figure 18-27) labeled "angle" in the upper right-hand corner of the Preview window. This display indicates which angle you are currently viewing. Make certain that the angles you have added display in the correct order.

Figure 18-26: Click the Angle button at the bottom of the Preview window to switch video streams.

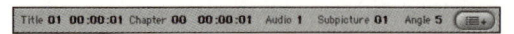

Figure 18-27: Notice that the display in the upper right-hand corner of the Preview window indicates the angle number you are currently viewing.

4. **After you are finished previewing the video track, click the Stop button at the bottom of the window to return to the Graphical View workspace where you may continue editing your track to make any additions, deletions, or corrections necessary to your multi-angle project.**

Summary

✦ Multiple angles can add greater interactivity to a DVD project by providing several viewing options for a user to choose from.

✦ A maximum of nine video streams may be utilized on a DVD.

✦ DVD Studio Pro allows you to use the maximum number of angles allowed by the DVD standard.

- Multiple angles can be used for many things, including the addition of supplemental interviews, behind-the-scenes material, and camera positions that run concurrently and synced to the main video stream.
- DVD Studio Pro allows you to manage multiple assets through the use of the Project window and other views.
- Final Cut Pro is a great video editor for the preparation of multiple angles.
- Final Cut Pro's Batch Export feature is ideal for exporting several video tracks at one time with the same settings.
- All of the multiple angles for a particular track must be created with the same MPEG-2 settings.
- The Angles container on a track allows you to add and organize multiple video streams.
- The Preview feature in DVD Studio Pro lets you test multiple angle tracks.

The next chapter explains how to add multiple languages to a DVD Studio Pro project. This is another DVD feature that lets you customize the experience for a viewer. By making use of special features, such as multiple angles and languages, you may be able to provide the user with an enhanced viewing experience that truly makes use of the capabilities that DVD has to offer.

✦ ✦ ✦

Working with Multiple Languages

CHAPTER 19

✦ ✦ ✦ ✦

In This Chapter

Understanding multiple language capabilities

Managing multiple languages in DVD Studio Pro

Making a multiple language DVD Project

✦ ✦ ✦ ✦

In this chapter, you will learn about the multiple language capability of the DVD format and various ways you can take advantage of this capability while using DVD Studio Pro. Whether you are working at a professional or casual level, investigating and implementing multiple languages can be worthwhile, allowing you to take a DVD to a new level, from Kansas all the way to the Land of Oz.

It's easy to take the beauty of languages for granted. Exposure to other languages can come primarily through movies, or occasional acquaintances. In a DVD project, even though the features are there, it's natural to make the assumption that working with more than one language is simply too difficult. If you don't speak a particular language, how can you work with it?

But think of working with one language in a DVD project like a black and white movie, like Dorothy in Kansas in the *Wizard of Oz*. Kansas is interesting, even entertaining, but when you reach the Land of Oz, there's a whole new world of possibilities.

Having more than one language in a DVD project gives you access to a wider audience, gives the project more credibility, and makes life interesting for the DVD author. If the client has the budget and is game, it allows you to bill more hours, and thus come closer to a (G5?) Mac or a Cinema Display. Or two Cinema Displays.

Admittedly, there was a storm on the way to the Land of Oz, and sometimes a DVD project can feel like a tornado. When you are working with more than one language, managing the interconnections of a host of audio, video, subtitle, and graphic elements can get complex.

Aim to get a handle on the related concepts and methods, and be confident that a systematic approach will allow you to build a multilingual DVD project with a minimum of hassle. Start with a review of the technical capabilities of the DVD standard, consider various approaches of implementing languages, and practice making a multiple language DVD project.

Understanding DVD Multiple Language Capability

The Multiple Language feature is a boon to both DVD marketing and authoring. It gives the people who market DVD players yet another thing to distinguish DVDs from VHS tapes for the consumer, and any disc that features multiple languages will be able to stand out. It gives the content provider the ability to market to a wider audience, expanding the potential market by degrees. Thinking cross-culturally and cross-lingually is especially useful in this day and age, as global marketing and distribution converge with the continuing rise of the Internet, causing the world to be both a larger and smaller place — a larger world, in the potentially wider audience that can be reached, and yet a smaller world, as people and companies perceive themselves as being closer to each other, through the interconnectedness of modern communications.

When the DVD standard was created, the entire globe was taken into consideration. For better or worse, concepts were developed such as Region Coding, where the planet is divided up into six regions, and discs created for one region won't work in another, which makes Hollywood rest easier. Region coding is a matter of copy protection, and has little to do with language, unless you are planning to make separate versions of a disc for various regions, where you also need to calculate which languages will be appropriate for each region. Good luck! Region Coding is discussed in Chapter 20.

A more attractive feature of the DVD standard is that it takes advantage of digital technology to allow for multiple languages in a video-related project. Technically speaking, DVD players are capable of supporting up to 8 separate audio tracks (each with 8 channels, for a total of 64 audio streams) and up to 32 separate subtitles for each project.

Note The multiple audio track capability of DVDs doesn't necessarily have to be used for different languages, but it can be. The capability is often used for things like narrative voice-overs in the same language, such as when a director is commenting on a movie, allowing the user to switch between the regular audio for a movie and the commentary. Even subtitles don't necessarily have to be in a different language; they could be used for commentary, or even comedy in the same language.

When it gets right down to it, a DVD project can be done in any language you choose, and technically, you are only limited by the storage capacity of the final disc. As long as you can create the appropriate graphics and audio, you can basically do whatever you want, instructing the player to display anything from Abkhazian to Zulu. (See Figure 19-1.)

Abkhazian	Estonian	Kirghiz	Russian	Tonga
Afan	Faroese	Kurundi	Samoan	Tsonga
Afar	Fiji	Korean	Sangho	Turkish
Afrikaans	Finnish	Kurdish	Sanskrit	Turkmen
Albanian	French	Laothian	Scots Gaelic	Twi
Amharic	Frisian	Latin	Serbian	Ukrainian
Arabic	Galician	Latvian	Serbo-Croatian	Urdu
Armenian	Georgian	Lingala	Sesotho	Uzbek
Assamese	German	Lithuanian	Setswana	Vietnamese
Aymara	Greek	Macedonian	Shona	Volapuk
Azerbaijani	Greenlandic	Malagasy	Sindhi	Welsh
Bashkir	Guarani	Malay	Singhalese	Wolof
Basque	Gujarati	Malayalam	Siswati	Xhosa
Bengali	Hausa	Maltese	Slovak	Yiddish
Bhutani	Hebrew	Maori	Singhalese	Yoruba
Bihari	Hindi	Marathi	Siswati	Zulu
Bislama	Hungarian	Moldavian	Slovak	
Breton	Icelandic	Mongolian	Slovenian	
Bulgarian	Indonesian	Nauru	Somali	
Burmese	Interlingua	Nepali	Spanish	
Byelorussian	Interlingue	Norwegian	Sundanese	
Cambodian	Inupiak	Occitan	Swahili	
Catalan	Irish	Oriya	Swedish	
Chinese	Italian	Pashto	Tagalog	
Corsican	Japanese	Persian	Tajik	
Croatian	Javanese	Polish	Tamil	
Czech	Kannada	Portuguese	Tatar	
Danish	Kashmiri	Punjabi	Telugu	
Dutch	Kazakh	Quechua	Thai	
English	Kinyarwanda	Rhaeto-Romance	Tibetan	
Esperanto	Klingon	Romanian	Tigrinya	

Figure 19-1: It's a small world after all. 139 languages are available in DVD Studio Pro, from Abkhazian to Zulu, all of which are accessible in the language options within DVD Studio Pro. These language choices are more a matter of organization than anything else; it is up to you to generate the appropriate content. This diagram actually has an additional language inserted to keep you on your toes; see if you can guess which one it is (compare it with Figure 19-2 if necessary) and then send an e-mail to Apple requesting the language be added in version 2.0 of DVD Studio Pro.

The DVD standard addresses a certain number of languages that are found throughout the planet, but not every single one. Each of these languages addressed by the DVD standard has its own language code, which can be used by the computer brain inside the DVD player to perform certain kinds of tasks.

When a DVD authoring program generates the digital instructions that a DVD player ultimately reads, language codes are passed along in the midst of all the other information, but the DVD author doesn't need to think about the codes, only the languages.

Choosing a default language

A DVD player, without a disc inserted, may give you the capability to determine a "default" language. Typical choices for DVD players purchased in the United States may include English, French, German, or Spanish. This does not mean the player *translates* any material; the content which appears is entirely dependent on how the DVD was put together.

DVD player default language

If a DVD disc has been created with a language that matches one of the default languages in the DVD player, it is possible for a person to have his or her chosen language automatically come up when he or she inserts the disc in the player. For example, say you have a DVD disc that is set to make use of the default language capability, and you insert it into a player set to a default language of Spanish, the Spanish version of the DVD content automatically comes up.

DVD disc default language

As a DVD author, you can choose a default language for the DVD project, which should theoretically be the one that comes up first when the disc is inserted in the player, depending on where the DVD player was purchased, what features it has, and so on.

Regardless of how the default languages are set, it is customary to give the user the ability to choose which language version they want from the main menu of a DVD.

Considering various multilingual approaches

There is more than one way to implement multiple languages on a DVD. You can take a variety of approaches to giving the user the option of choosing another language. It could be as simple as adding a subtitle stream in another language, or as complex as creating different language versions of the DVD. You can also create each of the different language versions to include menus, audio and subtitles customized for that particular language. The main question to keep in mind is who your audience will be, and if the menu choices will make sense to the user.

The following approaches are examples of ways to implement languages, suggested scenarios ranging from simple to complex. They are representative of typical methods, but are not exhaustive. There aren't really any rules, as long as you keep the audience in mind.

The Lite Approach

The lite approach consists of including either a subtitle or an extra audio stream in another language. The user accesses the additional language by clicking the Subtitle button or Audio button when instructed either on-screen or in print.

The Direct Approach

A direct approach bypasses the capability of a DVD player to automatically choose the language and, instead, allows the user to choose a language from a menu screen on the DVD. After the user selects a language, a video track appears with the appropriate audio and subtitles.

For example, an independent Klingon filmmaker decides to make an instructional DVD on how to fire plasma bursts at Federation starships, and he wants to make a Romulan and Borg-language version available throughout the galaxy. When his Romulan or Borg friends insert the DVD disc in their player, they see an opening animation of an inspiring space battle followed by a menu with language choices displaying Klingon, Romulan, and Borg buttons. (See Figure 19-2.)

Figure 19-2: The opening screen of the Klingon Plasma Training Video, with Romulan button preselected. Romulan, Klingon, and Borg are registered trademarks of their respective galaxies and Paramount. "Hab SoSlI' Quch!" is a Klingon insult meaning, "Your mother has a smooth forehead!" For more information on the Klingon language, including instruction on how to exclaim "Today is a good day to die!" or the proper way to ask "Where is the bathroom?" visit the Klingon Language Institute at `http://www.kli.org`.

Clicking on the individual language buttons leads directly to the video, with audio in the appropriate language, running simultaneously. If his friends want to view subtitles, they just use the Subtitle button on their remote control. For example, the members of the Borg collective can choose the Borg version of the training video so that they can listen in their native tongue. However, if they want to brush up on their Romulan, they can choose the Romulan audio version and Borg subtitles. Or if they are feeling frisky, they can choose the Klingon audio version and Romulan subtitles.

In essence, with the direct approach, a main screen leads to video segments with preselected audio tracks, but any audio track or subtitle incorporated into the DVD is a part of a palette that the user can select from, no matter which audio is preselected. In other words, when you start watching a video sequence and hear the audio, if there is more than one audio track to choose from, you can switch between them. Additional audio tracks can be recorded in a separate language.

To accomplish the intended user experience, the Klingon warrior/filmmaker convinces his DVD-authoring girlfriend to help out, and she creates a single main menu and video segment, three separate versions of the audio, and three separate subtitle files.

The Alternate Reality Approach

The next step beyond the direct approach is the alternate reality approach. In this scenario, you have the language choice menu lead back to a main menu, where all of the menus and content appear in that language.

For example, say you are infected with the desire to learn Esperanto, and it becomes your all-consuming passion. You work in a corporate environment where the creative Mac types are left alone, and though the project that came through was originally going to only be in English, it is now going to include Esperanto as well (because the DVD author met a cute co-worker who is fluent in Esperanto). So when the project is presented to the Marketing Director, a Language button at the bottom of the main menu leads to a language choice menu with the choice of either English or Esperanto. (See Figure 19-3.)

Figure 19-3: The Vortecs Corporation advertising DVD, with rural and urban video segment choices, and the Language button, which leads to a menu where the user can select either English or Esperanto.

The Marketing Director selects the language button, and then selects Esperanto on the language choice menu. The Marketing Director is one of the two million people who speak Esperanto, is a big William Shatner fan, and just purchased the newly released Incubus DVD. After choosing a language, the main menu appears in Esperanto. (See Figure 19-4.)

Figure 19-4: The Vortecs Corporation advertising DVD, in Esperanto, with the same urban and rural video segment choices, which now lead to video content with audio in Esperanto.

The Default Approach

When talking about using multiple languages, the DVD Studio Pro manual is assuming that you will be setting up a disc where the ability to choose a particular language is reliant upon the person's DVD player. In other words, if it's a bilingual English and Spanish disc and the person sets the player to Spanish, the Spanish menus and audio will come up. If they can't set the player to Spanish, there's no way for them to see the Spanish menu screens, but they should be able to click on the Subtitle button or Audio button to select any available Spanish audio or subtitle streams.

Apple doesn't use the term Default Approach (and probably no one else does), but thinking of using languages in terms of approaches can clarify what needs to be done within DVD Studio Pro to achieve the desired effect. The Default Approach is simply using languages in DVD Studio Pro in the way Apple describes in the manual.

Managing Multiple Languages in DVD Studio Pro

DVD Studio Pro makes it convenient to work with multiple languages when you are using the default approach. Essentially what you do is create the DVD project with your chosen primary language, such as English, and then add a new language to the project, such as Spanish.

You have the option of creating a separate Photoshop file for a menu appearing in your chosen additional language, as long as there are the same number of layers as the Photoshop file corresponding to the primary language. The layers have to be the named the same, and need to be in the same order. You may want to create a separate audio track and subtitles in another language as well.

Adding a new language to a DVD project causes new options to appear in certain places in the Property Inspector, most notably for menus. (See Figure 19-5.)

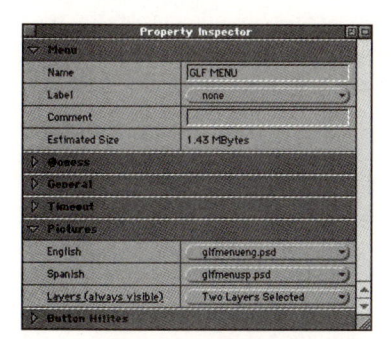

Figure 19-5: The Property Inspector with a GLF Menu tile selected.

> **Note**
> In the Pictures section of the Property Inspector normally you only have one line to choose the Photoshop file you are using for the menu. After adding the new language (Spanish), a new line appears (see Figure 19-5), allowing you to choose the Photoshop file which will be associated with the menu when the DVD player is set to the Spanish language. The third line is "shared," so the choice you make of which layers are always visible applies to both languages, and thus the respective Photoshop files. So in the context of using the default approach with multiple languages, a single Layers (always visible) setting applies to multiple Photoshop files. This is why additional Photoshop files for additional languages must have the same number of layers, with the layers named the same, and in the same order, as the Photoshop file for the primary language.

When you develop a multiple language project, it's nice to have all of the assets prepared beforehand, including any multiple audio streams for dialogue in a different language, subtitles, and graphics. But you can also go back into a previously developed project, identify what you need to make a different version of, and then import the new assets after they've been created.

Say you just want subtitles or audio in a foreign language available to the user without getting into separate versions of a menu or having the foreign language material come up automatically based on a person's DVD player. You can just add the subtitle and audio streams to an existing project. In this kind of scenario, the DVD player isn't set to a particular language and there is no Language key on a DVD remote control. A person can click on the Audio or Subtitle button as they are watching a video clip to "cycle through" the additional language choices.

> **Note**
> In DVD Studio Pro, when you associate an individual subtitle with a video track, it is always going to display with that track. When you have more than one subtitle, a user can switch between the subtitles. If you want to display a video segment where users can turn subtitles on and off, you can create a dummy subtitle (with nothing in it) that is accessed when the video track plays so that the user doesn't see any subtitles. However when the user presses the Subtitle key on the remote control, the actual subtitle is displayed. In this way, the Subtitle key has the effect of switching subtitles on and off. This technique is used later in this chapter.

Making a Multiple Language DVD Project

This section shows you how to add an additional language to a DVD project, resulting in a DVD with the "default approach."

To help you get acquainted with working with languages in DVD Studio Pro, try doing the following project with resources from the DVD-ROM, located in the CH19 folder in the Tutorial section of the disc. Not only does it involve a different language, it actually involves another species!

Adding a new language

Initially, the way you add a new language is through the Project View window in the DVD Studio Pro workspace. You may have already done several DVD projects and have never had to access the Languages tab in the Project View window. If you go back and look at the project, you'll probably find an Untitled language asset there; it can be named whatever you want. (See Figure 19-6.) If you click on it, the Property Inspector will show the actual language that is associated with the language asset, one of the 139 available in DVD Studio Pro.

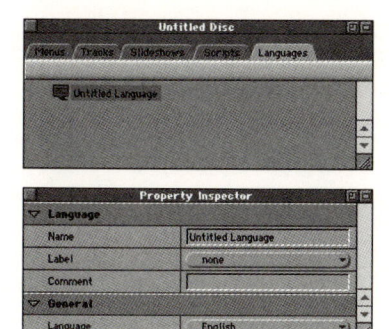

Figure 19-6: The Project View with Languages tab selected, and the respective contents of the Property Inspector, showing the Language drop-down menu.

The item you click in the Languages tab in the Project View is called a *language,* and the choices on the drop-down menu in the Property Inspector are also called *languages.* If this is confusing, think of the item in the Project View as an asset. Like other assets in a DVD Studio Pro project, the asset itself can be renamed, and you can change what it is associated with. A graphic asset in the Assets Container can be named examplefile.psd and be linked to examplefile.psd, or the asset can be renamed to Menu File and still link to examplefile.psd. Similarly, a "language asset" can be named anything you like, and you can switch the language it is associated with to any one of the 139 languages. (See Figure 19-7.)

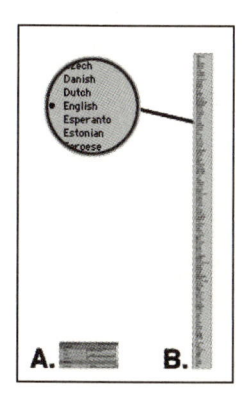

Figure 19-7: The relative height of the Languages section in the Property Inspector (A), next to the height of the Languages drop-down menu (B). This menu could get in the *Guinness Book of World Records* for being one of the longest drop-down menus in recorded history.

To get the ball rolling with the example project, go ahead and copy the CH19 folder over to your hard drive from the Tutorial section of the DVD-ROM, and from the File➪Open function in DVD Studio Pro, open up the GLF generic DVD project.

Note Project files copied over from the DVD-ROM should open up fine, but if you get a message asking you if DVD Studio Pro should try to relink any Photoshop layers, click on the Re-link button in the dialog box. The Log window will appear, indicating whether the Photoshop layers were successfully relinked. If the layers were not successfully relinked, review Chapter 3 and choose Item➪Asset Files to relink the assets in the project.

The project you are going to add a new language to is a simple DVD, which has a main menu, two music videos linked from the main menu, and four additional screens whose only function is to display information. When the file loads up, you should see the five menu tiles and the two track tiles in the Graphical View window. (See Figure 19-8.)

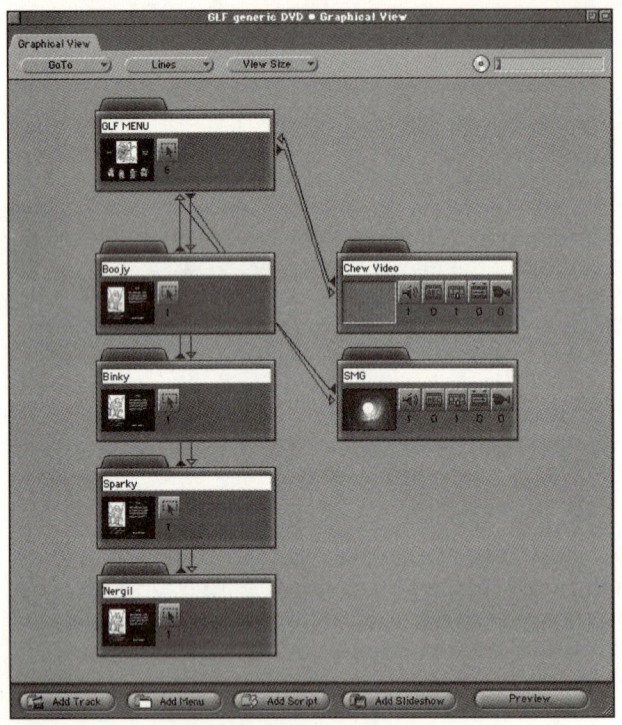

Figure 19-8: The GLF Generic DVD, with the Lines drop-down menu set to the Always position, making the connections between the items visible.

To add the new language, follow these steps:

1. **Click on the Languages tab in the Project View window.** You should see an Untitled Language asset. (See Figure 19-9.)

Figure 19-9: The Project View window with Languages tab selected.

2. **Click on the Untitled Language asset, and rename by typing** English **in the Property Inspector.** Check to make sure that the Language drop-down menu is set to English.

3. **Click on the Languages tab in the Project View window again to make sure the Languages tab is selected, and choose Item ⇨ New Language to add a new language asset to the Project View window.** Select the new language asset and name it Spanish in the Property Inspector, and select Spanish from the Language drop-down menu. (See Figure 19-10.)

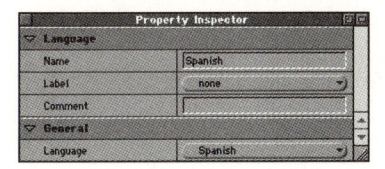

Figure 19-10: The Property Inspector showing the new language asset.

4. **Now that you have the new language added to the project, save the project!** Always a good idea. Especially if there's a scheduled power outage coming that you forgot about!

You are ready to import the previously prepared Spanish-language assets to the DVD.

Note: As soon as you add the new language, you won't be able to preview individual menu tiles until new assets are assigned in the additional language version of those tiles. When you add a new language to a project, nothing will look different, and the text in the title area of any menu tiles initially stays the same, but if you save the project file and re-open it, you will notice that the title text becomes italicized, and indication that an asset is needed.

To import the assets, follow these steps:

1. **Choose File ⇨ Import, and locate the Spanish assets folder, which is in the assets folder in the CH19/GLF ML DVD folder in the Tutorial section of the DVD.** There are two Photoshop files there — one corresponding to the main menu, and one for the submenus. (See Figure 19-11.)

2. **In the Import Assets dialog box, select each file and click the Add button.** Then click the Import button.

3. **Choose File ⇨ Import.** Locate the Subtitle Files folder, which is also located in the Assets folder. Import the smgsp.spu subtitle file. This is a Spanish subtitle file, destined to be used with the SMG video track. It was prepared using the Subtitle Editor application that comes with DVD Studio Pro. More information on subtitles can be found in Chapter 7.

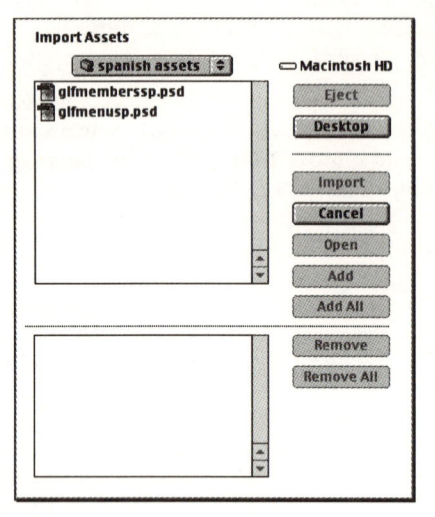

Figure 19-11: The Import Assets dialog box showing additional language versions of the Photoshop files, ready to be associated with the new menu tiles that DVD Studio Pro automatically creates when the new language is added.

You have now imported all the necessary assets to develop the Spanish version of the DVD.

Assigning assets for the new language

When you are using the default approach for developing additional languages for a DVD project, you don't need to create any additional track or menu tiles. This is automatically done for you when you add the new language asset in the Project View.

Using the Language menu

You may notice that after adding a new language, a new Program Menu choice — Language — appears at the top of the screen in between the Windows and Buttons menus. Clicking on the Language menu allows you to switch languages. At first, you won't notice anything changing, until you start assigning assets. Once you have things set up, the Languages menu at the top of the screen allows you to easily switch between language versions of your DVD; it is useful for previewing the additional language versions of a menu, for example.

If the concept seems confusing, think of the Language program menu as an elevator. In this project, there are two floors, English and Spanish. Each floor has the same basic layout, but when the Spanish floor is customized, you will need to "ride the elevator" to see the Spanish floor.

Assigning the graphic assets

It's easy to assign the new assets to the project. As mentioned earlier in this chapter, one of the effects of adding a new language to a project is that the available choices in the Property Inspector change for certain items, such as menu and track tiles. You then assign the newly imported assets to the additional language version of a particular menu or track.

First you need to set up the main menu.

> **Note** If you speak Spanish, please forgive the translations in this sample project; they were done using the Translation feature of http://babelfish.altavista.com, intended as a sample only.

To set up the main menu, follow these steps:

1. **Click on the GLF Menu tile in the Graphical View window. In the Property Inspector, notice how both of the languages you named now appear in the Pictures area (see Figure 19-12).** Initially the asset for Spanish will not be set.

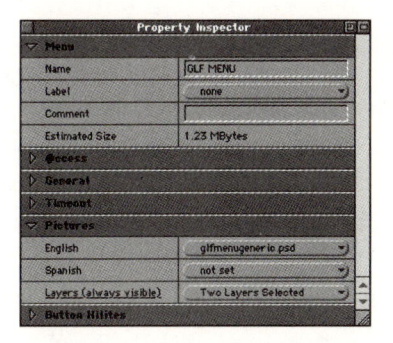

Figure 19-12: The Property Inspector, showing the Pictures section, with English and Spanish areas.

2. **Click on the not set drop-down menu and select the glfmenusp.psd asset that you imported earlier (in the Adding a new language section).**

The layer selection for the Spanish version of the Main Menu is the same as the English version; they are tied together. This is why the alternate language versions of a Photoshop file need to be arranged the same way.

If you like, to preview the Spanish version of the Main Menu, choose Language⇨ Spanish, click on the GLF Menu tile and click the Preview button in the Graphical View window. You will see the Spanish version of the menu. (See Figure 19-13.) Try switching back and forth between the English and Spanish versions, by exiting the Preview area, selecting the Menu tile, and using the Language program menu.

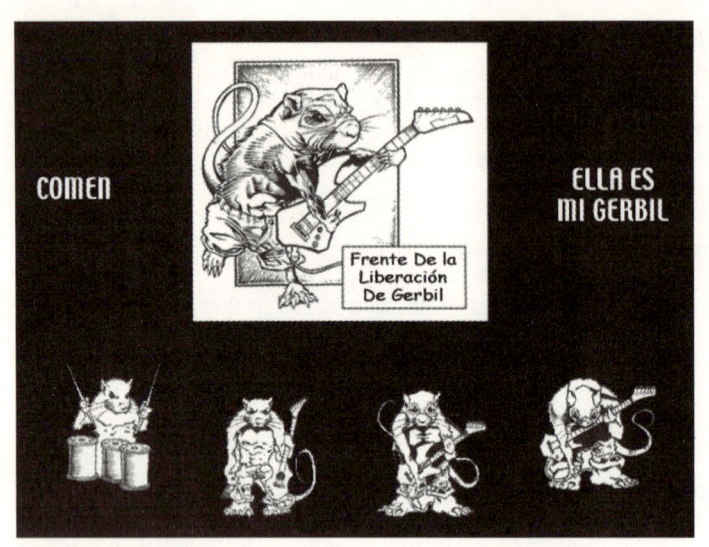

Figure 19-13: The Spanish version of the GLF Menu tile.

Now you need to set up the additional menus. Each additional menu (Boojy, Binky, Sparky, and Nergil) all share the same Photoshop file. As with the Main Menu, the layer selection for the Spanish version stays the same as the English version.

To set up the additional menus, follow these steps:

1. **Click on the Boojy Menu tile in the Graphical View window.** In the Property Inspector, set the Spanish picture to glfmemberssp.psd.

2. **Repeat the Picture asset selection for the Binky, Sparky, and Nergil tiles, setting them all to glfmemberssp.psd.**

Now that the menus are all set up, the final element in this project is to associate the Spanish subtitle file that you imported earlier with the SMG video track.

Associating the subtitle file

The subtitle was created with the Subtitle Editor, using a project movie, which was a reduced-size QuickTime movie of the original video. The Subtitle Editor application has project files, just like DVD Studio Pro does, ending in a .PRJ file extension. When you "compile" the subtitle, it generates a file with the .SPU file extension; this is the one you import into DVD Studio Pro, and associate with the video track containing the MPEG-2 video file.

To associate the subtitle file, follow these steps:

1. **Click on the Asset View window and scroll to where you can see the smgsp.spu subtitle file that was imported earlier.**

2. **Click on the smgsp.spu asset, and drag it onto the SMG Track tile in the Graphical View window.** To verify that it worked, click on the Subtitle button in the track tile (the one with ABC on it). This will open up the Subtitle Streams window, which should look like Figure 19-14.

Figure 19-14: The Subtitle Streams window, activated by clicking the Subtitle button on the track tile, with the smgsp.spu asset selected. If you like you can open this window first and then drag a subtitle asset onto it directly from the Asset View window.

3. **Select the smgsp.spu subtitle asset.** In the Property Inspector, set the Language to Spanish. (See Figure 19-15.)

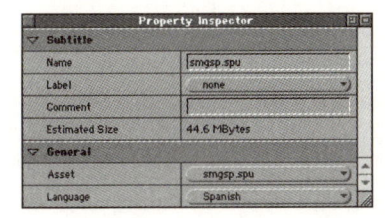

Figure 19-15: The Property Inspector, after the smgsp.spu subtitle asset is selected in the Subtitle Streams window. The Language has been set to Spanish. I know it is tempting to set it to Bashkir or Shona, but don't do it, it will only confuse the computer.

You have now completed the main tasks for adding an additional language to a DVD Studio Project.

Adding additional audio streams

You can add additional audio streams in another language in much the same way as you did the subtitle file: by importing them, clicking on the audio streams button in the respective track tile, selecting the audio file, and setting the particular language you want in the Property Inspector. If you work with more than one audio stream, be sure to set all of the audio streams to a particular language, to keep things straight.

Considering a dummy subtitle

What "should" happen when the disc is inserted in a DVD player where the language has been set to Spanish is that the Spanish versions of the menus will display, and when the SMG track is played, the Spanish subtitle will run beneath the video. Murphy's Law says that the Spanish subtitle may even run underneath the video when the disc is inserted in a player where the language is set to English. Because of this potentiality, you may want to consider making a "dummy" subtitle.

Another reason to make a dummy subtitle is to make certain that the user can switch a subtitle on and off with the subtitle key on their player. A dummy subtitle can be used with any of the different approaches of implementing multiple languages.

The approach for making a dummy subtitle is to open the Subtitle Project file for a subtitle that has already been created and where the Subtitle Project file is still associated with a Project Movie.

> **Cross-Reference**
> There is more information on subtitles in Chapter 7.

Save a copy of the Subtitle Project file. Then you can basically delete all the subtitles in the newly created file, leave one blank subtitle, and save it.

Adding a dummy subtitle

If you don't have it open, open up the GLF generic DVD project file that you just completed (or you can open up the finished version in the Finished Files folder in the CH19/GLF ML DVD folder).

To add the dummy subtitle, follow these steps:

1. **Choose File ⇨ Import, and import the dummy.spu subtitle file, located in the GLF ML DVD/assets/subtitle files folder.**
2. **In the Graphical View window, click the Subtitle Streams button (the button with ABC on it) in the SMG Video Track tile.** Click and drag the dummy.spu asset from the Asset View window into the open Subtitle Streams window.
3. **In the Subtitle Streams window, click and drag the dummy.spu subtitle so that it appears first in the list.** The technique is to click on the filename or icon, and drag it slowly toward the top of the window, until a horizontal line appears above the file you want to be in "second place," which in this case is smgsp.spu subtitle. When you are done, the Subtitle Streams window should look something like Figure 19-16.

Figure 19-16: The Subtitle Streams window for the SMG Track tile, showing the dummy.spu subtitle before the smgsp.spu file in the list. The order here from top to bottom determines which subtitle will run first with the track.

Now when you preview the SMG Track tile, the subtitles will only run when you click on the simulated Subtitle button in the Preview area. Clicking the Subtitle button once again will turn them off.

Having described this technique, you could simply have one menu choice lead to a track with the subtitles, and a different menu choice lead to a different track with no subtitles, where both tracks use the same video and audio. Some may prefer this approach; it's just a question of whether you would like people to be able to turn the subtitles off.

Making an Alternate Reality Multiple Language DVD Project

Now that the built-in language feature of DVD Studio has been covered, these features will be set aside so you can look at other options.

In the alternative reality approach, you manually create the multiple language versions of menu and track tiles, using the Duplicate function and then using the additional language versions of the graphic assets to customize the manually duplicated tiles. Unlike the default approach, you see both language versions at the same time in the Graphical View window. Furthermore, you provide a menu item on the screen so that users can switch back and forth between the languages. (In a case where you have only two languages, this will work; with more than three, you probably want a separate screen as a "jumping off point" for choosing which language you want.)

The phrase "alternate reality" refers to the user experience in which the primary language appears when the disc is inserted in the DVD player but then a menu enables the user to switch languages. When the language switch button is selected, the same menu appears, but now it is in a different language. This is in contrast to the default approach, where the user's player determines which language comes up, and there is no menu choice in your DVD project that allows them to switch languages.

> **Note:** If you run into trouble as you go through the turorial, you can always check the finished project file, which is called GLF Alternate DVD, located in the CH19 folder in the GLF ML DVD/finished projects folder.

To create the alternate reality version of the GLF DVD, follow these steps:

1. **Open up the GLF Generic DVD project file, located in the CH19/GLF ML DVD folder in the Tutorial section of the DVD-ROM.** If you went through the previous Tutorial, you may want to copy over a fresh copy of the project file or the whole CH19 folder to your hard drive. There will be five menu tiles and two track tiles in the Graphical View window.

2. **Click toward the bottom of the Graphical View window on the background, hold the mouse button down, and drag up and to the left, until all the tiles are selected.** Then choose Edit ⇨ Duplicate to make copies of the tiles. You could also select each tile and use the Duplicate function individually. The Graphical View window might look a little messy to begin with, especially if you have the Lines drop-down menu set to Always, but you can re-arrange the tiles so the Graphical View window looks something like Figure 19-17. (When you duplicate a series of tiles at once, they will remain selected until you click on one of them or on the background of the Graphical View window. So if you want to move them individually, click on the background first.)

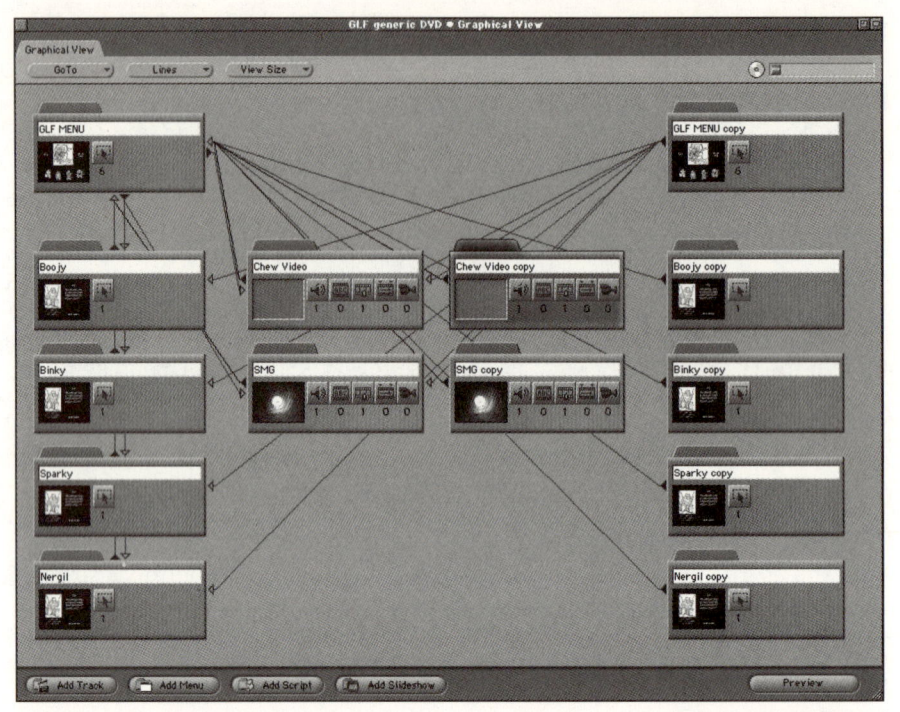

Figure 19-17: The Graphical View window, showing the original tiles on the left and the copied tiles on the right with the Lines function set to Always, showing the interactive paths.

3. With each of the tiles on the right side of the window, change their names, replacing the word *copy* in each case with *Spanish*, so it is Chew Video Spanish, and so on.
4. **Now that you have a GLF Menu Spanish Menu tile, click on each of the two Spanish track tiles, and in the Property Inspector, set the Jump When Finished value to GLF Menu Spanish.** If you have the Lines function set to Always when you select this function, you will see that the lines from these track tiles now lead to the GLF Menu Spanish Menu tile.

Now that you have set up the tracks, it's time to import the Spanish assets and associate them with your newly created tiles. Unlike the default approach, with this alternate reality approach, you could have the additional language versions of the menus in the same Photoshop file, but we will just use the same files as before.

Setting up the Spanish assets

To set up the Spanish assets, follow these steps:

1. **Choose File ➪ Import, and in the CH19 folder, locate the GLF ML DVD/assets/alternate folder, and import the glfmenu.psd file.** Then import glfmemberssp.psd from the GLF ML DVD/assets/Spanish assets folder, and import the smgsp.spu subtitle file from the GLF ML DVD/assets/subtitle files folder.

2. **Click on the GLF Menu Spanish tile. In the Pictures area of the Property Inspector, reset the Asset from glfmenugeneric.psd to glfmenu.psd, and set the Picture Layers (always visible) value to include the Foreground, Background and Spanish layers.** This changes the Menu tile from the English version of the menu to the Spanish version. If you like, click on the GLF Menu Spanish tile and preview it. You will notice that the buttons all still work but that they need to be reset to lead to the newly created Spanish tiles, which you will do in just a moment.

3. **Click on each of the "band member" Menu tiles and reset the Picture Asset from glfmemberseng.psd to glfmemberssp.psd.** You will notice that the appropriate layers automatically display. This is one reason why having a separate Photoshop file for the alternate reality approach may be nice: When you are assigning assets for the duplicated tiles, the layers will be more likely to automatically come up correctly. (If you were using one Photoshop file for both languages, you would need to manually go through each tile, and set the appropriate layers to be visible.)

Now that you have set up the asset files, you are ready to redirect the button "jumps" in the Spanish language versions of the menus, so that the main Spanish menu will lead to the right tiles, and the "member menus" will lead back to the main Spanish menu.

Redirecting buttons in duplicated tiles

This process is fairly easy. You need to go into each Menu tile and open up the Menu Editor (by double-clicking on the thumbnail area of the individual tiles). You could also take this opportunity to investigate the Jump Matrix, which is mentioned in Chapter 2. The Jump Matrix comes in handy in situations like this, where you need to make adjustments to a number of individual buttons.

Redirecting buttons with the Menu Editor

Start with the GLF Menu Spanish Menu tile. Double-click on the thumbnail area of the tile to open up the Menu Editor. You will be looking at the Spanish version of the Main Menu. (See Figure 19-18.)

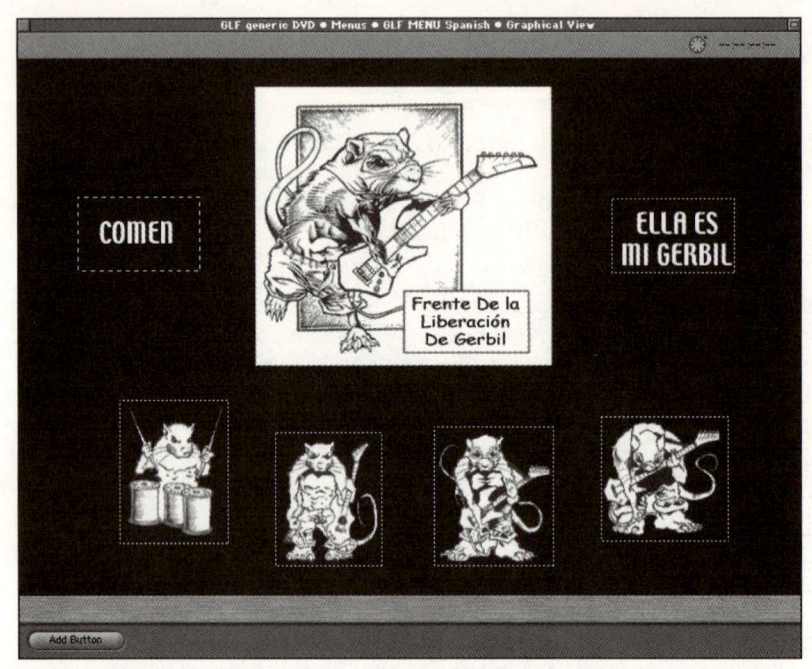

Figure 19-18: The Menu Editor, activated by double-clicking on the thumbnail area of the GLF Menu Spanish Menu tile. Each of the buttons in this duplicated tile still retains the values that point them to the English tile. Only the Jump When Activated setting needs to be changed; the Button Link settings, which determine what happens when the arrow keys are pressed on the remote, can stay the same for now.

> **Note** Throughout this process, if the linking gets confusing, you can go back to the English version of the tile on the other side of the screen, and see where the corresponding button is pointing to (by clicking in the rectangular region of the button in question in the Menu Editor and looking at the Jump When Activated value, in the Property Inspector).

Each of the buttons on this Spanish version of the menu will need to be redirected to the Spanish language version of the tile you duplicated in earlier steps. This is where naming the new tiles the same as the old ones but with "Spanish" added can help to keep things straight.

To redirect the buttons, follow these steps:

1. **In the Menu Editor for the GLF Menu Spanish Menu tile, select the Chew Video button, which appears over the word "Comen" ("Comen" corresponds

Chapter 19 ✦ **Working with Multiple Languages** 435

to the English word chew), and in the Property Inspector, set the Jump when activated value to Chew Video Spanish (see Figure 19-19).

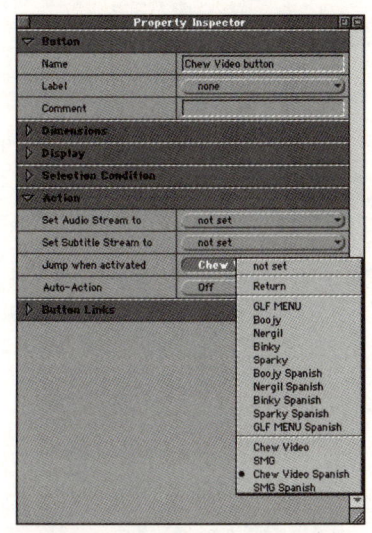

Figure 19-19: The Property Inspector, with the Chew Video button selected in the Menu Editor (the one that appears over "Comen"). The Jump when activated value is now set to lead to the Chew Video Spanish track tile.

2. Set the SMG Video button (Ella Es Mi Gerbil) to lead to SMG Spanish, and set each "band member" button to go to its Spanish equivalent.

3. Close the Menu Editor by clicking in the square at the very top left part of the Menu Editor window.

4. Open up each of the "band member" menu tiles, and set the Jump When Activated value for the Back to GLF Menu button ("Vaya Detras al Menu") to **GLF Menu Spanish.** If you are using the Lines function as you make these adjustments, you will notice the lines now lead back to GLF Menu Spanish.

At this point you have done just about everything required to make the project multilingual. The Graphical View window (with Lines function set to Always) should now look something like Figure 19-20.

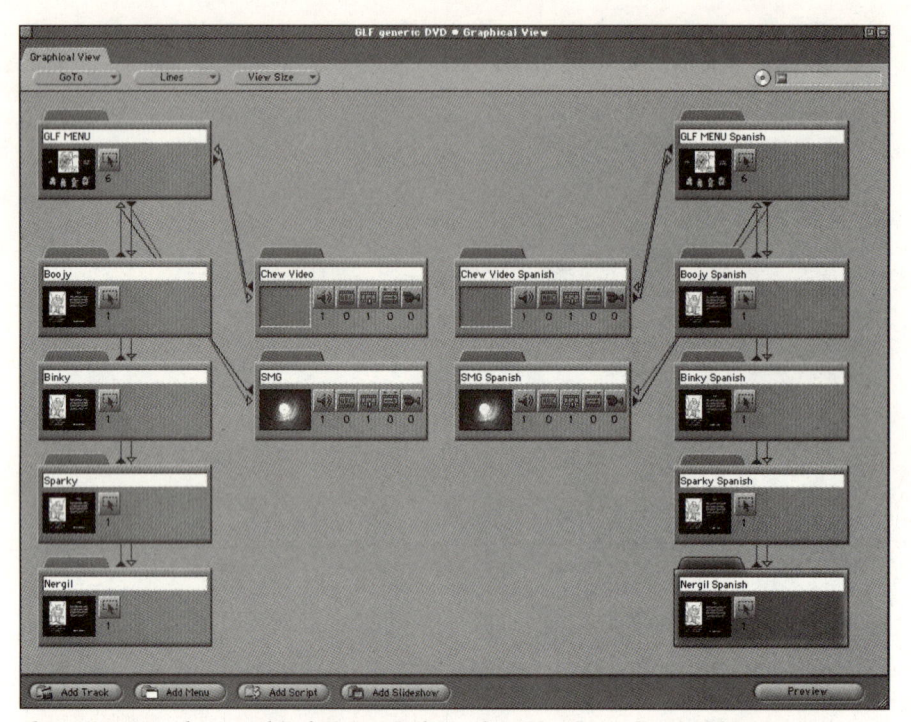

Figure 19-20: The Graphical View window, showing the redirected buttons.

Making the language bridge

The two separate language versions of the project are created, and you could click on either the GLF Menu or GLF Menu Spanish tiles to preview the entire project. Based on the experience you gained from the last section, feel free to add a Spanish subtitle to the SMG Spanish video track or even add a dummy subtitle.

But as for the disc's general properties, the preview will start off with the English version, since the startup action is currently set to GLF MENU. So at present there is no way for the user to select the Spanish version, or switch between the two.

So you need to go into the GLF Menu and GLF Menu Spanish Menu tiles, and activate the layers in the Photoshop files for a "switch" button, create a new button in each tile, and direct it to the other.

To set up the bridge, follow these steps:

1. Click on the GLF Menu tile, and in the Pictures area of the Property Inspector, reset the Asset from glfmenugeneric.psd to glfmenu.psd, and set the Layers (Always Visible) value to include the Foreground, Background, and English layers.

2. **Double-click on the thumbnail area of the GLF Menu tile to open up the Menu Editor.** Create a button for the bridge by clicking the upper left of the phrase Change Language and dragging down and to the right, to make a rectangular area. (See Figure 19-21.) (If you don't get it quite right, you can click on the edge of a button and resize, or you can click in the button and move it around a bit.) In the Property Inspector, name the button Change to Spanish.

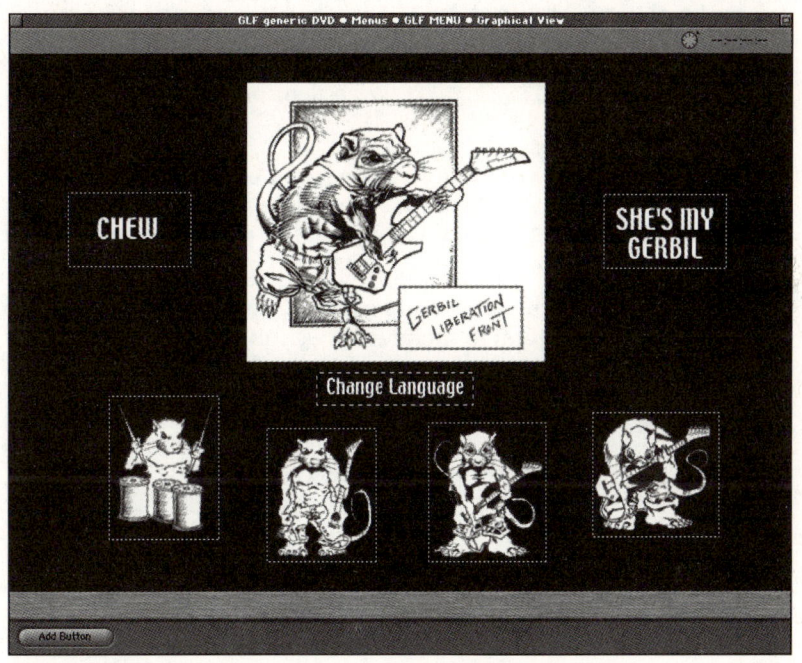

Figure 19-21: The Menu Editor for the GLF Menu tile, with the new Change to Spanish button (identified as Change Language).

3. **With the Change to Spanish button selected, set the Jump when activated value to GLF Menu Spanish in the Property Inspector.** Then adjust what the arrow will do on the remote control with this new button by adjusting the values in the Button Links section of the Property Inspector as follows:

   - **Up:** Not Set
   - **Down:** Nergil
   - **Left:** Chew Video Button
   - **Right:** SMG Video Button

   See Figure 19-22 for details.

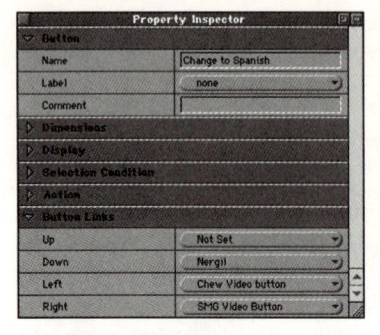

Figure 19-22: The Property Inspector, featuring the Button Links section, with the Change to Spanish button selected in the Menu Editor. The Button Links values are set for what buttons the users jump to when they press the various arrow keys on the remote control. You can set them any way they make sense; it's usually a good idea when setting the Button Links values to preview the menu and experiment, pretending you're the user, to see if the arrow key actions make sense.

4. **Click on the Nergil and Binky buttons, and adjust their Up Button Link values to Change to Spanish, so that when you are on the Nergil and Binky buttons, you can get to the new button that was just created.** You can now exit the Menu Editor.

5. **Select the GLF Menu Spanish tile in the Graphical View window, and set the Layers (Always Visible) to include Foreground, Spanish, and Background layers.**

6. **Open up the Menu Editor for the GLF Menu Spanish tile and create a button over the phrase Cambie El Linguaje, name it Change to English, and set the Jump when activated value to GLF Menu.** Remember to set the Button Links values for this new button, the Nergil and Binky "Up" values, and the Default Button value, as you did for the English version of this menu.

Whew! You're done! When you preview the correctly-prepared project, you can enjoy the English menus or switch to the Spanish versions, check out the SMG video with Spanish subtitles, get to know the band members, and start thinking about making your own multilingual project.

Thinking about multilingual workflow

We basically followed these steps to create this multilingual project:

1. **Make a copy of the original Photoshop file.**

2. **Copy and paste the editable text directly from the Photoshop layers into the Babelfish Web page at** http://babelfish.altavista.com.

3. **Translate the English text.**

4. **Paste the Spanish text back into the Photoshop document and make adjustments.** There aren't Spanish remixes of the GLF songs as of yet, but incorporating multiple audio tracks would have been an option as well.

For the subtitles, we followed a similar procedure: First we made an English version of the subtitle file and then saved the project file (using the Save As command) so that the in and out points of each individual subtitle line were preserved. Then we tried a couple different techniques, such as copying the text out of each subtitle line and adding it to a SimpleText document and then copying that text into the Babelfish Web page. Then we translated and pasted the Spanish text into the Subtitle Editor at the appropriate line.

(Microsoft) Word to your mother

(In the spirit of neo hip-hop multilingualism, Word to Your Mother is an affirmation/agreement. It can also be used as a greeting. Like, "Yo G, word to your mother!" Variations include "Word up!" or simply, "Word!")

Anyway, the Babelfish site is fun, but not necessarily reliable in all situations. But with any kind of translation, it might be helpful to have a two column Word document, where you put the original dialogue or lyrics in the left column, and have the translator put the translation in the right column. If you know you are doing subtitles, you may want to go ahead and do subtitles in the original language first, and then adjust the "line breaks" in the left column, to reflect how the subtitles display with each successive video segment.

For example, if you have a sentence where the subtitle starts on one screen and ends on another, you could go into the Word document so that each line accurately reflects a particular screen. For example, the movie is running, and the actor asks, "Where the heck is my orange juice? Did you take it, cousin Cletus?" So one subtitle is the first question, and the next subtitle is the second question. The Word Document might look something like Figure 19-23.

English	French
Where the heck is my orange juice?	L'estacade à claire-voie est mon jus d'orange??
Did you take it, cousin Cletus?	Vous le prenez, cousin Cletus??

OR

Where the heck is my orange juice?	L'estacade à claire-voie est mon jus d'orange??
Did you take it, cousin Cletus?	Vous le prenez, cousin Cletus??

Figure 19-23: A simple example of a way to work with line-by-line translations, which can be copied and pasted into the Subtitle Editor. The example is based on the Table function in Word. (Choose Table ⇨ Insert ⇨ Table and create two columns.) You could keep expanding one row with new lines, add additional rows, or add additional tables.

So the idea is, you set up subtitles in an original language, and make some kind of record of how the subtitles appear, screen by screen. The subtitles can then be translated line by line, so that it's easier to insert the translations back in the appropriate spot, keyed to the original video. Using the Save As command, you make a copy of the original subtitle file and then insert the new translation, line by line, lickety split!

Imagining additional fun

You may have noticed that there is an additional fun folder in the CH19 folder. It contains a chew subtitle files folder and an smg subtitles folder, which includes the subtitle project file and QuickTime project movie that was used to generate the subtitles. You might want to investigate the files and play around with the Subtitle Editor and add the English sing-along subtitles for the SMG video track.

But the chew subtitles folder contains a QuickTime project movie along with a Word document with English and Spanish versions of the lyrics to Chew (a living example of the two-column approach for translation and subtitling). Your mission, should you choose to accept it, is to make the Chew video multilingual, by generating both English and Spanish subtitles, and incorporating them into one of the GLF DVD project files. If you are working in an office setting, you might want to turn the sound down on your speakers; if your colleagues ask you what the heck you are doing, you can reply that you are learning how to generate and compile subtitle files for a multilingual DVD project.

Patting yourself on the back

Congratulations! You made it through the chapter!

(It's time to consider a lesson you may have already learned from *Saturday Night Live*, now that you are a multilingual-capable DVD author.)

Take a break, go look in a mirror, and say "I'm good enough, I'm smart enough, and doggonit, clients like me!"

Summary

This chapter wraps up the section in Advanced Interactivity, where you've been exposed to scripting, multiple angles, and multiple languages. All of these features are worth some experimentation, as they have the potential to bring a DVD project to a new level. Becoming familiar with multiple language features allows the DVD author to make a unique contribution to collaborative projects right from the beginning. Awareness of what you can do with a DVD can open up a new world of brainstorming possibilities, in terms of the way video is shot, the way audio is recorded, and the way graphics are prepared.

- ✦ A DVD project can include up to 8 separate audio tracks (each with 8 channels, for a total of 64 audio streams), and up to 32 separate subtitles for each project.
- ✦ If a DVD disc has been created with a language that matches one of the default languages in the DVD player, it is possible for a disc to automatically come up with the content in the right language, such as audio, subtitles, and menus.
- ✦ The lite approach for a multilingual DVD is simply to add a subtitle or additional audio stream to an existing project and give the user instructions on-screen or in documentation on how to activate the subtitle/audio stream.
- ✦ The direct approach for implementing a multilingual DVD bypasses the capability of a player to automatically choose the language. The choice of languages is incorporated into a menu screen on the DVD, which takes a person directly to the video track, with the appropriate audio and subtitles.
- ✦ The alternate reality approach involves ignoring the language detection capability of DVD players. The user chooses the additional language version of the project from a main menu or as part of an options subscreen, and is returned to the main menu. The main menu now appears in the new language, and leads to the additional language version of tracks and other menus.
- ✦ Unless the subtitle and audio keys are disabled, no matter how many different language versions you have and how they are implemented, a person will only be able to "cycle through" the various subtitle and audio streams that are associated with an individual track tile.
- ✦ To allow the user to turn a subtitle on or off, it may be necessary to create a dummy subtitle so that it can be the first active subtitle, giving the appearance that the actual subtitle is "off."
- ✦ Besides William Shatner, there are (allegedly) two million people who speak Esperanto. (See www.esperanto.org for more information.)
- ✦ In DVD Studio Pro, a language is really a language asset. As with other kinds of assets, you can rename the language asset to anything you like and switch the language that it is associated with.
- ✦ When you add a new language asset to a project, you won't be able to preview individual tiles until assets have been assigned.
- ✦ When a new language asset is added to a project and a language is chosen for the asset, the Language program menu becomes active, allowing you to access the additional-language versions of tiles in the Graphical View window. It works like an elevator, where each language is like a different floor, with the same layout.

- ✦ If you add multiple subtitles to a project, the subtitle that appears first in the Subtitle Streams window will appear first when the track is played.

- ✦ If you are developing additional language subtitles, it can be helpful to create a subtitle in the original language first, setting in and out points to get a sense of what appears in each scene. A copy of this subtitle file can be made using the Save As feature (Find ⇨ Save As), and when the translation is made, the translated lines can be matched up to the original lines and inserted in the right spot with the original in and out points.

The next section, Exploring DVD Content Delivery, discusses various methods for delivering DVD content once a project is created, includes coverage of hybrid DVD/DVD-ROM development, and introduces the concept of simultaneously developing content for delivery on DVD, CD-ROM, and the Internet. The section starts out with Chapter 20, including a review of copy protection methods and the phenomena of Regional Zone Coding.

✦ ✦ ✦

Exploring DVD Content Delivery

PART VI

In This Part

Chapter 20
Evaluating DVD Player Options

Chapter 21
Developing DVD-ROM Content

Chapter 22
Cross-Developing for DVD, CD-ROM, and the Internet

Evaluating DVD Player Options

CHAPTER 20

◆ ◆ ◆ ◆

In This Chapter

Discussing methods for copy protection

Examining Contents Scrambling System (CSS)

Examining Macrovision

Looking at Region Coding

Evaluating the Apple DVD Player

◆ ◆ ◆ ◆

A DVD project cannot be appreciated until someone puts it into a DVD player. Until then, it is nothing more than a shiny disc with lots of potential. In this chapter, we discuss the topics of copy protection and region coding, in order to give you an idea of some of the options you have when manufacturing a DVD project. In addition, there is a closer look at Apple's DVD player, an example of a software-based DVD player.

In the effort to maximize profits and reduce illegal duplication of copyrighted material, movie studios have come together to develop methods for making piracy a thing of the past. Through various copy protection schemes and region coding, DVD creators and manufacturers are able to greatly decrease the number of pirated videos and increase the potential for profits worldwide.

CSS and Macrovision are the methods currently preferred by studios to make the attempt of copying a disc difficult. Region coding ensures that discs, which are made available for sale in one area of the world, are not sold in markets ahead of release schedules or in areas where the price of discs is much lower. Both of these methods are made possible through the use of special digital coding, and are available for use in DVD Studio Pro.

Investigating Copy Protection

When the DVD platform was first developed, movie companies made certain that their investment would be properly protected. Without such assurances, many companies would never have agreed to release their material on a format that provides the potential for exact digital copies to be made. Every year, the entertainment industry loses millions of dollars to pirates who illegally copy and distribute their product. This has been a problem since video tape recorders were introduced and is even more of a problem now, in our digital age, where data (including video and music) can be re-recorded with incredible quality and transmitted across the world at the speed of light.

Some may argue that being able to make digital copies is not necessarily a bad thing — although the movie studios, record companies, and most artists would tend to disagree. With DVD, the policy was to develop copy protection schemes to prevent, or at least deter, would-be video pirates.

Debating the use of copy protection

Unfortunately, copy protection does have its problems. Opponents of these measures point to the conflicts caused by discs that alter or stretch the standards followed by disc makers and player manufacturers. Although copy-protected discs are solely intended for use by the person purchasing them, there are times when the discs themselves do not work in perfectly legal players. This leaves the consumer, who paid money to rightfully own the disc, debating the merits of copy protection. By going to great lengths to protect their products, companies can sometimes interfere with the consumer's ability to enjoy them. Still, this only occurs in very rare cases and discs usually work without any problems whatsoever.

As a consumer, it is your right to assume that a product that you have paid for should work in the system it was intended for. For example, most people would be surprised if they purchased a CD and found that they couldn't play the disc in one of their CD players. But DVD compatibility is not quite as good as CD, because of copy protection. Certain older set-top players may not support some newer copy protection schemes, and some computers with DVD-ROM drives may not support the coding either.

Few content delivery issues have caused as much protest as DVD copy protection methods like CSS or Macrovision. Litigation arose, pitting independent hackers and protesters against the money and influence of large corporations (like the case of DeCSS, an underground program that was designed to "rip" a disc without the CSS encryption).

Intellectual property rights and the fair use of products by a consumer are issues that may never be resolved to everyone's satisfaction. In the end, it remains evident that companies and many artists have no intention of backing down from protecting their works by every means necessary. If sufficient copy protection methods did not exist, the wait would have been even longer for movie studios to release their works on DVD. In addition, many jobs may have been lost to piracy, due to less income and demand for the production of studio titles.

If you intend to manufacture a DVD project, consider whether you are going to need copy protection. If you are distributing a commercial disc available for retail sale, you may want to protect your investment by using either CSS or Macrovision. If your disc is for corporate presentations or self-promotion, it may not require any special means to protect it. In either case, decide from the start because there are going to be several issues to contend with, especially when using DVD Studio Pro to create the project.

Contents Scrambling System

Contents Scrambling System (CSS) uses a form of encryption to prevent the copying of data from a disc. By using two different types of encryption keys (disc key and title key), it scrambles sections of data on the disc, producing a signal that can only be played back on a player with the proper decryption hardware. Fortunately, this decryption mechanism is a standard feature on set-top DVD players.

For the most part, encryption is done by a disc manufacturer who purchases a special license to obtain a set of keys that allows them to properly add CSS to a project. When authoring a disc, the producer can set "flags" (special bits of data that mark sectors on a disc) by selecting CSS protection for the project. The replication house reads these flags and activates the encryption process when writing the discs based on the information the flags provide.

Before attempting to produce a disc with CSS copy protection, you must make certain that the disc replicator you are working with is capable of adding CSS to disc. A simple phone call to the replication house beforehand can save you a lot of trouble. Also, it is very important to note that DVD-R General media, required by Apple's internal SuperDrive, Pioneers DVR-103, and other external models, does not allow for the copying or inclusion of disc projects with CSS.

DVD-R Authoring media, on the other hand, prewrites a section of the disc where CSS information is stored. By doing this, it is impossible to make direct copies of discs with CSS encryption, protecting movie studios from illicit copying. Unfortunately, this prewritten section also prevents DVD authors from using the general media drives for creating projects with CSS. If you want to create a DVD project with CSS copy protection, you must record your disc to DLT, removable hard drive, or other non-protected media before sending to the replicator. (See Figure 20-1.)

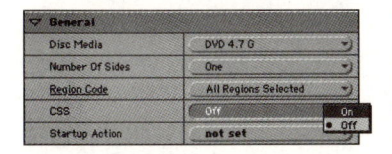

Figure 20-1: Enabling CSS for a DVD Studio Pro project does not work unless you have the right media, such as a DLT, for writing discs that are then enabled for copy protection at the replication house.

For more information on CSS, visit the Motion Picture Assocation of America (www.mpaa.org), or the DVD Copy Control Association (www.dvdcca.org). To investigate the alternative viewpoint, try typing in DeCSS in Yahoo! (www.yahoo.com).

Macrovision

Another form of copy protection that can be added to a disc at a manufacturing plant is called Macrovision. It is a process that prevents the direct copying of material from one device to another, such as from a DVD player to a VCR. On the way to a television or recording device, the signal first passes through a digital-to-analog conversion process. Macrovision copy protection modifies the video signal through the component that does the conversion, yielding copies that play back fine on a television or monitor but are distorted when captured by a recording device.

Macrovision has been around for a while and has been added to all sorts of video formats, from videocassettes to pay-per-view television. Due to its long proven history of copy protection, it is probably the most trusted option for many media producers today. The process is updated on a regular basis and has shown little sign of aging. In fact, with the arrival of DVD, Macrovision has gained even more popularity and widespread use. If you want to add Macrovision to your next DVD project, you should first contact the plant where you are planning to replicate your discs to make certain that they offer Macrovision and to acquire tips for preparing the master DLT or disc that you plan on replicating. Disc manufacturers who offer Macrovision are required to hold a license for the process and charge customers on a per unit basis for activating the capability.

Note For more information, check out www.macrovision.com.

In the next section, we discuss the usage of region coding as a method of copy protection.

Understanding Region Coding

Region coding consists of assigning a special code to a DVD disc that limits playback of a disc to a specified area of the world, by making it incompatible with any DVD player except one that is specifically for that region. The vast majority of DVD players are tied to a particular region code, but there are "region-free" players that can be obtained on the Internet for playing discs from any region.

DVD Studio Pro allows you to assign region coding to a project, as shown in Figure 20-2, but unless your project is intended for commercial release, this option may not be necessary. In fact, if you do not specifically require region coding then it might be good to ignore it. Otherwise, you are limiting areas of the world where your disc could be played.

Note If you are intending to set region codes for a disc project, you are required to add CSS if you want them to function properly. If you do not add CSS with your region codes, the player may not adhere to the coding that you have selected.

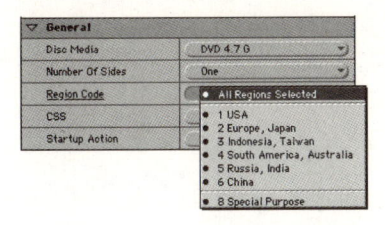

Figure 20-2: The DVD Studio Pro Property Inspector allows you to choose from several different regions for coding your disc project.

If you want to add region coding to a project in DVD Studio Pro, there are a couple ways to set it. The most direct method is to simply select the region that you want from the drop-down menu under the General section of the Property Inspector. Select a code (or codes) that corresponds to the country or region of the world that you want the disc to play in. For example, choose 2 if you want the disc to work in Europe and Japan or 5 if you want it to work in regions encompassing Russia and India. For most of your projects, you should choose the default All Regions Selected for your disc to work everywhere. In the Property Inspector, you can also click the Region Code link to open the window for selecting codes shown in Figure 20-3. And region codes can be set for your project's playback by selecting a code from the Preferences menu in DVD Studio Pro. (See Figure 20-4.)

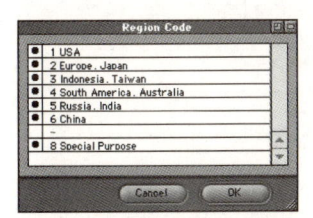

Figure 20-3: By clicking on the Region Code link in the Property Inspector, a separate window opens that allows you to select or unselect specific region codes.

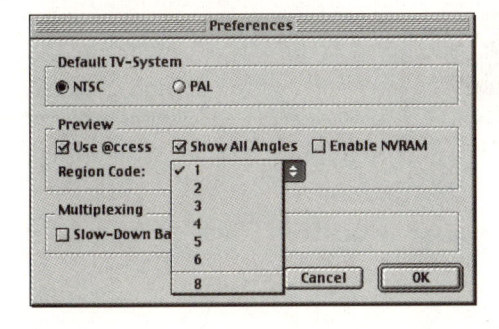

Figure 20-4: Region codes can be set for your project's playback by selecting a code from the Preferences menu in DVD Studio Pro.

In DVD Studio Pro, regions are designated with the following codes:

1. USA, Canada
2. Europe, Japan

3. Indonesia, Taiwan
4. South America, Australia
5. Russia, India
6. China
7. Special Purpose (only available for special players in airplanes)

> **Note** Remember, if you are setting a region code then you must also set the CSS copy protection to "On" or the coding will not work. And in order to use CSS you must have a suitable means for exporting the file, such as DLT, which allows for the inclusion of CSS information for the manufacturing process. The best approach for the development of a CSS or region-coded title is to start by consulting with the manufacturer to make sure you have a way to get the project to them properly.

> **Note** If you are using a Mac to play DVD videos, your DVD drive may prompt you to select a region code if you are attempting to play back a disc from a different region code other than the one already selected for your drive. While this may seem like a nice option to have, you are only allowed to change the region code a set number of times before the drive is permanently locked. Be careful that you do not accidentally lock your player for a region other than the one you use most often. Once you have locked the drive, there is no going back.

Justification for Region Coding

In order to protect their investments in other markets of the world, movie studios and video content makers and distributors decided that they needed to add region coding to their DVDs. By doing this, they ensure that someone isn't purchasing a disc from another country. The reasons for doing this are clear. Movies are released with different schedules and different content depending on where they are located geographically. One reason is due to the inability of studios to produce enough release prints to show movies simultaneously in every market around the world. By using region coding on discs, studios are able to limit which markets have access to the movies, limiting the accessibility to the home video in countries where the movie has not yet been released.

To appreciate the benefits of DVD you must have a player. Apple provides a free software player with its new operating systems.

Evaluating the Apple DVD Player

Any Macintosh computer with a drive capable of playing back DVDs can use the Apple DVD software player. (See Figure 20-5.)

The latest Mac operating systems come with a copy of the player already installed. In fact, if you meet the requirements to use DVD Studio Pro software, you must have it on your system. (If you don't have it for some reason, it can be installed with system software, or the latest version can be downloaded from www.apple.com.) DVD Studio Pro uses some of the player's functionality to preview projects before and after building the disc project.

Figure 20-5: The Apple DVD Player allows you to view DVD discs on your Mac.

Playback is accomplished through windows of varying sizes, and the on-screen remote control allows you to make changes to a movie during playback, similar to a set-top player. Other player preferences may be manually selected such as parental controls and language options.

> **Tip** When you insert a DVD disc into your drive, the Apple DVD Player may automatically start. To disable the automatic launching of the player, use the Extensions Manager (easily accessed through the Control Panel in the Apple menu) to turn off the DVD AutoLauncher extension.

Modifying Player Preferences

The Apple DVD Player allows you to specify many different options for the playback and function of a disc. By selecting Edit ⇨ Preferences a window with tabbed options appears. From this window you can make several choices that affect the playback of discs, as shown in Figure 20-6.

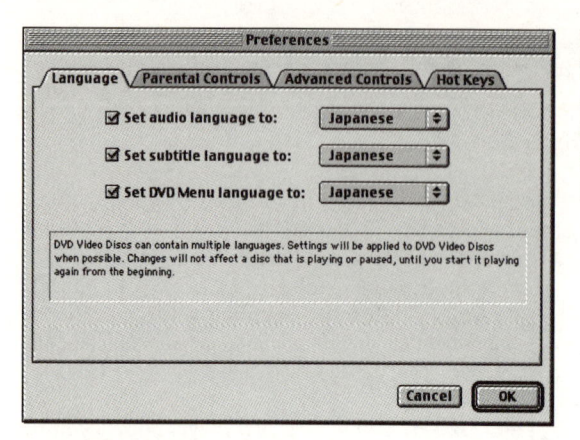

Figure 20-6: Preferences for the Apple DVD Player include choices for the language that is automatically selected for multiple audio, subtitle, and menu options.

The first set of choices is for languages, which is relevant if the disc contains multiple audio, language, or subtitle tracks. By checking the appropriate box and choosing a language from the drop-down list, the player knows which languages to play automatically. For instance, if you set the DVD menu language to Spanish, only Spanish menus appear for discs that contain them. If the disc contains Spanish menus, they are displayed when you start up the disc.

You may also set parental controls if you wish to limit the type of movies that children have access to. By clicking on the Parental Controls tab and selecting the Parental Control checkbox, you can select an MPAA rating (if the disc has one) that you want to lock from the drop-down menu. (See Figure 20-7.) After you have selected the rating that you do not want your children to see, click the Set Password button and type in a code. When you wish to unlock the player for viewing a "mature" disc, you need to come back to this menu, set the DVD parental lock to Off in the drop-down menu and enter your password when prompted.

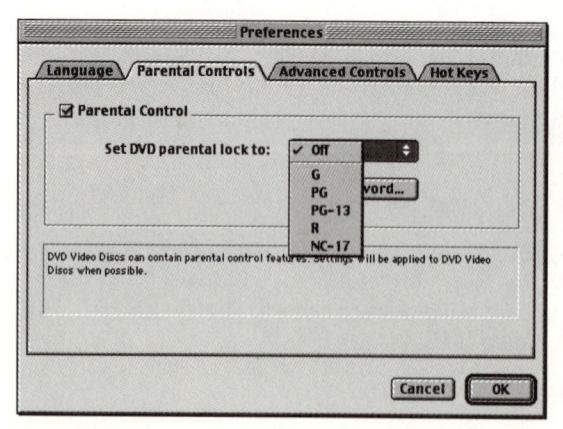

Figure 20-7: Preferences for the Apple DVD Player includes choices for parental controls, which limit what movies (based on the MPAA rating system) children are allowed to see.

Chapter 20 ✦ **Evaluating DVD Player Options** 453

The Advanced Controls tab offers you a few unique choices. From this menu, you can choose whether to activate Dolby Surround Pro Logic audio if your system supports it. In order to make the best use of Dolby, you need to have an external decoding device (many recent stereo systems have this capability) and a set of speakers. If you do not have special Dolby hardware, choose None from the menu.

If you have an external amplifier that requires line level input, select Enable Line Mode. Also, you can choose to enable DVD@ccess Web Links, which automatically launches your Internet browser when a special Web link is encountered on a disc. Figure 20-8 shows where DVD Studio Pro offers you the option of creating these types of links.

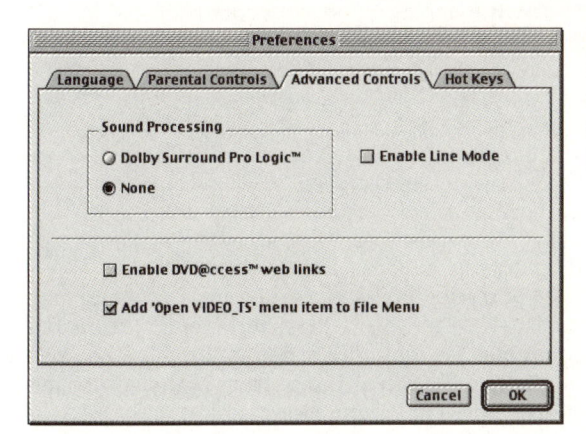

Figure 20-8: Preferences for the Apple DVD Player include choices for advanced controls, such as sound processing and DVD@ccess Web Links.

Hot keys allow you to quickly change the display properties. In the Hot Keys menu, you can choose to create hot keys for showing or hiding the Apple DVD Player's controller (remote control), as well as special keys for switching to Presentation mode. (See Figure 20-9.) You may want to hide the controller if it is obstructing your view of the window, or if it seems to be slowing down the playback of your video (this may happen at times, particularly if you drag the controller around the window a lot).

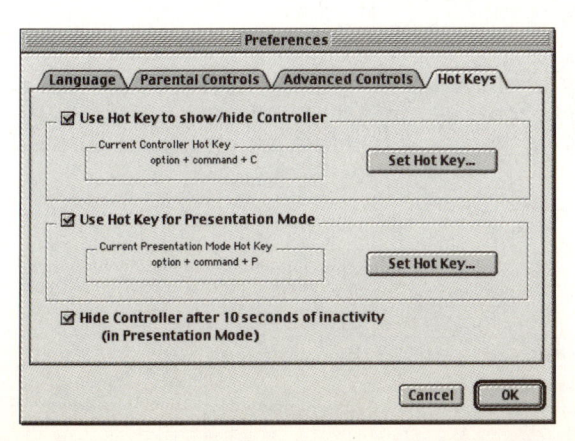

Figure 20-9: Preferences for the Apple DVD Player include choices for hot keys that can show or hide the controller.

Using the controller

The controller (remote control) for the Apple DVD Player, shown in Figure 20-10, is a convenient visual method for accessing disc features. Using the controller, you can accomplish functions similar to those offered by traditional set-top player remote controls. If it is not already showing, select Window ➪ Show Controller to display the controller. It can be dragged around the screen and placed in a more convenient location, especially if you plan on having it visible while playing a video. If you wish to hide the controller, select Window ➪ Hide Controller.

Figure 20-10: The Apple DVD Player's controller is a convenient device for selecting disc options.

Disc operations that are normally performed on a physical remote control can be performed by pressing the Forward, Reverse, Stop, Play, and Menu buttons to name a few. More complex functions, such as switching between multiple audio and subtitle tracks, can be performed by clicking the tab at the bottom of the controller to reveal an additional set of buttons, shown in Figure 20-11.

Figure 20-11: The Apple DVD Player has many control buttons for advanced functions, such as the selection of multiple language tracks, neatly hidden in the bottom of the controller.

The controller can also display information about your position in a disc, such as time elapsed and time remaining. You can change the information that is displayed by clicking the small clock symbol on the right side of the controller. (See Figure 20-12.)

Figure 20-12: By clicking on the small clock symbol to the right of the counter, you can access more time display options and view your location on the disc.

Note DVD discs cannot be navigated while in Pause mode. Make sure the disc is playing if you want to make changes to the subtitles, audio tracks, or other disc options, including forward and reverse functions.

Tip You can optimize DVD playback for the Apple software player by making a few simple adjustments to your system. First, in the Energy Saver Control Panel, set the system to Never Go to Sleep. This ensures that playback is not disrupted by your system's hard drive and display suddenly shutting down. Next, make certain that your Find by Content indexing is turned off by deselecting the relevant system extension. In addition, it is best if no other applications or windows are running when using the player for ordinary DVD playback. Checking these things should imporve overall performance of your Apple DVD Player.

You have gained some insight into the methods available for affecting the playback and usage of DVDs, and have gotten to know Apple's own DVD player better.

Summary

In this chapter, we learn what methods are available for limiting access to contents on a disc. For your reference, it may be helpful to summarize a few of the points we have discussed.

- ✦ Copy protection methods are intended to prevent the illegal duplication of commercial discs.

- ✦ The implementation and ongoing success of copy protection affects the willingness of movie studios and other companies to release their valuable projects on the DVD format.

- ✦ There are methods for circumventing copy protection, although most of them are deemed illegal by current legislation.

- ✦ CSS, or Contents Scrambling System, is a popular copy protection method that uses encryption to prevent the copying of disc content.
- ✦ Macrovision is another widely used copy protection method that prevents the successful copying of video by distorting the signal to the recording device.
- ✦ Region coding prevents the playback of DVD discs in areas of the world they were not intended for.
- ✦ Region coding allows movie studios to limit the availability and compatibility of discs around the world, thus protecting the value and timing of releases in certain markets.
- ✦ The Apple DVD Player is a software solution for the playback of DDS on your Macintosh computer.

In the next chapter, we jump to a different topic — authoring content for hybrid DVD/DVD-ROM discs. Discussion includes system requirements and the standards used for creating hybrid projects.

✦ ✦ ✦

Developing DVD-ROM Content

CHAPTER 21

In This Chapter

Understanding DVD-ROM

Preparing DVD-ROM content

Exploring DVD-ROM options

In this chapter, you will learn about the various ways you can include DVD-ROM content that people can access with their computer.

More and more commercially released DVD titles have some kind of DVD-ROM content, sometimes known as "special features," that people can experience when they put the DVD disc in a DVD-ROM drive on their computer. DVD-ROM content utilizes the advanced power of the desktop computer, so the interactive possibilities are far more advanced than what you have with a DVD player.

For movie-related DVDs, DVD-ROM content often includes things like games, animation, additional video footage, or commentary on the main DVD. The content is typically created with applications such as Macromedia Director or Flash, but you can put any kind of file you like, as long as it will run on a computer with a DVD-ROM drive.

In a nutshell, DVD-ROM possibilities are based on whatever you do with the space you have left over from your DVD project.

What is DVD-ROM?

Technically speaking, DVD-ROM content can consist of any file(s) that you can run on a computer. The ROM part of DVD-ROM stands for Read Only Memory, and thus it is the DVD equivalent of CD-ROM.

How CD-ROM influenced DVD-ROM

Ever since computers started including CD-ROM drives, people have been experiencing interactive content. Most people's first experience with CD-ROM discs probably consisted of installing a program when they got their computer. And for millions of people, their most memorable CD-ROM experience was probably with Broderbund/Cyan's megahit Myst, a sleep-depriving game that burst out of nowhere, featuring astonishingly realistic 3D art.

Aside from the millions of AOL CDs and other installation discs out there, CD-ROM content has typically been some form of interactive multimedia, a file that you can click on that opens a window and enables you to experience some combination of video, audio, animation, text, and graphics.

DVD-ROM and CD-ROM content often consists of some kind of *standalone application*, which means that you can run the file without having to have a particular program on your computer. In other words, instead of opening a file within a particular program, you click on the standalone application and run it directly.

CD-R discs

A CD-R is like a CD-ROM, but it starts out as blank, and it is *recordable*. Once you put data on a CD-R disc, it essentially becomes a CD-ROM disc. And when you write audio data to a CD-R disc, it becomes the equivalent of an audio CD, if your CD player is compatible. Nowadays, the vast majority of audio CD players play CD-R discs, but there may still be a few out there that have trouble.

Likewise, a DVD-R is a recordable disc format. If you burn a VIDEO_TS folder to it, it will play in most DVD players, but the compatibility is not as good as audio CDs created with CD-R media.

DVD-ROM Capabilities

The DVD format is based on a standard known as UDF Bridge, which combines the UDF and ISO 9660 formats. These formats were agreed upon in the CD-ROM and DVD/DVD-ROM industry as standard ways of writing data to the physical disc media.

Essentially, a DVD player can read the disc, and the only thing it knows how to do is look for the VIDEO_TS folder. A computer with a DVD-ROM drive can read the disc as well, and you can run files directly off the disc or copy them to your hard drive.

Standalone content

Pragmatically speaking, if you wish to include DVD-ROM content on a DVD, you need to author the content in a program other than DVD Studio Pro. DVD Studio Pro is designed for authoring DVD content exclusively. Applications that are typically used to generate CD-ROM/DVD-ROM content include Macromedia Director and Flash, both of which can generate standalone files.

Both Flash and Director can be used to create content with animation, interactivity, and Web links, where users can jump to the Web if they have an active Internet connection. (See Figure 21-1.)

Dynamic CD

The move from CD-ROM to DVD-ROM marks a major step forward in technology, but both formats are rather static compared with the Internet's ability to deliver content dynamically. That is until ad2 developed a new technology called *DynamicCD*. While originally developed for CD-ROM applications, the underlying programming code functions the same on DVD-ROMs. Here's how it works . . .

The DVD can be programmed to make a connection to the Internet without the use of a browser. In fact, the connection and the delivery of updated content can be completely transparent to the user. *DynamicCD* works behind the scenes by making the Internet connection, retrieving files from a server and storing them on the user's hard drive. The programming on the DVD causes the new content to be displayed within the DVD interface creating the impression that the information was on the DVD from the start. Depending on file sizes and the user's connection speed, the *DynamicCD* process can happen in just a few seconds. This is an important step toward converging the rich media experience of DVD-ROM and the immediacy of the Web. Does this represent the first example of a total media convergence? From a branding perspective, it does.

DVD-ROMs are a 'call-to-action', with DVDs you can find the audience rather than having them find you. The packaging and the face of the DVD can carry the 'brand,' whether it's a feature film or a corporate marketing message. The use of MPEG-2 video represents the delivery of broadcast quality video. The fact that it's a ROM means that it can deliver a rich, multi-media experience with audio, games and many other interactive components. Moreover with the *DynamicCD* technology, the last layer of convergence is present, which allows the immediacy of the Internet.

The *DynamicCD* technology means a lot more, too. It ensures the user stays connected with your brand. ROMs have long been a popular format because the perceived value means that people are reluctant to throw them away. By making a disc dynamic, some of the content may be new everyday, or updated as frequently as information can be changed on a server. This makes the disc 'sticky' and ensures it is close at hand; therefore, the brand is creating an impression even when it's out of the computer.

There are a variety of applications for *DynamicCD* technology. A few of these include: daily, weekly or monthly promotions; the latest news or product announcements; customer profiling in a less intimidating environment than on the Web; and gaming takes on a whole new dimension with *DynamicCD*. Other applications are waiting to be discovered but one thing is sure . . . it delivers on the promise of convergence.

Having a standalone file allows someone to experience multimedia without necessarily having to install anything; he can just click on a file. Exceptions to this may be if the overall presentation incorporates some kind of file that requires a program like QuickTime. For example, a Director presentation might include video files in QuickTime format, so users are either instructed to download the free QuickTime installer, or arrangements are made with Apple to include the QuickTime installer on the disc.

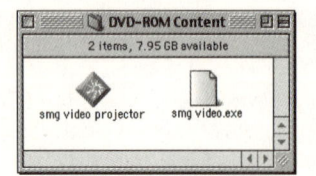

Figure 21-1: A simple example of standalone content, the Mac and Windows versions of a Flash animation, exported from Flash in the self-contained projector format, so that you can just double-click on the files and they run. The characteristic Mac Flash icon may be familiar. Notice how the Windows version has the `.exe` file extension (which stands for executable), the same as any other Windows program file.

Player-based content

There's no reason why you necessarily have to put standalone application files on the DVD-ROM. Depending on your audience, other potential formats for conveying content might include HTML documents, which can be used by anyone with a Web browser, and Acrobat (PDF) format, which is becoming increasingly popular for storing information. Both HTML and PDF are increasingly being used on CD/DVD-ROM discs for documentation and other kinds of additional information. For example, this entire book has been converted into PDF format and is available on the disc accompanying this book.

There is also the potential for a DVD author to include bare video or audio files on the disc in formats such as mp3, `.WAV`, `.AIFF`, QuickTime, `.AVI`, or even Web-based plug-in formats like RealVideo or Windows Media Player.

For dependent content, it all really depends on what "player" applications you expect your audience to have. (OK techies, give us some artistic license, think of HTML as "playing" in a browser.) Most people these days who have a DVD-ROM drive probably have the ability to play most of the common audio and video formats, whether they are using a Mac, Windows, or Unix/Linux operating system. The most typical use of a player application would be something like listening to a downloaded mp3 audio file, or watching a video stream on a Web site with the QuickTime player or RealPlayer.

Yes, you heard right. There are an increasing number of people using the Linux operating system who also have DVD-ROM drives on their computer that can read the universal ISO 9660 format, and they would probably appreciate having some dependent DVD-ROM content so they could pull out the Linux versions of the various media players or browsers.

Considering platform compatibility

When you include DVD-ROM content, you have to consider what kind of computer (platform) the audience will be using. Let's say the audience is small, that you're burning DVD-R discs, and including DVD-ROM content on the discs. If they all have Macs, chances are they can run whatever you can on your computer. But if the audience is large or you don't really know what kind of computers they have, if any, then you will have to enter into conversation with Bill Gates. That is, you will have to account for the possibility that someone may access your DVD-ROM content on a machine equipped with one of the variants of the Windows operating system.

Before you start going into shock or developing a slight twitch, consider that you don't necessarily have to get a Windows machine to generate Windows-compatible content. This would especially be the case with player-dependent files such as straight audio, graphic or video or PDF files, where you can play them on whatever computer you like, as long as you have the right player.

So don't be discouraged. (See Figure 21-2.) The following is an excerpt from the May 15, 2001 section of a news thread at `www.macintouch.com` (a well-regarded source of information on all things Mac):

> Apple currently has around 5% market share in personal computers. This means that out of one hundred computer users, five of them use Macs. While that may not sound like a lot, it is actually higher than both BMW's and Mercedes-Benz's share of the automotive market. And it equals 25 million customers around the world using Macs. But that's not enough for us. We want to convince those other 95 people that Macintosh offers a much simpler, richer, and more human-centric computing experience. And we believe the best way to do this is to open Apple stores right in their neighborhoods.

You can easily generate generic player-based content for both Mac and Windows platforms, where a single file will play on either computer as long as a person has the right player. But you can also use applications such as Macromedia Director and Flash, which have the capability of generating platform-specific files.

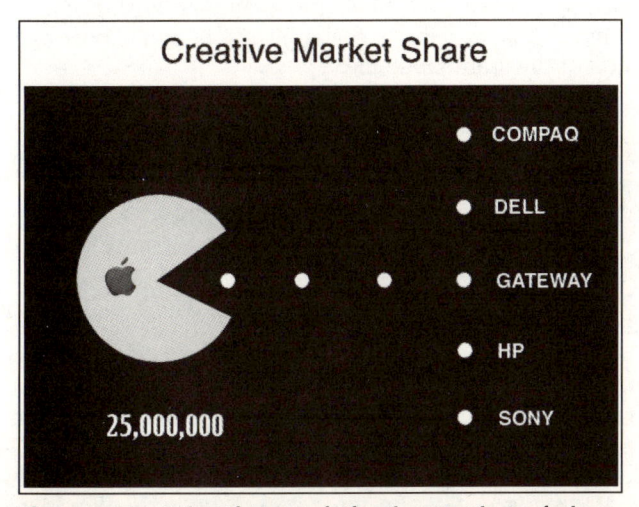

Figure 21-2: Point Abstracted Charting-Matrix Analysis Nexus. This is a top secret, very scientific diagram showing an abstracted pie chart overlaid against a statistical point chart. The high score displays in the lower left hand corner.

> **Note:** If you are including player-based content on a disc (such as straight mp3, PDF, or QuickTime files) for use on Windows, Linux, or other operating systems, remember to make sure the file extensions are on the files; otherwise they might not be recognizable to the player applications on the other computers. For example, if a QuickTime file on a Mac is saved with a filename of roneyrules with no file extension, the QuickTime player on a Windows machine might not recognize it. So add a .MOV on the end, and you get roneyrules.mov. Voilà! (mp3 = .mp3 // Acrobat (PDF) = .pdf // avi = .avi // wave = .wav // um // Quicktime = .qt or .mov // HTML = .html or .htm // JPEG = .jpg // GIF = .gif // um // ah! // AIFF = .aif // MPEG or MPEG-2 = .mpg // GLF = .glf // etc.)

Cross-Platforming Standalone Content

Cross-platforming. OK, so finally we get to what most Mac-based DVD authors will probably end up doing if they want to put interactive content on a DVD disc. The answer? Macromedia Director or Flash. (Or possibly, Adobe's LiveMotion.)

> **Note:** GLF is not actually a file extension, it is an acronym for the Gerbil Liberation Front. Yet it is admitted the .GLF could be a file extension if the GLF made its own compressed image format, as rumors indicate they have plans to do for revenue generation, modeled after CompuServe's .GIF format. CompuServe actually owns the .GIF image format, and can theoretically collect licensing fees from anyone who uses it on a Web page or disc.

Both Director and Flash can generate standalone files for Mac and Windows operating systems. These standalone files are also known as projectors. With Flash, you end up authoring content in the Mac version of the application, and exporting both a Mac and Windows version of the standalone file.

With Director, in order to make a Windows-compatible projector, you may have to reopen the project file in the Windows version of Director. When you purchase the program, it may come with both Mac and Windows versions of the program, so you could conceivably install the Windows version on a friend's computer, or try something crazy like running the Windows version of Director 8.5 in Virtual PC, for the express purpose of burning Windows Director projectors.

Now say that five times fast. Windows Director projectors. Windows Director projectors.

In Figure 21-3, you see examples of Mac and Windows standalone Flash, as seen through the eyes of Bill, circa Windows 98. These are the same files as Figure 21-1, where you are looking at the Mac and Windows standalone Flash projector files. Notice how this time, the Mac version gets the generic icon, while the Windows .EXE file has the characteristic Flash diamond. Notice the file size as well.

Depending on what you do, a Flash file can be extremely small, even when it's exported as a projector, with the integrated Flash player. The She's My Gerbil Flash animation was rendered into full-screen video and converted to MPEG-2 for the GLF DVD (not very long, but still 36MB), but it was also saved as Flash for the Web, and saved as standalone Flash to go on the DVD as a party favor that can be e-mailed to friends. The size of Flash also permits files to be placed on (gasp) floppy disk, which in this case means a cheap, experimental Electronic Press Kit that can be given away in mass quantities and help the GLF in the quest for world domination.

Figure 21-3: Examples of Mac and Windows standalone Flash, as seen in the Windows 98 operating system.

Comparing Flash and Director

If you have the money and time, Director is pretty powerful, and is the standard for working with CD-ROM/DVD-ROM interactive multimedia. It is expensive, well over a thousand dollars, but it has more robust support for working with video than Flash.

For example, in Flash 5, you can integrate a QuickTime movie within a Flash document, the Flash content then becomes a QuickTime track, and the Export option generates a QuickTime file. It is fun, and works nicely for Mac owners or Windows owners with QuickTime installed. But Director presentations can support both QuickTime and the Windows-native AVI format, and you could have a standalone file for Windows users with AVI video that would not require an installation, and a Mac version using QuickTime that would probably not require an installation, since many Mac users would probably have a recent version of QuickTime installed. (See Figure 21-4.)

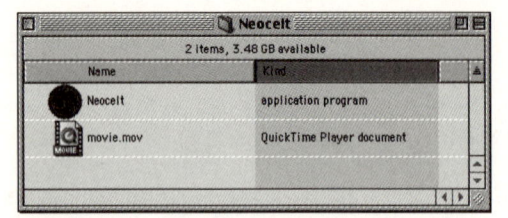

Figure 21-4: The projector and movie file. This is a Mac version of an interactive piece created in Director.

In this case the application icon is customized with a logo. A Director file does not have to use video, but it can. The QuickTime movie is not integrated into the Director presentation—it remains a separate file. (In a very rudimentary way, this setup with an interactive file referencing a video file externally is the ancestor of the InterActual/PCFriendly system mentioned later in this chapter. InterActual/PCFriendly DVD-ROMs allow interactive presentations that access the encoded MPEG-2 files within the VIDEO_TS folder of the DVD itself.)

To examine Macromedia's commentary on the question of what the differences are between Flash and Director, try visiting `www.macromedia.com/software/flash/productinfo/faq/` and look for the "What is the difference between Macromedia Director and Macromedia Flash?" link.

Tip Director allows you to import Flash, so you can theoretically have the best of both worlds.

The approach for DVD-ROM content might be a question of resources, and the purpose of your DVD-ROM content. Flash is only a couple hundred dollars, is ideal for cross-purposing certain kinds of content (mentioned in the next chapter), and if you don't plan to author the content yourself, there are a lot more people out there who know Flash. Director ultimately still has advantages for disc delivery, with its support of video, the level of sophistication you can achieve with scripting, and the many third-party "Xtras" which can be incorporated into presentations to achieve additional functionality.

Figure 21-5 shows one of the screens from the Neocelt multimedia presentation. One of the nice things about Director is its ability to take over the entire screen.

> **Tip**
>
> For people who use Flash — taking over the entire screen can be done in standalone Flash, too, using the FScommand action in a frame or button, by selecting the Fullscreen command with the True argument. The result is that the Flash is scaled larger — whereas in Director, the rest of the screen outside a presentation can become black without scaling the content. If you are using vector art in a scaled Flash presentation this re-sizing is not a problem, but bit-map art may be a different story.

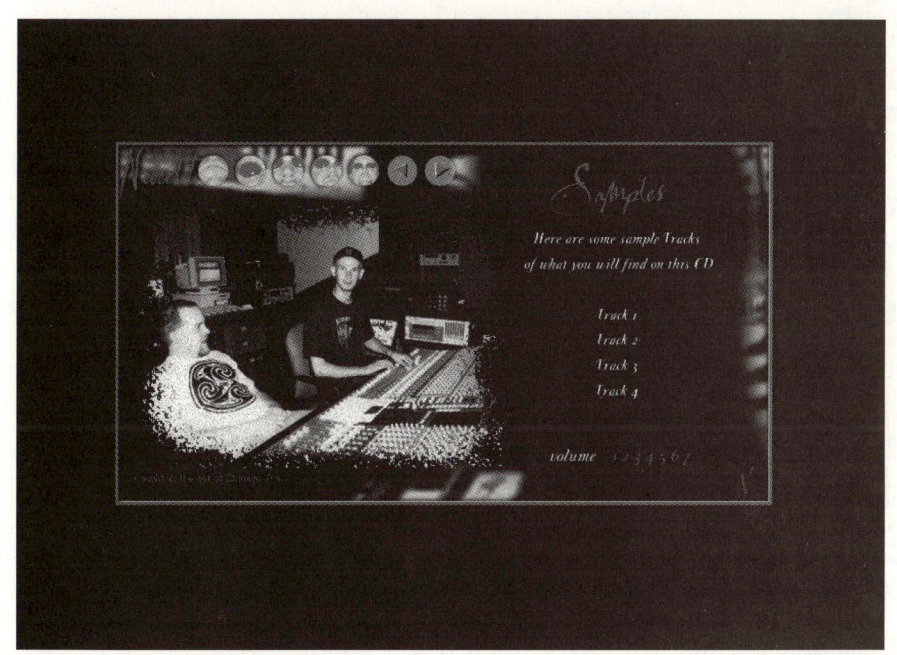

Figure 21-5: One of the screens from the Neocelt multimedia presentation, an interactive project that was developed to go with the Neocelt music/poetry CD. Some of the related files are in the Goodies section of the DVD-ROM for this book. (Todd Kelsey on left, producer Howie Beno on right. Photo by Audrey Cho, interface design by Jason Reeves. `www.neocelt.net`).

Director has introduced robust 3D support in version 8.5 for either disc or Web delivery. Director does have powerful Web delivery features through the Shockwave format. Not as many people have the Shockwave player as the basic Flash player, but Shockwave can have advantages in certain cases. For more information on the relationship between Flash and Shockwave, visit `www.macromedia.com`.

Don't be concerned if your eyes are glazing over. The best thing to do is have some fun with Flash and then start investigating what Director and Shockwave can do for you. Try tinkering in Flash for a bit with the trial version on the DVD-ROM that comes with the book, make something fun, make a standalone version of your project, and imagine some of the things you can do. (See Figure 21-6.)

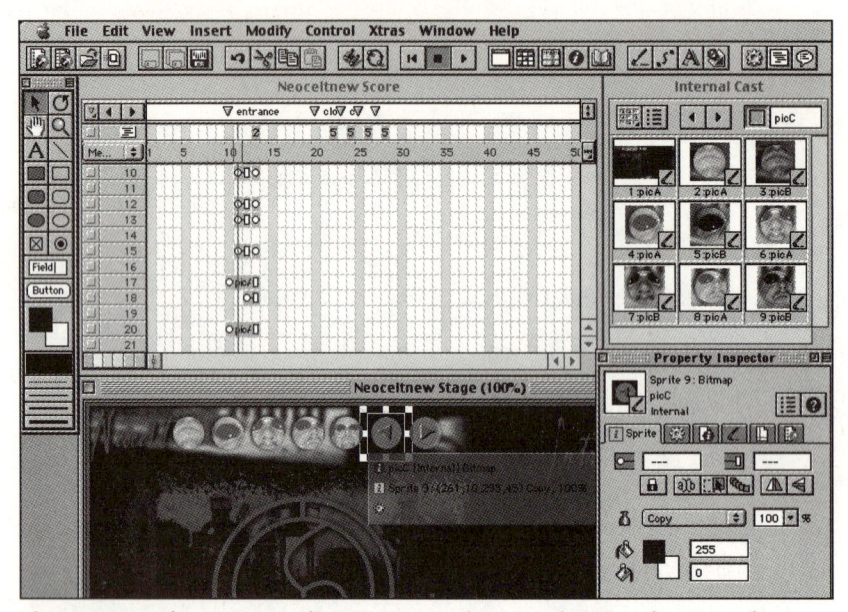

Figure 21-6: The Macromedia Director workspace, showing the Neocelt multimedia project. Director has similarities to Flash, with the Score window, Property Inspector, and Stage areas. The Internal Cast in Director roughly corresponds to the Library in Flash. In addition, in Director, you have Sprites, whereas in Flash, you have Graphics/Symbols.

If you try Director, don't be intimidated by the lengthy manuals. Just work through the tutorials, and think of a fun project you can do to give you inspiration to look up what you need to. You can also use the manuals as a convenient tool for building upper body strength.

Fireworks

Macromedia Fireworks is an image editing program that is commonly used in conjunction with Dreamweaver, Director, or Flash. It can create vector or bitmap art, and can even be used to generate .PSD files. It is fun to create a multilayered file Photoshop file, import it into Fireworks, and export it as .SWF so that you can effectively bring it into Flash. (Adobe's LiveMotion imports .PSD directly, but then again, it's not perfect either.) Macromedia sells Fireworks alone, and also offers bundles with Flash or Director.

At some point you will probably sell your remaining secretly held Microsoft shares and buy a radio control helicopter and Director Shockwave Studio. Don't be dismayed when you find that Director 8.5 comes with Fireworks and Peak LE (a sound editor), if you already own both programs. You can give the extra copies to a friend and offer to help him learn the programs. And of course, under the right conditions, fireworks and shockwaves can be the natural result of mutual interactive development. But remember, while personal digital assistants may be acceptable, public displays of affection could get you in trouble.

LiveMotion

LiveMotion is Adobe's answer to Flash, and has some interesting features such as tighter integration with other Adobe applications. It generates .SWF content, but one thing it can't do as of version 2 is generate a standalone file.

The best thing to do is to imagine what you would use them for, try both Flash and LiveMotion, and then choose either or both.

There are some interesting comparisons of LiveMotion and Flash on the Web. At the time of writing, Adobe has an aggressive comparison on its site, Macromedia feels like Flash speaks for itself, and some independent sites have interesting feature by feature comparison.

Adding player-based Flash on a DVD-ROM

If you generate standalone interactive content with Flash for a DVD-ROM, you may still wish to include the bare .SWF file. If you have the extra space, an easy way to score points and look impressive is to create an extra series of folders for additional operating systems that the Flash player can be used with, such as Linux or Solaris. To examine this brave new world, visit the Macromedia Flash Player Download Center at www.macromedia.com and click on the Need a Different Player? link. (It could be amusing to create a folder for the Pocket PC operating system, too.)

Preparing DVD-ROM Content

When you are developing a project in DVD Studio Pro that you want to distribute with DVD-ROM content, the approach is probably going to be the same, whether you are distributing the project on DVD-R discs that you burn yourself, or whether you are getting the project manufactured.

Ultimately what you end up with on the final disc is a VIDEO_TS folder, and alongside it you have whatever other folders you decide to put on the disc, which may include standalone projector files, HTML documents and other kinds of files.

The first step would be to multiplex the project in DVD Studio Pro using the Build Disc command, so that you end up with the VIDEO_TS folder. Then you would create a folder structure for the DVD-ROM content, separating it into Windows and Mac if you have both. Finally, you burn it to a disc with a program like Roxio's Toast Titanium. (See Figure 21-7.)

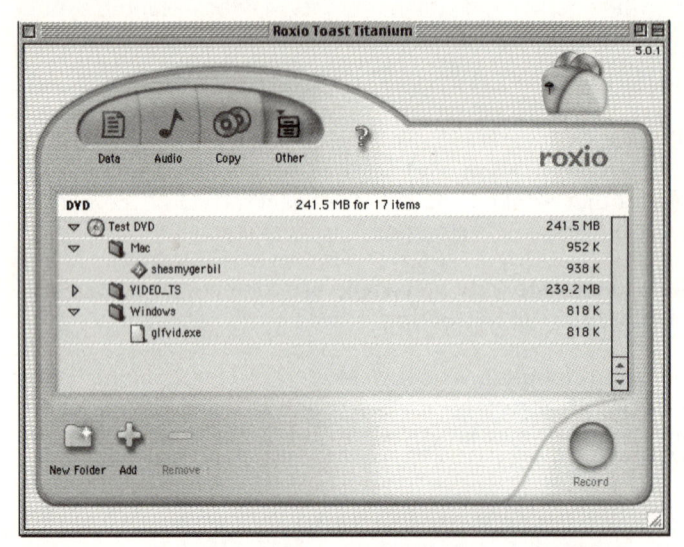

Figure 21-7: Burning a DVD in Toast. Clicking on the Other button and holding it down gives you the option of burning a DVD disc. In this sample, separate Mac and Windows folders are created for the DVD-ROM content.

Manufacturing a DVD with DVD-ROM content

If you are manufacturing a DVD with DVD-ROM content and wish to send the project on DVD-R media, first make sure the manufacturer accepts DVD-R.

If you have a DLT drive, there is no way to include DVD-ROM content on the DLT tape through DVD Studio Pro. In this scenario, talk to the manufacturer/replicator and see if you can turn the DVD-ROM content in separately, either on DVD-R or DLT.

Finally, if you have a DLT drive and Toast Titanium, you should be able to burn a DLT tape with the VIDEO_TS folder and the DVD-ROM content, but check with the replicator to make sure what kind of format it can accept!

Enhanced DVD-ROM?

Before there was DVD, there was CD. And with a CD-ROM, you can have the disc set up so that the Mac files only show up on a Mac, and the Windows files only show up on a Windows machine. You could even have an audio CD that had CD-ROM content, with separate Mac and Windows sections that would show up on the appropriate computer. This enhanced format for CD is technically known as the Blue Book standard. (It doesn't have anything to do with the infamous UFO-related Blue Book legends. Or does it?!?!)

Toast Titanium has an excellent Help resource that describes all the different possible CD formats. And while Toast can be used to create DVDs, it can't write the CD formats to DVD-R — you are stuck with the UDF Bridge format.

UDF Bridge

UDF Bridge format is the standard file format of the DVD disc. It is a combination of UDF (Universal Disc Format), and the ISO 9660 format. Basically ISO 9660 is a format that has been around for a while, and is readable by most computers, whether you are talking about CD-ROM or DVD-ROM drives. UDF is newer and gaining in popularity.

Exploring DVD-ROM Options

In this section, several scenarios are discussed that may be of interest at one time or another in your exploration of DVD development.

Previewing cross-platform content with Virtual PC

If you're developing DVD-ROM content for both Mac and Windows machines, you might want to consider purchasing Virtual PC from Connectix. Virtual PC is a program that emulates a computer running the Windows operating system. (See Figure 21-8.)

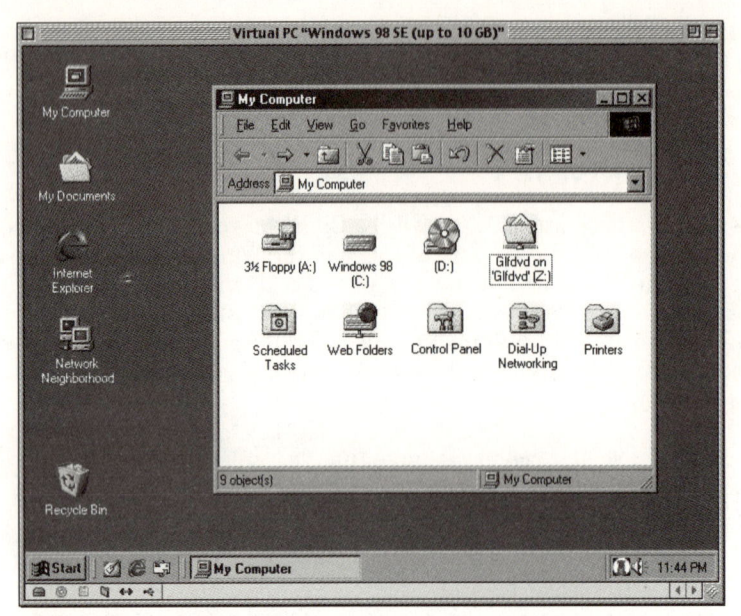

Figure 21-8: Windows 98 on a Mac, courtesy of Virtual PC 4.0. Basically you can run Windows and Mac simultaneously — the Windows operating system is contained in its own Mac window. Virtual PC supports access to CD-ROM drives and printers, as well as other kinds of removable storage.

> **Caution** The current version of Virtual PC does not support access of DVD-ROM drives directly, but you can share folders between Virtual PC and your Mac, so you can treat a DVD-R disc as a shared folder, as in the example above, where the DVD appears on the Mac desktop, as well as Drive Z: in Virtual PC. (You can get Virtual PC with Windows 2000 or Windows Me, or even Linux.)

The relative performance of Virtual PC is not as good as if you had an entirely separate Windows computer sitting there, but it is a convenient way to run Windows programs on a Mac. (See Figure 21-9.) It is amazing that it works at all, given that what the software is doing is simulating the actual Windows computer. Virtual PC can't really compete with actual PC hardware, and it's not meant to; it's more of a convenience issue. Yet even so, performance has increased over previous versions with support of the G4's Velocity Engine.

Virtual PC is not an accurate measure of the *performance* of the Windows versions of DVD-ROM content, but it can be used to test content *functionality* on Windows. And there are Windows-only programs out there that might justify the purchase in and of themselves. One such program is Swish, an excellent, inexpensive Flash-related application that makes certain kinds of Flash animation a lot easier to implement. (To find out how to get Swish, go to www.swishzone.com.)

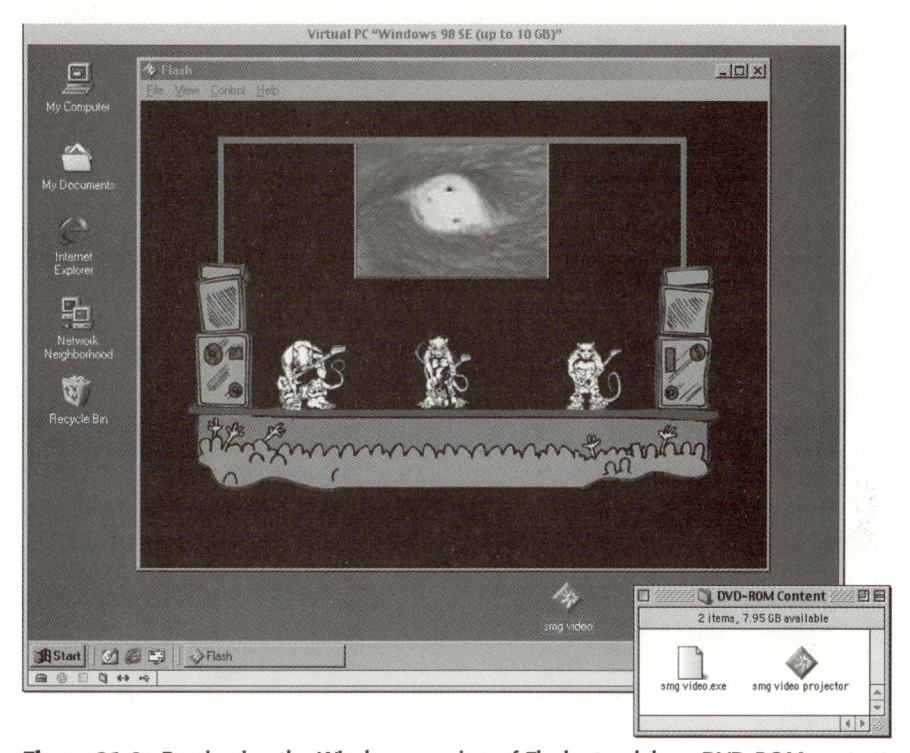

Figure 21-9: Previewing the Windows version of Flash standalone DVD-ROM content, running in Virtual PC on a Mac. It is nice how you can actually drag and drop files from the Mac to the Virtual PC desktop. In the lower right hand corner, you can see the smg video.exe file, which has a generic icon in the Mac window. When it is dragged to the Virtual PC desktop, you see the icon in its native state as if it were on a Windows machine.

Evaluating PCFriendly/InterActual Player

Ah, PCFriendly . . . alas it is not yet Mac-friendly, but theoretically they're working on it.

PCFriendly is pretty much the standard for delivering entertainment-related DVD-ROM content, and is used by nearly every major Hollywood DVD release that also has DVD-ROM content. Walk in to a Blockbuster or Hollywood video store and you'll start to see the PCFriendly logo on a lot of the DVD cases. The company which created PCFriendly is called InterActual, and the most recent version of the software is called the InterActual Player. (www.interactual.com)

InterActual's PCFriendly consists of custom software that works with several of the high-end DVD authoring systems out there to allow the DVD-ROM content to access the video on the DVD from the DVD-ROM side. That is, PCFriendly provides a shell, which is now called the InterActual Player, that allows the user to access special interactive features such as Web links as well as giving them a way to watch the DVD from the computer.

The way PCFriendly achieves interactivity is through a customized implementation of a Web browser, so much of the content is HTML, and often involves Flash or Director/Shockwave elements. (See Figure 21-10.)

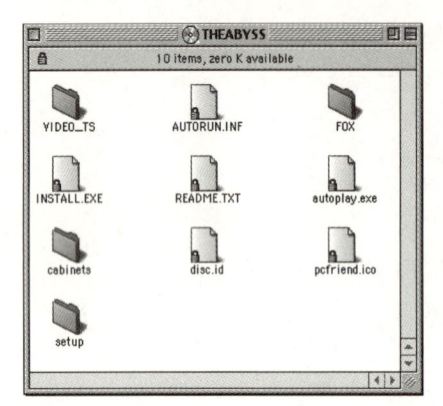

Figure 21-10: The way that a PCFriendly-enabled DVD disc looks when you open it up on a Mac. You have your characteristic VIDEO_TS folder in the upper left-hand corner. The autorun.inf file is a simple text file that tells a Windows machine what file to run when the disc is inserted; in this case, the computer is instructed to run autoplay.exe, and the user is prompted to install the PCFriendly software.

In certain cases, without even running Virtual PC, you may be able to preview certain kinds of Windows-based DVD-ROM content. If the content is HTML based, which could definitely be the case, try wandering around the folders and looking for documents ending in .HTML or .HTM, such as index.html, double-clicking on them to bring up a browser window — if you get that far, you may need the latest version of the Flash/Shockwave plug-in.

Experimenting with Multiple DVD

Adventurous DVD authors may wish to try burning multiple VIDEO_TS folders to allow a person to preview additional DVD content with a software-based DVD player.

This is the one technique for achieving a form of DVD-ROM content using DVD Studio Pro. True, anything you can burn in a VIDEO_TS folder can be merged into an existing DVD Studio Project. But what the heck, it could be fun.

The merge command in DVD Studio Pro allows you to combine several projects, but it can get messy in the Graphical View window, especially if you start trying to delete certain project elements that are dependent on other project elements. If viewing the projects on a computer is acceptable, having multiple VIDEO_TS folders may be a convenient way of previewing several projects on one disc, or taking a look at multiple multiplexed versions of a project.

The idea is that you have the option to leave a single VIDEO_TS folder at the top level. If the disc is inserted in a player, the player will look at that folder. Any additional VIDEO_TS folders are placed inside other folders so that the set-top player doesn't get confused. (See Figure 21-11.)

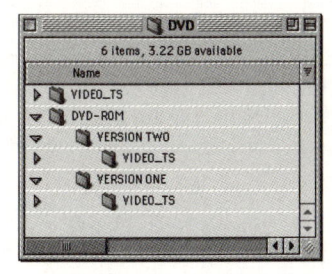

Figure 21-11: Concentrically Arranged Recursive Points of Entry for Disc Integration En Masse, also known as Carpe Diem. A highly complex technique for distinguishing folders of the same name from each other.

If you plan to incorporate multiple VIDEO_TS folders for DVD-ROM, be kind to yourself and name the folders something you will be able to remember.

That just about wraps up the whirlwind tour of DVD-ROM.

For more detailed information of data formats, read Lee Purcell's *CD-R/DVD Disc Recording Demystified,* or Jim Taylor's *DVD Demystified*. You might also wish to visit Jim Taylor's `www.dvddemystified.com` and search through the DVDFAQ. For more information on these and other DVD-related books, visit `www.dvdspa.com`.

Jim Taylor's DVDFAQ is also available in HTML format in the Goodies section of the DVD-ROM that came with this book. When you have a DVD-related question, it can be helpful to search the DVDFAQ using a particular keyword with the Find feature on your Web browser (probably accessible from the Edit menu of your browser).

Summary

In this chapter, you learned about some of the ways you can include DVD-ROM content with a DVD project. Adding DVD-ROM content requires additional effort, but the investment of time and resources can significantly enhance the DVD experience.

- ✦ DVD-ROM content can consist of any file(s) that you can run on a computer. The ROM part of DVD-ROM stands for Read Only Memory, and it is the DVD equivalent of CD-ROM.
- ✦ DVD-ROM and CD-ROM content often consists of some kind of *standalone application*, which means that you can run the file without having to have a particular program on your computer.

- Applications that are typically used to generate CD-ROM/DVD-ROM content include Macromedia Director and Flash, both of which can generate stand-alone files.

- Both Flash and Director can be used to create content with animation, interactivity, and Web links, where a person can jump to the Web if he has an active Internet connection.

- Player-based content can be included on the DVD-ROM portion of a DVD disc. This could include audio or video files, or HTML and PDF documents; in short, you can put any kind of file you want on the disc, if you think a person is going to have the appropriate player, or if you can instruct him on how to download the player.

- You can easily generate generic player-based content for both Mac and Windows platforms, where a single file will play on either computer as long as a person has the right player. But you can also use applications such as Macromedia Director and Flash, which have the capability of generating platform-specific files.

- The main advantage of Director is the ability to work with video, its new 3D support, and some of the extended functionality you can integrate within multimedia presentations using third-party "Xtras."

- The main advantage of Flash is that it is ideal for cross-developing content for Web and disc delivery, it is cheaper than Director, and easier to learn.

- When you are developing DVD-ROM content, ultimately what you end up with on the final disc is a VIDEO_TS folder, and alongside it you have whatever other folders you decide to put on the disc, which may include standalone projector files, HTML documents, folders with files inside them, and so on.

- Virtual PC from Connectix is a program that allows you to run Windows programs on a Mac and is a convenient way of previewing standalone platform-dependent DVD-ROM content, such as the Windows version of a Flash or Director projector.

In the next chapter, we move on to the wild wild world of cross-developing content. There's no reason you couldn't do a CD-ROM and Internet version of a DVD, or a DVD version of something that has been developed for Internet and CD-ROM — it all depends on your goals and requirements. So we take you on a tour of some interesting possibilities, providing you with a potentially endless supply of fun!

✦ ✦ ✦

Cross-Developing for DVD, CD-ROM, and the Internet

CHAPTER 22

◆ ◆ ◆ ◆

In This Chapter

Investigating delivery options

Considering disc delivery

Considering Internet delivery

Managing workflow between applications

Simulating DVD — real world examples

◆ ◆ ◆ ◆

In this chapter, you find out about the various directions in which you can take your content when you consider going beyond the DVD disc.

Alternative delivery methods can help you extend the audience for the original source material you have, whether it is video, audio, animation, or a collection of still images and text. Becoming familiar with the various ways you can deliver content can help you to expand the possibilities for reaching your audience, whether the purpose is entertainment, education, or marketing. You may want to take excerpts from a DVD project and deliver them on the Web to market the DVD project. Or you may want to develop both DVD and CD-ROM versions of a project to reach as wide of an audience as possible.

Familiarity with content delivery options can also spur thought about how you can recycle items in your archives. For example, you may have access to some Photoshop images and Flash/LiveMotion animations that have been developed for the Web, and you realize you can take the images and use them in a DVD menu. Then you can recut the Flash animation into full video files and put them on a DVD!

Investigating Delivery Options

Content delivery options revolve around what you can do with the content. With so many converging technologies, the question of how to deliver content can be difficult to answer, especially if you're not familiar with all the options.

Think of cross-developing content as expanding your repertoire.

You can start with what you're familiar with. Find a project that inspires you (or drives you) to learn how to work with alternative delivery methods. You can then find people who know how to work with the programs that these delivery methods rely on, or you can learn the programs yourself, building on the existing familiarity you have. Skills that you have developed in one program can often be applied in another program.

For example, if you know Adobe Photoshop but you've never developed a DVD before, you quickly find out how important Photoshop is for making DVDs in DVD Studio Pro. Or you may be familiar with using Adobe Premiere for editing video, and you want to use the program for DVD. Then you realize that you can use Premiere to simultaneously cut versions of the same footage for DVD and Web delivery.

There's more than one way to arrive at your goals, but the seeming complexity of all the different technologies and programs can be simplified, when you purposely make content the central aspect of your project.

You can easily get distracted or intimidated by all the programs and delivery methods available, but don't worry about the technology. Keep focused on the content, and ask yourself what you want to do with it. Then you can find the appropriate method to achieve your goals.

And if you don't know what you want to do, you can learn about the delivery options, so that you know what you *could* do. The resulting familiarity can help you to answer the question of what you *want* to do.

Paying passenger and stowaway content

When you are planning the cross-development of content, one question to ask yourself is whether the alternate versions of a DVD project are paying passenger or stowaway, or a combination of both.

Paying passenger content is where you are attempting to faithfully reproduce a project in a variety of mediums, as a product that you are actually selling, or because it is important for some other reason to have as much as possible of the original project in each alternative version. For example, if you cut a DVD of a movie and want to also be able to offer it for sale on Video CD for an extended audience, you want the whole film on there. When you consider various delivery options from this perspective, some may be ruled out simply because they won't allow the kind of quality you're looking for, or the storage capacity and other limitations may be factors.

In contrast, *stowaway* content is likely to be more promotional in nature and doesn't necessarily need to reflect the entire project or meet the quality of the main presentation. One delivery strategy for stowaway content is to get it on other discs or Web sites. The content is getting a free ride. The stowaway content is not the reason that the disc or Web site is created, but it goes along for the ride, hoping for exposure.

For example, if you're developing a DVD, you can develop a standalone Macromedia Flash or Director presentation that integrates some of the content. You can then negotiate to place the resulting file on any DVD-ROM or CD-ROM disc you can get your hands on, such as musical albums that have CD-ROM content, DVDs that have DVD-ROM content, any disc-based project where the company or entity that is releasing it can be approached to give you space. For Web versions, the possibilities for stowaway content are limitless, if the file is downloadable or e-mailable, or an address is promoted to point to a central location.

With stowaway content, you have more freedom to take content that is also appearing in a DVD, and hint at the original project. The limitations of alternative delivery formats are not as much of an issue, because you don't necessarily want to threaten the sales or distribution of the main project. Often stowaway content can take the form of excerpting the main project, like a sneak preview or sampler of some kind.

Perception and delivery

When you're thinking of reusing content or cross-developing content for a variety of mediums, separating the content from all of the surrounding technologies is helpful. The essence of the project does not consist of MPEG-2 video or RealMedia or Shockwave or Flipbook or Scratch 'n' Sniff. The true identity of what you are delivering is in the ideas and stories and information, and the purpose behind them.

Focusing on the essence of your creations frees you to better harness the strengths of available technology. Your delivery is then built around your own goals, rather than simply because a particular method is available.

In other words, don't necessarily let the technology carry you along. Figure out what you want to do and put the technology to work for you.

Considering perception

Consider the basics of perception. Consider how people use their sight and hearing to experience what you create. Thinking of a project in terms of how it is ultimately experienced can help you to see the essence of the content, or the soul of your project.

Some people believe that the human race descended from apes and/or monkeys. Others believe that primates and humans evolved separately. Whatever you think about monkeys, they serve as an insightful example of sensory perception. They are also a reminder of the essential value of play. Remember play? Both humans and monkeys like to monkey around.

Human see

The central aspect of any interactive experience is probably the visual one. The quality and enjoyability of a visual experience translates into impact. The better the images look, the better it matches the person's style or expectations. In modern society, we experience the world through a barrage of digital images, and the most enjoyable and memorable digital experiences are often the best-looking ones.

Human hear

Sound adds another dimension to an interactive experience. Even if the content didn't originally have sound, it may be worth considering, as long as it does not become annoying. Repetition can be the pitfall of sound on a DVD or on the Internet. But if a person enjoys the sound, the experience is more memorable and deeper.

Human do

If a person has the option to play with the way things look or sound, so much the better. Being able to explore is fun, as long as it is not confusing.

Establishing goals

You know that whatever you do with the content and whichever directions you take it, you want it to look good, sound good, and be fun, or at least functional. Establish goals for your delivery by figuring out the whos, whys, hows, whats, and whens. If you don't know, try starting by asking questions, to stimulate thought.

Following are sample questions:

- Who is the audience? Primarily Web surfers who also own DVD players? Surfers who own surfboards and DVD players? Players who own DVDs?

- Why do you want to develop a Web version of a DVD? Simply because you can?

- How do you want to use the Web version of the project? To market the DVD? If so, do you want to have the whole DVD represented or just an excerpt?

- What is the purpose of putting your project on the Web? If you have video footage to work with, do you really need to put the video on the Web, or would stills from the video footage serve just as well?

- What if you could make a version of the DVD on CD-ROM, where the quality of the video is not quite as good, but where you reach a wider audience? Does most of your audience have DVD players or DVD-ROM drives?

- When does the project need to be launched? Can you have a successive release with aspects of the project on the Web followed by the release of the disc?

When you have considered the essence of the content, move on to considering the ways you can deliver it.

Considering Disc Delivery

Disc delivery offers four main options for the content creator: DVD, DVD-ROM, CD-ROM, and Video-CD, or a mixture of the four.

DVD

DVD gives you a lot of space to work with, a high level of video quality, and limited interactivity. If you are developing a DVD, then you will have video and graphic resources to work with that can be formatted for other mediums.

DVD-ROM

If you have the need, you could develop a disc that is exclusively DVD-ROM, that is, which is designed specifically to play in computers. In that case, think of the DVD-ROM as a 4.7-gigabyte CD-ROM — it's just like a CD-ROM, in the sense that you're storing data, yet a person needs to have a DVD-ROM drive to access the content.

Consider using DVD-ROM if you need a level of interactivity that DVD players do not support. You can do more with a Director-based multimedia presentation or a series of HTML documents than you can achieve on DVD. For example, the scripting capability in Director is quite advanced, allowing for a wide variety of customization, as well as data input and processing, whereas DVD scripting is relatively limited. If you have the resources, planning a project around DVD-ROM could make sense if you require advanced interactivity, and you could still have a simplified version of the project in DVD format. Remember, DVD-ROM and DVD can exist on the same disc — it's all a matter of space.

For developing DVD-ROM, you first need to decide what kind of program you are going to use to program the interactivity, and you want to consider Director or Flash. Then you need to decide what kind of format you want the video files to be in. If the audience is a mix of Windows and Mac people, you'll probably want to use QuickTime and give an indication of how to download and install it for those who don't have it or make arrangements to include the installer on the disc.

Another option for DVD-ROM video is to use straight MPEG-1 video — it's approximately VHS quality but takes up a lot less space than MPEG-2.

CD-ROM

If you are building a project around DVD, making a CD-ROM version may enable you to reach a wider audience. The issues with developing a CD-ROM version of DVD content are similar to the issues when you are developing additional content to go with a main DVD presentation.

Capturing an entire DVD presentation on CD-ROM can be a bit tricky, given that you only have about 650 megabytes to work with, but if you don't mind going to a smaller size for the video and making some sacrifices in quality, you can do it.

You can thus take the video that had been developed for DVD, save it out to smaller, more compressed file sizes, and create an interactive simulation of the DVD presentation, in a program such as Director. The main advantage of Director is that the audience member can often run the Director presentation without having to install anything.

Sonic Solutions, the granddaddy of DVD authoring software, has a format they call cDVD, which is basically the same level of interactivity as you would have on a DVD, but the format allows you to save a DVD project to CD. Essentially what it does is take advantage of the ability of software-based DVD players to read a VIDEO_TS folder.

Simulating cDVD if you can fit your DVD project on a CD-ROM is easy. Basically, you write the VIDEO_TS folder to CD-ROM. As long as a person has a software program, such as Apple's DVD Player or Intervideo's WinDVD, that person can access the content. (In the case of the Apple DVD player, if the CD-ROM doesn't come up automatically, you may need to go into the preferences in the program to change the setting that enables you to open up the VIDEO_TS folder from the File menu.)

If you attempt to do a cDVD, you may also consider using MPEG-1 video in the original project file. For example, you can create your regular DVD with full MPEG-2 video, and so on, in DVD Studio Pro, and burn to DVD-R or get it manufactured. At the same time, you can save MPEG-1 versions of each video segment and create another version of the main project file by using the MPEG-1 video. This version probably has a greater chance of fitting on CD-ROM.

When you are considering the various options for CD-ROM, think of your audience. If you don't know who your audience is or what kind of computers they have, then the majority of them probably have Windows, and a minority has Macs. If you use Director, you can create separate versions of a project for each platform with video files tailored to the operating system; for example, if you use Director to create a version of the project for windows that used AVI video files, chances are that people would not have to install anything to run it. And if you created the Mac version that used QuickTime, chances are they wouldn't have to install anything to run it.

With either DVD-ROM or CD-ROM, the ideal situation is to determine how you want to deliver the project on disc and then seek to make arrangements to license/include all the necessary installers that a person may potentially need to run the presentation, or at least include Web addresses where they can download them. Some companies may allow you to license/include the installer for a free player application on your disc; contact the software manufacturer for more information. Having the installer on the disc would be convenient for someone, though giving a Web link for downloading would give access to the latest version of the player application. A good compromise may be to do both — make arrangements to include the installer, and provide a link if a person chooses to download.

If the audience is captive (they have purchased the disc), or if it is a closed audience like members of a group or company, you probably have more freedom to use a delivery format that is more likely to result in people having to install a file. With promotional presentations, you should structure delivery around the principle of making it as easy as possible for a person to access the DVD-ROM/CD-ROM version of your project. In this case, a standalone delivery format, such as Director or Flash, may be ideal.

Video CD

Video CD is an interesting disc delivery option. This option is based on the standard CD/CD-ROM disc and uses the MPEG-1 video format. Although it has its limitations, Video CD is nice because it's cheap. You can burn it with a CD-R burner, and the discs can play in most DVD players. You can fit a little less than an hour of video on a disc, and the audio is 44.1Khz, just like a regular CD. When you work with video files in QuickTime Pro, you can save to MPEG-1 format and use a program like Roxio's Toast to burn the CD. You have to make sure that you're burning in Video CD format. You can't simply put an MPEG-1 video file on a CD-ROM disc.

Now that you've considered various disc delivery formats, Table 22-1 provides a sense of perspective.

Table 22-1
Typical Disc Delivery Formats

Disc Delivery	Video File Format	Video Format	Typical Quality	Interactivity Tools	Compatibility
DVD 4.7 gigabytes	MPEG-2	High	DVD Studio Pro for DVD project QuickTime Pro For encoding Video Photoshop for menu graphics Final Cut Pro or Premiere for editing video	DVD Studio Pro encodes interactivity into VIDEO_TS folder, interactivity is limited by capability of players	1. Can run in set-top DVD player 2. Can be accessed by computer with DVD-ROM drive and appropriate software-based DVD player
	MPEG-1	Medium			
DVD-ROM 4.7 gigabytes **CD-ROM** 650 megabytes	QuickTime, AVI, etc.	Dependent on how much you compress the file	Director for project Adobe for Premiere or Terran's Cleaner to compress video files	Dependent on what program you use	Computer with DVD-ROM drive; Interactivity or video may require install of appropriate software, such as QuickTime
Video CD 650 megabytes	MPEG-1	Medium	QuickTime Pro or Cleaner can generate Video-CD compatible MPEG-1. Roxio's Toast burns Video CDs nicely.	None; video is presented in a linear format. Any stopping or searching is dependent on the player	Video CDs will play in most set-top DVD players and in many computers

Considering Internet Delivery

Trying to cross-purpose DVD content over the Internet can be fun or a nightmare, depending on how you look at it. The level of nightmare may be in direct proportion to how close you are trying to get the Web version to appear and function like the original.

DVD and Internet parallels

Because of DVDs limited interactivity, certain aspects of a DVD interface are easy to reproduce for Internet delivery.

Rollovers — selection and activation

Web developers who try DVD authoring may notice the interesting parallel between the use of hilites to represent selections in a DVD menu, and the corresponding use of rollovers on a Web page. For example, in a DVD menu, you may have three button choices to select from on-screen, and when you move from one choice to the other, a hilite appears above each button choice. Similarly, on many Web pages, when you roll the mouse over an item, something happens to indicate that you are about to select, depending on how the page is programmed. For example, on the sample page (www.psrecords.net/wake), rolling over each main navigation choice changes the logo at the top of the screen. As you can see in Figure 22-1, there are also parallels with button states between DVD and Flash, which helps make it easier to develop a Flash version of DVD interactivity.

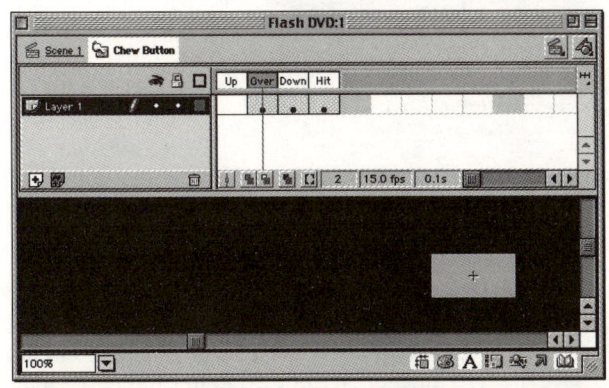

Figure 22-1: A comparative look at buttons. The Property Inspector in DVD Studio Pro on top, Flash below. At the top, the Selected Set 1 setting is set to 33%, so that when a button in the DVD menu is selected, a gray hilite appears.

With a reasonable knowledge of the respective programs, simulating the buttons of a DVD menu on the Web is not too difficult and working with the graphics is fairly straightforward, too.

Graphics resolution

The resolution of a graphic that you see in a DVD or on a Web page is associated with the dpi, the *dots per inch* of that image. This refers to the number of individual pixels or dots that appear in an inch of spaces.

The resolution or dpi of an image has something to do with the quality of the image, but this is usually more of an issue when you are printing images on paper. In that case, you often want as high of a dpi, or as many dots per inch as possible.

On-screen however, whether you are talking about video, DVD, or the Web, in most cases when you work with an individual image, you can work at 72 dpi. So when you're cross-purposing content, you can usually convert in either direction. For example, if you have developed graphics in Photoshop for a DVD menu, you can use the same graphics on the Web without necessarily having to change the resolution. (See Figure 22-2.)

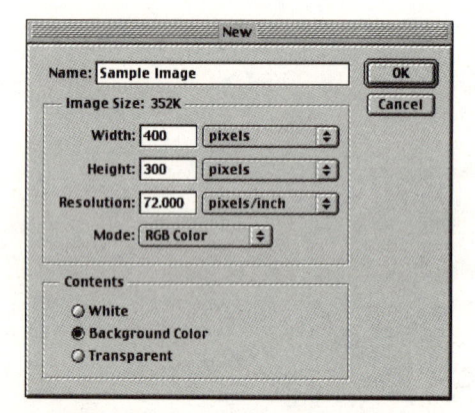

Figure 22-2: The New File dialog in Photoshop. With the resolution set to 72 dpi, the image can easily be used in a DVD menu or a Web page.

You'll encounter other issues when converting graphics from one medium to another, but resolution is a fundamental one. Note that in this respect, the Web is a close cousin to DVD.

> **Note** When I was developing the graphics for the menu of the GLF DVD, I sometimes took a look at 72 dpi graphics that had been developed for the Web site as potential files to be used. After developing the menus for the DVD at 72 dpi, I sometimes found myself using the Save for Web function in Photoshop to capture the look of the DVD menu in a single jpeg file and posting it on the Web in a simple page to share the menu design with colleagues and friends. I figured that if I got in the

habit of saving out Web versions of the various menu screens, at some point stringing them together with links to simulate the DVD on the Web would be easy. (The Gerbil Liberation Front DVD is mentioned further in Chapter 19 and Chapter 27, and excerpts appear in the DVD that is included with this book.)

The parallels between the Internet and DVD are nice, but you need to keep some differences in mind, too.

In general, the Internet can handle a lot less information at one time because the transfer of information is limited by the speed of a person's Internet connection. One of the only exceptions to this rule is when DVD-ROM or CD-ROM content combines an Internet connection with content that is accessed from the disc. Otherwise, video and audio are very limited. When considering the cross-purposing of content, keep in mind that moving images on a DVD are virtually impossible to reproduce at the same size on the Internet. In most cases, putting regular video from a DVD on the Internet means that the video has to be reduced significantly in size and quality.

Streaming media

Streaming media is a term that refers to the process of delivering audio and video on the Web. Most of the time, streaming media relies upon plug-ins, which are mini-programs that are installed and added to an Internet browser, such as Netscape or Internet Explorer. (Current versions of America Online incorporate a customized version of Internet Explorer.)

The most common streaming media plug-ins are RealPlayer, Windows Media Player, and QuickTime. Statistically speaking, most people have some version of Windows Media Player, and many people have RealPlayer. Many of the big sites that have video online offer video streams that are compatible with all three formats.

As with other forms of Web content, to deliver streaming video one has to have a place to put it, and an entire industry has evolved around hosting streaming media and caters to every size of client.

In regards to the discussion at the beginning of the chapter between paying passenger and stowaway content, streaming media is clearly most compatible with the latter, where you may want to deliver excerpts of your DVD material for promotional purposes, but otherwise the quality is fairly low by comparison. Great advances have been made in the technology associated with streaming media, and if your audience has very fast connections, cross-purposing content between DVD and the Web may be more of an option.

Generating streaming media usually consists of taking the original video and encoding it in various Web compatible formats. The makers of each plug-in have free tools that can be downloaded to encode the video, but more advanced tools may be necessary to achieve the quality and compatibility you may want.

An excellent tool to seriously consider getting in this area is Terran's Cleaner, which you may already be using to generate MPEG-2 video. Cleaner is especially suited for someone who intends to cross-purpose content for a variety of mediums. You can drag and drop an original media clip and then output it directly in a great variety of formats. (See Figure 22-3.)

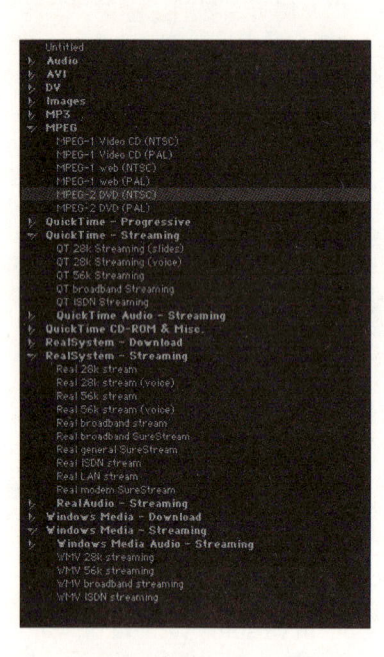

Figure 22-3: An excerpt from the Settings window in Terran's Cleaner, showing some of the main categories for preset Streaming media formats, as well as some of the DVD-related MPEG formats. Each of the categories has a number of individual settings that can be further customized if you want. One of the advantages of using Cleaner is its batch processing capabilities, where it can kick out the desired formats automatically for any number of individual video files. It's all drag and drop, baby!

To implement streaming media, you need to develop some experience in making a Web page and using the tools to encode the video, but another option is to get someone else to actually put the Web page together for you. You still may be able to generate the appropriate files for Web developer with the downloadable tools from Microsoft, Real, or Apple, but if you have the money, get Cleaner!

When you have considered streaming media, you may also want to consider streaming animation formats, such as Flash and Shockwave.

Streaming animation

Streaming animation is another form of streaming media. The most common form of streaming animation is Flash. Flash can be mixed with streaming video and audio, and vice versa, to a certain extent.

Using Flash may be a nice way to promote your DVD on the Web, by taking graphics that had been developed for the DVD, and incorporating them in a Flash animation for the Web.

Streaming Flash

At this point, a majority of Web surfers have the Flash plug-in installed on their computer, and many new computers have Flash already installed. In this chapter, the use of Flash is considered from the standpoint of developing a standalone file that can be included on a disc. That is, you can develop Flash as a file that you just click, like any other file on your computer, and it's like a jack-in-the-box, which opens up and starts playing. This standalone capability is a more recent development, but you have always been able to make splash with Flash on the Web.

Streaming Shockwave

Before there was Flash, there was Shockwave. Earlier generations of Macromedia Director let you create Web versions of multimedia projects through a special program called Afterburner. Afterburner compressed Director files into Shockwave format; this format included the ability to have audio and video. And before there was Flash, there was Splash. Before Macromedia came along and absorbed/renamed it, Splash introduced the ability to have Web animation with very small file sizes.

So eventually what happened is that the use of Flash skyrocketed, while the use of Shockwave grew, but not at the same pace. It's easy to get confused, but basically Shockwave goes a step beyond Flash and introduces the ability to achieve new levels of complexity in interactive features — it also has more capabilities with video. To do Shockwave, you need Director. If you want to use Shockwave, consider that not as many people have the full Shockwave plug-in — most people have just Flash.

If you are developing DVD-ROM or CD-ROM content for your DVD project using Macromedia Director, Shockwave may be a nice way to deliver it on the Web. Or an experienced Director person could potentially go further than you could with Flash simulating a DVD on the Web with Shockwave.

HTML delivery

All delivery of streaming media involves HTML to some extent, because the World Wide Web is based on the Hypertext Markup Language. But you don't necessarily need to use streaming media to give someone a taste of a DVD interface.

Another option to consider for introducing an audience to a DVD is to take simple screen shots of the DVD interface (or to export the screens directly from the original Photoshop files), and put them on the Web by using a program like Dreamweaver, and incorporating simple rollovers (mentioned earlier in this chapter) to simulate DVD buttons, allowing the user to click back and forth between the screens.

Whether you are delivering by disc or by the Internet, you have choices to make about how to develop and refine the content; these choices revolve around the programs you can use.

Managing Workflow Between Applications

If you decide to experiment with cross-purposing content, you will end up using a variety of applications, or at least working with a variety of specialists.

If you are doing a project alone or working on a team, consider what programs you are using, and how content is created, exported, and utilized from one program to another. Draw a simple flowchart, listing the formats that the original source material is in, how the material needs to be encoded to be able to use it in the various formats, and what programs are necessary to accomplish the overall plan.

Looking at an overview of how content flows from one application to another can help you notice patterns and may give you ideas of how to approach a project with a particular goal in mind. You can find out how to maximize quality and resources or discover how to work within a particular budget. You may also see the value of thinking ahead when creating the content. Imagining the various ways the content could be used outside of DVD may have an influence on which method you use to create it.

For example, if a DVD involves animation, Flash may be one program to consider generating the animation with, because you can save the animation in a variety of formats that are ideal for different delivery options. (See Figure 22-4.)

Note: In Figure 22-4, I looked back at the GLF She's My Gerbil music video, a cartoon animation that had been originally developed in Flash. It was intentionally made for video, and incorporated in a DVD project. I traced the various paths that the content took between applications. It is an example of how the original source material was specifically planned so that it could easily be delivered in a variety of formats from square one.

Now that you've taken a look at some of the various delivery options, consider a few ways of how you can simulate a DVD interface.

```
                    ┌─────────────────────┐
                    │   She's My Gerbil   │
                    │ music video animation│
                    │   created in Flash  │
                    └─────────────────────┘
                 ┌───────────┼───────────┐
                 ▼           ▼           ▼
```

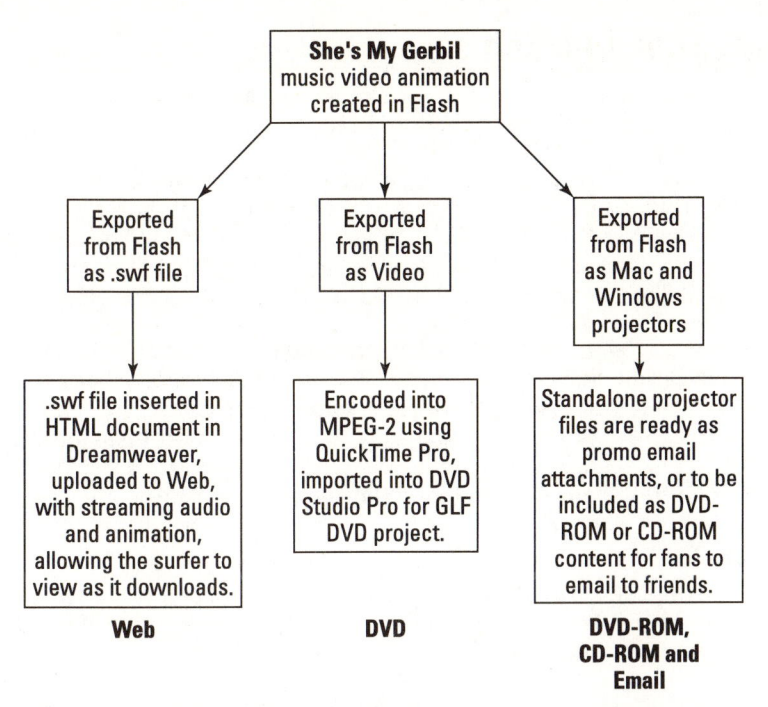

```
┌──────────────┐  ┌──────────────┐  ┌──────────────┐
│   Exported   │  │   Exported   │  │   Exported   │
│  from Flash  │  │  from Flash  │  │  from Flash  │
│ as .swf file │  │   as Video   │  │  as Mac and  │
│              │  │              │  │    Windows   │
│              │  │              │  │   projectors │
└──────────────┘  └──────────────┘  └──────────────┘
        │                 │                 │
        ▼                 ▼                 ▼
```

.swf file inserted in HTML document in Dreamweaver, uploaded to Web, with streaming audio and animation, allowing the surfer to view as it downloads.	Encoded into MPEG-2 using QuickTime Pro, imported into DVD Studio Pro for GLF DVD project.	Standalone projector files are ready as promo email attachments, or to be included as DVD-ROM or CD-ROM content for fans to email to friends.
Web	**DVD**	**DVD-ROM, CD-ROM and Email**

Figure 22-4: The flexibility of Flash. The cartoon animation was intentionally created for DVD; yet Flash allowed for easy delivery of content in a variety of ways. The Web version of the animation is accessible at www.psrecords.net/glf/shegerbil.

Real World Examples — Simulating DVD

Now for some fun!

Warning: This section assumes some familiarity with various applications such as Flash. The resource files are available to open up in the Tutorial/CH22 folder on the DVD-ROM.

The goal with the Gerbil Liberation Front DVD is to reach as wide of an audience as possible. The idea with cross-purposing content meant trying as many different ways as possible to convey aspects of the content. The limitations in quality with some of the delivery options was not a concern, because the alternative versions of the DVD content were promotional in nature, and the desire was not to have them so good or so complete that they would compete with the actual DVD.

Simulating DVD in Flash

Flash was appealing because it offered the ability to do completely standalone files, but the rub was how to incorporate video into Flash.

It was understood that in Flash 5 the only "legal" way to integrate video with Flash is to import QuickTime files, build interactivity, and then publish the whole thing as QuickTime, which would save the Flash content as a QuickTime track. The disadvantage of this option is reliance upon the QuickTime Player, as the final file is QuickTime. The advantage would be video quality, and easier to put together than in Director.

Another approach was also considered, called *frame exporting*, which is based on the ability Premiere gives you of exporting a video file in individual frames. The essential principle for the video was to export the original video clip at a very small size, 160 × 120 pixels, as a series of PICT files for each frame, then import them into Flash.

A simulated DVD interface was developed for use in either version.

Simulating the DVD interface

To begin with, the original Photoshop file for the DVD was opened up, and I used the Save for Web function to save each screen in the DVD as a separate JPEG file, so that I could easily import them into Flash.

I then created a layer for each screen, and started developing buttons for the main menu screen. (See Figure 22-5.)

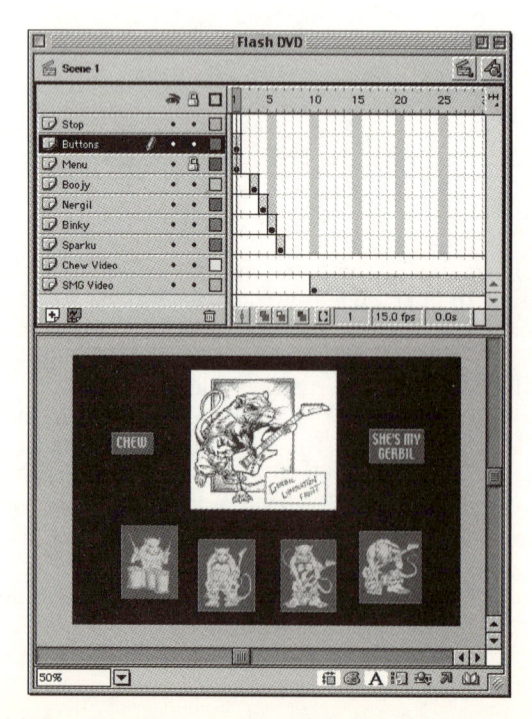

Figure 22-5: The GLF Flash DVD. In the Score window, layers were set up according to the main screens that a person could access, equivalent to the separate menu tiles in the main DVD Studio Pro project. Two additional layers were set up below for the "video" frames. The Preview area shows the buttons created in Flash, which lead to the separate selections.

The approach for the buttons was to simulate the hilite feature of DVD players, so I first went to each character along the bottom of the screen, and in a separate layer I would draw a rectangle, size it so it would roughly correspond to the size of the character underneath, and then use the Edit⇨Cut function to save the rectangle in the clipboard memory. Then in the Library window I chose New⇨Symbol, made it a button, and in the button editing area I pasted the rectangle in the Over frame and subsequent frames, because I didn't want anything to appear until a user rolled over the button area. (See Figure 22-6.)

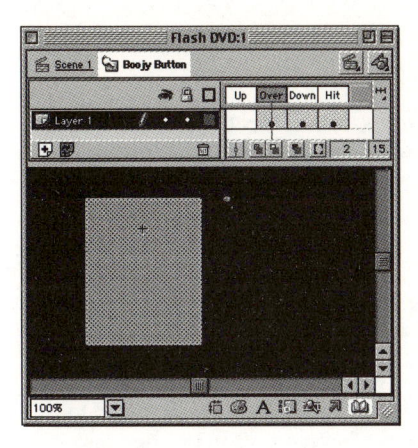

Figure 22-6: The simulated DVD button; a gray rectangle, drawn in Flash and then set to Arrange⇨Group to make it easy to move around.

To achieve transparency when the button was placed over the individual choices on the main menu screen, I set the button's Effect setting to Alpha, and chose 71%. (See Figure 22-7.)

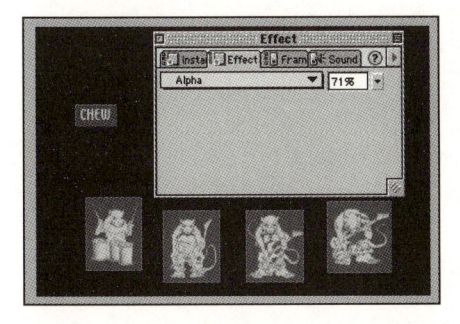

Figure 22-7: The main menu screen in the GLF Flash DVD, with the button's Effect set to Alpha, to achieve transparency.

After I created the Effect, I set each button so that it would jump to the frame where I had dropped the screen I had saved from the original Photoshop file that I used in the DVD Studio Pro project. Instead of a Menu tile leading to other Menu tiles, a screen in one frame jumped to another frame where the other screen was represented. In order to prevent the Flash animation from cycling through every frame, I set the Action for the buttons to go to a particular frame and stop. (See Figure 22-8.)

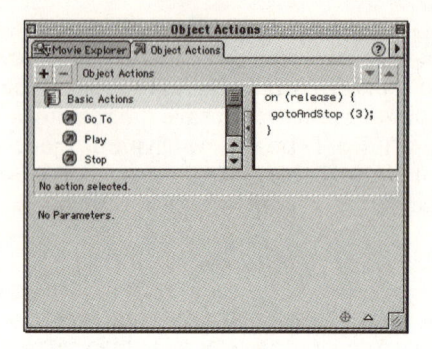

Figure 22-8: A sample of a simulated DVD button's Object Actions setting, going to a particular frame and stopping on that screen.

Flash makes it easy to add actions to an object — you select the object, open the Actions window, select an action, and customize it if you need to. Setting up was fairly straightforward. (See Figure 22-9.)

As soon as the subscreens were set up, I created a button that led back to the main screen, and I was able to use the same button in each of the subscreens. Again, a Go To And Stop Action was used instead of Play, so that Flash would stop at the first frame.

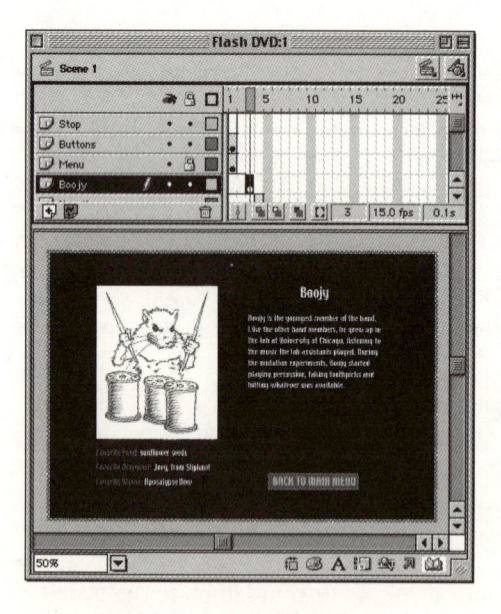

Figure 22-9: The Action setting in this figure is for a button off the main screen that leads to the page about Boojy, drummer for the GLF. The button jumps the movie to Frame 3, shown in this screen.

Simulating the DVD video

After getting all the basics set up, I then opened up Premiere and investigated options for exporting a QuickTime version of the videos I had created for CD-ROM delivery. (See Figure 22-10.) I had imported the QuickTime files, and then basically used the Export feature to generate a series of numbered PICT files. (When you export as PICT, Premiere automatically numbers the files, so that smart programs like Flash can recognize a sequence and automatically import a large range of files.)

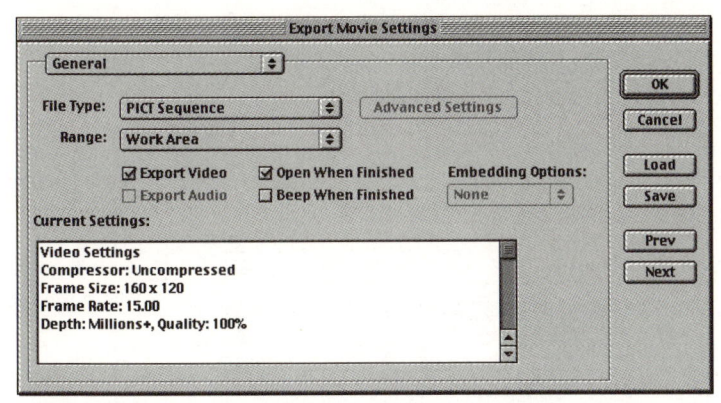

Figure 22-10: The Export Settings dialog in Adobe Premiere. In the General Settings, the File Type is set to PICT Sequence, which generates a series of numbered PICT files.

Because I only wanted a representative sample of frames from the video sequence, I actually cancelled the Export option when there was about 100 frames exported. I tried exporting as GIF, but doing so didn't turn out too well, so I went with Uncompressed and a bit-depth of millions of colors, and decided to allow Flash to do the compression on the images.

After I had a folder on my hard drive with the numbered PICT files, instead of importing the files directly into a layer of the Flash file, I chose to make a Movie Symbol and import the files there. (See Figure 22-11.) I also imported the appropriate audio file and let it run underneath. I think my motivation was to keep the audio with the individual frames and to be able to use one layer in the main Flash score for the video simulation.

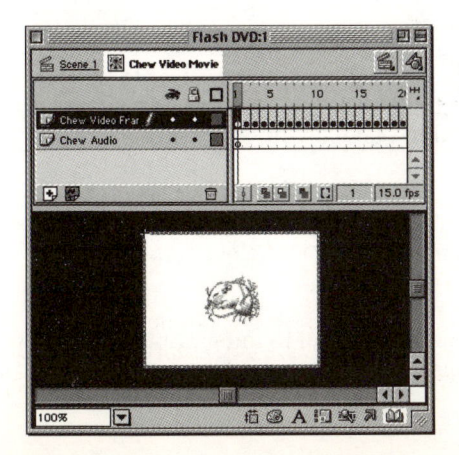

Figure 22-11: The Movie Symbol in Flash, showing a single frame of the miniature version of the video used in the DVD.

After setting up the Movies, I inserted them in the main layers, set keyframes up and let them run until their last frame, and then set an action to bring the Flash movie back to the main screen after the simulated video was done playing. (See Figure 22-12.)

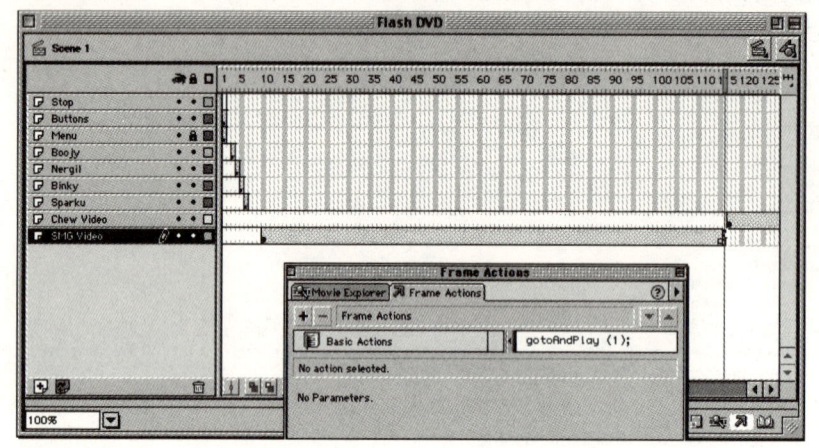

Figure 22-12: A zoomed out view of the GLF Flash DVD, with the frame at the end of the video sequence, showing the Action setting which takes things back to frame one.

The way that the files had been shaping until this point in the library was pretty straightforward, some JPEG files and the buttons I'd created. When I started importing the video files, I ended up creating folders in the library to hold all the bitmap images for the individual frames, so that it didn't become a mess. (See Figure 22-13.)

Essentially it worked, and if you're curious about the file size, check it out in the Tutorial/CH22 folder on the DVD-ROM. Flipping through 15 frames per second at 160 × 120 pixels was not that big of a deal, and it was nice to be able to publish a standalone projector that was completely self-sufficient, requiring no plug-in, no player, no nothing.

With tweaking of the JPEG compression and using other tricks, the file size for something like this probably wouldn't be too bad. A tad larger than may be normal for e-mailing, but still something that wouldn't take up much space as a promo CD-ROM giving fans a taste of the DVD.

I knew that Flash couldn't be as efficient at displaying video frames as RealVideo, but I figured that being able to have a completely self-sufficient standalone file would be worth it, and I was pleased with the result.

But to compare the quality that QuickTime would lend to integrating Flash with video, I had to try importing true video files into Flash, as well.

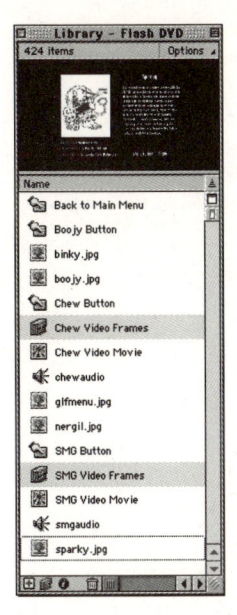

Figure 22-13: The Library window for the GLF Flash DVD, showing buttons, the main screen jpeg files, and a glance at the future, with the folders created to hold all the individual frames that are used by the Movie Symbols.

Using QuickTime with Flash to simulate DVD

Continuing with the basic DVD interface simulation that I had set up in Flash, I proceeded on the basis of being able to use the Flash interactivity as a track in QuickTime, with the QuickTime Publish option in Flash 5.

I went back and picked up the 320 × 240 pixels QuickTime versions of the GLF music videos that had been used in the DVD, and imported them directly into the Flash project file.

Figure 22-14 shows Frame 3860, representing the last frame of the QuickTime file that had been integrated with Flash, with an action set to return to frame one when the video is done playing.

The kooky thing about working with video in Flash is that you end up way out in the boondocks as far as frames go. The number of frames I used was even more than I usually use because I had misgivings about having two QuickTime files sitting directly on top of each other.

The initial result was not the most attractive — partly because I had developed the Flash interface at 640 × 480 pixels but was exporting from Flash at 320 × 240 and then 400 × 300. Resizing and adjusting graphics with the end size in mind would help in that respect. But there it was before my eyes, DVD in QuickTime. (See Figure 22-15.) The result can be found in the Tutorial/CH22 folder on the DVD-ROM that accompanies this book.

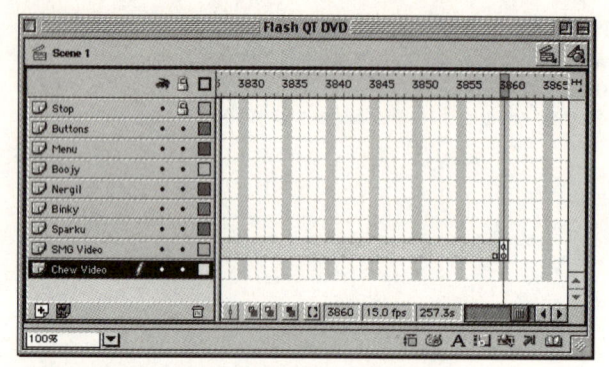

Figure 22-14: Frame 3860, the end of the line for a QuickTime file in Flash, with an action set to return to frame one when the video is done playing.

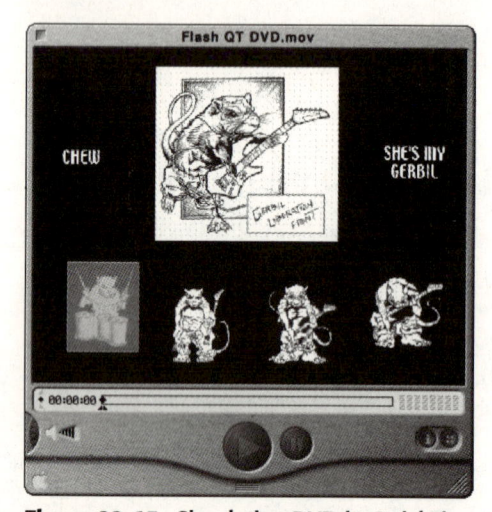

Figure 22-15: Simulating DVD in QuickTime. The video file has a Flash track. Both music videos and the Flash interactivity have all become one contiguous file.

When you play a simulated DVD in QuickTime, the QuickTime Player enables you to click on the buttons, so the rollovers and button jumps from screen to screen still work.

After completing most of the settings and testing the simulated DVD, I realized I had no way for the viewer to get back to the beginning of the video, no simulated Return button for the DVD. Of course a person can use the controls in the QuickTime player to do so, but I wanted it to be a closer simulation, so I set up a button layer above each QuickTime layer in Flash. I wanted a Back option to be available. I set the Back button to be semi-transparent so that it would interfere less with the video. (See Figure 22-16.)

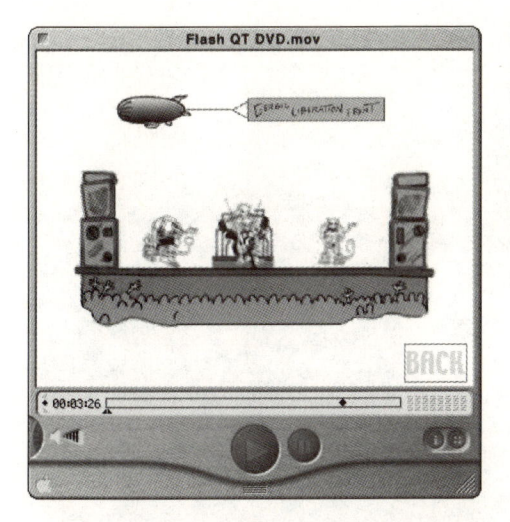

Figure 22-16: The simulated DVD, during the music video; in the lower-right corner, the rollover state of the button is shown, with a blue rectangle. Normally the button is a semi-transparent gray. The Action setting on this button brings the viewer back to frame one, where the Main Menu screen is.

I was pleased with the results in QuickTime. The look and feel of the DVD was preserved, and having everything bundled in one contiguous QuickTime file was nice. The final file size was about 50 megabytes, which is suitable for CD-ROM or DVD-ROM delivery.

Simulating DVD with Adobe Acrobat (PDF)

Not content with just doing Flash and QuickTime versions of a DVD? Perhaps you want to take advantage of the QuickTime feature of Adobe Acrobat. Acrobat Exchange enables you to create Acrobat documents from the print function in a file menu in a program like Microsoft Word or by piecing individual elements together from various sources such as Photoshop PDF, and so on.

Acrobat is a common format for transferring printed documents, but the Acrobat reader is becoming more popular as a way for including electronic versions of printed manuals on disc, for example.

Acrobat has a lesser-known feature — the ability to integrate QuickTime movies. The QuickTime movies remain as separate files, but you can actually embed the movie so that it plays within the document.

Simulating a DVD with Acrobat and QuickTime results in a file that is cross-platform compatible, as long as the user has QuickTime and the Acrobat Reader.

Yes, using PDF to simulate a DVD is a little crazy, but it works.

In the context of the ongoing attempt to simulate the GLF DVD, I wanted to try Acrobat, as a means of integrating the video directly within a larger document. Kind of like using Macromedia Director but about $1000.00 less costly. After you get the hang of it, it's pretty simple.

First, I opened up the original PSD file that I used for the DVD Studio Pro project file and saved each screen as a Photoshop PDF file; then I created a special startup page for the overall PDF DVD simulation. (See Figure 22-17.)

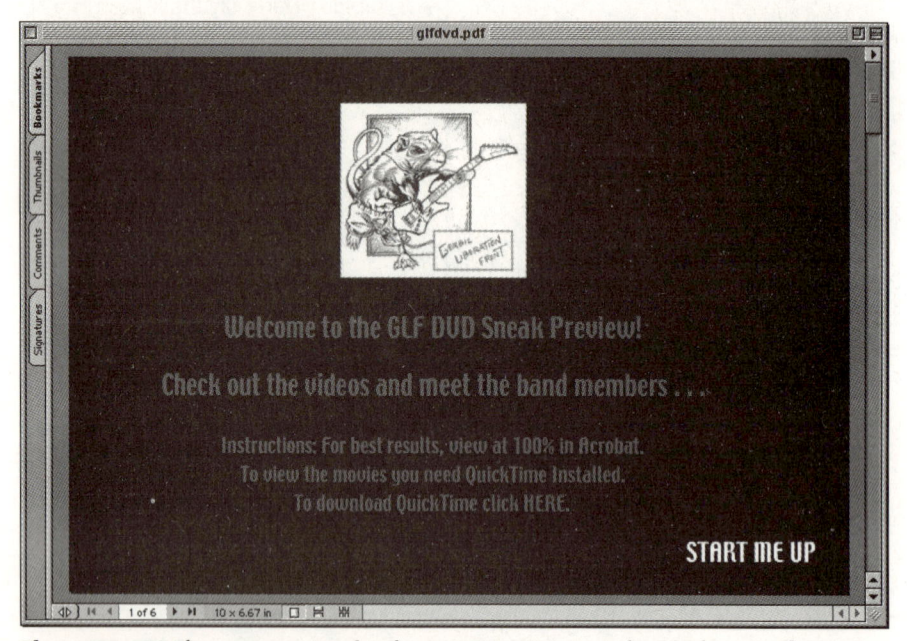

Figure 22-17: The startup page for the GLF PDF DVD. Acrobat Exchange allows you to insert hotspots in a PDF page, which can lead to a Web link, or to another page.

On the start-up page for the GLF PDF DVD, the reader is instructed that QuickTime is necessary to have to view the videos, and a hotspot is given to allow the reader to click directly for a link to download QuickTime. I felt like it was necessary to have some simple instructions before getting into the nitty-gritty of the PDF DVD.

You can go from page to page by using the normal arrow buttons in the Acrobat Reader to navigate, as though you were reading a book, but most people go for the hotspots.

I created a slightly different version of the main DVD menu in Photoshop and placed thumbnail images of the videos sized to 160 × 120 pixels. Clicking the thumbnail images or the text below the images is the equivalent of selecting the DVD button. (See Figure 22-18.) In PDF, you can't do rollovers, but when you do roll over a hotspot, the hand changes to a hand with an index finger.

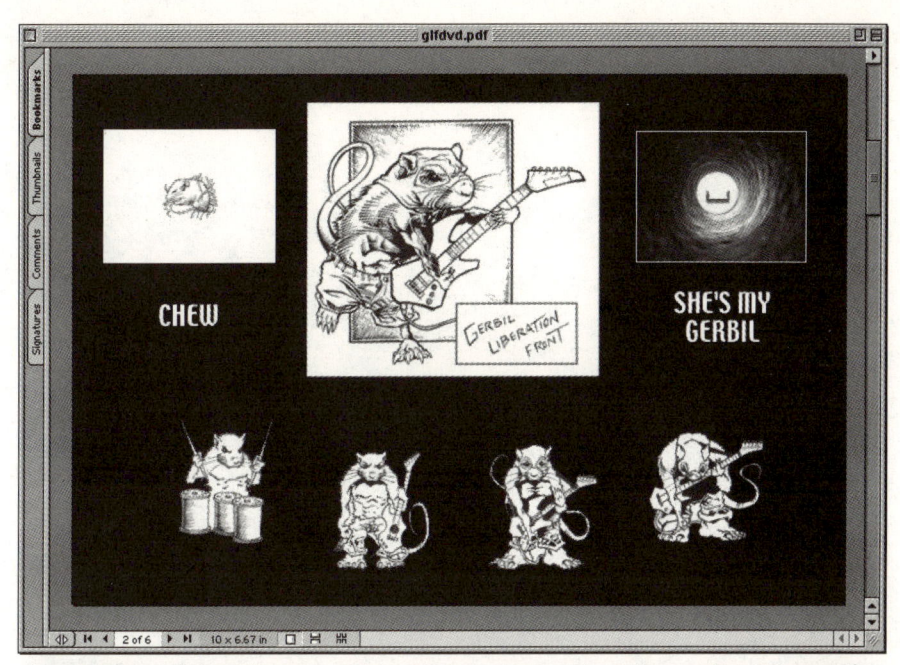

Figure 22-18: The main menu of the GLF PDF DVD. When the reader clicks on the movie thumbnails, they start playing. Clicking on any other hotspot takes you to a different page.

Acrobat has an easy to use Movie tool that I used to define the area I wanted the QuickTime movies to play, right over the thumbnail images. (See Figure 22-19.)

Figure 22-19: The Movie Properties dialog in Acrobat enables you to select some basic settings for displaying QuickTime movies in PDF documents.

After I tweaked the QuickTime movies, I set about inserting the additional pages that represented the submenus for the DVD, and it went smoothly. Much the same as the Flash DVD simulation, simple hotspots were set up to go to and from the submenu pages, using the link tool in Acrobat. (See Figure 22-20.)

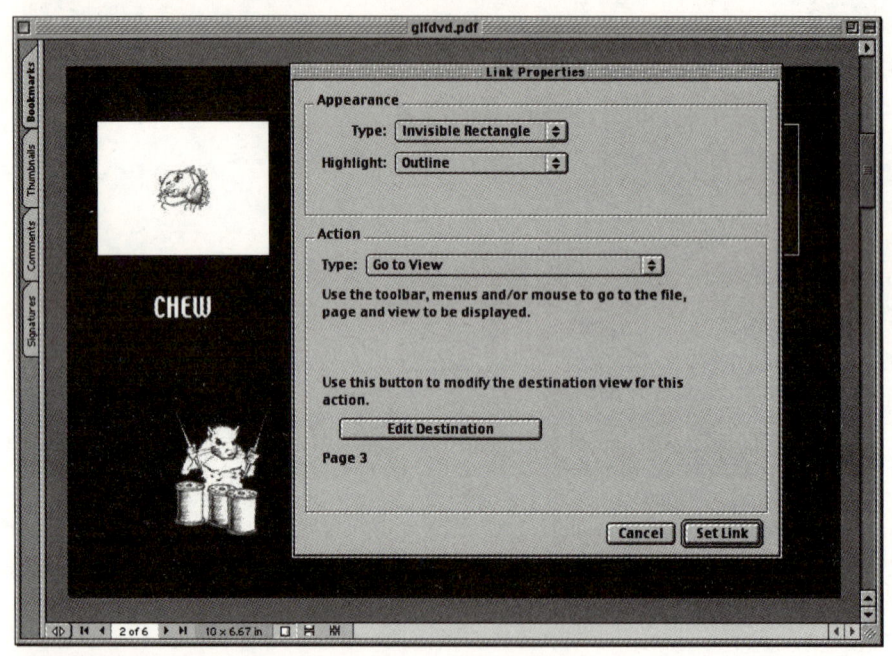

Figure 22-20: The Boojy hotspot on the main pdf page, where a hotspot is defined leads to page three of the overall document.

Overall the whole process went pretty smoothly, and the only thing that tripped me up is the annoying habit of Acrobat documents appearing as fitting to the size of your screen, rather than appearing at actual size, 100% of original size. In other words, when I tested it, the GLF PDF was blown up to fill the entire screen, and I wanted it to show up at a specific size.

To do this in Acrobat, set a Page Action by choosing Document➪Set Page Action. (See Figure 22-21.)

You can execute a menu item through a page action, meaning you can have a hotspot perform an action, such as if someone had accessed the menu to resize the document. I added the Execute Menu Item Page Action and set it to Actual Size. (See Figure 22-22.) Now, when the PDF document is opened, even when Acrobat Reader tries to display it to fit the screen, it goes back to 100%.

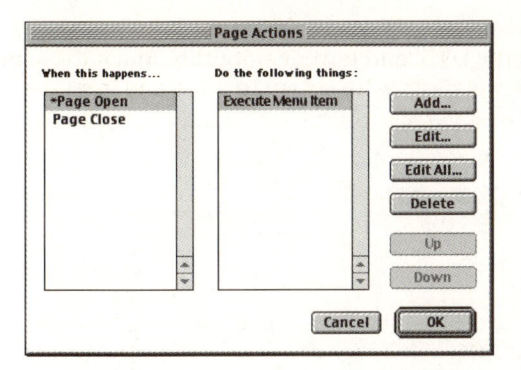

Figure 22-21: The Page Actions dialog in Acrobat with the Execute Menu Item option added.

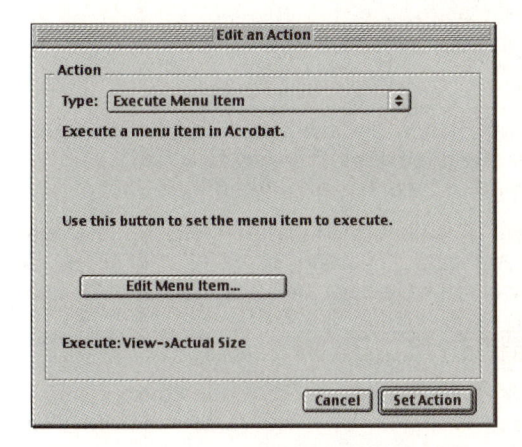

Figure 22-22: The Edit an Action dialog, where you can click Edit Menu Item to have Acrobat automatically set the document to Actual Size. Acrobat Reader typically expands any PDF document to fill the entire Acrobat window, which means it might re-size your document to larger than 100%. Using this method, when a person launches the document, you can ensure that the opened document retains its original dimensions.

After going through the process of figuring out how to simulate a DVD in Acrobat, this method seems like an easy way to get a nice integrated QuickTime movie on a screen with a minimum of hassle. It seems like it could be an easy way to prototype simple DVD interactivity as well (See Figure 22-23).

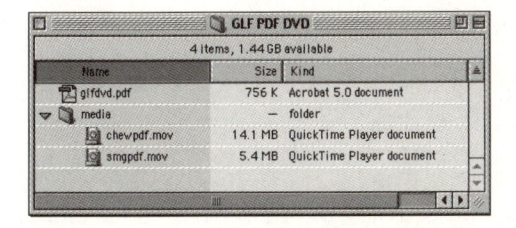

Figure 22-23: The GLF DVD in PDF format. Truth is stranger than fiction. The QuickTime files are external but run within the document. Wacky stuff!

I found the process of simulating DVD in Acrobat to be an enjoyable and useful exercise. I think that this PDF technique may be most useful when there is a need to incorporate such content into a larger document, such as documentation that may already be in PDF format. I think that this technique has possibilities for exposure of certain content on CD-ROM and DVD-ROM delivery as well, especially where other PDF documents are being conveyed.

Acrobat gives you an easy way to have a simple, nice looking functional simulation of DVD interactivity.

Summary

In this chapter, we looked at the varied possibilities of what you can do to take content beyond the DVD disc. We investigated delivery options, became acquainted with typical possibilities for disc and Internet delivery, and considered some of the programs people often use. Finally, we visited the very edge of DVD civilization, the outpost known as Simulated DVD, where rules have not developed quite yet.

- Paying passenger content is content that needs to be complete, either because expectations are higher or because the content is being marketed to a person for potential purchase. An example relative to DVD may be releasing the equivalent of the DVD on CD-ROM for people who don't have a DVD player.

- Paying passenger pigeon content is content that is printed out, rolled up into a tiny scroll, and sent by passenger pigeon. A passenger pigeon may not be able to carry a DVD, though African swallows have been known to do so on occasion.

- Stowaway content is content that is borrowing space from some other project, whether it is on a disc or on the Internet. The purpose of stowaway content can be to provide an excerpt of another project. In the context of DVD, various delivery methods that can't match the video quality or storage capacity of DVD may still be viable options for promotion.

- The four typical options for disc delivery are DVD, DVD-ROM, CD-ROM, and Video-CD.

- Internet delivery allows for streaming media, which can contain video. The quality is very limited but is ideal in situations where the desire is to reach a wide audience.

- Flash is an ideal format for cross-purposing content between disc and Internet delivery formats because it can generate high quality video, Web-ready streaming animation files, and standalone interactive files for disc or e-mail delivery.

- DVD can be simulated effectively in such interactive formats as Flash, QuickTime, and even Acrobat. The experience of a DVD can be simulated through a combination of basic interactivity and compressed video clips.

In the next several chapters, we look at case studies of some DVD authors who have used DVD Studio Pro in one way or another and encountered interesting possibilities and challenges along the way.

✦ ✦ ✦

Case Studies

PART

VII

In This Part

Chapter 23
Case Study: Warm Blankets

Chapter 24
Case Study: Atomic Paintbrush Studios

Chapter 25
Case Study: Pioneer Electronics' SuperDrive

Chapter 26
Case Study: Simple Motion Menus

Chapter 27
Case Study: Metatec — Manufacturing a DVD

Chapter 28
Case Study: RealWorld Promotional DVD

CHAPTER 23

Case Study: Warm Blankets

Background

Warm Blankets is an organization that helps orphans in Cambodia by building orphanages and providing caregivers to teach and raise children in a loving environment (www.warmblankets.org). As another part of its mission, it provides outreach to communities on day-to-day issues related to quality of life, such as basic hygiene.

For several years, Warm Blankets has been traveling to Cambodia to document the conditions in which people are living and to teach Cambodians about health topics, such as disease — a matter of great concern in that part of the world — because Cambodia has some of the highest rates of infection in southeast Asia. In addition, Warm Blankets makes regular visits to Cambodia to record the progress of the homes and schools it has built and supported in various communities.

At its home base just outside Chicago, Illinois, Warm Blankets produces short videos and multimedia programs to solicit potential donors and for teaching purposes. In both cases, using the right medium to do the job and to reach the desired audience is important. (See Figure 23-1.)

To reach a wider audience of investors, Warm Blankets has considered producing a DVD that can be shown in portable presentations as well as in homes. Ordinarily, presentations are given with laptops in a conference room environment, which draws a limited, albeit a desirable, range of interested people. With a DVD presentation, potential donors or volunteers can also peruse a disc at their leisure in the privacy of their homes, exploring aspects of the company that are of interest to them. A DVD project for a non-profit organization with these goals serves as a good example, because it provides insight into planning for a particular audience and medium while touching upon questions of budget and distribution. The following is a case study for Warm Blankets describing some of the details that went into planning and producing the DVD.

Figure 23-1: Warm Blankets supports the needs of orphans and communities in Cambodia.
Image courtesy of Warm Blankets

Project Planning

The first planning stage of this project began with a meeting between the DVD project producers, Todd Kelsey and Chad Fahs, and the directors of the Warm Blanket organization. Establishing the exact needs and requirements of the project from the beginning is important.

At issue were the benefits of DVD. Until now, the majority of Warm Blankets' presentations were given directly from the hard drive of a portable computer. This gave Warm Blankets the flexibility to display presentations in video, Microsoft PowerPoint, and other formats. However, given the amount of video that had to be shown and the relative quality levels of their current MPEG-1 presentations over the MPEG-2 of DVD, it was determined that DVD was a desirable option in terms of storage capacity and quality.

Warm Blankets found the interactive aspects of DVD particularly interesting. Because Warm Blankets had a large variety of video clips that they wanted to both show in presentations as well as archive in a single location, DVD was a logical choice for putting these materials together in an attractive interface. For example, Warm Blankets has video tours of many homes that they support in Cambodia. With DVD, a map image could be used for navigation, with each dot or home on the map linked to the appropriate video clip. This format would provide a logical, elegant means of selecting a particular location or video — something that certain formats, such as VCD (video CD), can't offer. (See Figure 23-2.)

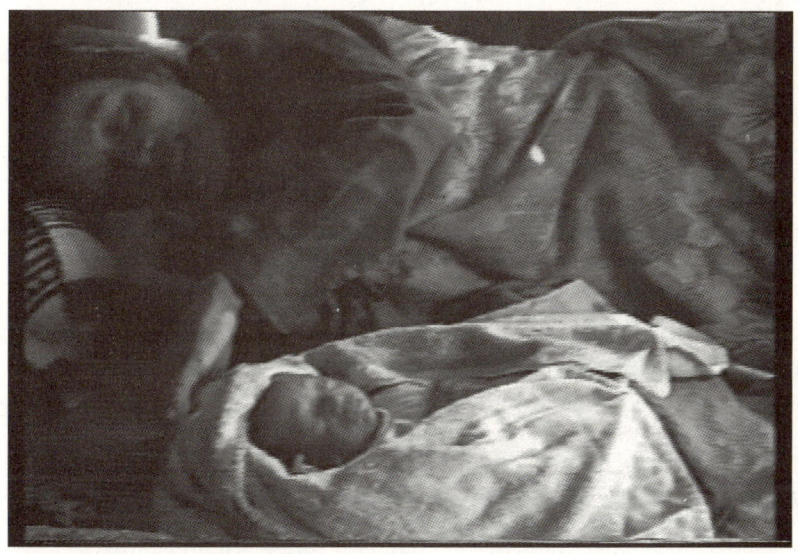

Figure 23-2: "The Good Khmer Mother" video was converted to video CD format. It provides information on hygiene, ranging from the preparation of food to the disposing of waste. The above image is a mother from the Khmer tribe and a doll that is used for educational purposes.
Image courtesy of Warm Blankets

However, for some of Warm Blankets' more immediate needs — where issues of quality or interactivity were not important — VCD remained an option for instructional videos that could be shown to communities in Cambodia. Many people in Cambodia possess VCD players, which are more popular than VHS.

One particularly important factor that had to be decided was whether DVDs would be "one-off" duplicated DVD-Rs (burned as needed), or manufactured, replicated DVDs. This was an important decision, because presentations are often given on portable computers, which may be equipped with DVD-ROM drives but not necessarily compatible with DVD-R media. (Most, but not all, DVD-ROM drives will play DVD-R media). To make use of the DVD-Rs, Warm Blankets would need to test the discs out first to find a portable that works with the DVD-R media. Replicated DVDs, on the other hand, play on any DVD or DVD-ROM capable player. The only drawback to manufacturing these discs is the price, which may be viewed as somewhat prohibitive for some. In the short term, Warm Blankets decided that DVD-Rs would suffice until a particular need for distributing the discs was determined. Also, the option remained to seek out a manufacturer who could donate a short run of discs.

Development Issues

Many of the issues facing the development of any DVD project remain the same. First, the layout and navigation of a disc must be determined. This is one of the most important considerations, as it affects a viewer's ability to access the material on a disc. Without proper navigation, a project could fail to keep a viewer's interest or may even frustrate the viewer. With this project, Warm Blankets wanted to provide access to material in a simple, attractive, and logical manner, which would hold the viewer's interest.

As mentioned in the previous section, one form of navigation considered was creating an interactive map, providing access to video clips associated with a particular location. This could be accomplished by first obtaining a copy or picture of the map, scanning it into Adobe Photoshop, and then retracing it to produce a simpler design. The map could also be recreated in other graphic applications such as Adobe Illustrator, Macromedia FreeHand, or even Flash. The map just needs to be big enough and simple enough to be easily navigable. The outlines should be more than 1 pixel high to avoid flickering, and the colors should be within video safe limits (discussed in Chapters 5 and 8). Also, we needed an overlay graphic from Photoshop to trace circles to mark locations on the map and to act as buttons for access to video clips or other menus. After the map graphic was created, it could be brought into Adobe After Effects to composite some motion elements, such as moving pictures of faces or locations to add interest.

Motion Menu Loop Points with QuickTime Pro

For other menus, Warm Blankets wanted to use video clips that would loop (meaning, sections of video that repeat once they have played). This was accomplished by using the Loop function for a motion menu (discussed in Chapter 6), located in DVD Studio Pro's Property Inspector, shown in Figure 23-3.

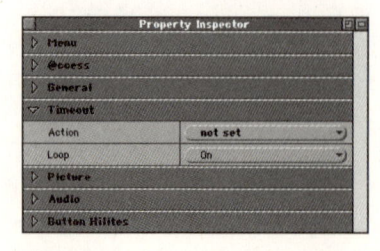

Figure 23-3: The Loop function in the DVD Studio Pro Menu Property Inspector enables you to loop video for a motion menu.

We created some of these loops by using QuickTime Pro — without the need for a separate video editor. Creating and testing loops with QuickTime Pro is fast and easy.

1. **To create loops with QuickTime Pro, we first found the DV footage file that we originally imported for our main video program.**
2. **After locating the file, we double-clicked it to launch the QuickTime player.**
3. **After the video opened in the QuickTime window, we searched for an appropriate portion to loop for a video menu.**
4. **After finding a good portion, we dragged the in and out point indicators at the bottom of the Timeline (the line directly beneath the main video window) to mark that portion of the video that we wanted to loop (see Figure 23-4).** This process of marking points on a video clip is similar to Apple's iMovie.

Figure 23-4: Selecting a portion of video to loop is quick and easy with QuickTime Pro. Here's a clip of a man boating down a river in Cambodia.
Image courtesy of Warm Blankets

5. **With the video selected (now marked in gray), we chose Edit ➪ Copy (⌘+C) to copy the marked portion.**
6. **We chose File ➪ New Player to open another QuickTime window, and selected Edit ➪ Paste (or ⌘+V) to paste the video clip into a new player.**

7. **After the video was in a new window, we tested the loop by selecting Movie ⇨ Loop to see how well it would work.**

8. **We made a copy of the entire loop by selecting Edit ⇨ Select All (⌘+A), choosing Edit ⇨ Copy (⌘+C), and pasting it back into the same window a few times by choosing Edit ⇨ Paste (⌘+V) in order to create a longer, smoother clip.** By repeating a loop several times like this, you decrease the chances of a viewer experiencing a pause in the video while it loops back on itself when it reaches the end of a sequence. After we had a usable file, we decided to export it for use as a background element to composite in After Effects. For a few other menus, we created a background image by extracting frames from the main video program. Then we used Photoshop to add buttons, which could be exported to After Effects to place over a motion background. A stroke was also created for the button layer, which could be used as an overlay for the highlights. (See Figure 23-5.)

Figure 23-5: Simple buttons were combined with the logo and a digital picture for the beginnings of a simple interface idea. Buttons were composited onto a background image taken from the main video program.
Image courtesy of Warm Blankets

We also played with the logo taken from the main video and made overlays on certain elements. Using the Magic Wand tool in Photoshop, we selected portions of the graphic that would make interesting buttons for activating video or other features. In this case, we decided to forgo the logo as a button idea and stick to menus with clearly defined buttons. (See Figure 23-6.)

Figure 23-6: Using the Magic Wand tool in Photoshop, you can select portions of a graphic to be used as overlays in a motion or still menu — such as the "squiggly" line in this logo.
Image courtesy of Warm Blankets

Summary

The Warm Blankets Foundation is doing important work and changing the lives of many people in Cambodia. Hopefully, the use of a medium like DVD will bring its message more readily into the hearts and minds of people who are willing to contribute their time and resources. By educating people and presenting the facts of what is happening in Cambodia, Warm Blankets hopes to aid children and families who desperately need assistance. To find out more about the organization, visit www.warmblankets.org.

✦ ✦ ✦

CHAPTER 24

Case Study: Atomic Paintbrush Studios

In this chapter, you have the opportunity to look at a DVD project created to showcase Atomic Paintbrush Studios, an independent illustration and animation company based in the New York City area.

The project was created as a portfolio item to show potential clients by Atomic Paintbrush, with the intention of demonstrating as many of the features of DVD Studio Pro as possible. The results are viewable on this book's DVD-ROM when you insert the disc in a DVD player.

Conversation with Atomic Paintbrush Studios

This interview with Dennis Calero, illustrator and DVD author, co-founder with Kristin Sorra of Atomic Paintbrush Studios focuses on the intent and creativity Calero and Sorra used to create a DVD project.

Who are you?
Kristin and I graduated from Pratt Institute in 1994 and '95. We studied illustration and design and started Atomic Paintbrush, which means different things to different people. To some, it's a graphic design and video production studio; to others, it's the official face of our individual illustration projects. We've done everything from computer coloring comic books and package illustration, to storyboard and video sleeve design.

Figure 24-1: The main DVD menu for the Atomic Paintbrush showcase DVD.
Image courtesy of Atomic Paintbrush, Inc.

What was the purpose of the project?
The purpose of creating this DVD was two fold: to take some of the properties that we were trying to market and put them in the context of a sophisticated presentation, and also to try our hand at DVD authoring using an affordable new tool. (See Figure 24-1.)

What features were used?
I think we used just about every feature in the program, from still and motion menus, galleries, subtitles, and multiple audio streams, to scripts. Some successfully, others . . . not so much. (See Figure 24-2.)

What scripts did you use?
We used scripts to allow viewers to switch between audio streams and to elect to turn subtitles on or off, which they can do with a DVD remote anyway, but we wanted to mimic features found on commercial DVDs as closely as possible.

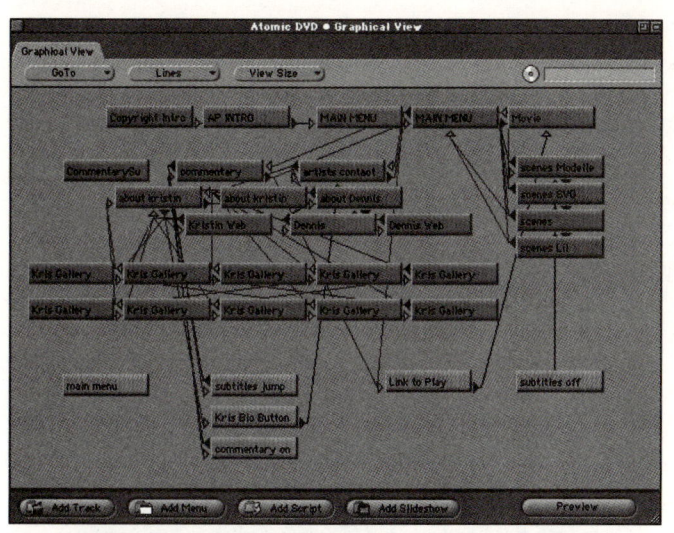

Figure 24-2: The Atomic Paintbrush DVD project in the Graphical View window of the DVD Studio Pro project file. The Lines function is set to display the interactive relationship of various project elements.

What hardware did you use?
For the DVD itself, we used a Mac G4 733 with the SuperDrive.

What third-party applications did you use?
We used Adobe Photoshop to do all of the layered files (Figure 24-3) for the still menus.

The content itself was created using Macromedia Flash and Adobe After Effects, and we used After Effects again to create the motion menus.

Do you plan to release on DVD-R or manufacture a short run at some point?
Not at this point, no, but we have done some research into long form production of glass masters from a DVD-R or a DLT and it seems we would have no problem.

What was the typical workflow?
Each project began with the writing and storyboarding of the cartoon. We had to be concerned with giving a taste of the show, without being too concrete on details. On one level, people want to see the material as finished as possible. After all, they want to know what they're getting. But an another level, there are going to be a lot of cooks wanting to throw their individual little morsels into the pot, so there had to be room for that, too.

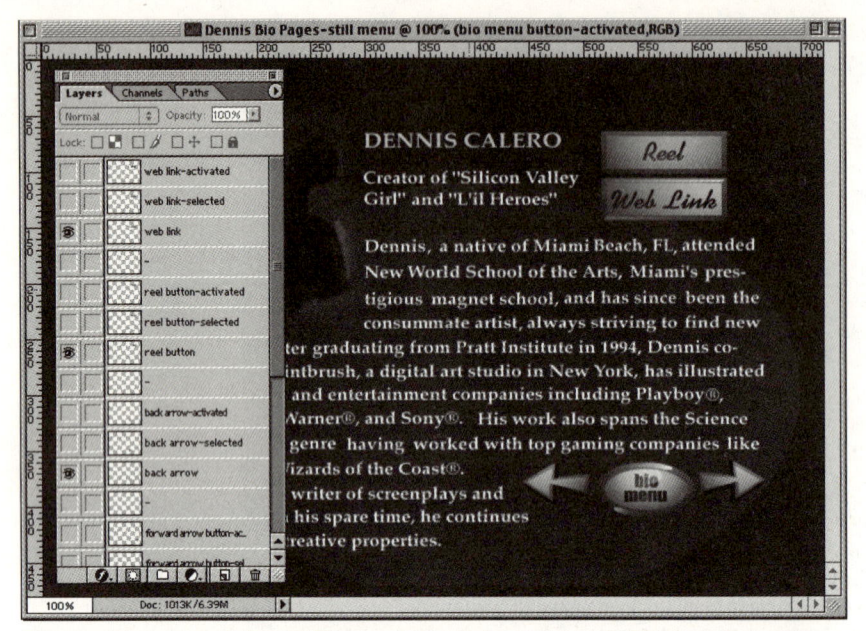

Figure 24-3: One of the multilayered Photoshop files used for a menu screen in the Atomic Paintbrush DVD project, along with the Photoshop Layers palette.
Image courtesy of Atomic Paintbrush, Inc.

For *Modelle*, we decided on animating using Flash, because it was an easy tool to use for tweening animation, which basically means that you do a limited number of drawings and let the computer interpolate frames in between, much like animatic work done for advertising. Also, animating in Flash would let us put it on the Web should we ever decide to do so.

Once the basic animation was done in Flash, we exported it as a QuickTime file and then worked on it in After Effects, adding special effects and sound, which we recorded in SoundEdit. While the concepts were developed individually, all the art and animation and storylines were developed concurrently.

Modelle basically came out of Kristin's experience with the fashion industry, which is its own strange sort of world. Fashion business logic seems to have very little to do with logic in the larger sense of the word. So we took a lot of that and created a character to be our point of view, someone who was pretty much totally normal, with the exception of having some ambition. (I mean, she's supposedly from some podunk town in the midwest and she gets on a bus and goes to New York with a definite goal in mind, which certainly is *not* normal.) So she's dropped into this world of insanity and quickly becomes indispensable in her environment, not because of her skills, which she's still developing, but because she refuses to have her even-headed view of the world tainted by the day-to-day nuttiness that goes on around her.

Silicon Valley Girl basically came from the title, a pretty obvious play on words. So it obviously worked best as a satire. I guess the ultimate goal with something like this is to market it to the *Barbie* girls who've discovered their computers. After all, the Barbie CD-ROM series is among the best-selling in history. Be that as it may, I went in for the satirical bent of it, because that's what interests me.

Could you comment on how you did the motion menu?

We only used motion menus in the scene selection menus, which might seem odd, because each feature is only 4 or so minutes long. But we felt, if we were going to show this as an example of DVD authoring as well as a medium for our properties, that motion menus were an important feature to demonstrate what we could do.

The animation was created in After Effects with the overlays created in Photoshop. (See Figures 24-4, 24-5, and 24-6.)

Figure 24-4: A Photoshop file was prepared for importing into After Effects, where video was added, so that each of the green frames contained separate clips. The resulting file was used as an asset for a motion menu in the DVD Studio Pro project file.

Image courtesy of Atomic Paintbrush, Inc.

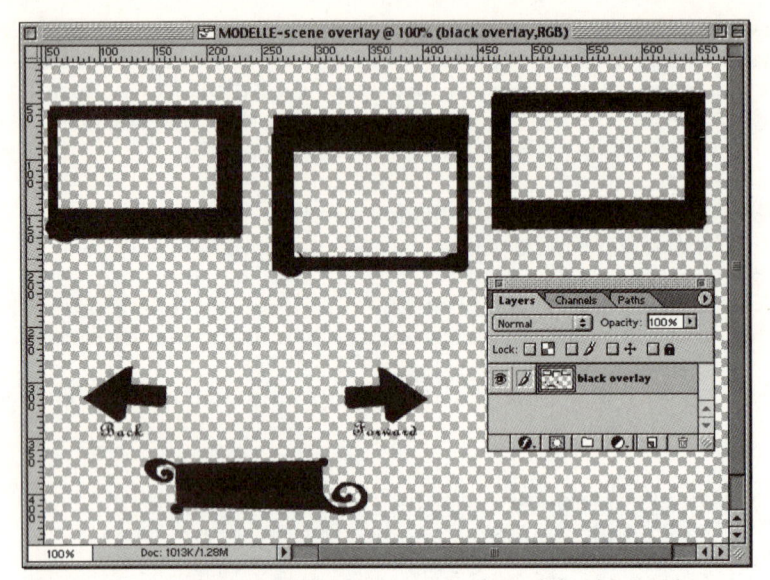

Figure 24-5: A Photoshop file representing the solid color overlay, which was used in the motion menu. The foreground elements are black, and everything else is transparent. (You can create a new Photoshop file with the background set to transparent.) The foreground solid areas define where the hilites will appear when a user is selecting menu choices on the screen. Compare this figure with Figure 24-4.

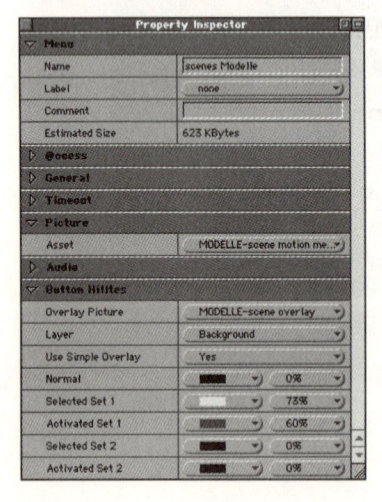

Figure 24-6: The Property Inspector for a menu tile in the Atomic Paintbrush project that has been set up to be a motion menu. The Picture/Asset has been set to the prepared video clip, and the Overlay Picture has been set to reference the Photoshop file shown in Figure 24-5. The Selected Set value in Button Hilites determines the color and transparency of the overlay image.

Could you comment on why and how you made the video one contiguous track, using story and chapter features, rather than using separate MPEG-2 files?
It was simply the more elegant way to go. I also got the impression that separating the features into separate MPEG-2 files would make the total file size larger, which may or may not be true. It was mostly just the convenience of not having to deal with 4 files instead of one. Author's Note: Dennis made one video file and used markers within the video file allow users to jump to various locations.

@ccess Web Links

The Atomic Paintbrush project uses the @ccess Web Links feature in DVD Studio Pro. Using a Web link in a DVD Studio Pro project makes it possible to jump directly to a Web site while watching the DVD on a computer with an active Internet connection.

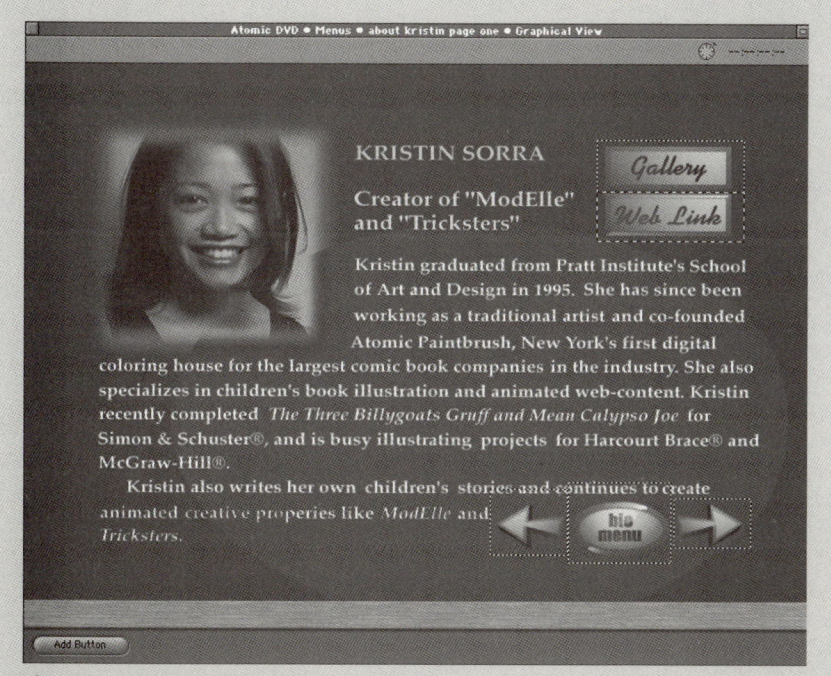

The Kristin Sorra bio page, as seen in the Menu Editor in DVD Studio Pro. The Web Links button leads to a separate menu, and when the next screen appears, the @access link is activated.
Image courtesy of Atomic Paintbrush, Inc.

Continued

Part VII ✦ Case Studies

Continued

When you add the Web links, DVD Studio Pro includes the Installers when you burn the disc, so a person needs to go through the installation process to be able to use the feature. It may be more hassle than the DVD author or DVD user wishes to go through, unless the DVD project would include a great number of Web links, in which case the installation might be worth it to the user.

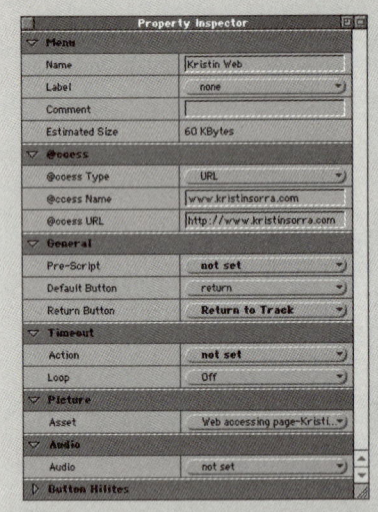

The Property Inspector settings for a menu that has been created for the purpose of launching an @ccess link. In this case, the menu is actually a motion menu. The Picture/Asset value is set for a movie that has been created to run while the Web is being accessed.

When the Web link is activated, the DVD presentation continues to run on the computer, while a browser opens.

Do you have any thoughts on how to achieve the best balance of quality in setting bit-rate for MPEG-2 compression? Did you use QuickTime Pro?
I used QuickTime Pro and had some advice from some guys who encoded in MPEG-2 pretty regularly. It seemed that a bit-rate of 4-5 was pretty average and if you could squeeze out more, then great. That's why it was so important to encode the audio into AC-3 files, because it really does save a lot of room that you'd rather conserve for the video.

What are your favorite features of DVD Studio Pro?
The scripts (in theory) allow a lot of customization, letting the programmer do many neat things.

What kind of supporting materials did you find on the Web or in book form that were helpful to you in investigating DVD?
Well, of course there was the DVD Demystified site, which delved deeply in the basic mechanics of the DVD, the standards, the manufacturing, and so on.

> **Note** The DVD FAQ is accessible at www.dvddemystified.com, and the entire DVD FAQ HTML document is on the DVD-ROM that accompanies this book. It is helpful to use the browser's Find feature to find particular information.

✦ ✦ ✦

CHAPTER 25

Case Study: Pioneer Electronics' SuperDrive

Apple's revolutionary SuperDrive DVD-R burner is based on technology developed by Pioneer Electronics. This case study focuses on the SuperDrive, Pioneer Electronics, and other related issues that came up in a question and answer session with Andy Parsons, Senior VP of Product Development at Pioneer. Before looking at the SuperDrive, consider what came before it.

Before the SuperDrive: Pioneer DVR-S201

Before the SuperDrive came out, the cheapest DVD-R burner — the Pioneer DVR-S201 — cost about $4,000.

Using DVD-R (authoring) media and a cousin to the SuperDrive, the DVR-201 is used by many DVD authoring companies. The DVD-R (A) discs can be used with certain kinds of DVD manufacturing systems and actually serve as a master disc for manufacturing; other manufacturing projects are typically turned in on DLT tape.

The SuperDrive A03 Mechanism

A majority of DVD Studio Pro users will probably use a SuperDrive or one of the external FireWire equivalents at some point in their DVD authoring career. (See Figure 25-1.)

Figure 25-1: The heart of the SuperDrive. Looking at the front of the DVR-A03 mechanism.
Photograph courtesy of Pioneer, Inc.

Most DVD-R burners at present use the same internal drive mechanism as the SuperDrive, whether you are talking about a FireWire DVD-R burner or an internal DVD-R burner that might ship with a Compaq Presario or Sony Vaio.

A *drive mechanism* is the innards of a storage device, usually provided to a computer manufacturer by a third party such as Pioneer. The SuperDrive is technically known as the DVR-A03 mechanism, and Pioneer's DVR-103 is the same drive mechanism. (See Figure 25-2.)

Figure 25-2: DVD in the nude: the DVR-103 drive mechanism. Just the drive; no external case.
Photograph courtesy of Pioneer, Inc.

DVD-R media

The SuperDrive and any other DVD-R drive built around the A03/103 mechanism can write to either DVD-R or DVD-RW media (see Figure 25-3). DVD-RW media enables you to write more than once to the same disc.

Figure 25-3: A brave new world: the DVD-R disc. The prices of both drives and discs will continue to go down as the technology becomes more common.
Photograph courtesy of Pioneer, Inc.

External FireWire DVD-R burners

The popularity of external FireWire drives will continue to increase. (See Figure 25-4.) I wouldn't be surprised if iMacs have DVD-R burners someday, if someone comes out with a slot-loading DVD-R mechanism.

Figure 25-4: An external DVD-R drive. Doesn't necessarily have to have a FireWire connection, but most external DVD-R drives do.

Photograph courtesy of Pioneer, Inc.

Conversation with Pioneer Electronics

The following is an interview with Andy Parsons, Senior VP of Product Development and Technical Support at Pioneer Electronics. Andy's group is responsible for product planning and technical support of industrial DVD and CD products, including DVD-R and DVD-RW drives, DVD Video players and recorders, multi-disc library systems, and professional DJ products. His group works closely with Pioneer Japan and customers to help define, market, and sell products in these categories. Over the past few years, he has presented frequently at industry conferences, usually on recordable DVD technology.

Who invented the A03 drive? Was it a team of engineers? Is it patented?

The DVR-A03 is Pioneer's third generation DVD-R writer. It was designed and developed by Pioneer Corporation in Japan. The drive uses a variety of optical disc technologies, many of which were developed by Pioneer. As with all things DVD, a number of patents are licensed by many companies that are members of the DVD Forum.

Will the A03 drive affect sales of Pioneer's higher-end DVD authoring drives?

Interestingly, it hasn't had much of an impact so far. This may be because the A03 represents new market trends. It is the first Pioneer DVD-R drive that uses 650nm (nanometer) recording lasers, which are less expensive. 650nm lasers also have a better ambient temperature tolerance — the ability to operate well at higher temperatures. This is one reason that we were able to achieve a lower price point for the drive, as well as package it in an internal IDE form factor. Because of this lower price, the A03 has allowed a movement of DVD-R products into the consumer market, where we think there is significant interest in recording personal DVD-Video discs. So this is a market expansion as opposed to a market replacement. As a result, we continue to see demand for the DVR-S201 authoring drive, most likely because it supports both 3.95GB and 4.7GB media and because of its good reputation in the pro authoring industry. So as long as demand for the S201 continues, we will keep it in our product line.

What are the strengths of the A03 mechanism?

There are several. The A03 is the first DVD-R drive to achieve 2X recording on DVD-R media. This is a big deal, because recording a complete disc only takes 30 minutes instead of 60. Also, we introduced lossless linking in this drive, which had previously only been used in our set-top video recorders. This allows a very effective buffer underrun protection — if the host computer cannot sustain the 22+ megabits per second that the drive needs to record at 2X, the drive will pause momentarily until the bit stream resumes and then it will seamlessly continue recording onto the disc. Then of course we have the ability to write CD media, which is currently a unique capability. Incidentally, we believe that CD-R will be around for a long time, with DVD-R/RW adding a new layer for video applications and large-scale data uses. Finally, we managed to do all of the above at a much lower cost.

How did Apple manage to score the deal with you?

Well, I guess it's best to say that we both scored. We don't normally comment on our OEM relationships, but I will say that it was clear that Apple understood the importance of video in computer systems. Video, in my opinion, is the "killer application" for writable DVD — the mass market enabler, if you will. Video also happens to be the most likely type of information right now that can easily fill up 4.7GB of capacity per disc in a typical user environment. No other consumer-generated data source I can think of, with the possible exception of high res digital photography, needs all this space on removable media. So I think the timing was right for Apple and Pioneer to link up because Apple was obviously very clued into the importance of a computer's ability to add value to personal video (editing, special effects, and the like), but Apple had nowhere to put the finished work except back to videotape. Our drive came along in late 2000 and everything just fell into place. I should also say, however, that we were working with other OEM customers as well. I'm sure you know about our relationships with Compaq, Sony, and Packard Bell in Europe.

Did manufacturers approach you, or did you approach them in marketing the drive?

We maintain business relationships with all the major computer manufacturers because we have supplied DVD-ROM drives to many of them for years. Naturally, we always present our customers with our newest product offerings.

Were there any particular hurdles to overcome in bringing it to the market?

There were many. Writing and reading both DVD and CD media in one drive — DVD R, DVD-RW, CD-R, CD-RW, DVD-ROM, and CD-ROM — had never been done before, but this is what ultimately made it "super," to borrow Apple's name for it.

Doing all this while lowering the price so dramatically was a very big challenge. I'm very proud of our engineers in Japan for doing such a great job with it.

Where do you see things going with DVD authoring?

I think that there are two types of users. One type wants to make sophisticated, fully-featured titles with menus, subpictures, multiple angles, and the like. These are people who want to know everything there is to know about DVD authoring and how it all works. Then there is the other type of user who doesn't want to know anything about DVD authoring at all. They just want to focus on the creative aspects of the video content, and maybe they'll want to make a simple menu, and that's it. A variation of this second user is someone who doesn't even want to edit video — he just wants to dub content to a DVD disc much like duplicating a videotape. So I think authoring tools have begun to divide into two fundamental kinds, targeted for each type of user — the full-featured do-it-all tools and the simple, easy-to-use type that isolates the user from all the authoring jargon. For the most authoring-averse user, we have developed a set-top video recorder that works just like a VCR. In this case, we completely removed authoring from the equation.

Do you think that consumers will adopt DVD-R as a format more rapidly than it took with CD-R, since a precedent has already been set?

Yes, and that is because DVD-Video has been so phenomenally successful. The installed base of DVD-Video players continues to grow dramatically, so this is the engine that drives the demand for making personal DVD-Video discs. My way of thinking is that there are three attributes that determine the likely success of a particular writable DVD format to the mass market, in order of importance:

1. **Compatibility with existing video players and DVD-ROM drives.** This is of paramount importance, because without compatibility with a large installed base of players and computer drives, you effectively have a niche product.

2. **Low cost recording media.** If the recording media can be inexpensive enough, then it makes practical sense to use it for everyday applications. As an example, CD-R media is replacing the floppy disk because it's so cheap and, per item number one, compatible and interchangeable with ubiquitous playback devices.

3. **The number of rewrite cycles.** This is a distant third place, in my opinion, because other than applications like time shifting television programs or computer system backups, rewritability is not so critical to most removable storage applications. As an example, I always like to ask people how often they recycle floppy disks. Most of us just grab another new disk to avoid erasing something that may be irreplaceable, which effectively wastes the format's rewritability. The bottom line is that the low cost of the new disk outweighs the potential loss due to accidental erasure. Interestingly, one research analyst, Strategic Marketing Decisions, reports that only about three to four percent of writable CD media consumption is rewritable — historical evidence that rewritability by itself is not of overriding importance.

Do you think that tools like iDVD will make it so easy for people to make their own DVDs that it will become as pervasive as home movies on VHS?

Again, it depends on the user. I think iDVD is a brilliant product, and I think it's just the beginning. But remember that there are also people who don't even want to use a computer when they archive their existing tape content. This is one reason we support DVD-R recording in our standalone video recorders.

Do you see the emergence of consumer standalone DVD recorders being more likely to appeal to the consumer than the combination of a DVD authoring program and a DVD-R burner in the computer?

Many people may want both capabilities, but I view them as different functions. In my house, for example, I can see myself doing the video editing of our old camcorder tapes on a workstation, while the rest of my family would love to use DVD-RW for temporarily storing all their TV programs onto removable media. We have a Tivo/satellite combination unit that we all love because it's so simple and convenient to use. But it has one major limitation — it can store "only" 35 hours of content. Removable media would tremendously expand this product's practical value to us.

I should also say that our set-top boxes allow for some very rudimentary editing in that you can easily cut scenes out — all those shots of the sidewalk because someone forgot to turn the camcorder off — and avoid using up disc space accordingly. This may be the extent to which a set-top user is willing to go for this function. Anything more sophisticated than simple deletions becomes more of a workstation task rather than a living room task.

How does the Panasonic mechanism differ from Pioneer's?

ALL DVD Forum disc formats use error correction (you couldn't use any of these discs if they didn't!). DVD-RAM adds defect management to the equation, which is a feature that some enterprise users may be interested in. The notion that DVD-RW has no error correction and that ANY error will result in lost data is not true. This is only true if the error correction algorithm — the same one that other DVD formats use, including the DVD-R format that Panasonic has included in their new drive — cannot correct for an unusually large (unrecoverable) error.

It's true that you must write data to a DVD-RW disc sequentially - but only during the first recording of the disc. Once data has been written to a -RW disc, you can write randomly anywhere that data exists. This feature is called restricted overwriting, and is spelled out in the DVD-RW white paper. So if you write 4.7 GB of data the first time, you can subsequently rewrite to any data block on the disc regardless of its position.

(Author's Note (TK): The Panasonic mechanism in question is a dual DVD-R/DVD-RAM drive. It writes at DVD 1x, as compared with the SuperDrive/AO3/103 2x, and Pioneer's mechanism can burn CD as well as DVD.)

How long do you think it will take for the price of DVD-R burners and DVD-R media to approach the level of cost of the CD-R equivalents?

It will probably happen much faster than many people might think. Everything in the DVD world seems to move much faster than it did in the CD world. Consider that we have seen a 94 percent price reduction from our first to our third generation DVD-R drives in about three years. Also, as more manufacturers begin to support DVD-R, this will help the format achieve greater production volumes, which will in turn help to encourage lower prices in future generations of product.

Do you see DVD-R eclipsing CD-R media, or sharing the field?

I think that DVD-R and CD-R will co-exist for the foreseeable future. It's my understanding that the majority of CD-R media consumption is for audio recording, and this application is not going away anytime soon. CDs aren't really well suited for high-quality video recording, however, so that's where DVD-R and DVD-RW come in. DVD adds a video layer on top of the existing audio layer, and both can be used for data applications as well depending on the capacity requirements. This neatly defines the DVR-A03, because it supports both families of media in one drive.

Any URLS you think people should check out?

`www.pioneerelectronics.com`

I wondered if you knew of any sources of information on how widely compatible DVD-R media is with internal DVD-ROM mechanisms in desktop and portable computers?

I assume by "portable," you mean laptop computers with DVD-ROM drives in them. The problem with tracking this is that most PC companies change drive brands all the time, so it's hard to say that such and such laptop model is the best to use (the same model might use more than one brand of drive). I do know that some laptop users actually prefer to use the handheld portable DVD players instead, because they are quicker to launch and more dependable (no operating system, player software, crashes, and so on). They are also getting quite inexpensive now. If you are looking for anecdotal reports of video player compatibility, there are several charts posted on the Internet. A good example of these is on the `www.dvdmadeeasy.com` site (you need to register — it doesn't cost anything). I have not seen any that specifically address computer drives, though.

Would you happen to know if authoring media burned out of a DVR-S201 would be more compatible with DVD-ROM or DVD players than general use media?

The 4.7GB varieties should all have the same compatibility. The recorded discs look identical to a player. Some people report that 3.95GB authoring media plays better, especially in older players, but this may not work for you if the programs are larger than 3.95GB. The 3.95GB media spins a little faster and has a bit more space between tracks, making it easier to play. Meanwhile, we published a white paper that talks about the differences between 4.7GB authoring and general media, which can be found on our Web site at `www.pioneerelectronics.com/pioneer/files/dvdr_whitepaper.pdf`.

You may hear claims that there are playback differences between authoring and general flavors of 4.7GB media, but this has more to do with different companies' manufacturing processes than with the differences in recording wavelength, which is irrelevant to playback. Also, some people compare 3.95GB authoring media with 4.7GB general media, which is comparing apples and oranges.

Note: Three white papers from Pioneer are in PDF format on the DVD-ROM accompanying this book, covering DVD-R media in the Goodies folder.

✦ ✦ ✦

Case Study: Simple Motion Menus

CHAPTER 26

♦ ♦ ♦ ♦

This case study is a brief account of the preparation of a simple video file for a motion menu — to satisfy the curiosity of those who wonder if they can piece something like it together without necessarily being experienced in compositing. The sequence of events cycles through Terran's Cleaner 5, Adobe After Effects, QuickTime Pro, and DVD Studio Pro, but there's no reason other programs couldn't be used in addition to or in place of these.

Creating a Motion Menu Using Adobe's After Effects

Before I sat down and tried to create a motion menu, I had never used After Effects. Despite a couple of frustrating situations (admittedly due to not reading the manual beforehand), I found After Effects to be a fairly easy, and satisfying program to use to finally achieve my motion menu dream.

Prepping the clips

The first step in creating a motion menu is prepping the video clips you want your menu to include. I wanted to have a brief fade in and fade out on each of the clips before I tried compositing them against a version of the original DVD menu, so I used Terran's Cleaner 5 to quickly export a couple of clips. (See Figure 26-1.)

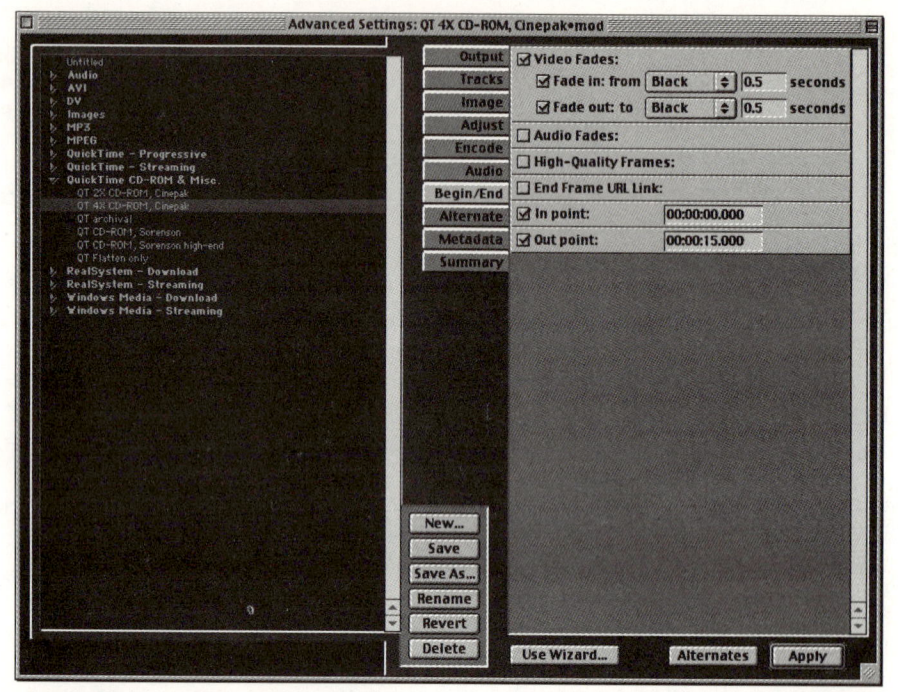

Figure 26-1: Output settings in Cleaner 5 — applying fades and setting an in and out point for the video clips.

Creating the After Effects composition

I started up Adobe After Effects and created a new composition. I imported the two video clips I wanted to use as well as a flattened version of the DVD menu, saved in `.pct` format. The overall After Effects file is called motionmenu.aep, and a composition is a kind of subproject. (See Figure 26-2.)

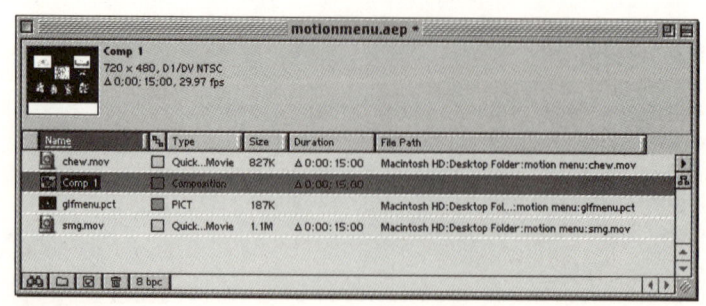

Figure 26-2: Files imported into an After Effects project file.

I had to go back and double-click the composition to adjust the duration of the composition, and set it to 15 seconds, because it originally was set to 10 seconds. I spent about 10 minutes trying to figure out why I couldn't get past 10 seconds in the Timeline, until I realized I had to adjust the duration of the composition. (See Figure 26-3.)

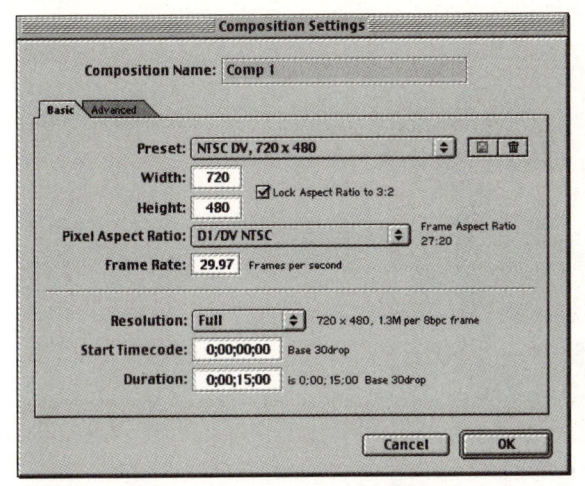

Figure 26-3: The Composition Settings window, where you can change a host of settings, as well as give the composition a name, unlike the author, who was naughty and didn't bother.

Adjusting layers in the Timeline

I dragged the imported elements down into the Timeline and realized that I had to have the movie clips in the uppermost layers above the background screen, so I clicked near the filename icons and dragged them into the proper order. (See Figure 26-4.)

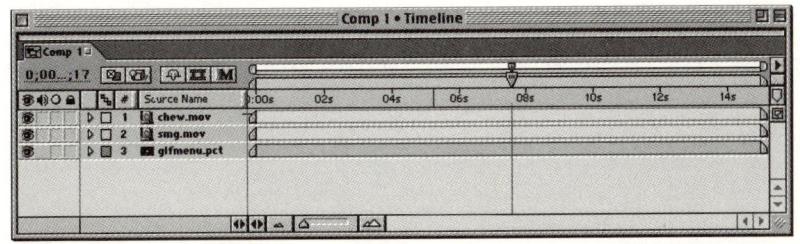

Figure 26-4: The After Effects Timeline, with multiple layers of video fun.

Positioning the clips

I found it easy to reposition the 160-x-120 clips over the spots I created for them in the underlying `.pct` file. (Although admittedly, I had to go back into Photoshop several times to re-adjust the background image, re-import it in After Effects, and re-drag it into the timeline.) (See Figure 26-5.)

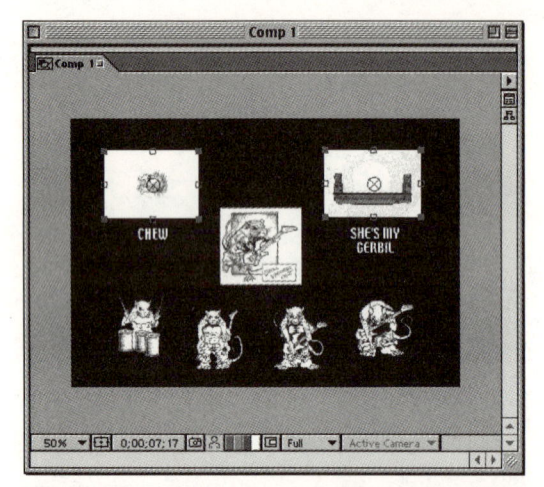

Figure 26-5: The Preview window in After Effects. The two superimposed clips have selection points and can be resized or repositioned with ease.

Exporting as QuickTime

The next step was to export. The first few times I tried QuickTime with no compression for the heck of it, but the effect was unpleasant, so I remembered the wisdom of Yoda, who once said "Do or do not, there is not try!" So I exported from After Effects as a QuickTime file with Animation compression, Millions of Colors, and Best quality selected — which results in a fairly large file, but excellent individual frame to frame quality. (See Figure 26-6.)

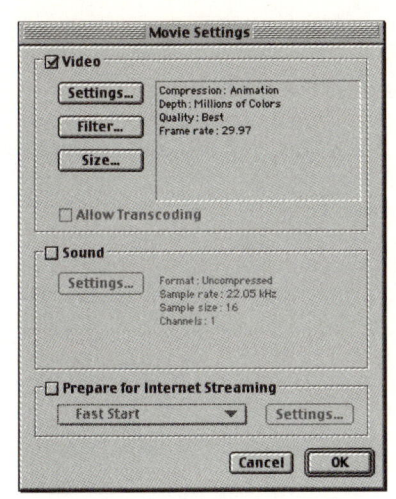

Figure 26-6: The Movie Settings window in After Effects.

The export was reasonably zippy, which kind of surprised me. (See Figure 26-7.)

Figure 26-7: The Export window, giving a frame by frame progress report on the export.

Encoding into MPEG-2

Good old QuickTime Pro. Just open it up, and off you go! I opened up the resulting composited file in the QuickTime player and exported it as MPEG-2, clicking the Options button to set the bit-rate at a safe value of 5.7. (See Figure 26-8.)

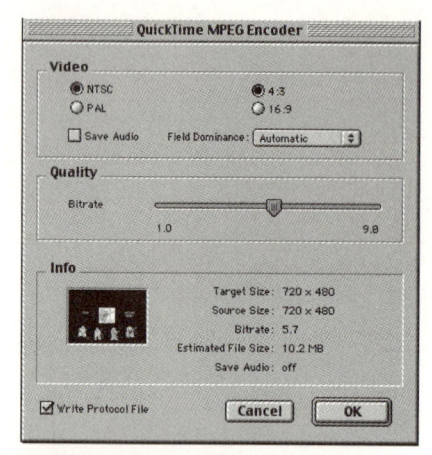

Figure 26-8: QuickTime Pro. Who would have thought the QuickTime player could do so much when upgraded to Pro? Apple rocks.

Creating the motion menu in DVD Studio Pro

My excitement steadily increased as I realized that I was on the verge of a multiple clip motion menu — opening up DVD Studio Pro, creating a simple menu tile, and simply importing the MPEG-2 clip, with no sound but the slithering of my roommate's python.

Alas, it was as easy as selecting the clip as an asset in the Property Inspector, which then brings up the Loop option in the Timeout area, which I immediately turned on so I could hypnotize myself in a mesmerizing haze of composited visual ecstasy. I carelessly set the Button Hilites to a Selected Set 1 value of gray, 33%, but I quickly forgot as the cursor was drawn inexorably to the Preview button. (See Figure 26-9.)

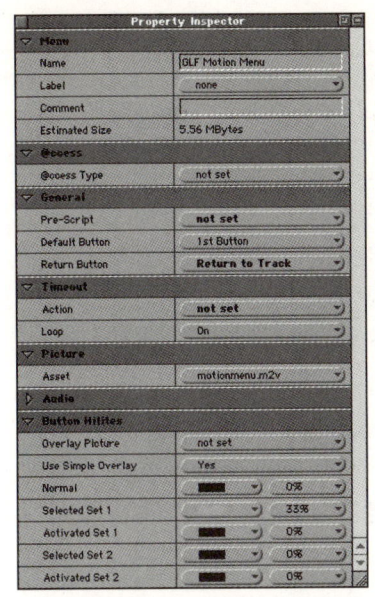

Figure 26-9: The Property Inspector in DVD Studio Pro, with menu tile selected and Picture/Asset chosen.

Jubilation

After experiencing the delights of a multiclip motion menu, I opened up the familiar Menu Editor, which to my horror showed black as the blackest soul, head like a hole. But I simply advanced the clip a few frames using the slider, to bring up the images, and began to add buttons as usual. The menu was there! (See Figure 26-10.)

Figure 26-10: The Motion Menu Editor. Much the same as the regular Menu Editor, except you get a little slider at the bottom to do wacky things like advance the clip a few frames.

Summary

Apple, Adobe, and Terran are wonderful companies. Special thanks to all the work they put into a suite of excellent programs. Semi-complex motion menus with multiple video clips are within the grasp of the ordinary insomniac-narcoleptic DVD author! The following are questions Todd is asking himself, for fun and a minor amount of candid insight.

What would you have done differently?

Actually prepared the video clips from original DV NTSC footage instead of lifting them from 15fps Cinepak QuickTime files. Shhh! Don't tell anyone!

What was your favorite feature of After Effects?

Probably the little button you can click on in the Preview window that allows you to have an overlay of video safe and title safe areas.

What's your next goal with a motion menu?

To make the rest of the menu an animation, do some wacky things with a Flash file, generate 30fps QuickTime at 720 × 480, leave space for the video clips, re-composite the whole thing, maybe make the motion menu longer than the videos, possibly put some secret buttons on the page somewhere that appear at certain times in the video segment.

Author's Note: There may be alternatives to After Effects. If After Effects is out of the question, you might be able to use Premiere. Premiere is an entirely different tool, but you could theoretically use the Motion Control feature to superimpose a scaled-down version of a clip against a backdrop and just not move it, and then go back in and re-render another clip in another place, and so on. Premiere also has some support for After Effects plug-ins, so there might be some possibilities there. But do try to get a copy of After Effects; it could be argued as an indispensable tool for the DVD author. If you are still paying off a PowerBook Titanium and need to conserve, try hunting on eBay, you don't necessarily need the latest version.

✦ ✦ ✦

Case Study: Metatec — Manufacturing a DVD

CHAPTER 27

◆ ◆ ◆ ◆

This case study provides an overview of the DVD manufacturing process from the manufacturer's perspective. This study also includes a brief visual account of the process of getting a DVD manufactured through Metatec International, a provider of optical media manufacturing services in the United States and Europe. The project — the GLF DVD — was a short run of 100 discs.

Background

Metatec International is a provider of optical media manufacturing services including CD-ROM, DVD-ROM, and CD-R. The company offers these services from strategically located ISO-9000 certified plant operations in California, Ohio, and Europe. Metatec integrates its core business of CD-ROM and DVD manufacturing with Internet-based electronic commerce, secure online software and information distribution technologies and supply chain management solutions to provide businesses with a range of information distribution services. The company maintains sales and customer support offices throughout the United States and in Europe.

Note Metatec (www.metatec.com) is aggressively pursuing the DVD replication market, and specifically targets the short-run market with its pricing and the E-Store. As of the time of this writing, Metatec accepts DVD-R General media, and does not charge a mastering fee. Keep in mind though, that a DVD-R General project cannot incorporate CSS or Region

Coding. Some replicators might be able to rebuild a project and add it in but this is not likely. This wasn't an issue with the GLF DVD, because the desire is to get it out there as much as possible.

The following article comes from Metatec:

DVD Manufacturing

The process for making DVDs consists of premastering, mastering, replication, printing, and packaging.

Premastering

Premastering consists of compiling all of the files together and building a DVD image. (See Figure 27-1.) This image is then written to a DLT (digital linear tape) or a DVD-R, depending on the software being used to create the image. During this phase, the data must be saved in a UDF Bridge format. DVD uses the Universal Disc Format (UDF) for its file structure. While many new operating systems use this format, some use the ISO 9660 format. Because there are still very few systems using UDF, it is necessary to make a UDF Bridge when building the disc image. This enables the disc to meet the DVD specification as well as make the disc compatible with older 9660 systems.

Figure 27-1: Premastering
Photograph courtesy of Metatec International, Inc.

Another form of premastering is *authoring*. This term generally refers to the joining together of audio, video, and navigation menus to form an interactive full motion experience. Examples of authoring include corporate video, full-length movies, and point-of-purchase sales and training videos. Other enhancements might include links to Web sites and live updates. Once all the video and graphic assets are put together in an organized fashion, a DLT or DVD-R that contains the DVD disc image is created.

When the source or master DVD-R (or DLT) is submitted to the replication facility, a quality control process is performed. (See Figure 27-2.) Is the data readable? Is the image complete? The incoming data is analyzed to ensure data integrity. After the input media is verified to be compliant, a DLT copy is made. For DVD-R (authoring) and DLT, a direct transfer can take place. During this process, the data is checked for uncorrectable errors. If any are found, the media is sent back to the client for review and submission of new data. For DVD-R (general), the contents are first transferred to a hard drive and then the image is written to a DLT tape. During this conversion, it is not currently possible to add additional features such as CSS, region coding, or Macrovision copy protection. After a DLT has been made, the tape is sent to mastering. Note that for a DVD-9, 2 tapes must be made, one for each layer.

Figure 27-2: Quality control
Photograph courtesy of Metatec International, Inc.

Mastering

Mastering refers to the process of making a master stamper that will be used as a mold for making the discs. Various methods exist, but in general, a piece of glass is coated with a special material. (See Figure 27-3.) The data is "cut" into the coating by using an LBR (laser beam recorder). The coated glass is then coated with metal, usually nickel. This forms a rigid metal stamper. All of the mastering work is performed in a clean-room since even a small spec of dust could render the disc unreadable if it were to get into the master.

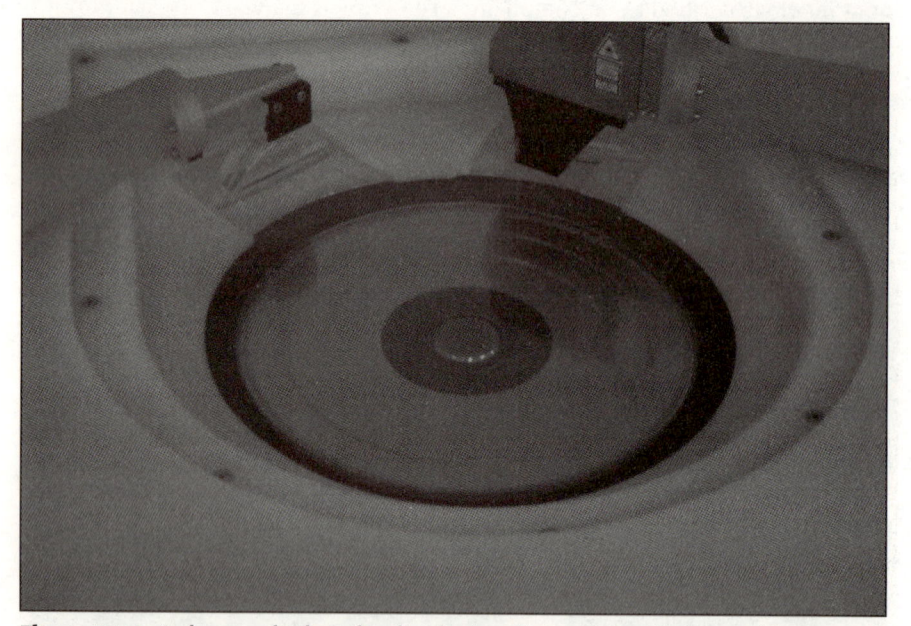

Figure 27-3: A close-up look at the developing disc, which will ultimately serve as the master disc for making a large number of copies
Photograph courtesy of Metatec International, Inc.

Remember that the features on a DVD disc are much smaller than on a CD so cleanliness is very important. The use of glass and nickel in the making of a stamper leads some to refer to this as glass-mastering or a nickel master. For small quantities the mastering charge is usually separate from the replication charges. For longer runs, the cost of the master is generally included.

Replication

A metal stamper or master is necessary for replicated discs. *Replication* is the act of actually molding the information into the disc, where *duplication* involves writing or recording the data on the disc.

The master stamper is then mounted in the replication or molding machine. The stamper acts as a mold. Polycarbonate is injection molded against the stamper, resulting in a plastic disc with the data molded into it. Subsequent metallization forms the reflective layer. The metallization is done immediately following the molding process. A DVD disc is comprised of two halves, each being half the thickness of a finished DVD. For a simple DVD-5 disc, one layer contains data and one is black, or a "dummy." The two halves are then bonded together to form a finished disc. It is possible to have data on just one or both layers. A dual layer (DVD-9) or dual sided disc (DVD-10) requires two stampers but makes the capacity of the disc essentially double that of a DVD-5. The data surface(s) are trapped in the center of the two halves, making the disc a very robust assembly and very tolerant of surface scratches. The finished DVD disc is the same diameter and thickness as a standard compact disc. The replication process for DVD discs is done on machines that have HEPA filters for cleaning the air and have positive air pressure to keep dust from entering the machines environment. (See Figure 27-4.) Dust could get molded into the disc and cause it to be unusable. In-line inspection equipment checks the quality of the metallization, the thickness of the bond layer, and for tangential and radial tilt or warpage.

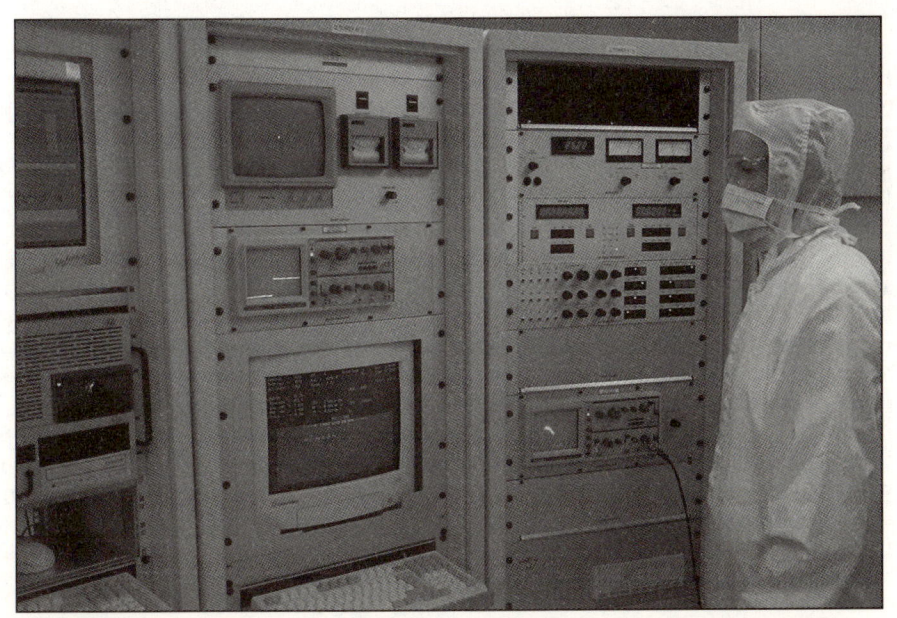

Figure 27-4: A controlled atmosphere is required at certain stages in the DVD manufacturing process.
Photograph courtesy of Metatec International, Inc.

Inspection is the next phase of the process. The inspection process for DVD is different from that of compact disc. Various signal levels of the disc are checked and verified against the DVD specification. Additional signal related tests include random errors (PI errors), jitter, scanning velocity, and track pitch. However, because the DVD disc is made up of two disc halves, additional tests are required to verify the physical nature of the disc. These physical tests include the disc weight, eccentricity when rotated, reflectivity, disc thickness, center hole diameter, tangential and radial deviation (warp), and bond layer thickness. In addition, a byte for byte comparison is done against the client's original data to ensure all data has been replicated correctly. For DVD-video discs, basic navigation is verified using commercial grade DVD players. This is to ensure that the disc will function on players that people have in their homes. While errors cannot be fixed at this point, the customer is notified so that they are aware of the issues. The customer can choose to proceed with the process or send in a new master and begin the process all over again.

For large runs, a "check disc" package is available. This includes the mastering and replication of 10 discs. The purpose is to allow the client samples to verify their authoring and navigation functionality. This limits the exposure in case there are any issues. If everything checks out, the master is simply remounted and the discs are pressed.

Printing

After the discs are tested and verified to be in spec, disc decoration is applied using silkscreen processing. Customers have the option of preparing their own disc and packaging artwork, or having our art technicians prepare the designs for them. If the customer has sent in his own graphic files, we open the files and QC (Quality Control) them. We make sure all fonts and graphic images have been included with the file. We also will size the image(s) to ensure the printing will fit in the printable area. Once the artwork has gone through QC, it is posted on our secure Internet site for the customer's approval. An e-mail is sent to the customer letting him know the artwork is available. The customer can then view the `.pdf` version online to make sure it looks correct. The customer is given the option to approve or reject the artwork. Hardcopy proofs are also available to ensure color correctness. Films are then generated from the art files, one film per color. Screens are produced and ink is mixed on-site. In-line inspection equipment verifies the color of each disc to ensure consistency throughout the run. An alternative method of disc decoration is pit art. Recall that a DVD-5 has one half-disc that contains data and one that is a blank or "dummy." A second stamper can be made in place of the dummy half that will contain data, but will look like a picture when viewed. Basically, a standard bitmap can act as the input media and, using a modified stamper making process, will yield a metal stamper that makes a picture or graphic on the disc. This option is only available for DVD-5 discs.

DVD STANDARDS

A set of DVD standards exists, much like the Yellow Book specification for CD-ROMs. The DVD specification actually contains five books. The books are:

Book A: DVD-Read only memory (DVD-ROM)

Book B: DVD-Video

Book C: DVD-Audio

Book D: DVD write once (DVD-R)

Book E: DVD Re-Writable (DVD-R/W, DVD+RW)

(The set of standards is available from Toshiba in Japan (one of the patent holders) for approximately $5,000 a set. These standards are purchased by manufacturers as a license to manufacture discs according to the standard.)

Packaging

The packaging of DVD discs can vary compared to compact discs. DVDs, for the most part, are packed in standard DVD-Video packaging and held in place by the disc's center hole. For DVD-ROM, a standard jewel box is common. DVD-Video has a standardized box, which is about as wide as a standard jewel box, but taller, approximately the height of a VHS cassette box. These dimensions are not pure coincidence. Retailers do not want to alter their shelf space by introducing larger packaging. Thus, the width is the same as a jewel box. Producers and retailers did want to differentiate DVD-Video in some way to reduce the level of confusion on the consumer's part. By making the case taller, the package has the look and feel of a VHS movie box, something the consumer is familiar with. Packaging continues to be an issue with DVD. It is conceivable that a title can be released in three different formats. For example, a Hollywood movie can be released as a DVD-Video, the movie's soundtrack can be released as a DVD-Audio disc, and screen savers and trivia games from the movie can be released as DVD-ROM. How does the producer differentiate these three discs that look the same and share the same title? This has yet to be defined.

For shorter runs (1,000 units), digital printing is an option. This is a good alternative for four-color process printing. No film generation is necessary and no inks need to be mixed. It is important to note that solid PMS colors will be made from CMYK when digitally printed. Customers should not expect an exact color match.

Pictorial Case Study: The GLF DVD

The following visual sequence represents the life cycle of a DVD manufacturing project, from the ordering to the preparation process.

This project, the GLF DVD, (Gerbil Liberation Front) was prepared using DVD Studio Pro. The VIDEO_TS and AUDIO_ST folders were burned to DVD-R General disc from Apple using Roxio Toast Titanium. Art templates from Metatec were downloaded in Freehand, and submitted separately to the Metatec art department. Turnaround time for a quantity of 100 discs was less than a week. For more information on the Gerbil Liberation Front, see www.gerbilfront.com.

Step 1: Ordering online — Getting a quote

DVD replicators, such as Metatec, are making the process of ordering a manufactured DVD easier, using the Web to handle quotes and ordering. (See Figure 27-5.)

Figure 27-5: Metatec's E-Store. DVD replicators are making the process of ordering a manufactured DVD manufactured easier, using the Web to handle quotes and ordering.

Step 2: Choosing a quantity

Short DVD runs are now a possibility. The GLF DVD was manufactured (rather than burning DVD-R discs) to get a more professional appearance, as well as ensuring 100 percent compatibility with all DVD players.

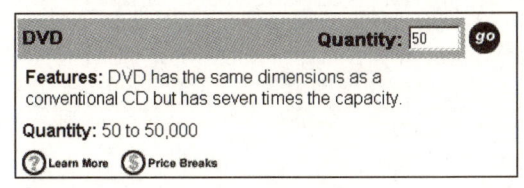

DVD   **Quantity:** 50   go

Features: DVD has the same dimensions as a conventional CD but has seven times the capacity.

Quantity: 50 to 50,000

? Learn More $ Price Breaks

Figure 27-6: Choosing a quantity

Step 3: Choosing disc capacity

Most people choose DVD-5 or DVD-9. The GLF project takes up a relatively small portion of the disc, so DVD-5 was fine.

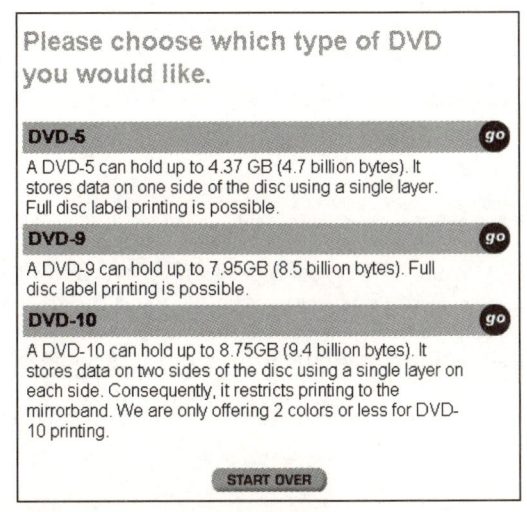

Please choose which type of DVD you would like.

DVD-5   go
A DVD-5 can hold up to 4.37 GB (4.7 billion bytes). It stores data on one side of the disc using a single layer. Full disc label printing is possible.

DVD-9   go
A DVD-9 can hold up to 7.95GB (8.5 billion bytes). Full disc label printing is possible.

DVD-10   go
A DVD-10 can hold up to 8.75GB (9.4 billion bytes). It stores data on two sides of the disc using a single layer on each side. Consequently, it restricts printing to the mirrorband. We are only offering 2 colors or less for DVD-10 printing.

START OVER

Figure 27-7: There is a variety of disc capacities available. The higher capacities involve greater cost.

Step 4: Choosing art colors

One-color and two-color art is generally used for logos or text that you want to appear in specific, individual solid colors. Four-color art work is based on the CMYK color system, so if you have photographs or other complex color ranges, the manufacturer will take the files and convert them to CMYK, which allows them to use the four primary ink colors in varying ranges to represent your image. Four-color printing is usually more expensive, whether you are talking about printing on the disc, or printing supplementary material such as booklets. The GLF project is based on black and white, so a total of two colors were used. (See Figure 27-8.)

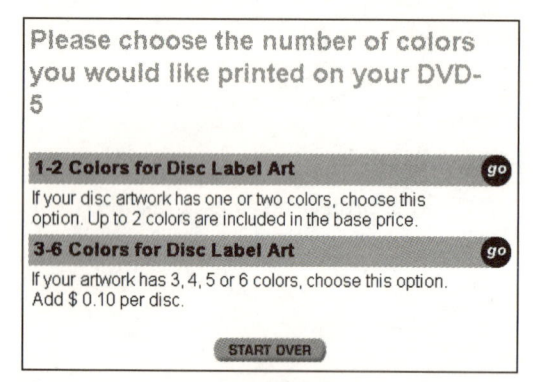

Figure 27-8: DVD is like CD, in the sense that you can print in several colors.

Step 5: Choosing packaging

For the GLF project, we went with a standard DVD video case and elected not to order any printed supplement, because of the cost. However, we used the templates for supplementary printed material to guide our own design, and printed them out on a color printer. For a short run, some form of digital printing can be a lower-cost alternative. (See Figures 27-9 and 27-10.)

How you would like your DVD to be packaged?

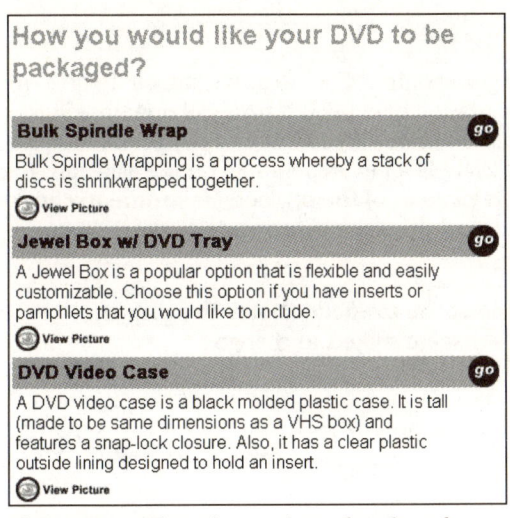

Bulk Spindle Wrap `go`
Bulk Spindle Wrapping is a process whereby a stack of discs is shrinkwrapped together.
View Picture

Jewel Box w/ DVD Tray `go`
A Jewel Box is a popular option that is flexible and easily customizable. Choose this option if you have inserts or pamphlets that you would like to include.
View Picture

DVD Video Case `go`
A DVD video case is a black molded plastic case. It is tall (made to be same dimensions as a VHS box) and features a snap-lock closure. Also, it has a clear plastic outside lining designed to hold an insert.
View Picture

Figure 27-9: There is a variety of options for DVD packaging.

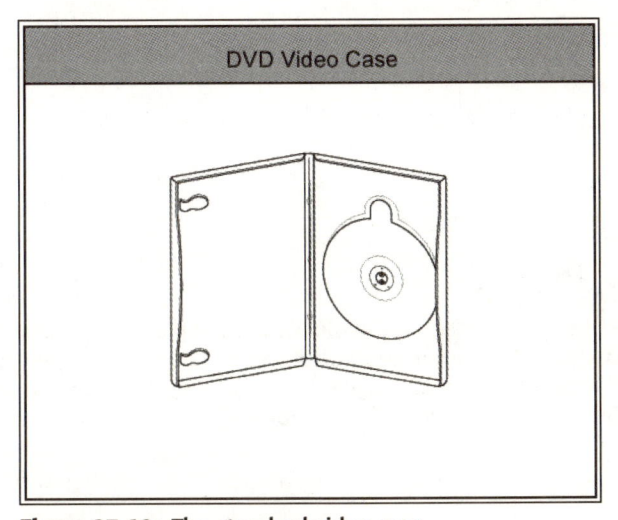

Figure 27-10: The standard video case

Step 6: Finishing up

DVD manufacturing is increasingly going online. For sites that are set to give quotes online, the process ends with a calculation of expected totals. Remember, it is always a good idea to talk to the manufacturer if you have any special requirements or questions. Don't assume. Figure 27-11 is an example of an online assistant that calculates the individual costs for each element of the project. Make sure to check with a manufacturer to see if there are any additional mastering or conversion costs.

Figure 27-11: The online assistant. (Note: This diagram represents prices that are subject to change.)

Step 7: Preparing the art

To prepare the art for the GLF DVD project, I downloaded a template in Freehand from Metatec's Web site, opened it, saved it as a copy, and did the layout. (See Figures 27-12 and 27-13.) It's best to use the template provided by a particular manufacturer, to make sure it is compatible with their equipment.

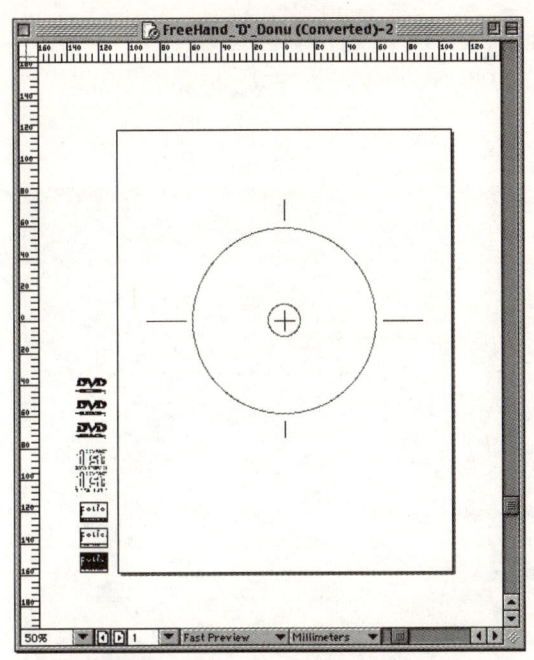

Chapter 27 ✦ **Case Study: Metatec — Manufacturing a DVD** 555

Figure 27-12: DVD manufacturers often have some way for you to download appropriate template files in a variety of file formats.

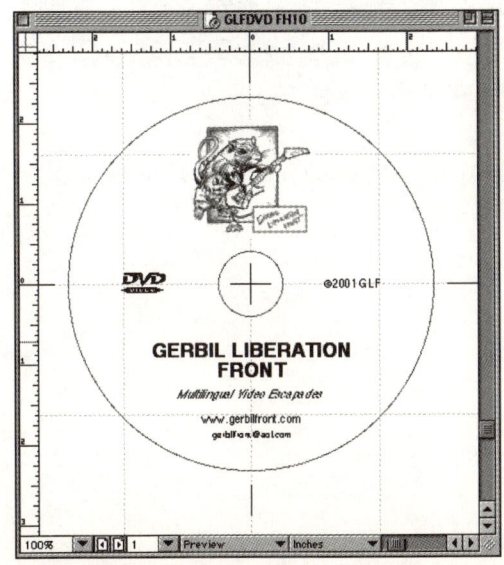

Figure 27-13: Close-up view of the disc art. The cartoon logo was a bitmap image, and the text was generated in Freehand.

Step 8: Submitting the project and waiting

I sent off the DVD-R master and art files, waited about a week, and got the DVD discs back. (See Figure 27-14.)

Figure 27-14: The finished disc, in a standard DVD video case

I immediately tested a sample disc and found it to be fine. (See Figure 27-15.)

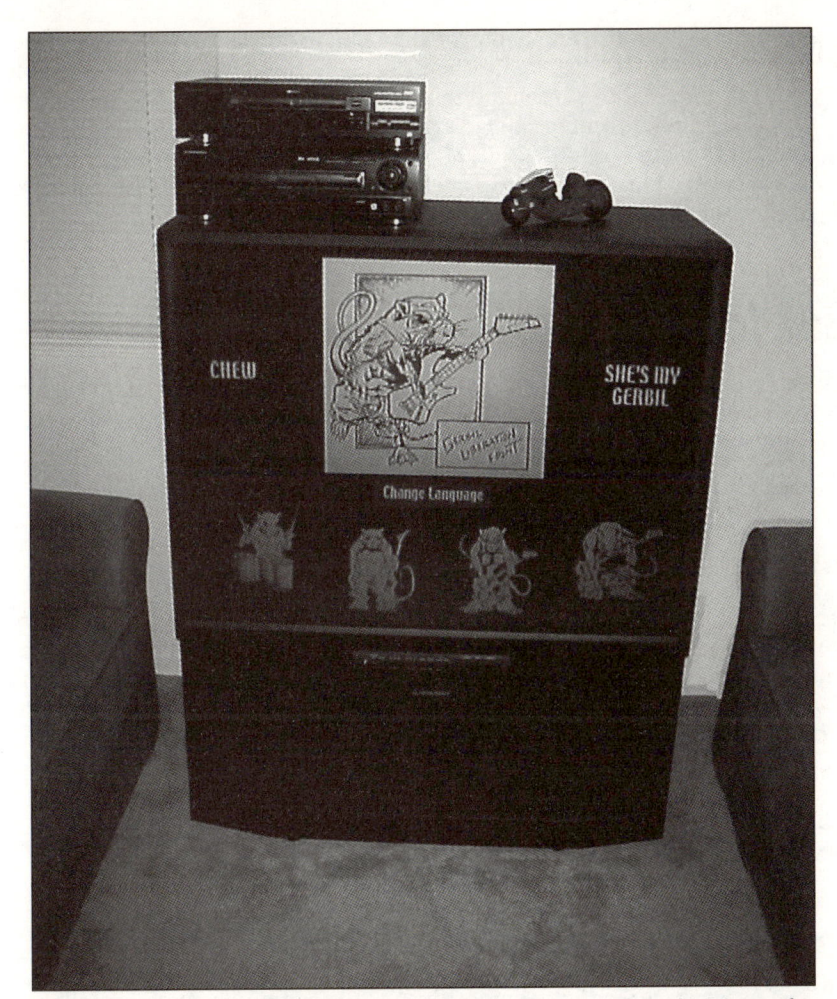

Chapter 27 ✦ **Case Study: Metatec — Manufacturing a DVD** 557

Figure 27-15: Testing the DVD on a television. See how the television only shows the "video safe" area. For example, the gerbil on the far right (Sparky) is much closer to the edge than in Figure 19-21.

It is always fun to try out the finished project. As an exercise in do-it-yourself cost savings, as an alternative to paying for the printing of the outside cover, a couple weeks later I downloaded the Freehand template for the outside cover and started developing my own art. I printed the covers out on an inkjet printer, went to a local Kinko's, and cut them to size (10 3/4" × 7 3/16"). After inserting the covers in the packaging, I was very pleased with the overall results, and the professional look and feel of the manufactured DVD.

Discount for Readers

Negotiations are under way for a discount on the cost of DVD manufacturing for readers of the Macworld DVD Studio Pro Bible. For more information, visit www.dvdspa.com and look for the DVD Manufacturing Discount link.

Summary

Because the DVD project was designed to get publicity and hopefully gain the interest of record labels and/or television shows, DVD-R wasn't really an option, due to concerns that a person might happen to have a DVD player that is not compatible with DVD-R. I didn't want to take that risk with this project, and it gave me confidence to know these manufactured discs are 100 percent compatible with DVD players.

So I've begun to look for opportunities to shop the Gerbil Liberation Front as a creative property for television and music. And I'm armed with a professional broadcast-quality DVD demo, made using a variety of programs, including DVD Studio Pro, Premiere, Flash, Photoshop, and Freehand. Thanks Apple, Adobe, Macromedia, and Metatec!

All in all, the process worked very smoothly. As long as you cover the details and ask questions of the DVD replicator, the process of manufacturing a DVD is straightforward. Always remember to plan what you want to have happen and check with your manufacturer to make sure they can do what you want, accept the master in the format that you want to submit it, and accomplish any additional features or packaging requirements you have.

✦ ✦ ✦

Case Study: Real World Promotional DVD

CHAPTER 28

This case study gives a unique insight into the first project that Real World Interactive did with DVD Studio Pro. The project was a promotional DVD associated with Real World Records, Peter Gabriel's record label (see Figure 28-1). Real World Records has an international roster of a variety of interesting musical artists, and the DVD was released by Virgin in France as part of a special campaign where people could purchase several Real World records and receive a free DVD.

The first part of this chapter includes an interview with York Tillyer, the director of Real World Interactive. The second part of this chapter includes a brief visual tour of some of the related files and techniques used in the project.

Note For those who are curious, several of the files used in the promotional DVD are included on the DVD-ROM that accompanies this book.

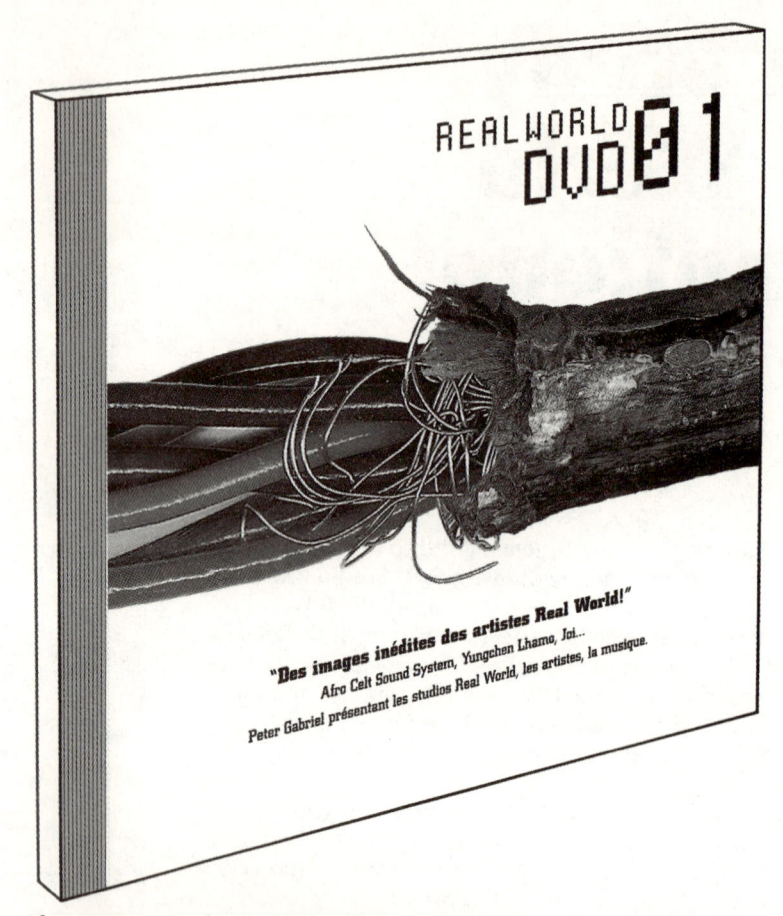

Figure 28-1: Real World's first DVD release in France, made with Macs. The cover art is a 300 DPI Photoshop file, saved in `.tif` format. The design was created by Marc Bessant.
Image courtesy of Real World, Inc., 2001.

Visual Tour of the Real World DVD Project

The following screen shots were taken from DVD Studio Pro and Photoshop files used to create the Real World promotional DVD. Some of these files are included on the DVD-ROM that accompanies this book.

Interview with York Tillyer, Director of Real World Interactive

Background

Peter Gabriel is a famous international recording artist, pioneer in interactive multimedia, and Macintosh enthusiast. Some may remember the Xplora interactive CD-ROM, or the more recent Eve CD, which featured world-class design and cutting-edge techniques. Real World Interactive is Peter Gabriel's interactive company, which designed the Real World Web site. It also develops new media for a variety of related enterprises, such as Real World Records, W.O.M.A.D., and so on. To find out more, visit http://realworld.on.net.

Author's Preamble (TK)

I visited Real World when I was in a band called Sister Soleil, supporting a solo artist named Stella Katsoudas, who was signed to Universal Records. The first release on Universal was recorded at Real World Studios in England, and while I was there, one of my great pleasures was seeing all of the Macs they use in the various studios and offices. I visited the interactive folks, and we chatted about Cyan's interactive game Riven, which had just been released in the United States. More recently, when DVD Studio Pro came out, knowing that Real World was cutting edge, I had a hunch Real World would be getting into DVD Studio Pro, and sure enough, they had the software and hardware on order. Later on, York Tillyer agreed to comment about DVD Studio Pro, and the first project they created with it — a special promotional DVD released in France, a compilation of Real World related music videos that packed out the disc at 4.6GB.

Who are you? What do you do?

I've been here for 6 years, joining whilst working for my Masters in 3D computer-aided graphical technology applications. We have a small team working mainly in-house for Real World, W.O.M.A.D., and Peter Gabriel, producing Web sites, CD-ROM, ECD video, and DVD content. It's a multidisciplinary team that works on both the technical and creative side of projects.

What is the trend in DVD going to mean for Real World?

Maybe the main thing is that the software and hardware now available have let us take stuff from camera to DVD distribution very quickly and cheaply — we can distribute old material and get new content made very easily. It opens up a lot of possibilities.

How are Macs used at Real World?

We use Macs as our primary development tool and to serve our Web site. It's just the best tool for the job, with DVD Studio Pro and Final Cut Pro being up there now as some of the best reasons for using this platform. The internal SuperDrive and the software brought DVD authoring within our grasp.

Continued

Continued

What was the purpose of the project?

This was our first commercial DVD project; it went out as part of a marketing campaign for Virgin France. I think if you brought two Real World Record titles in the campaign, you got a free DVD. We had to make it quickly and to a tight budget, so we stuck with a simple project design.

Audience for project?

Discerning music buyers in France.

As a video professional, what impressions have you formed about DVD Studio Pro?

Really just that it is so quick and easy to set up menu options and navigate through the disc content.

What hardware did you use?

Mac G4 733 with SuperDrive building the DVD project, and G4 500 running Final Cut with Canon XL1 grabbing and editing the video content and mirroring content on TV.

What third-party applications did you use?

We used Adobe Photoshop to build the graphical elements and Illustrator and Cinema 4D to make the intro sequence. There is also probably a fair amount of After Effects in the video content as well as the two Apple apps mentioned.

Do you have any thoughts on how to achieve the best balance of quality in setting bit-rate for MPEG-2 compression?

Well, it was interesting to play about with the compression settings (for bit rates when compressing video). The initial temptation was to put everything to the maximum, but playback seemed to glitch. In the end, we had to lower the settings on some clips (there is not a lot of guidance on the best settings here — you only have the one slider to play with). The basic lesson was that 4.7GB can soon fill up.

How did you do the motion menu?

We ended up with a motion intro to a static menu. The art seems to be in finding a good loop point (like the white-out flash in the DVD menu for *A Perfect Storm*. It is important to note that the loop point must be the same for the audio and video — I guess because nothing is loaded into memory. The limitations were such that we had to ditch motion in the menu. It's the challenge for the next one.

What kinds of things do you see yourself doing in the future with DVDSP?

Really we just want to build more interesting projects; this first one was just proof that we could turn it around, but we really want to explore and push what people think is possible with the design of the motion menus and interaction.

Any DVD-ROM content that came with the disc?

No.

Do you plan to make DVD-ROM content in the future?

Yes, given time we would have added it to this project, but I think we would in the future, largely to give Web links and ways for people to give feedback on the content.

Have you considered using Flash for interactive content, or will you stick with Macromedia Director?

Our last ECD used Flash extensively, but this was embedded in Director to give us more control and the ability to play back QuickTime. Director is also much better at handling bitmaps. There does not seem to be any reason to limit it to one or the other when both work together.

ECD stands for Enhanced CD and is a regular audio CD that you can put in a CD-ROM drive for interactive multimedia. Real World is tied in with a number of recording artists and does a fair amount of ECD titles.

What kinds of things do you see people putting on DVD-ROM content?

I think it's going to be marketing material, linking to Internet services like retail and feedback.

What programs were used to generate the art for the printed materials that came with the project?

I didn't handle this, but I'm pretty sure it would be Photoshop, Illustrator, and Quark.

How did you submit the material for premastering?

On DLT tape — it seems manufacturers are not yet geared up to support DVD-R masters; they want it on tape. Interestingly this means the SuperDrive is not really crucial, particularly since the Apple DVD player will open Video_TS folders. Being able to burn a disc is useful for testing and viewing on domestic players, but it's not really crucial.

Did the manufacturer give you a test disc?

No, time was tight on this deal, so Virgin France checked it against a DVD-R pre-master copy we sent along with the DLT tape.

Is the DVD available for order online somehow?

No, it's exclusive to this campaign in France.

What are your favorite features of DVD Studio Pro?

Just basically that it is quick and easy to use. The question is a difficult one because I don't have any other DVD authoring systems to compare it with. I suppose that means its best feature is that it has enabled a small studio like ours to move into this field.

Main menu

The main navigation screen for the DVD project was designed as a straightforward selection of each music video. The screen appears after an initial animation where a 3-dimensional logo flies through the screen, and then the same logo appears in the background for the main screen. (See Figure 28-2.)

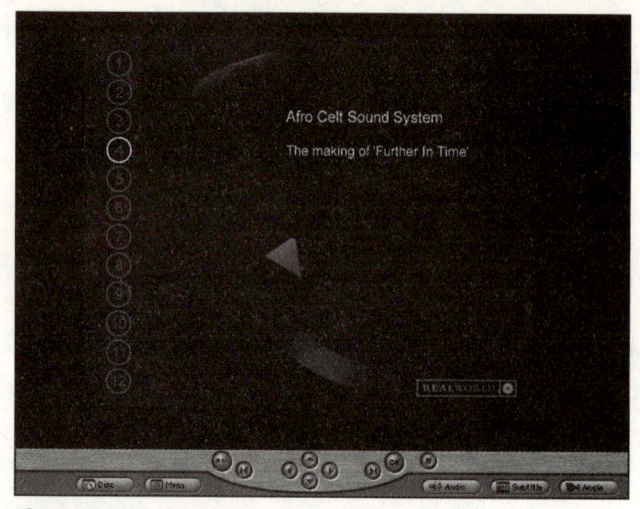

Figure 28-2: The main navigation screen.

Graphical View window

Even though the project was created in the PAL format, it looks the same in the Graphical View area. A track tile was created for the animation intro, and then the user is taken to the main menu, where they can branch out to the individual videos. (See Figure 28-3.)

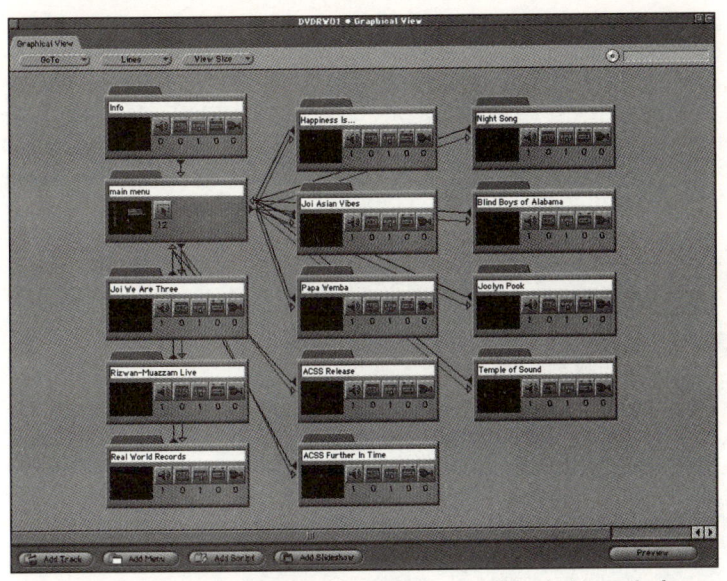

Figure 28-3: Looking at the project in the Graphical View window.

General settings

The general disc settings for the project would be much the same as any other DVD project, but in this case, the Video Standard was set to PAL. The project was designed to fit within the 4.7GB capacity of a disc, but the material just barely fits onto a DVD, because the material's size turned out to be 4.6GB. (See Figure 28-4.)

Figure 28-4: Looking at the general disc settings in the Property Inspector.

Photoshop layer arrangement

When a person is viewing the main menu on the DVD, for each button they select, there is corresponding text which appears on-screen. In the Photoshop file, there is a separate layer with a bright white circle to represent the "on" state of each button. (See Figure 28-5.) (This might be a good time to check out the PDF version of this book on the DVD-ROM, and look at the full color version of this chapter.)

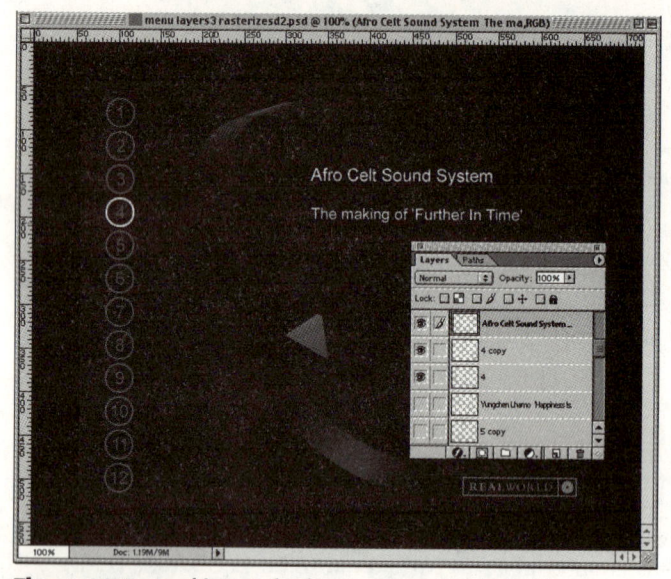

Figure 28-5: Looking at the layer arrangement for the main menu file.

Design guides in Photoshop

Design guides are a helpful way of ensuring accurate placement of images when designing screens in Photoshop. Typically the Photoshop user sets the rulers to display and drags these lines out from the rulers, to help align various elements on the screen. In this case, the vertical and horizontal lines helped to ensure accurate, consistent placement of the text layers in the Photoshop file. (See Figure 28-6.)

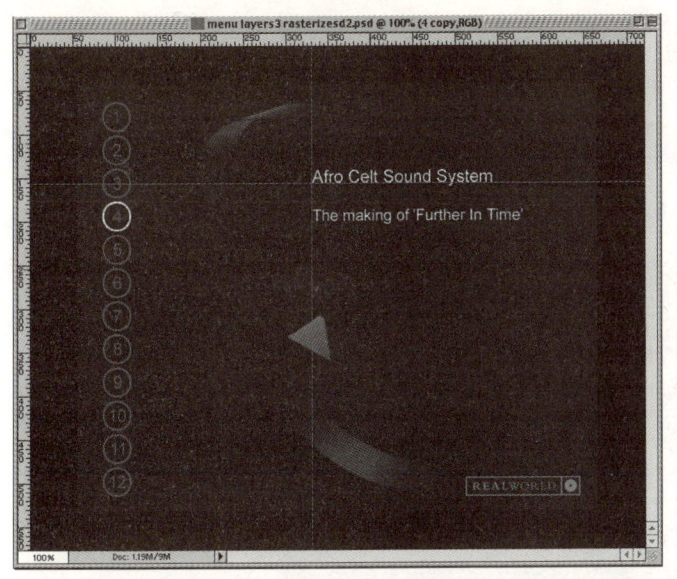

Figure 28-6: Design guides in the Real World DVD Photoshop file.

Buttons in the Menu Editor

In the Real World DVD menu screen, the area around each circle is defined as a button (see Figure 28-7). When a button is activated, a specific Photoshop layer appears. The exact alignment of the button outlines is not important, because DVD player-generated hilites are not being used. Normally, the area defined by a button outline is used to define the hilite. In this case, the only real purpose of defining the button outlines is to make it easy to program the remote control settings that determine which button leads to another. It also defines the regions for each button that will be "hot" if the DVD is accessed in a software-based DVD player that allows the mouse pointer to selct DVD buttons.

In theory, the button outlines could be drawn in any part of the screen, because the Photoshop layers themselves are in a fixed position. Aligning the buttons in a logical order makes it easy to remember which button leads to another. In other words, by stacking the outlines in an easy to understand order, it is easier to go and set the action for button A, the down arrow on the remote-control should lead to button B, and so on, using the button outlines as a visual guide.

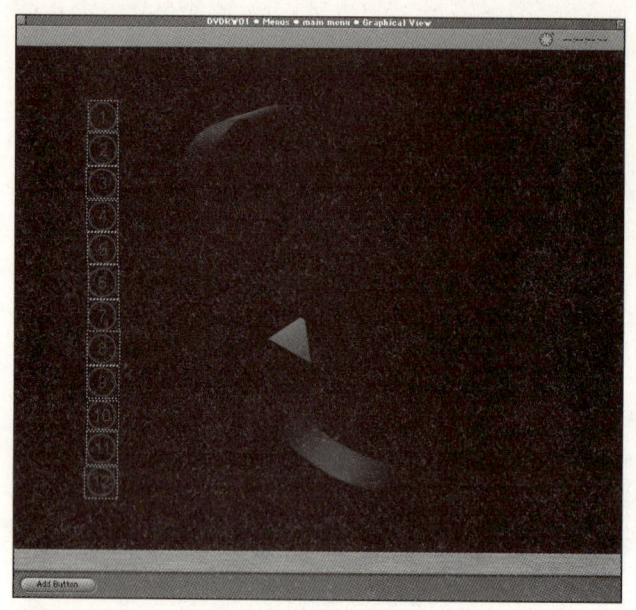

Figure 28-7: Looking at the main Real World DVD menu in DVD Studio Pro in the Menu Editor window.

Adjusting button display settings

The Real World DVD follows a steady pattern with the way the button display settings have been adjusted to make respective Photoshop layers appear. For example, with a representative button, the Normal State is set to Layer 4 (see Figure 28-8), which represents the unselected state of the button, and adjacent Photoshop layers represent other button states.

When the button is selected from the menu on the DVD screen, two layers are set up to display. One layer is the white circle which appears above the particular number of the video the user is on, and the other layer is the text which describes that particular video. (See Figure 28-9.)

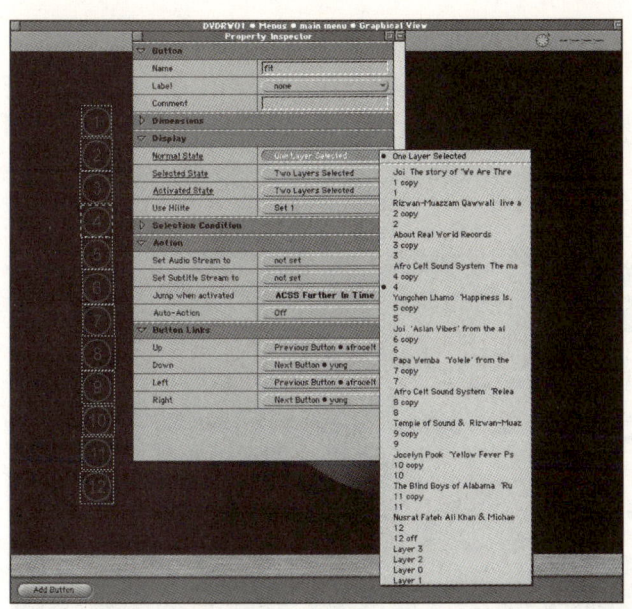

Figure 28-8: Adjusting the button settings in the Property Inspector. The drop-down menu represents all the potential Photoshop layers that are available to be turned on and off through DVD buttons.

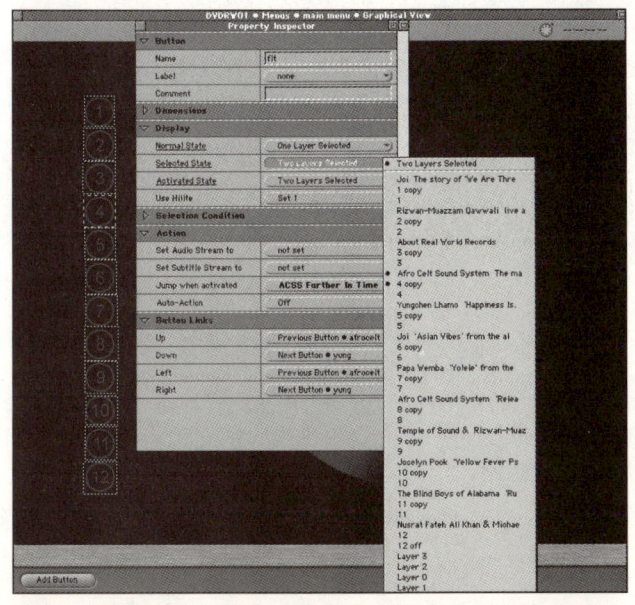

Figure 28-9: The Selected State for the button. When the button is selected, the 4 copy layer and the Text layer (Afro Celt Sound System) will show up in the diagram.

Preview

Even though the design is simple, there are a lot of settings to individual buttons and layers in this project. It is an example of a project where the Preview window becomes especially helpful, to verify that the correct layers are displaying with the correct buttons, and to ensure that the user experience makes sense. (See Figure 28-10.)

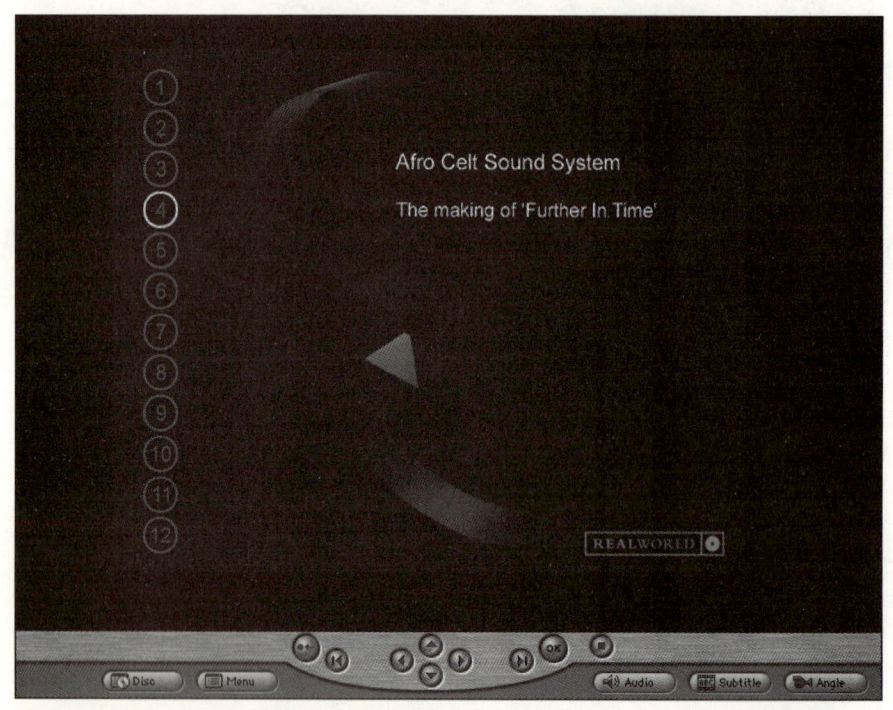

Figure 28-10: The preview of the project after adjusting settings, representing the Selected State of the button in Figure 28-9. In the preview, the two Photoshop layers are activated because the button is selected.

Summary

This case study represents a straightforward menu design — making use of the capability of DVD Studio Pro to use layers for button states, rather than the hilites, which draw an overlay on the television screen. Thanks, Real World!

✦ ✦ ✦

Appendixes

PART

VIII

♦ ♦ ♦ ♦

In This Part

Appendix A
DVD Studio Pro Installation and System Requirements

Appendix B
Resources

Appendix C
Glossary and Terms

Appendix D
About the DVD-ROM

♦ ♦ ♦ ♦

DVD Studio Pro Installation and System Requirements

APPENDIX A

System Requirements for DVD Studio Pro

In order to run DVD Studio Pro, your system must meet a few basic requirements. The most important requirement is that you are using a G4 Macintosh. Without the Altivec processing provided by a G4, you cannot even run the application. Besides, processing times for MPEG-2 compression would be considerably longer without the power of Altivec processing — the key to software encoding (without extra hardware) on the Mac.

This section covers system requirements as listed by Apple:

Hardware requirements

- Macintosh with a G4 processor and an Apple-supplied AGP graphics card
- At least 128MB of random-access memory (RAM)
- DVD-capable drive
- Display software and hardware capable of 1024 × 768 pixels at thousands or millions of colors

- ✦ For writing to disc: DVD-R recorder (such as the Pioneer A03, 103, or S201), DVD-RAM drive, or DLT tape drive
- ✦ Sufficient hard drive space to hold at least twice the size of your DVD project — for the source files and multiplexed project
- ✦ Hard drives capable of handling video for multiplexing — such as Firewire drives or (preferably) ultra wide A/V drives

Software requirements

- ✦ Mac OS 9.04 or 9.1
- ✦ QuickTime Pro 4.1 or later (included on the DVD Studio Pro CD)
- ✦ Apple DVD Player 2.3 or later (installed with the DVD Studio Pro software)
- ✦ MPEG encoding software, such as the QuickTime MPEG Encoder included with DVD Studio Pro
- ✦ Adobe Photoshop (Version 4.0 or later) for creating menus and buttons
- ✦ Video creation and editing software, such as Final Cut Pro, for preparation of video and audio files

After you have the necessary hardware and software, it is time to install the applications.

Installing and Configuring DVD Studio Pro

The installation process for DVD Studio Pro is simple and straightforward. By following the steps that you would use for other Macintosh applications you should be on your way to creating DVDs in no time at all.

Tip: Before you actually install DVD Studio Pro, we advise you to follow a few steps that may help guard against potential problems. First, you should turn off any security software and virus-protection software that may be running on your computer. Also, make sure you install and unlock QuickTime Pro (included with the application) before installing DVD Studio Pro.

QuickTime Pro is included as a part of the overall DVD Studio Pro package. Follow these steps to install QuickTime Pro software on your Macintosh computer:

1. **Insert the DVD Studio Pro CD into your computer's disc drive.**
2. **Open the QuickTime Pro folder.**
3. **Double-click the QuickTime Pro Installer icon to begin installation.**

4. Follow the on-screen instructions to install the software to the appropriate location on your hard drive.

5. Select **Custom** when the option for Choose Installation Type appears.

6. Choose **Select All** and click **Continue**.

7. Restart your computer when you have completed the on-screen installation process.

8. After your computer has restarted, locate the QuickTime Pro application on your hard drive and open it.

9. Complete the registration information and enter your serial number, which is found on the QuickTime Pro sheet that came with your copy of the DVD Studio Pro CD.

Now that you have installed QuickTime Pro you can proceed with the installation of DVD Studio Pro. Follow these steps to install DVD Studio Pro software on your Macintosh computer:

1. Insert the DVD Studio Pro CD into your computer's disc drive.

2. Double click the DVD Studio Pro Installer icon to begin installation.

3. Follow the on-screen instructions to install the software to the appropriate location on your hard drive.

4. Locate the DVD Studio Pro application on your hard drive and open it.

5. Complete the registration information and enter your serial number, which is on the sleeve of your DVD Studio Pro CD.

This concludes the installation of your DVD Studio Pro software. If you have any additional questions regarding installation, you may check the official Apple Web site at `www.apple.com/dvdstudiopro` for more information or updates. Also, make sure you check out the Readme file located in your DVD Studio Pro folder.

In addition to properly installing your software, make certain that you have enough memory on your system to adequately run the application. If you need to assign more memory to the DVD Studio Pro or QuickTime Pro application, you can do this by opening the Memory window.

Follow these steps to assign more memory to your DVD Studio Pro or QuickTime Pro applications:

1. Select the application icon (not an alias) by clicking it once. Make sure you don't open the application.

2. Choose **File**⇨**Get Info**⇨**Memory** to open the Memory Information window for the application.

3. **Enter a new value for Preferred Size and Minimum Size based on your memory requirements and the memory you have available.** Remember not to assign too much memory if you plan on running more than one application at a time.

4. **Close the Memory Information window and start up your application.**

New memory values can also be assigned for the A.Pack and Subtitle Editor applications, which were installed along with DVD Studio Pro. You can change the amount of memory assigned to an application as often as you like and as your needs and system capabilities change.

✦ ✦ ✦

Resources

APPENDIX B

The following is a list of Web sites that may be useful if you are seeking additional information regarding DVD Studio Pro or related topics. The official DVD Studio Pro Bible Web site is www.dvdspa.com, which includes up to date resources and links, as well as information on DVD manufacturing. (More information on the site is available in this book's Preface.)

Apple Links

www.apple.com/dvdstudiopro	General information about DVD Studio Pro
www.apple.com/dvd/compatibility	News on compatibility of DVD-R media with DVD players
www.apple.com/dvdstudiopro/update	Get the upgrade to Version 1.1, for free!
www.apple.com/quicktime	Get the latest version of QuickTime.

DVD-Related Sites

www.dvddemystified.com	Home of the extensive DVD FAQ
www.dvdmadeeasy.com	General information of interest to DVD authors
www.dvda.org	Worldwide organization for DVD community
www.2-pop.com	Lots of discussion on DVD and video coverage of DVD Studio Pro and Final Cut Pro
www.interactual.com	The leaders in DVD-ROM content
www.dvdspa.com	Up-to-date links to DVD manufacturing and authoring services

Video Companies

Name	URL	Services
Angle Park Productions	www.anglepark.com	Full range of video production services
Timeline Video	www.timelinevideo.net	Full range of video production services
Skipwave	www.skipwave.net	Music video production and related services
OVT Visuals	www.ovtvisuals.com	Video artists extraordinaire
Alien Arts	www.alienarts.com	Audio and video experts

DVD Manufacturing

URL	Services
www.metatec.com	Full service manufacturing, e-store
www.emvusa.com	Full service manufacturing
www.sanyolaserproducts.com/dvd/index.htm	Full service manufacturing, general DVD info

Web Hosting

URL	Services
www.metric-hosting.com	Host of the official DVDSP Bible site
www.playstream.com	Good pricing on hosting streaming video online

Author-Related Sites

Chad likes the following:

URL	Description
www.tokyomix.com	Devoted to Japanese culture and media and includes video, audio, and text on a variety of topics
www.dvd5.net	Information on topics related to DVD

Todd likes the following:

URL	Description
www.psrecords.net	Independent label
www.neocelt.net	Irish musical project
www.gerbilfront.com	Gerbil Liberation Front HQ
www.cftw.net	Change for the World HQ, info on changing the world, incrementally

Additional Links

www.detholz.com	Band site for examples in this book
www.warmblankets.org	Nonprofit organization, mentioned in Chapter 23, which builds and maintains orphanages in Cambodia
www.atomicpaintbrush.com	Home of animators/illustrators Dennis Calero and Kristin Sorra, covered in Chapter 24
www.realworld.co.uk	General site for Peter Gabriel's Real World Records and associated entities, covered in Chapter 28

Software Manufacturers

www.adobe.com	Information about Photoshop, After Effects, and other Adobe products
www.macromedia.com	Information about Flash, Dreamweaver, and other Macromedia products
www.terran.com	Information about Cleaner and other encoding solutions
www.digieffects.com	Information about Cinelook, Delerium and other After Effects plug-ins
www.roxio.com	Information about Toast and other disc recording software
www.connectix.com	Information about Virtual PC and other Mac utilities
www.matrox.com	Information about RTMac and other video hardware solutions

✦ ✦ ✦

Glossary and Terms

APPENDIX C

AC-3
Also known as Dolby Digital, an audio format capable of producing 5.1 surround sound channels with relatively small file sizes and data rates for DVD.

AUDIO_TS
The AUDIO_TS folder is created along with the VIDEO_TS folder when you generate a project in DVD Studio Pro. It is a vestigial organ. The AUDIO_TS folder is designed for the DVD-Audio standard, but DVD players never look at it. Some experts recommend leaving it when you burn a disc, even though it's empty.

BUP Backup Files
On a DVD disc, the BUP, or Backup, files contain duplicate information for the IFO files in case the IFO files fail to function due to a scratch or corruption of a file.

CBR
An acronym for constant bitrate encoding. This process sets a constant bit rate for the entire project, which it does not deviate from while encoding.

DLT
Digital Linear Tape. DLT tape is the most common way to submit a DVD project for manufacturing.

DTS (Digital Theater Systems)
A competing 5.1 channel audio format that offers less compression than AC-3 and what some experts consider higher quality sound.

DVD-Audio
The latest high quality sound format with terrific audio specifications but questionable appeal at this point in time.

DVD-RAM
A DVD-RAM drive allows the user to write data to a DVD-RAM disc. DVD-RAM allows data to be written several times to the same disc. An alternative to DVD-RAM is using the new DVD-RW (rewritable) drives such as the SuperDrive, which can write several times to DVD-RW media.

HTML
An acronym for Hypertext Markup Language. The basic building block for the Internet.

IFO files
Information Files. IFO files contain all of the formatting information for the VOB files in your project. They are responsible for giving a DVD player the information it needs to enable navigation of a disc and to set up display options such as the aspect ratio, subtitles, languages, and menus. The VIDEO_TS.IFO file actually contains all of the data that is required to navigate the overall DVD.

Internet
A large network of computers. When you go to an address such as `www.psrecords.net/wake/index.html`, the browser is going over the network and retrieving the index.html document.

Linear PCM (Pulse Code Modulation)
The highest quality audio option available for DVD, since it is uncompressed and delivers the highest bit rates. Due to the uncompressed nature of linear PCM, the resulting file sizes are very large and take up a great deal of space on a DVD disc.

Menus
Visual, interactive elements of a project that ultimately allow a person to navigate a finished DVD. Menus consist of backgrounds and buttons that are linked to tracks.

Motion Menus
Menus that incorporate animation or video loops, as background imagery.

Multiplexing
The process of encoding the project files in a DVD authoring program into a special DVD format that is based around a VIDEO_TS folder.

SDDS (Sony Dynamic Digital Sound)
An audio format primarily for movie theatres that can produce eight channels of surround sound and uses ATRAC compression — similar to mini-discs.

Slideshow
Sequences of still images or video with the ability to advance automatically or manually when the user clicks the remote.

Still Menus
Motionless menus, where individual still images are used as buttons that lead to playing a selection or that lead to another menu.

Subtitles
A special feature of DVDs, which utilizes the overlay feature of DVD players to place layers of text over video.

Tracks
Collections of video and audio that play as a whole, containing the assets, which become the core of a DVD's material.

VBR
An acronym for Variable bit rate encoding. Variable bit rate encoding overcomes the problem of encoding long, memory-hungry videos by first looking at a scene to determine the complexity and then setting an appropriate bit rate to encode the scene properly. Variable bit rate encoding is often superior to constant bitrate encoding (CBR) since it can fit as much video as possible on a disc, while maintaining higher quality when encoding complex scenes.

VIDEO_TS
The VIDEO_TS folder is the standard for all DVD projects; it contains all of the multiplexed DVD video files. Within the VIDEO_TS folder you find Video Title Sets (VTS) and three main types of subfiles with extension labeled .VOB, .IFO, and .BUP.

VOB files
Presentation Files. VOB files, or Video Objects, contain the multiplexed audio, video, and subtitle streams from your project. They are the fundamental elements in a DVD project.

✦ ✦ ✦

APPENDIX

D

About the DVD-ROM

This appendix provides you with information on the contents of the DVD-ROM that accompanies this book. The following programs are included on the DVD:

- ◆ Director Shockwave Studio
- ◆ Adobe Photoshop
- ◆ Adobe Photoshop Elements
- ◆ Adobe Premiere
- ◆ Adobe After Effects
- ◆ Adobe GoLive
- ◆ Adobe LiveMotion
- ◆ Fireworks
- ◆ Flash
- ◆ Acrobat Reader
- ◆ Deck LE (Audio recording software)
- ◆ Peak LE (Audio recording software)

What's on the DVD?

The disc is a DVD-ROM, featuring tutorial files, applications, several additional case studies, miscellaneous goodies, and a full-color electronic version of the book in PDF, all of which can be accessed with a DVD-ROM drive.

Tip: This disc is also a regular DVD and can be inserted in a standard DVD player to display video and other information created for this book, as well as some showcase material such as music videos.

Following is a summary of the contents of the DVD-ROM arranged by category.

Tutorial files

The tutorial section of the disc has a series of folders that contain example files for individual chapters.

Applications

The following applications are on the DVD-ROM:

- Trial and tryout versions of Adobe Photoshop, Adobe Photoshop Elements, Adobe After Effects, Director Shockwave Studio, Premiere, Adobe GoLive, Adobe LiveMotion, Fireworks, Flash, Dreamweaver, Deck LE, and Peak LE.
- Freeware program of Acrobat Reader.

Trial, demo, or *evaluation versions* are usually limited either by time or functionality (such as being unable to save projects).

Freeware programs are free, copyrighted games, applications, and utilities. You can copy them to as many computers as you like — free — but they have no technical support.

Case studies

The DVD-ROM includes several case studies that demonstrate projects created with DVD Studio Pro. Some of these case studies are discussed in the text, and there are additional case studies that appear only on the disc where there are descriptions of a project, resource files, and so on. The disc is also a standard DVD — some of the case studies have DVD video that can be viewed by placing the disc into either a set-top DVD player or DVD-ROM drive with DVD video playback software.

Goodies

- Jim Taylor's DVD FAQ in searchable HTML format, courtesy of www.dvddemystified.com
- Three white papers on DVD technology from Pioneer Electronics

- ✦ DVD-related art templates in a variety of formats, including Quark, Illustrator, Freehand, used to submit art when manufacturing a DVD (courtesy of Metatec)
- ✦ Additional last-minute treats

Full-Color electronic version of Macworld DVD Studio Pro Bible

(An excellent way to follow along with diagrams and examples in the book.)

The complete (and searchable) text of the *Macworld(r) DVD Studio Pro Bible* is on the DVD-ROM in Adobe's Portable Document Format (PDF), readable with the Adobe Acrobat Reader (also included). For more information on Adobe Acrobat Reader, go to www.adobe.com.

Using the DVD with the Mac OS

To install the items from the DVD to your hard drive, follow these steps:

1. **Place the disc in your computer's DVD-ROM drive.**
2. **Click the disc icon on your desktop to open a window listing the file folders on the disc.**
3. **If you wish to install or copy a particular file from the disc, locate the appropriate folder and click it to open.**
4. **Select the file and either drag it from its location on the disc to a folder on your hard drive or double-click it to open it directly from the disc.**

Troubleshooting

If you have difficulty installing or using the DVD-ROM programs, try the following solutions:

- ✦ **Turn off any antivirus software that you may have running.** Installers sometimes mimic virus activity and can make your computer incorrectly believe that it is being infected by a virus. (Be sure to turn the antivirus software back on later.)

✦ **Close all running programs.** The more programs you're running, the less memory is available to other programs. Installers also typically update files and programs; so if you keep other programs running, installation may not work properly.

If you still have trouble with the DVD, please call the Hungry Minds, Inc. Customer Care phone number: (800) 762-2974. Outside the United States, call (317) 572-3994. Hungry Minds provides technical support only for installation and other general quality control items; for technical support on the applications, consult the program's vendor or author.

✦ ✦ ✦

Index

Symbols
<> (angle brackets), 88
?= (assignment), 375
@access option (Property Inspector), 137
@access Web Links feature, 530
⌘+A (Select All), 510
⌘+K (New Button), 325
⌘+V (Paste), 509
= (equal sign), 387
(pound) sign, 384
/ (slash), 88

A
absolute value, 166
Abyss, The, DVD, 174
AC-3 format. *See also* A.Pack application
    experimenting with, 227–228
    mixing for, 216–218
    monitoring audio, 242
    overview of, 54, 210–213, 223–224
Acrobat (PDF) format, 460, 497–502
action safe area, 258, 259
Action section (Property Inspector), 143
Action setting (Flash), 492, 494, 497
activated state (button), 118
ad2, Inc., 178, 459
Add Menu button (Graphical View window), 5, 35–36
Add Script button (Graphical View window), 36
Add Slideshow button (Graphical View window), 36–37
Add Track button (Graphical View window), 10, 34–35, 407
Adobe
    Acrobat (PDF), 460, 497–502
    After Effects. *See* After Effects (Adobe)
    ImageReady, 111–113
    LiveMotion, 467
    Photoshop. *See* Photoshop (Adobe)
    Photoshop Elements, 100
    Premiere, 24, 25, 226, 271, 493, 541
    professional media development products, 115
    Web site, 226
Advanced Controls tab (DVD Player), 348, 453
Advanced Settings window (Cleaner), 275–276, 280, 281
aesthetics, 176

After Effects (Adobe)
    alternatives to, 541
    benefits of, 193
    compositing, 23
    composition, creating, 534–535
    exporting file as QuickTime, 536–537
    features of, 540
    Motion menu, creating, 132, 190, 533, 534–538
    Movie Settings window, 537
    overview of, 187–189
    plug-ins, 189, 190, 191–195
    Preview window, 536
    safe colors, determining in, 183
    Timeline, 535
    video clips, positioning, 536
    video clips, preparing, 533–534
AIFF (Audio Interchange File Format), 54, 207–208, 224
Aldrich, Nika, 212–213
alias, 60
aligning buttons, 567
all4dvd Web site, 356
anamorphic video, 257
angle brackets (<>), 88
Angle Park, 262
angles, 23. *See also* multiple angle project
Angles container, 408, 409
animations
    enjoyability and, 174
    Flash and, 195–199
    overview of, 135
    streaming, 486–487
    transitional, 180
    tweening, 526
    wire-frame, 285
anti-aliasing, 198
Antitrust (movie), 355
A.Pack application
    assigning more memory to, 575
    Batch Encoder window, 235–236, 240–242
    default bit-rate, 213
    encoding with, 211, 236–237, 238–239
    Instant Encoder window, 228–235, 238–239
    mixing audio, 237
    monitoring audio, 242
    overview of, 228
    preparing audio for import, 237

Apple
    compatibility list, 366
    DVD Player. *See* DVD Player (Apple)
    Final Cut Pro. *See* Final Cut Pro (Apple)
    iMovie, 303
    Pioneer Electronics and, 527
    Power Mac G4, 364, 573
    Sonic and, 358, 369
    Spruce Technologies and, 369
    SuperDrive, 83, 367, 523
    Web sites, 3, 353, 363, 366, 575, 577
applications. *See also specific products,* such as After Effects (Adobe)
    assigning more memory to, 575–576
    on DVD-ROM, 585, 586
    managing workflow between, 488–489
    manufacturer Web sites, 579
    multitrack recording, 225–227
    standalone, 458
    3D, 135
    video preparation, 270–277
APS Technologies Web site, 361
Arrange⇨Group (Flash), 491
art for DVD manufacturing, preparing, 554–555
aspect ratio, 42, 54, 254–255
Asset Files window, 66–69
Asset Matrix, 49, 63–64, 65, 397
Asset View window, 7. *See also* Assets Container
Asset window, 29, 30
assets
    assigning, 65, 426–430
    assigning file to, 67–68
    associating, 319–322, 422
    creating, importing by, 59
    deleting, 63, 245, 287
    description of, 21, 53, 66
    importing, 6–8, 56–59
    linking to menu buttons, 140–144
    maximum number per slideshow, 307
    renaming, labeling, and commenting, 60–61, 244, 286–287
    selecting, 59–60
    sending to Land of Oz, 62
    sorting, 62–63
    viewing, 63–64
Assets Container
    deleting assets, 63
    importing file, 57–58, 243
    managing file, 244–245
    New Asset option, 59
    overview of, 38–39
    renaming, labeling, and commenting assets, 60–61
    selecting assets, 59–60
    sorting assets, 62–63
    video assets, 286
Assign button (Asset Files window), 66, 69
assigning
    assets, 65, 426–430
    file to asset, 67–68
    language property to audio asset, 247
    memory, 575–576
    scripts, 378
    state to button, 118–119, 143–144
    value to variable, 387–388
assignment (?=), 375, 387–388
associating assets, 319–322, 422
Atomic Paintbrush Studios
    @ccess Web Links feature, 519–520
    bit-rate, 520
    content preparation, 515
    main menu, 514
    menu screen, 516
    motion menu, 517–518
    overview of, 513, 514, 515
    scripts used, 514
    workflow, 515–516
ATRAC (Adaptive Transform Acoustic Coding) compression, 215
audience
    captive vs. free, 78
    for CD-ROM, 480–481
    distribution and, 285
    enjoyability concept, 76
    expectations of, 78–79
    identifying, 77–78
    non-profit organization, 505–506
    overview of, 75–76
    re-purposing content and, 79–80
    responsibility to, 185
audio. *See also* audio format; audio track
    adding to track, 10
    assets, associating, 321–322
    importing, 243
    low-pass filter, 235
    menu, including with, 139
    mixing for multi-channel, 216–218
    monitoring channels, 218–220
    multitrack recording, 225
    overview of, 205, 223
    preparing, 224–225
    as Project Movie, 152

Audio Coding Mode options (A.Pack), 230
audio format
    AC-3/Dolby Digital, 210–213, 216–218, 223–224, 227–228, 242
    compatibility issues, 54
    DVD-Audio, 215–216
    linear PCM, 54, 206–208, 264
    MPEG audio, 214
    overview of, 205–206
    QuickTime Pro and, 209–210
    SDDS, 215
    support for, 19, 224
Audio Interchange File Format (AIFF), 54, 207–208, 224
Audio section (Property Inspector), 138–139
audio stream, 22, 245–247, 313, 429
Audio Streams folder, 246
audio track
    locking, 402–403
    multiple, capability for, 416
Audio value (Slideshow editor), 305
AUDIO_TS folder, 15, 16
authoring, 545
authoring environment, 21, 89
authoring media, 531

B

background
    interface, for, 175, 177, 179, 181
    Motion menu, for, 190–191
    Web page, for, 88–89, 94
    working with, 126–130
backing up file, 26
Backup (BUP) file, 265
backup storage system, 365
Barbie CD-ROM series, 527
Batch Encoder window (A.Pack), 235–236, 240–242
Batch window (Cleaner), 272–273, 281
Batch⇨Flag for Encoding (A.Pack), 241
Batch⇨New Job (A.Pack), 241
Baumgaertner, Martin, 262–263
Beno, Howie, 465
Berg, Randy, 357
Bessant, Marc, 560
Best Buy Web site, 227–228
Bias Web site, 225
bidirectional frames, 267
Bin window (Final Cut Pro), 404, 405
bit-depth, 297

bit-rate
    adjusting for multiple angle project, 411
    limitations on, 157, 279
    MPEG vs. Dolby AC-3, 214
    setting, 562
    specifying for audio format, 213
Bit Stream Mode options (A.Pack), 231–232
Blair Witch Project, The (movie), 262
Blue Book standard, 469
blurring image to remove moiré pattern, 298–300
bookmark, 84
border for subtitle, 153–154, 163, 170
Boris Continuum Looper plug-in, 190
buffer underrun protection, 527
build and format multiplexing, 346
Build Disc and Format command (File menu), 15, 352–353, 354
Build Disc command (File menu), 346–348, 352
build multiplexing, 346
building project
    button creation phase, 322–328
    linking phase, 328–338
    overview of, 337–338
    preparation phase, 317–322
burning
    avoiding errors when, 346
    DVD, 14–16
    DVD-R, 354–356, 360
    hybrid disc, 351, 356
button creation phase of project
    band member menus, 327
    Detholz Menu, 324–326
    Main Menu, 322–324
Button Hilites, 120–122, 131–132, 329, 483–484
Button Hilites section (Property Inspector), 45, 121, 139–140, 329
Button Links (Property Inspector), 143, 340, 343, 438
buttons
    aligning, 567
    assigning state to, 143–144
    checklist, developing, 344–345
    colors, matching, 105–106
    creating, 12–14, 111–113, 322–327
    interactive marker, as, 122–123
    layer effects for, 102
    linking assets to, 140–144

Continued

Index ✦ B–C

buttons *(continued)*
   linking to tracks, 144
   maximum number of, 142
   minimum number of, 100
   naming, 130, 142
   overlay image, creating, 119–120
   overview of, 99
   Property Inspector options, 142–143
   Real World Interactive project, 567–569
   redirecting, 433–439
   setting properties, 118–119
   shapes, 106–109
   still image and, 130–131
   still menu and, 100–105
   testing, 119
"buzz" of lines, eliminating, 185

C

Calero, Dennis, 523–524
camera angle, 23
case studies, 579, 586. *See also* Atomic Paintbrush Studios; Metatec International; Motion menu case study; Pioneer Electronics' SuperDrive; Real World Interactive; Warm Blankets Foundation
CCD (charge-coupled device), 296
CD-R disc, 458
CD-R/DVD Recording Demystified (Purcell), 367
CD-R media, 528, 530
CD-R technology, 527
CD-ROM
   disc delivery option, as, 480–481
   DVD-ROM and, 457–458
   outputting to, 367
cDVD (Sonic Solutions), 480
Center Downmix settings (A.Pack), 232
checklist, developing, 344–345
Cho, Audrey, 465
chroma values, 260
CineLook plug-in (DigiEffects), 189, 190, 191, 192, 193, 251
CineMotion plug-in (DigiEffects), 192–193, 251, 253, 255, 256
Cleaner (Terran), 271–276, 279–282, 486, 533–534
cloning tile, 327–328
CMF (Cutting Master Format) format, 362–363, 364
color
   choosing, manufacturing process, 552
   hexadecimal code for, 117
   matching, 105–106
   NTSC video and, 113
   reproduction of on television, 184
   RGB model, 100–101
   safe, determining, 116–117, 183
   subtitle overlays and, 156, 157
   for Web page, 88–89, 94
Color Menu Settings button (Subtitle Editor), 155
Color Picker dialog box, 116–117
Color Settings section (Project Settings window), 153–154, 163–164
combining projects, 472–473
Command+P (preview menu), 104
commands, adding to script, 385–387, 392–393
commenting, 60–61, 136, 383–385
communication and design issues, 180, 181
compatibility
   copy protection and, 446
   DVD-R and DVD-ROM, 530–531
   DVD-R media and DVD player, 366–367, 528
   DVD-ROM content and, 460–462
   format and, 54–55, 270
   prototype and, 83
   Web site on, 16
Composite, 184
compositing, 23, 24, 191
Composition Settings window (After Effects), 535
Compound shape options (Photoshop), 108
compression
   DVD-Audio, 215–216
   MPEG and, 18–19, 263, 266
   SDDS and, 215
Compression settings (A.Pack), 234
Computer Discount Warehouse, 365
computer-generated (CG) image, 135
Configure Lines dialog box, 146
configuring, steps for, 574–576
connection
   displaying for menu, 145–146
   Matrix Views and, 146–147
   menu and asset, between, 64
   project elements, between, 10–11
constant bit-rate encoding, 267
container, 398. *See also* Asset window; menu; track
content issues, 284, 285, 475–477, 527. *See also* cross-platforming standalone content
   cross developing, 24, 476–479
   cross-purposing, 78–79, 485, 489

Contents Scrambling System (CSS), 41, 358, 359, 447, 448, 450
copy protection
    CSS, 41, 358, 359, 447, 448, 450
    DVD-R media and, 365, 416
    Macrovision, 448
    overview of, 445–446
    problems with, 446–447
copying, 3. *See also* duplicating
Create Warped Text button (Photoshop), 110
cross developing content, 24, 476–479
cross-platforming standalone content
    Fireworks, 466–467
    Flash vs. Director, 463–466
    LiveMotion, 467
    overview of, 462–463
    previewing with Virtual PC, 469–471
cross-purposing content, 79–80, 485, 489
CSS (Contents Scrambling System) encryption, 41, 358, 359, 447, 448, 450
Custom shape tool (Photoshop), 108
customizing labels, 61–62
Cutting Master Format (CMF), 362–363, 364
Cyclone (all4dvd), 356

D

Data Rate options (A.Pack), 231
DCT (discrete cosign transformation), 266
DDP (Disc Description Protocol) format, 362–363, 364
Deck (Bias), 225
DeCSS, 446, 447
Default button, 137, 342
default language, choosing, 417–418
Delete option (Marker window), 160
deleting
    assets, 63, 245, 287
    slide from slideshow, 309
    tile from Graphical View window, 35
Delirium plug-in (DigiEffects), 190, 194–195
delivery options
    CD-ROM, 480–481
    disc formats, 479–482
    DVD, 479
    DVD-ROM, 479
    goals, establishing for, 478–479
    HTML, 487
    Internet, 482–487
    overview of, 475–476

paying passenger vs. stowaway content, 476–477
    Video CD, 481
design guides in Photoshop, 566–567
design issues. *See also* interface design
    media and, 114–117
    visual quality, 284–285
Dialog Normalization options (A.Pack), 231
Digidesign Web site, 226
Digital Linear Tape (DLT), 351, 358, 361–363
Digital Performer (Motu), 227
digital printing, 549
digital slideshow, 22
Digital Theatre System (DTS) format, 212–213, 214–215
digital video, 113, 249–250. *See also* video
Dimensions section (Property Inspector), 142–143
Director (Macromedia)
    CD-ROM and, 480–481
    cross-platforming and, 462–463
    DVD-ROM and, 458
    Flash compared to, 463–466
    interface design, 182
    overview of, 200–201
    scripting capability of, 479
    Shockwave and, 487
    York Tillyer on, 563
    workspace, 466
disabling automatic launching of DVD Player, 451
disc
    capacity of, 551
    previewing, 348–349
    specifying number of sides for, 41
Disc Description Protocol (DDP) format, 362–363, 364
disc media settings (Property Inspector), 40
disc menu settings (Property Inspector), 42
Disc properties (Property Inspector), 7–8, 39–40, 336–337, 341
disc space indicator (Graphical View window), 37–38
Discmakers, 366
discrete cosign transformation (DCT), 266
Display area (Property Inspector), 119, 143, 342–343
displaying
    asset information, 286–287
    connections between project elements, 10–11
    connections for menu, 145–146
distribution and copy protection, 446–447
DLT (Digital Linear Tape), 351, 358, 361–363
Do Nothing command, 386
Document⇨Set Page Action (Acrobat), 500, 501

Dolby AC-3 audio format. *See also* A.Pack application; surround sound
   DTS compared to, 214–215
   experimenting with, 227–228
   mixing for, 216–218
   monitoring audio, 242
   overview of, 54, 210–213, 223–224
Dolby Pro Logic, 212
Dolby Surround Mode settings (A.Pack), 232–233
domain name, 90
dpi (dots per inch), 484
drag-and-drop method
   associating assets, 319–320
   importing assets, 58–59
   importing file, 38
Dreamweaver (Macromedia), 86, 89–90, 94–97
Dressel, Brian, 190–191
drive, DLT vs. DVD-Authoring, 358
drive mechanism, 523–524
driver, 353, 369
DTS (Digital Theatre System) format, 212–213, 214–215
dummy subtitle, 429–430
Duplicate function, 327–328
duplicating
   folder, 3
   menu, 333–334
   service bureau and, 40
duplication, 547
duration
   of subtitle, setting, 166
   of video track for multiple angles, 404
Duration value (Slideshow editor), 306, 310
Dust & Scratches filter (Photoshop), 300
DV camera and PCM audio, 206–207
DVD-5 disc, 40, 547, 548, 551
DVD-9 disc, 545, 547, 551
DVD-Audio, 215–216
DVD authoring
   Apple and, 364
   Baumgaertner on, 262–263
   levels of, 369
   Mac vs. Windows, 358
   Parsons on, 528
DVD Copy Control Association Web site, 447
DVD Creator system, 357
DVD Demystified (Taylor), 357, 367
DVD FAQ Web site, 473, 531, 586
DVD format, 359–360, 479, 506–507
DVD menu, 8

DVD player
   default language, 417–418
   DVD-R media and, 366–367
DVD Player (Apple)
   controller, 454–455
   disabling automatic launching, 451
   File⇨Open VIDEO_TS, 352
   overview of, 348–349, 451
   Player preferences, 451–453
DVD-R burner
   Apple SuperDrive, 523
   cost of, 530
   demand for, 368
   external, support for, 353
   FireWire, 523–524, 525–526
   Pioneer DVR-S201, 523
   product comparison, 368
   prototyping and, 83
DVD-R disc, 14, 354–356, 360, 525
DVD-R format
   adoption of by consumers, 528
   DVD-ROM compatibility and, 359
   Hollywood and, 359
   overview of, 83, 361, 458, 507
DVD-R media
   A03 mechanism and, 525
   CD-R compared to, 530
   cost of, 530
   CSS encryption and, 447
   DVD players and, 366–367
   types of, 363, 364–365
DVD-ROM
   adding player-based Flash on, 467
   case studies, 586
   combining multiple VIDEO_TS folders for, 472–473
   cross-platforming content, 462–463
   designing interface for, 182
   developing content for, 457, 464
   as disc delivery option, 479
   DVD-R compatibility and, 359
   DynamicCD technology and, 459
   electronic version of book on, 587
   goodies on, 586–587
   overview of, 457–458
   PCFriendly and, 471–472
   platform compatibility, 460–462
   player-based content, 460, 462
   preparing content, 467–469
   previewing content, 469–471

programs included on, 585, 586
standalone content, 458, 459–460
troubleshooting, 587–588
tutorial files, 586
UDF Bridge and, 458, 469
using with Mac OS, 587
DVD standards. *See* standards
DVD video, simulating, 492–495
DVD-Videosize section (Project Settings window), 152, 162
DVD@cess Web Links (DVD Player), 453
DVDSP manual, 388
DVR-A03 mechanism, 523–524, 526–527
DynamicCD technology, 459

E

Easter egg, 62, 175
eBay, 365
ECD (Enhanced CD), 563
Echo Fire (Synthetic Aperture), 184
Edit and Action dialog box (Acrobat), 501
editing software, 270–276
editing video, 19
editors, 46. *See also* Marker Editor; Menu Editor; Script Editor; Slideshow Editor; Subtitle Editor
Edit⇨Clear, 63
Edit⇨Color Settings (Photoshop), 101
Edit⇨Cut (Script Editor), 382
Edit⇨Duplicate, 333
Edit⇨Paste (QuickTime Pro), 509
Edit⇨Paste (Script Editor), 382
Edit⇨Preferences, 61
Edit⇨Preferences (DVD Player), 348, 451–453
Edit⇨Select All (QuickTime Pro), 510
eliminating
  "buzz" of lines, 185
  lag at end of loop, 135
  lag before subtitle appears, 166
Ellipse tool (Photoshop), 107
Emagic Web site, 226
encoders, 19, 133–134, 285
encoding. *See also* A.Pack; multiplexing
  description of, 54
  MPEG-2, 277–283, 537–538
  variable bit-rate vs. constant bit-rate, 267–268
  video content, 19
  visual quality and, 284–285
encoding time, 364

encryption, 447
enjoyability concept, 76, 174–175
equal (=) sign, 387
error correction, 529
errors
  avoiding when burning disc, 346
  syntax error, 384
Esperanto, 19
Estimated Size option, 39–40, 136–137
evaluating
  image quality, 261
  interface design, 181–182
  screen size, 254–260
  sound quality, 261, 264
EventStream authoring (Cleaner), 272
evolution|bureau, 184, 185–186
.exe file extension, 460
expectations of audience, 78–79
Export dialog box (QuickTime Pro), 278
Export Queue window (Final Cut Pro), 405
Export QuickTime dialog box (Flash), 198
Export Settings dialog box (Premiere), 493
Export window (After Effects), 537
exporting
  After Effects, from, 536–537
  QuickTime, to, 195–199
  QuickTime Pro, from, 55–56
external drive
  burning and, 356
  FireWire, 14, 523–524, 525–526
  Format window and, 16
  manufacturers of, 369
Eyedropper tool (Photoshop), 105–106

F

Fahs, Chris, 506
fair use of product, 446
feedback, providing to user, 175, 178
Feith, Paul, 284–285
Fenster, Dan, 180–181
file. *See also* file extension; format
  backing up, 26
  creating, 5
  folders for, 27–28
  importing, 6–8, 38
  locating missing, 67–68
  managing, 244–245, 286–287, 396–399
  nondestructive, 272
  sizes of, 18, 26

file extension
   description of, 55
   .exe, 460
   HTML document, 87
   player-based content, 462
   video, 54
   working with, 55–56
file structure, 264–265
File➪Batch Export (Final Cut Pro), 405
File➪Build Disc, 16, 346–348
File➪Build Disc and Format, 15, 352–353, 354
File➪Export Movie (Flash), 197
File➪Export (QuickTime Pro), 278
File➪Export➪QuickTime (Final Cut Pro), 129, 133
File➪Get Info➪Memory, 575
File➪Import, 6, 56–58, 243
File➪Import➪File (Final Cut Pro), 133
File➪Import➪Subtitles (Subtitle Editor), 170
File➪New, 5
File➪New Batch List (A.Pack), 240
File➪New Player (QuickTime Pro), 509
File➪New➪Bin (Final Cut Pro), 404
File➪Open VIDEO_TS (DVD Player), 352
File➪Preferences (Subtitle Editor), 155
File➪Save, 5
File➪Save for Web (Photoshop), 92
film. *See also* movie
   aspect ratio for, 254–255
   video compared to, 250–253
FilmDamage module (DigiEffects), 190, 191–192
Filter➪Blur➪Gaussian Blur (Photoshop), 298
Filter➪Noise➪Dust & Scratches (Photoshop), 300
Filter➪Sharpen➪Unsharp Mask (Photoshop), 299
Filter➪Video➪NTSC Colors (ImageReady), 113
Final Cut Pro (Apple)
   capturing still from video clip, 128–130
   cutting clip into loop, 133–134
   overscan lines, 258
   overview of, 132, 186, 271
   preparing angles, 399–406
   rendering time and, 193
   safe areas, defining, 258, 259
   widescreen filter, 255, 256
Find feature (Marker window), 160
FireWire DVD-R burner, 14, 523–524, 525–526
Fireworks (Macromedia), 24, 100, 466–467
flagging job, 241

Flash (Macromedia)
   adding player-based on DVD-ROM, 467
   animation and, 488, 526
   cross-platforming and, 462–463
   Director compared to, 463–466
   DVD-ROM and, 458, 460
   Movie Symbol, 493
   overview of, 195–199
   Score window, 490
   simulating DVD in, 490–497
   streaming animation and, 486–487
   Tillyer on, 563
   transparency, achieving in, 491
   uses of, 178–179
flatbed scanner, 296
flattening
   image, 24, 128, 130, 302, 303
   movie, 276
floppy disk, 26
flowchart, 488
focus group, 78
folder
   copying, 3
   for project, 27–28
font, subtitle and, 169–170
foreground elements. *See* buttons
format. *See also* audio format; video format
   disc delivery, 479–482
   DVD-ROM content and, 460
   graphics compatibility, 54–55
   importing files and, 6–8
   overview of, 53–54
Format Disc window, 15, 353, 354–355, 362
Format window, 15
frame exporting, 490
frame rate, 54, 250–251
frame size, 54
Freehand (Macromedia), 199–200
freeware program, 586
frequencies
   Dolby AC-3 format, 211
   DVD-Audio, 216
   linear PCM, 206
   SDDS, 215
Full Bandwidth Channels settings (A.Pack), 235
fun, building into project, 337
function, 389–391

Index ✦ G-I

G

G4 Macintosh, 364, 573. *See also* Mac OS
Gabriel, Peter, 559, 561
GameCube, 177
Gaussian Blur filter (Photoshop), 298
General properties (Property Inspector), 40–42, 137, 410, 449, 565
General settings (A.Pack), 234–235
.GIF format, 462
glass-mastering, 546
GLF (Gerbil Liberation Front) DVD, 485, 489, 550–558
Global Variable, 380
goals, establishing for content, 478–479
GoLive (Adobe), 115
GoTo drop-down menu (Graphical View window), 32
goto label command, 385
Gradient tool (Photoshop), 127
graphic overlay. *See* overlay image
Graphical View window. *See also* Assets Container
    Add Menu button, 5, 35–36
    Add Script button, 36
    Add Slideshow button, 36–37
    Add Track button, 10, 34–35, 407
    associated tiles in, 320
    Atomic Paintbrush Studios case study, 525
    button settings, 342–343
    description of, 29, 30
    Disc Properties, 341
    disc space indicator, 37–38
    drop-down menus, 32–33, 34
    Lines feature, 145–146, 338
    Lines function set to Always, 435–436
    menu settings, 341–342
    multiple language project, 431–432
    overview of, 4
    Preview button, 11, 14, 37, 340–341, 352, 412
    project element buttons, 34–38
    Real World Interactive project, 564–565
    Subtitle Streams button, 430
    tiles in, 30–32, 319
graphics compatibility, 54–55

H

halftone screen pattern, 297
hardware encoder, 282–283
hardware requirements, 573–574
HDTV, 257, 261
Helpers drop-down menu (Script Editor), 381–389
Helpers⇨Play⇨Play Menu, 382
hexadecimal code for color, 117
Hot Keys menu (DVD Player), 453
hotspots, 90–91, 94–96, 498, 500. *See also* roll-overs
HTML, 86–91, 487. *See also* HTML document
HTML document
    Dreamweaver and, 89–90, 94–97
    DVD-ROM and, 460
    naming, 87
    naming files and folders in, 91
    setting background color for, 88–89
Hungry Minds, Inc. Customer Care number, 588
hybrid disc, burning, 351, 356
hyperlink in Word document, creating, 84–85
Hypertext Markup Language. *See* HTML

I

ICE hardware board, 193
identifying audience, 77–78
iDVD, 529
IF-THEN statement, 378–379, 386
image. *See also* overlay image; still image
    computer-generated, 135
    flattening, 24, 128, 130, 302, 303
    interlaced video, 129
    quality of, evaluating, 261
image editing program. *See* Photoshop (Adobe)
image map, 90–91
Image⇨Image Size (Photoshop), 120
Image⇨Mode⇨Grayscale (Photoshop), 120
ImageReady (Adobe), 111–113
IMG file, 353
iMovie (Apple), 303
Import Assets dialog box, 7, 56, 57, 318, 426
Import Assets window, 286
Import button (Import Assets dialog box), 6–7, 57
importing
    assets, 6–8, 56–59
    audio, 243
    file, 6–8, 38
    multilayered image file, 24–25
    Photoshop file, 24, 25
    subtitle, 167–168, 170–171
    video, 283, 286
Incubus (movie), 19, 311–312
Information Architecture, 185
input channels, 229–230, 238
Insert⇨Bookmark (Microsoft Word), 84

Insert⇒Image (Dreamweaver), 95
Insert⇒Picture⇒From File, 81
inspection of disc, 548
installing, 573–575
installer, including on disc, 481
Instant Encoder window (A.Pack)
    audio settings, 230–232
    bitstream settings, 232–234
    encoding with, 238–239
    input channels, 229–230
    overview of, 228–229
    preprocessing settings, 234–235
intellectual property rights, 446
interactive marker, button as, 122–123
interactivity
    DVD-ROM and, 182, 479
    interface design and, 174, 178
    perception and, 477–478
    reviewing by using Lines, 33
    testing, 339–345
InterActual
    PCFriendly, 471–472
    Player, 357–358, 471
interface design. *See also* specific software products, such as After Effects (Adobe)
    backgrounds, 175, 177, 179, 181
    DVD vs. DVD-ROM, 182
    enjoyability and, 174–175
    evaluating, 181–182
    Fenster on, 180–181
    Internet delivery and, 483–485
    Kuzmanich on, 178–179
    Martin on, 176–177
    overview of, 173
    Requiem for a Dream (movie) example, 79
    technical considerations, 183
    usability, 175
    Zada on, 184–186
interlaced scanning, 253–254
interlaced video image, 129
Internet, 87
Internet delivery285, 482–487
intra frames, 266–267
inverse telecine, 252–253
ISO 9660 format, 544
Item⇒Duplicate, 328
Item⇒New Angle, 408, 409

Item⇒New Audio Stream, 321
Item⇒New Button, 325
Item⇒New Language, 425
Item⇒New Story, 291
Item⇒New Subtitle (Subtitle Editor), 168
Item⇒Preview Marker, 293
Item⇒Preview Menu, 104, 292
Item⇒Sort Assets, 62–63

J

jewel box packaging, 549
joining subtitles, 161
Jolt Cola Web site, 373
JPEG format, 92
Jump Matrix, 50, 123, 146, 397–398
Jump when activated action, 343

K

Katsoudas, Stella, 561
Kelsey, Todd, 465, 506
key frame, 151, 161
Klingon Language Institute, 419
Kuzmanich, Justin, 178–179

L

LaBarge, Ralph, 357
labeling
    assets, 60–61
    customizing, 61–62
    scripting and, 385
lags, eliminating, 135, 166
Language menu, 426
languages. *See also* multiple language project
    adding new to project, 421–422
    adding new to slideshow, 311–313
    alternate reality approach to, 419–420
    assigning assets to, 426–430
    default, choosing, 417–418
    default approach to, 420
    direct approach to, 418–419
    lite approach to, 418
    support for, 19–20, 417
    working with multiple, 415–416
lasers, 526
Layer Matrix, 50–51, 123, 146–147
layered image file. *See* multilayered image file
Layer⇒Duplicate Layer I (ImageReady), 112

Layer⇨Duplicate Layer (Photoshop), 103
Layer⇨Flatten Image (Photoshop), 120, 302
Layer⇨Layer Style (Photoshop), 102
Layer⇨Merge Down (Photoshop), 128
Layer⇨Merge Visible (Photoshop), 128
Layer⇨New⇨Layer (ImageReady), 111
Layer⇨New⇨Layer (Photoshop), 101
layers
    multilayered image file, 24–25, 100, 126
        menu and, 24–25
        naming, 101
        organizing, 131
        preparing for use in still menu, 100–105
        selecting, 9
Layers (Always Visible) menu, 9, 138
Layers palette (ImageReady), 112
Layers palette (Photoshop), 24, 526
LBRDF (Laser Beam Recorder Data Formatter), 363
letterboxing, 255
LFE Channel settings (A.Pack), 235
Library window (Flash), 495
Line tool (Photoshop), 108
linear PCM, 54, 206–208, 264
Lines drop-down menu (Graphical View window), 33
Lines feature (Graphical View), 145–146
linking, 140–144. *See also* linking phase of building project
linking phase of building project
    Detholz Menu, 331–332, 334–335
    Disc properties, setting, 336–337
    Main Menu, 329–331
    overview of, 328–329
    Rick menu, 332–334
    Track properties, setting, 335–336
Linux operating system, 460
LiveMotion (Adobe), 115, 467
load times and interface design, 176
Locate button (Asset Files window), 66, 68
locating missing file, 67–68
locking audio tracks, 402–403
Log window, 391–392
Logic Audio (Emagic), 226–227
looping video, 134–135, 508
lossless linking, 527
low-pass filter for audio, 235
luminance values, 260

M

Mac OS
    alias, 60
    DVD-ROM and, 587
    file extensions in, 55
MacMall Web site, 227
Macromedia
    Director. *See* Director (Macromedia)
    Dreamweaver, 86, 89–90, 94–97
    Fireworks, 24, 100, 466–467
    Flash. *See* Flash (Macromedia)
    Flash Player Download Center, 467
    Freehand, 199–200
    interactive design products, 115
    Shockwave, 465, 487
    Web site, 464
Macrovision, 448
MacWarehouse Web site, 227
Magic Wand tool (Photoshhop), 510–511
Main Menu
    creating buttons for, 322–324
    linking, 329–331
managing
    file, 244–245, 286–287, 396–399
    workflow between applications, 488–489
manufacturing DVD. *See also* Metatec International
    discount for readers, 558
    DVD-ROM content and, 468–469
    overview of, 361
    Web sites, 578
Marken, Andy, 368–370
marker
    assigning script to, 378
    Atomic Paintbrush Studios case study, 530
    creating, 159–160
    creating stories with, 290–292
    description of, 22, 46
    interactive, 122–123
    maximum number of, 289
    navigating with, 158–159, 160
    placing in video track, 288–290
Marker Editor, 47, 48, 289
Marker window (Subtitle Editor), 158–160
market for DVD author, 360, 370
Martin, Alan, 176–177
masking, 210

master for replication, producing, 364–366
mastering, 546
Matrix, The, DVD, 49, 50, 175, 190
Matrix Views
    Asset Matrix, 49, 63–64, 65
    Jump Matrix, 50, 123, 146, 397–398
    Layer Matrix, 50–51, 123, 146–147
    multi-angle assets and, 397–398
    overview of, 49
Matrix⇨Assets of Disc, 397
Matrix⇨Assets of Disc PS Records DVD, 63
Matrix⇨Jumps of Disc, 397
media for encoded video, 285
Median filter (Photoshop), 299
memory
    assigning more to application, 575–576
    size of project and, 157
Memory Information window, 575
menu. *See also* background; buttons; layers; Motion menu; still menu
    adding to Graphical View window, 35–36
    audio, including with, 139
    commenting, 136
    connection between asset and, viewing, 64
    creating, 5
    description of, 22, 125
    duplicating, 333–334
    elements of, 126
    foreground elements, creating, 130–131
    multilayered image files and, 24–25
    naming, 6, 136
    Property Inspector and, 136–140
    resizing, 103
    scripting and, 375–377
    standards, 18
    testing, 144–147
    types of, 23–24
Menu Editor
    @ccess link, 529
    assigning state to button, 118–119
    new buttons, 12–14, 324, 326
    opening, 46
    Photoshop layers, 325
    Real World Interactive project, 567–568
    redirecting buttons, 433–436
    still menu, 104, 105
    testing buttons, 119
    Untitled Button, 141–142

menu flow document, 180
Menu properties (Property Inspector), 45, 136–137
menu tile, 5, 8
Menus drop-down menu (Script Editor), 379–380
merge command, 472–473
merging layers, 128
Metatec International
    background, 543
    "check disc" package, 548
    E-Store, 550
    inspection, 548
    mastering, 546
    packaging, 549, 552–553
    pictorial case study, 550–558
    premastering, 544–545
    printing, 548
    quality control, 545
    quantity, choosing, 551
    replication, 547–548
Microsoft
    Windows Media Player, 285
    Word document, 81, 83–86, 439–440
MIDI, 226
MLP (Meridian Lossless Packing) compression, 216
moiré pattern, detecting and eliminating, 297–300
monitoring
    AC-3/Dolby Digital audio, 242
    audio channels, 218–220
Motion Control feature (Premiere), 541
Motion menu. *See also* Motion menu case study
    assembling, 131–132
    Atomic Paintbrush Studios case study, 527–529
    backgrounds, 190–191
    Button Hilites and, 120–122
    description of, 8, 23–24
    looping video, 134–135
    MPEG-2 format and, 132
Motion menu case study
    After Effects, using, 533, 534–538
    composition, creating, 534–535
    DVD Studio Pro, creating in, 538–540
    exporting file as QuickTime, 536–537
    overview of, 533
    summary of, 540–541
    Timeline, adjusting layers in, 535
    video clips, positioning, 536
    video clips, preparing, 533–534

Index ✦ M–N

Motion Menu Editor, 539–540
Motion Picture Association of America Web site, 447
Motu Web site, 227
movie. *See also* film
    flattening, 276
    interface design for, 178
Movie Properties dialog box (Acrobat), 499
Movie Settings section (Project Settings window), 151–152, 162
Movie Settings window (After Effects), 537
Movie Symbol (Flash), 493
Movie⇨Loop (QuickTime Pro), 510
moving water scenes, 284
MP3 format, 271–272
MPEG audio, 214, 224
MPEG (Motion Picture Experts Group), 18–19, 54, 265
MPEG video format, 261
MPEG-1 audio format, 54
MPEG-1 video format, 261, 270
MPEG-2 encoder, 19, 133–134
MPEG-2 video format
    bidirectional frames, 267
    compatibility, 54
    compression and, 263, 266, 562
    encoding, 277–283, 537–538
    exporting from QuickTime Pro, 55–56
    intra frames, 266–267
    Motion menu and, 132
    overview of, 261, 265–266
    predicted frames, 267
    quality and, 270
    standard, as, 18
    variable bit-rate vs. constant bit-rate encoding, 267–268
multilayered image file, 24–25, 100, 126. *See also* layer; menu
multiple angle project
    adjusting bit-rate, 411
    creating, 406–411
    managing files, 396–399
    overview of, 395–396
    placing on tracks, 410
    preparing, 399–406
    previewing, 411–412
Multiple Language feature, 416
multiple language project
    adding language, 422–426
    additional fun folder, 440
    alternate reality approach, 431–432
    assigning assets for new language, 426–430
    language bridge, making, 436–438
    redirecting buttons, 433–436
    Spanish assets, setting up, 433
    translations in Word document, 439–440
    workflow, 438–439
multiplexed project, 264–265
multiplexing. *See also* encoding
    Build Disc and Format command, 352–353
    Build Disc command and, 346–348, 352
    description of, 339
    overview of, 15, 352
    preparing project for output and, 345–348
Multiplexing Progress window, 16
multitrack recording, 225–227
Myst game series, 174, 458

N

naming. *See also* renaming assets
    angle, 409
    buttons, 130, 142
    channels, 238
    HTML document, 87
    layers, 101
    menu, 136
navigation
    GoTo drop-down menu, 32
    interface design and, 175, 176
    Marker window and, 158–159, 160
    Pause mode and, 455
    Warm Blankets Foundation case study, 508
Navigation (IFO) files, 265
Neocelt CD, 464–465
New File dialog box (Photoshop), 484
New Marker dialog box (Subtitle Editor), 160
New⇨Symbol (Flash), 491
nickel master, 546
Nielsen, Jakob, 76
non-profit organization. *See* Warm Blankets Foundation
nondestructive file, 272
normal state (button), 118
normalization, 231
NTSC, 54, 100–101, 113, 152, 162, 260

O

Object Actions setting (Flash), 492
Oklahoma (movie), 250
opening
    Marker Editor, 48
    Menu Editor, 46
    Script Editor, 47
    Slideshow editor, 47
organizing. *See also* organizing assets
    images for slide show, 304
    layers, 131
    slides in slideshow, 309
organizing assets
    Asset Files window, 66–69
    Asset Matrix, 63–65
    Assets Container, 59–63
Output Name window (A.Pack), 240–241
Output window (Cleaner), 282
outputting. *See also* preparing project for output
    CD-ROM, to, 367
    DLT, to, 361–363
    Web, to, 114–115
overlay image
    color for, 156, 157
    creating, 119–120
    maximum number of, 165
    subtitles and, 150
overlay picture, 139
overscan, 134, 258
Overwrite command (Final Cut Pro), 403

P

packaging disc, 549, 552–553
Page Actions dialog box (Acrobat), 501
PAL (Phase Alternating Line), 54, 152, 162, 214, 260
Panasonic DVD-R burner, 368–369, 529–530
pans and zooms, 284
Parental Controls tab (DVD Player), 452
Parsons, Andy, 523, 526–531
Pause mode (DVD Player), 455
Pause value (Slideshow editor), 306, 308, 310
paying passenger content, 476
PCFriendly (InterActual), 471–472
PCM audio, 54, 206–208, 264
PDF (Acrobat) format, 460, 497–502
Peak DV application (Bias), 207, 208, 225
Peak LE, 467
Peak Mixing Level settings (A.Pack), 233
perception, 477–478
personalization, 178
Photoshop (Adobe)
    adding new language to project, 421
    Atomic Paintbrush Studios case study and, 527–529
    background, creating, 127–130
    design guides, 566–567
    effects layers, 130
    Eyedropper tool, 105–106
    Gradient tool, 127
    ImageReady and, 111
    importing file from, 24, 25
    Layer Style option, 185
    Layers palette, 24, 526
    Magic Wand tool, 510–511
    matching colors, 105–106
    modifying image, 94
    moiré pattern, eliminating, 298–300
    multi-layered image file, 100
    New File dialog box, 484
    overlay image, creating, 119–120
    pixels and, 113–114
    preparing image, 91–94
    .PSD format, 54–55
    Real World Interactive project, 562, 566–567
    resizing image, 300–303
    Save for Web function, 484–485
    selected state of buttons, 324–326
    shape tools, 106–109
    slice tools, 116
    still menu, 100–105, 126
    styles, 109
    text, using, 110–111
    transparent background, 528
    versions of, 100
Photoshop Elements (Adobe), 100
PICT file, 126, 493
Picture section (Property Inspector), 45, 138, 141, 421, 427
Pioneer DVR-S201, 83, 368, 523, 531
Pioneer Electronics' SuperDrive
    A03 mechanism, 523–524, 526–527
    Apple and, 527
    CD-R vs. DVD-R, 530
    compatibility issues, 530–531
    consumer standalone DVD recorder, 529
    DVD-R format adoption, 528
    DVD-R media, 525
    external burners, 525–526

marketing issues, 527–528
Panasonic mechanism vs., 529–530
Parsons and, 523, 526–531
Pioneer DVR-S201, 523
piracy, 445, 446
pixels, 113–114, 184
planning project. *See also* prototyping
audience, considering, 75–80
distribution options, 114–117
storyboarding, 80–82
Warm Blankets example, 506–508
Play function, 379
playback
DVD-R media and, 365, 531
optimizing, 455
PlayStation2, 177
plug-in, 485
pointillism, 113
Polygon tool (Photoshop), 107
positioning subtitle, 168–169
post house, 271
pound (#) sign, 384
Power Mac G4, 364, 573. *See also* Mac OS
pre-script, 378
Pre-Script option, 137, 342
predicted frames, 267
Preferences window (A.Pack application), 229
Preferences window (Subtitle Editor), 155–156
premastering, 544–545
Premiere (Adobe), 24, 25, 226, 271, 493, 541
preparation phase of building project, 317–322
preparing art for DVD manufacturing, 554–555, 558
preparing project for output
disc previewing, 348–349
managing multiple files, 396–399
multiplexing, 345–348
overview of, 339
testing interactivity, 339–345
Presentation (VOB) files, 265
Preview button (Graphical View), 11, 14, 37, 340–341, 352, 412
preview file, 25
preview multiplexing, 346
Preview window
After Effects, 536
DVD Studio Pro, 11–12, 412, 570
Subtitle Editor, 157–158, 164
Preview Window Color (Subtitle Editor), 154, 164

previewing
cross-platform content, 469–471
disc, 348–349
DVD, 14
loop, 135
multiple angle project, 411–412
slideshow, 311
tile, 11–12
video, 293
printing disc, 548
Pro Tools (Digidesign), 226
Progress window, 355, 356
progressive scanning, 254
project file, 25, 26, 317–318. *See also* file
Project Movie
creating subtitle with, 161–167
overview of, 151–152
Preview window and, 157
Project Movie section (Project Settings window), 151–152
Project Settings window (Subtitle Editor)
Color Settings section, 153–154, 163–164
DVD-Videosize section, 152, 162
Movie Settings section, 162
positioning subtitle text in frame, 168
Project Movie section, 151–152
Subtitle Settings section, 152, 162–163
Project View
description of, 29, 30, 38
Languages tab, 422–423, 424
markers and, 292
Menus tab, 38
multi-angle assets and, 398–399
Tracks tab, 322
Project window (Cleaner), 272, 273, 275
projector, 462
projects. *See also* building project; multiple angle project; multiple language project; planning project
adding multiple audio streams to, 245–247
associating element with language, 19–20
burning DVD, 14–16
buttons, creating, 12–14
combining, 472–473
default language, choosing, 418
folder structure for, 27–28
importing assets, 6–8

Continued

projects *(continued)*
    menu, 8–9, 22, 23–25
    multiplexed, 264–265
    overview of, 21
    preparing for output, 339–349, 396–399
    saving, 5, 26
    with script, making, 375–377
    sharing, 367
    starting, 3–6, 25–26
    tiles, previewing, 11–12
    track, 10–11, 22–23
Property Inspector
    @ccess link, launching, 530
    @ccess option, 137
    active layers, 323
    Assets Container and, 59–60
    assigning file to asset, 69
    assigning language property to assets, 247
    associating assets, 320–322
    Audio section, 138–139
    Button Hilites section, 49, 121, 139–140, 329
    Button Links, 143, 330, 343, 438
    button settings, 142–143, 335, 568–569
    disc menu settings, 42
    Disc properties, 8, 39–40, 336–337, 341
    Display section, 119, 143, 342–343
    displaying asset information, 286–287
    General properties, 40–42, 137, 410, 449, 565
    Go button, 333
    information in, 141
    Jump When Activated setting, 13
    Languages section and Languages drop-down menu, 423
    Loop function, 508
    Menu properties, 6, 9, 45, 136–137
    menu tile, 528, 539
    naming angle, 409
    Normal State drop-down menu, 332
    opening and closing section of, 136
    overview of, 4, 29–30, 39, 244
    Picture section, 45, 138, 141, 421, 427
    Remote Control properties, 43, 44
    resizing, 391–392
    reviewing interactivity, 341–344
    Script properties, 45
    selecting image layers, 104, 333–334
    Slideshow properties, 45
    Timeout section, 45, 134, 138
    Track properties, 10, 44, 335–336, 344
    Variable Names menu, 43
prototyping
    compatibility and, 83, 367
    Dreamweaver and, 89–97
    HTML and, 87–88, 90–97
    Microsoft Word document, 83–86
    overview of, 77, 82
Purcell, Lee, 364–366, 367

Q

quality and medium, 114
quality control, 545
QuickTime
    Adobe Acrobat and, 497
    editing file in Cleaner, 272–276
    exporting file from After Effects, 536–537
    simulating DVD, 495–497
    streaming media and, 485
QuickTime format
    exporting to, 195–199
    file, creating, 270
    Project Movie and, 151–152
QuickTime MPEG Encoder window, 278
QuickTime Player, 496
QuickTime Pro
    encoding MPEG-2, 277–279, 537–538
    installing, 574–575
    loops, creating and testing, 509–511
    PCM audio and, 207
    viewing audio file information, 209–210

R

re-purposing content, 79–80
Real World Interactive
    background, 559
    Mac use at, 561
    manufacturing process, 563
    motion menu, 562
    project overview, 562
    software used, 562, 563
    Tillyer interview, 559, 561–564
    visual tour of project, 560, 564–570
RealPlayer (RealNetwork), 285, 485
RealVideo compression, 263
rearranging slides in slideshow, 309

recordable disc, 458
recorder, consumer standalone, 529
Rectangle tool (Photoshop), 107
Rectangular Hotspot Tool (Dreamweaver), 95
redirecting buttons, 433–436
Reeves, Jason, 465
Reeves, Keanu, 49
region coding, 41, 416, 448–450
registering domain name, 90
relative value, 166
remote control
    DVD Player, 454–455
    settings, reviewing, 340–344
Remote Control properties (Property Inspector), 43, 44
removing. *See* deleting
renaming assets, 60–61, 244, 286–287
Rendering Options (Subtitle Editor), 153, 163–164
rendering time, 193
replication, 40, 361, 365, 507, 543–544, 547–548
replication house, 447
Requiem for a Dream (movie), 79, 182
resizing
    Acrobat document, 500, 501
    images in Photoshop, 300–303
    menu, 103
    Property Inspector, 391–392
resolution
    DVD vs. Internet, 484–485
    PAL vs. NTSC, 260
    scanning and, 297
resources. *See* Web sites
Return button, 137, 342
`Return` from menu command, 386
reverb when mixing for surround sound, 217
reviewing remote control settings, 340–344
rewards and interface design, 174
rewrite cycle, 528
RGB color model, 100–101
Riven game (Cyan), 561
Robbins, Tim, 355
roll-overs, 99, 483–484, 498–499, 500. *See also* hotspots
Room Type settings (A.Pack), 233–234
Rounded Rectangle tool (Photoshop) (Photoshop), 107
Roxio Web site, 351
RTMac card (Matrox), 271, 399

S

S-Video, 184
sample rate, 260
sans serif font, 169–170
Save dialog box, 5
Save for Web dialog box (Photoshop), 93, 484–485
Save Optimized As dialog box, 93
saving project, 5, 26
scanning
    moiré pattern, detecting and eliminating, 297–300
    progressive vs. interlaced, 253–254
    resolution and bit-depth, 297
    still image, 296
Scenarist system, 357
Score window (Flash), 490
screen size, 254–260
script, assigning, 378
Script Editor
    exiting, 382, 385
    general drop-down menus, 379–381
    Helpers drop-down menu, 381–391
    opening, 47
    overview of, 48, 379
Script properties (Property Inspector), 45
script tile, 5, 36, 374
scripting. *See also* Script Editor
    assigning value to variable, 387–388
    commands, adding, 385–387, 392–393
    comments and, 383–385
    comparisons, 388–389
    DVD vs. DVD-ROM capabilities, 479
    functions, 389–391
    `IF-THEN` statements, 378–379
    menu, 375–377
    overview of, 373–374, 392–393
    Play statement, 382
    pre-script, 378
    technique, choosing, 382–383
    variable, 374–375, 392
Scripts drop-down menu (Script Editor), 379
scrolling text, 285
SDDS (Sony Dynamic Digital Sound), 215
Select a Folder dialog box, 15, 354
Select Effect⇨Video⇨Broadcast Colors (After Effects), 183
Select Input Channel dialog box (A.Pack), 229

selected state (button), 118, 324–326, 569
selecting
    assets, 59–60
    tile, 35
    Web-safe color, 116–117
Selection Condition (Property Inspector), 143, 343
sending assets to Land of Oz, 62
Sequence⇨Insert Tracks (Final Cut Pro), 401
serif font, 169–170
service bureau, 40
`setAngle` command, 387
`setAudioStream` number command, 386
`setSubtitleStream` number command, 386–387
setting
    aspect ratio, 42
    background color, 88–89, 94
    bit-rate, 562
    button properties, 118–119
    Disc properties, 336–337
    duration of subtitle, 166
    Startup Action, 7–8
    Track properties, 335–336
    Video Standard, 42
setting up
    speakers, 218–220
    still menu, 8–9
    track, 10–11
Settings Modifiers window (Cleaner), 274
Settings window (Cleaner), 486
Settings window (Final Cut Pro), 406
Seurat, Georges, 113
Seven (movie), 192
shadow for subtitle, 153, 163, 170
shape of menu button, 12, 13, 101–102
shape tools (Photoshop), 106–109
Shatner, William, 19, 312
Shockwave (Macromedia), 465, 487
showing
    connections between project elements, 10–11
    connections for menu, 145–146
    controller for DVD Player, 454
sides, specifying number of for disc, 41
simulating DVD
    Adobe Acrobat (PDF), in, 497–502
    Flash, in, 490–495
    overview of, 489
    QuickTime with Flash, in, 495–497
simulating finished product. *See* previewing

size
    font for subtitle, of, 170
    frame, of, 54
    menu button, of, 13
    original scan, of, 297
    project file, of, 26, 136–137, 157
slash (/), 88
slice tools (Photoshop), 116
slide scanner, 296
slideshow
    adding language tracks, 311–313
    creating, 307–311
    moiré pattern, detecting and eliminating, 297–300
    overview of, 295
    previewing, 311
    resizing image, 300–303
    still frame image, using, 296–300
Slideshow Editor, 47, 303–307
Slideshow properties (Property Inspector), 45
slideshow tile, 5, 36–37
Slideshows drop-down menu (Script Editor), 380
SMPTE (Society of Motion Picture and Television Engineers) time code, 274
software. *See* applications
software requirements, 574
Sonic Solutions, 358, 369, 480
Sorensen compression, 263
Sorra, Kristin, 523–524, 529
sorting assets, 62–63
sound card and surround sound, 212
Sound Designer II format, 208
sound quality, evaluating, 261, 264
speakers, 218–220, 227
speaking up, 181
special effects, 189
splitting subtitle, 161
Spruce Technologies, 369
standalone application, 458
standards
    audio format, 19
    Blue Book, 469
    DVD, 416–417, 549
    image quality, 261
    interface design and, 176
    language support, 19–20
    menus, 18
    Region Coding, 416
    setting, 42

sound quality, 261, 264
storage capacity, 18
subtitles, 20
UDF Bridge, 458
video, 18–19, 250–251, 258
starting
    DVD Studio Pro, 3–4
    projects, 3–6, 25–26
Startup Action
    A.Pack application, 228–229
    description of, 42
    setting, 7–8
    Subtitle Editor, 155
sticky pages, 174
still image
    capturing from video clip, 128–130
    combining button elements with, 130–131
    moiré pattern, detecting and eliminating, 297–300
    resizing, 300–303
    scanner resolution and bit-depth, 297
    scanning, 296
    working with, 296
still menu
    background, creating, 127–130
    description of, 23
    layered image file and, 126
    preparing layers for use in, 100–105
    setting up, 8–9
Stir of Echoes DVD, 190
Stop playback command, 386
storage capacity, 18
stories, 288, 290–292
storing project file, 26
storyboarding project, 80–82
stowaway content, 476–477, 485
streaming animation, 486–487
streaming media, 263, 485–486
style, subtitle and, 169
styles (Photoshop), 109
subclip, 133
subtitle. *See also* Subtitle Editor
    associating with language, 428–429
    associating with video track, 422
    border, 153–154, 163, 170
    creating, 161–167
    dummy, 429–430
    eliminating lag before appearance of, 166
    fonts for, 169–170

    importing, 167–168, 170–171
    key frame and, 161
    line limitation, 169
    overview of, 149
    positioning in frame, 168–169
    setting duration of, 166
Subtitle Editor
    assigning more memory to, 575
    Color Settings, 153–154
    DVD-Videosize option, 152
    fonts, 169–170
    Marker window, 158–160
    overview of, 20, 149–150, 156
    positioning subtitle, 168–169
    Preview window, 157–158, 164
    Project Movie, 151–152, 157, 161–167
    Project Settings window, 150–151, 168
    Subtitle Settings, 152, 162–163
    Subtitle Streams window, 167–168, 429, 430
    Subtitle window, 160–161, 165, 169
    Word document pasted into, 439–440
Subtitle Settings section (Project Settings window), 152, 162–163
subtitle stream, 20, 150, 166–167
Subtitle Streams window (Subtitle Editor), 167–168, 429, 430
Subtitle window (Subtitle Editor), 160–161, 165, 169
subwoofer, 210
SuperDrive (Apple), 83, 367, 523
Surround Channels settings (A.Pack), 235
Surround Downmix settings (A.Pack), 232
surround sound, 205, 210–213, 216–220
Sweetwater Sound, 212
SWiSH, 470
switching video streams, 412
syntax error, 384
system requirements, 573–574

T

tags for HTML, 88
Target System option (A.Pack), 230
Taylor, Jim, 357–360, 367, 473, 586
technical quality and DVD authoring, 262
telecine process, 251–252
television
    designing graphics for, 184–186
    as medium for DVD, 183
    testing disc on, 557

Terminator 2: Judgment Day (DVD), 175, 181
testing
    audio mix, 220
    disc, 556–557
    interactivity, 339–345
    loop, 510
    menu, 144–147
    prototype, 77
text, scrolling, 285
text-warping feature (Photoshop), 110–111
thinned mpeg, 285
3:2 pulldown, 252
3D applications, 135
3D capabilities, 187–188, 465
tile. *See also specific types,* such as menu tile
    adjusting view size of, 33
    cloning, 327–328
    creating, 318–319
    deleting from Graphical View window, 35
    overview of, 5, 30–32
    previewing, 11–12
    selecting, 35
Tillyer, York, 559, 561–564
time code, 158, 274
Time value (Slideshow editor), 305
Timeline (After Effects), 535
Timeline (QuickTime Pro), 509
Timeline window (Final Cut Pro), 400, 401, 402
Timeout Action, 342
Timeout section (Property Inspector), 45, 134, 138
title, 265
title safe area, 258, 259
Toast Titanium (Roxio), 276–277, 351, 356, 360, 369, 468, 469
track
    adding to workspace, 34–35
    description of, 22
    linking button to, 144
    maximum number of video, 401
    setting up, 10–11
Track properties (Property Inspector), 10, 44, 335–336, 344
track tile, 5, 8
Tracks drop-down menu (Script Editor), 380
transfer modes, 190
transferring
    film to video, 250–253
    video to film, 252–253

transitional animation, 174, 180
transitions, 285
transparency, Flash and, 491
troubleshooting DVD-ROM, 587–588
Troubleshooting Windows, 51
tweening animation, 526
24p (progressive) video, 252
two-pass variable bit-rate encoding, 267–268
2X recording, 527

U

UDF Bridge standard, 458, 469, 544
Unsharp Mask dialog box (Photoshop), 299
Untitled Button (Menu Editor), 141–142
upgrading, 3
usability, 76, 175
user. *See* audience
user experience, 78–79

V

value, assigning to variable, 387–388
variable
    assigning value to, 387–388
    comparing to other variable or number, 388–389
    description of, 374–375
    maximum number of, 380, 392
variable bit-rate encoding, 267–268
Variable Names menu (Property Inspector), 43
Variables drop-down menu (Script Editor), 380
VCD (Video CD), 270, 481, 507
versions of DVD Studio Pro, 3
video. *See also* video format
    chroma and luminance, 260
    digital, 113, 249–250
    film compared to, 250–253
    frame rates, 250–251
    hardware-based encoding, 282–283
    importing, 283, 286
    overview of, 249–250
    placing markers in, 288–290
    previewing, 293
    progressive vs. interlaced scanning, 253–254
    screen size, evaluating, 254–260
    software for preparing, 270–277
    working with, 269
Video CD (VCD), 270, 481, 507

video clip
  capturing still image from, 128–130
  positioning for Motion menu, 536
  preparing for Motion menu, 534
video company Web sites, 578
video editing software, 270–277. *See also* Final Cut Pro (Apple)
video format, 18–19, 54, 114, 261, 270. *See also* MPEG-2 video format
Video Safe area, 258–259, 557
Video Standard, setting, 42
video stream
  adding primary to track, 408
  maximum number of, 400
  switching, 412
  using as multiple angles, 406
Video Title Sets, 265
video track, associating subtitle with, 422
VIDEO_TS folder, 15, 16, 352, 354, 472–473
View Size drop-down menu (Graphical View window), 33, 34
viewing. *See also* Matrix Views
  assets, 63–64
  connections between menu and asset, 64
View⇨as List, 58
View⇨Gamut Warning (Photoshop), 101
View⇨Title Safe (Final Cut Pro), 134
Virtual PC (Connectix), 469–471
visual assets, associating, 320–321
visual HTML editor. *See* Dreamweaver (Macromedia)
visual quality, encoding for, 284–285

W

Warm Blankets Foundation
  development issues, 508
  Motion menu loop points, 508–511
  overview of, 505–506
  project planning, 506–508
WAV format, 208
Web hosting sites, 578
Web output, designing for, 114–115
Web page, 174. *See also* HTML document
Web-safe color, selecting, 116–117
Web sites
  Adobe, 226
  all4dvd, 356
  Angle Park, 262

Apple, 3, 353, 363, 366, 575, 577
APS Technologies, 361
author-related, 578
Best Buy, 227–228
Bias, 225
compatibility, 16
computer outlets, 365
DeCSS, 447
Digidesign, 226
Discmakers, 366
DVD Copy Control Association, 447
DVD Demystified (Taylor), 367
DVD FAQ, 473, 531, 586
DVD related, 577
Emagic, 226
Fenster, Dan, 180–181
Gerbil Liberation Front, 550
HTML tutorials, 89
InterActual Player, 471
Jolt Cola, 373
Klingon Language Institute, 419
Mac information, 461
MacMall, 227
Macromedia, 89, 464, 467
Macrovision, 448
MacWarehouse, 227
manufacturing, 578
manufacturing discount, 558
Marken Communications, Inc., 368
Martin, Alan, 176
Metatec International, 543
Motion Picture Association of America, 447
Motu, 227
MPEG-1 video format, 270
Neocelt, 465
Nielsen, Jakob, 76
Pioneer Electronics, 530, 531
Real World Interactive, 561
resources, 227–228
Roxio, 351
software manufacturers, 579
Sweetwater Sound, 212
Swish, 470
video companies, 578
Warm Blankets, 505
Web hosting, 578

white, use of on video monitor, 184
Whitney Houston's Greatest Hits DVD, 180–181
widescreen video, 255–257
window, creating for container, 398
Window⇨Hide Controller, 454
Window⇨Log, 391
Window⇨Property Inspector, 58
Window⇨Show Controller (DVD Player), 454
Windows (Microsoft)
    cross-platforming standalone content for, 462–463
    DVD-ROM content for, 460–462
    file extensions in, 55
    Media Player, 285, 485
    running OS on Mac with Virtual PC, 469–471
Windows⇨Project View, 398
wire-frame animations, 285

Wizard of Oz (movie), 415
W.O.M.A.D., 561
Word document. *See* Microsoft Word document
workflow, 488–489, 525–527
workspace, 4, 29–30, 51, 138, 392. *See also* Graphical View window; Matrix Views; Project View; Property Inspector
written log, 396

X
X-Box, 177

Z
Zada, Jason, 184–186
zooms, 284

Hungry Minds, Inc.
End-User License Agreement

READ THIS. You should carefully read these terms and conditions before opening the software packet(s) included with this book ("Book"). This is a license agreement ("Agreement") between you and Hungry Minds, Inc. ("HMI"). By opening the accompanying software packet(s), you acknowledge that you have read and accept the following terms and conditions. If you do not agree and do not want to be bound by such terms and conditions, promptly return the Book and the unopened software packet(s) to the place you obtained them for a full refund.

1. **License Grant.** HMI grants to you (either an individual or entity) a nonexclusive license to use one copy of the enclosed software program(s) (collectively, the "Software") solely for your own personal or business purposes on a single computer (whether a standard computer or a workstation component of a multi-user network). The Software is in use on a computer when it is loaded into temporary memory (RAM) or installed into permanent memory (hard disk, CD-ROM, or other storage device). HMI reserves all rights not expressly granted herein.

2. **Ownership.** HMI is the owner of all right, title, and interest, including copyright, in and to the compilation of the Software recorded on the disk(s) or CD-ROM ("Software Media"). Copyright to the individual programs recorded on the Software Media is owned by the author or other authorized copyright owner of each program. Ownership of the Software and all proprietary rights relating thereto remain with HMI and its licensers.

3. **Restrictions On Use and Transfer.**
   (a) You may only (i) make one copy of the Software for backup or archival purposes, or (ii) transfer the Software to a single hard disk, provided that you keep the original for backup or archival purposes. You may not (i) rent or lease the Software, (ii) copy or reproduce the Software through a LAN or other network system or through any computer subscriber system or bulletin-board system, or (iii) modify, adapt, or create derivative works based on the Software.
   (b) You may not reverse engineer, decompile, or disassemble the Software. You may transfer the Software and user documentation on a permanent basis, provided that the transferee agrees to accept the terms and conditions of this Agreement and you retain no copies. If the Software is an update or has been updated, any transfer must include the most recent update and all prior versions.

4. **Restrictions on Use of Individual Programs.** You must follow the individual requirements and restrictions detailed for each individual program in Appendix D of this Book. These limitations are also contained in the individual license agreements recorded on the Software Media. These limitations may include a requirement that after using the program for a specified period of time, the user must pay a registration fee or discontinue use. By opening the Software packet(s), you will be agreeing to abide by the licenses and restrictions for these individual programs that are detailed in Appendix D and on the Software Media. None of the material on this Software Media or listed in this Book may ever be redistributed, in original or modified form, for commercial purposes.

5. **Limited Warranty.**

    (a) HMI warrants that the Software and Software Media are free from defects in materials and workmanship under normal use for a period of sixty (60) days from the date of purchase of this Book. If HMI receives notification within the warranty period of defects in materials or workmanship, HMI will replace the defective Software Media.

    (b) HMI AND THE AUTHOR OF THE BOOK DISCLAIM ALL OTHER WARRANTIES, EXPRESS OR IMPLIED, INCLUDING WITHOUT LIMITATION IMPLIED WARRANTIES OF MERCHANTABILITY AND FITNESS FOR A PARTICULAR PURPOSE, WITH RESPECT TO THE SOFTWARE, THE PROGRAMS, THE SOURCE CODE CONTAINED THEREIN, AND/OR THE TECHNIQUES DESCRIBED IN THIS BOOK. HMI DOES NOT WARRANT THAT THE FUNCTIONS CONTAINED IN THE SOFTWARE WILL MEET YOUR REQUIREMENTS OR THAT THE OPERATION OF THE SOFTWARE WILL BE ERROR FREE.

    (c) This limited warranty gives you specific legal rights, and you may have other rights that vary from jurisdiction to jurisdiction.

6. **Remedies.**

    (a) HMI's entire liability and your exclusive remedy for defects in materials and workmanship shall be limited to replacement of the Software Media, which may be returned to HMI with a copy of your receipt at the following address: Software Media Fulfillment Department, Attn.: *Macworld® DVD Studio Pro™ Bible*, Hungry Minds, Inc., 10475 Crosspoint Blvd., Indianapolis, IN 46256, or call 1-800-762-2974. Please allow four to six weeks for delivery. This Limited Warranty is void if failure of the Software Media has resulted from accident, abuse, or misapplication. Any replacement Software Media will be warranted for the remainder of the original warranty period or thirty (30) days, whichever is longer.

    (b) In no event shall HMI or the author be liable for any damages whatsoever (including without limitation damages for loss of business profits, business interruption, loss of business information, or any other pecuniary loss) arising from the use of or inability to use the Book or the Software, even if HMI has been advised of the possibility of such damages.

    (c) Because some jurisdictions do not allow the exclusion or limitation of liability for consequential or incidental damages, the above limitation or exclusion may not apply to you.

7. **U.S. Government Restricted Rights.** Use, duplication, or disclosure of the Software for or on behalf of the United States of America, its agencies and/or instrumentalities (the "U.S. Government") is subject to restrictions as stated in paragraph (c)(1)(ii) of the Rights in Technical Data and Computer Software clause of DFARS 252.227-7013, or subparagraphs (c) (1) and (2) of the Commercial Computer Software - Restricted Rights clause at FAR 52.227-19, and in similar clauses in the NASA FAR supplement, as applicable.

8. **General.** This Agreement constitutes the entire understanding of the parties and revokes and supersedes all prior agreements, oral or written, between them and may not be modified or amended except in a writing signed by both parties hereto that specifically refers to this Agreement. This Agreement shall take precedence over any other documents that may be in conflict herewith. If any one or more provisions contained in this Agreement are held by any court or tribunal to be invalid, illegal, or otherwise unenforceable, each and every other provision shall remain in full force and effect.